This is the first book to describe thoroughly the many facets of doping in compound semiconductors. Equal emphasis is given to the fundamental materials physics and to the technological aspects of doping.

The author describes various doping techniques, including doping during epitaxial growth, doping by implantation, and doping by diffusion. The key characteristics of all dopants that have been employed in III–V semiconductors are discussed. In addition, general characteristics of dopants are analyzed, including the electrical activity, saturation, amphotericity, autocompensation, and maximum attainable dopant concentration. Redistribution effects are important in semiconductor microstructures. Linear and non-linear diffusion, different microscopic diffusion mechanisms, surface segregation, surface drift, surface migration, impurity-induced disordering, and the respective physical driving mechanisms are illustrated.

Topics related to basic impurity theory include the hydrogenic model for shallow impurities, linear screening, density of states, classical and quantum statistics, the law of mass action, as well as many analytic approximations for the Fermi–Dirac integral for three-, two- and one-dimensional systems. The timely topic of highly doped semiconductors, including band tails, impurity bands, bandgap renormalization, the Mott transition, and the Burnstein–Moss shift, is discussed as well.

Doping is essential in many semiconductor heterostructures including high-mobility selectively doped heterostructures, quantum well and quantum barrier structures, doping superlattice structures and δ-doping structures. Technologically important deep levels are summarized, including Fe, Cr, and the DX-center, the EL2 defect, and rare-earth impurities. The properties of deep levels are presented phenomenologically, including emission, capture, Shockley–Read recombination, the Poole–Frenkel effect, lattice relaxation, and other effects. The final chapter is dedicated to the experimental characterization of impurities.

This book will be of interest to graduate students, researchers and development engineers in the fields of electrical engineering, materials science, physics, and chemistry working on semiconductors. The book may also be used as a text for graduate courses in electrical engineering and materials science.

Cambridge Studies in Semiconductor Physics and
Microelectronic Engineering: 1

EDITED BY
Haroon Ahmed
Cavendish Laboratory, University of Cambridge
Michael Pepper
Cavendish Laboratory, University of Cambridge
Alec Broers
Department of Engineering, University of Cambridge

Doping in III–V Semiconductors

Doping in III–V Semiconductors

E. Fred Schubert
AT&T Bell Laboratories

CAMBRIDGE
UNIVERSITY PRESS

Published by the Press Syndicate of the University of Cambridge
The Pitt Building, Trumpington Street, Cambridge CB2 1RP
40 West 20th Street, New York, NY 10011–4211, USA
10 Stamford Road, Oakleigh, Melbourne 3166, Australia

First published 1993

Printed in Great Britain at the University Press, Cambridge

A catalogue record for this book is available from the British Library

Library of Congress cataloguing in publication data
Schubert, E. Fred
Doping in III–V semiconductors / E. Fred Schubert.
p.   cm. – (Cambridge studies in semiconductor physics and
microelectronic engineering: 1)
Includes bibliographical references and index.
ISBN 0 521 41919 0
1. Compound semiconductors. 2. Semiconductor doping. I. Title
II. Title: Doping in 3–5 semiconductors. III. Series.
QC611.8.C64S34 1993
621.3815′2–dc20   92-35998 CIP

ISBN 0 521 41919 0 hardback

To Jutta Maria
and to
Anne, Martin and Ursula

Contents

Foreword

Technology revolutions mark their origin from a single breakthrough such as the demonstration of the transistor. Subsequent technological advances make the revolution a reality. The historical evolution from Si transistor to integrated circuits to high speed computers and telecommunications is one excellent example.

Active workers in the field know each advance represents a major expenditure of time and effort (really 'blood and sweat'). Progress comes about in a competitive atmosphere involving creative ideas, personality forces and technology prognostications.

We are presently in the midst of the microelectronics revolution. Insatiable demands exist for greater data rates, for consumer electronics and for superior telecommunications. We envision a world of wireless communications, video transmissions and displays, and numerous applications of high speed data transmission. Satisfying this demand is the goal of current research.

For our field of microelectronics this means a systems analysis from the final product to the atomic configurations of the materials that make the product. This analysis itself is remarkable; we can precisely relate the macroscopic system properties – how fast will the system operate – to the microscopic atomic structure – where do the atoms sit in the solid.

Analysis of today's devices reveal the limitations of today's semiconductors – mostly silicon. Silicon has been the workhorse for the last thirty years. The 'silicon community' continues to squeeze all that is obtainable from this most robust and manufacturable semiconductor. Nevertheless an all silicon technology has foreseeable and predictable limitations.

An electrical engineering student can open almost any textbook and see that the path is clear – III–V semiconductors can do all that Si can do and more – they emit light! Not only do III–V materials have fundamental characteristics which make faster electronics, but they are efficient emitters and detectors of photons. Combined with a transmission medium – the optical fiber – these III–V semiconductor devices gave rise to the era of optical communications. The intrinsically high speed in an optical system is the basis of our vision of ultra high speed data transmission, yielding electronically controlled visual communications, high definition TV and personal video.

The 'cost effective' realization of this technology, on a worldwide basis, depends on our ability to fabricate the materials that make the system. Modern silicon technology demonstrates that basic materials manipulation has governed the advance of the technology. Near perfect crystalline growth, a well behaved passivating oxide, controlled doping, and advanced lithography made the Si revolution a reality. The extension of the semiconductor revolution to III–V devices depends on these materials manipulations, which eventually determine their manufacturability.

Tomorrow's technology will be based on today's research; the body of information that allows the 'systems evaluation' and meets the 'manufacturability requirements'. As a starting point a systems evaluation may assume an ideal structure; however, no material is perfect. Better systems evaluation includes far more realistic material configurations and the picture of this realistic material is one goal of the semiconductor materials scientist. Manufacturing requirements will also be determined by this research in terms of yields, process margins, etc.

Both systems and manufacturability evaluations require a deep understanding of the numerous steps that go into the fabrication of a semiconductor. Each of these 'steps' becomes a sub-field unto itself – with hundreds of investigators, numerous conferences and publications in an ever expanding network of scientific research.

This book focuses on such an element of III–V devices – the doping of these semiconductors. Basically it addresses the question of the behavior of dopants in III–V semiconductors – how are they incorporated, where do the atoms sit? As explained in the Preface 'doping' is an essential element to make an active device from a semiconductor. Clever and controlled manipulation of 'doping' can enhance the device properties. Doping processes will be one critical factor in determining manufacturability. Thus the knowledge represented in this book is the critical input needed to provide insight into our technology directions.

One further point, lest it get lost in the current world focus on applied research. Semiconductor science is a field that spans the space of exciting fundamental science to immediate and current applications. Unique solid state structures with controlled doping have revealed new phenomena of the most fundamental character. Such reports have led to recent Nobel prizes and exciting new concepts. These 'unique solid state' structures combine the latest in materials fabrication and dopant control. Thus, it is worthwhile to note that this slice of science is not only crucial to technology but enables forefront basic research. It has opened the fascinating world of quantum structures and low dimensional configurations.

Those of us who have lived through this semiconductor science revolution look back with pride at the accomplishments of the field and look forward with excitement at what is to come. The science and art of fabricating active III–V semiconductors is a critical element in determining the future.

L. C. Feldman
AT&T Bell Laboratories
January, 1993

Preface

It was realized in the infancy of semiconductor science that impurities are the essential ingredient in these materials. The first usage of the term 'impurities' dates back to 1931 (Wilson, 1931):

> Electrons on a foreign atom [in a semiconductor] do not take part directly in [electrical] conduction. They must be transferred by the effect of lattice vibrations to an atom of the pure substance. In this case, the main function of the 'impurities' is to provide electrons for the upper unoccupied energy bands of the crystal, while acting as scatterers is only a secondary function.

The knowledge of the significant consequences of impurities motivated researchers to, first, purify the semiconductor materials, for example by zone refining, and, second, deliberately add 'foreign atoms' to purified semiconductors, for example by thermal diffusion of such 'foreign atoms' into the semiconductor. The deliberate doping of semiconductors with impurities required that the usage of the term 'impurities' was broadened. Such an extended definition was offered by Shockley (1950):

> The conductivity arises from the presence of [   ] atoms which are termed 'impurities', even though added to the otherwise pure [   ] [semiconductor].

The term 'dopant' was non-existent in the English language until its introduction in 1963 by Seidman and Marshall (1963), who wrote about crystal growth from the melt:

> The melt consists of intrinsic polycrystalline Ge and Si mixed with

suitable 'dopants' such as P, which produces an n-type melt, or Ga, which produces a p-type melt.

Thus, the term 'impurities' is used for unintentional as well as intentional foreign atoms in the semiconductor, while the term 'dopants' is used exclusively for deliberately added impurities.

This book is devoted to the many aspects of doping in III–V semiconductors. The book covers, with equal emphasis, fundamental concepts and basic theory of impurities, as well as the practical and technological aspects of the doping process. The basic conceptual theory of shallow impurities and deep levels and their effect on the free carrier concentration is given in Chaps. 1, 2, and 3. Chapters 4, 5, and 6 cover the characteristics of individual impurity elements as well as the doping process during epitaxy with solid elemental sources as well as gaseous sources. Chapter 7 covers characteristics of impurities incorporated during growth, ion implantation, and diffusion, as well as impurity characteristics such as amphotericity, autocompensation, and the maximum attainable doping concentration. Redistribution effects during growth and thermal processing are covered in Chap. 8 and include diffusion, segregation, migration, and other effects. Chapter 9 describes deep levels that are relevant to III–V semiconductor technology, such as the DX-center, EL2, O, Cr, Fe, and rare-earth impurities. Doping in semiconductor microstructures, which require highly controlled doping profiles, is described in Chaps. 10 and 11. Finally, Chap. 12 summarizes electronic, optical, and structural impurity characterization techniques.

The book is written for graduate students, researchers, and development engineers of the disciplines electrical engineering, physics, materials science, and chemistry working at universities, industrial research laboratories, government laboratories and production facilities. It is suited as a textbook for graduate courses in electrical engineering and material science. The book is self-contained and covers all major concepts in detail. Units of the *Système International* (S.I. units) are used throughout the book. Boxes are drawn around those equations which are frequently used in semiconductor science.

The author has benefited significantly from suggestions, discussions and collaborations with his colleagues at AT&T Bell Laboratories, especially with G. A. Baraff, T. Y. Chang, A. Y. Cho, L. C. Feldman, G. H. Gilmer, A. M. Glass, H. J. Gossmann, R. D. Grober, T. D. Harris, W. D. Johnston, Jr., R. F. Kazarinov, R. F. Kopf, M. Lax, A. F. J. Levi, H. S. Luftman, R. M. Lum, R. J. Malik, D. P. Monroe, M. B. Panish, L. Pfeiffer, D. Ritter, S. Schmitt-Rink, H. L. Stormer, V. Swaminathan, W. T. Tsang, J. S. Weiner, and J. L. Zilko. I also thank Professor D. G. Deppe (Univ. Texas, Austin), Professor Y. H. Yao (Univ. Nebraska,

Lincoln), and Professor K. Ploog (Paul Drude Institute, Berlin) for many suggestions and valuable discussions. The encouragement of F. Capasso and D. V. Lang during the course of this work was highly appreciated. I am indebted to L. C. Feldman for writing the foreword.

E. F. Schubert
AT&T Bell Laboratories
February, 1993

Symbols

α	(cm^{-1})	absorption coefficient
α	(-)	Sommerfeld's fine structure constant
α		variational parameter
a_B	(Å)	Bohr radius
a_B^*	(Å)	effective Bohr radius
a_0	(Å)	lattice constant
A	(cm^2)	area
A_1, A_2, A_3		normalization constants
$\mathcal{B}$	(T)	magnetic induction
c	(m/s)	velocity of light in vacuum
c		constant
c_n (c_p)	(s^{-1})	electron (hole) capture probability
c_n^* (c_p^*)	(cm^{-3}s^{-1})	electron (hole) capture rate
C	(F/cm^2)	capacitance per unit area
δ		delta function
Δ	(cm^{-2})	delta operator $(\partial^2/\partial x^2 + \partial^2/\partial y^2 + \partial^2/\partial z^2)$
ΔE	(eV)	activation energy
ΔE_C (ΔE_V)	(eV)	discontinuity in the conduction (valence) band
ΔE_{PF}	(eV)	lowering of emission barrier (Poole–Frenkel energy)
ΔR_L	(cm)	longitudinal straggle
ΔR_P	(cm)	projected straggle
D	(cm^2/s)	diffusion constant (also called diffusion coefficient)
D_i	(cm^2/s)	diffusion constant in intrinsic semiconductors
D_0	(cm^2/s)	pre-exponential factor of diffusion constant
ε	(A s/V m)	permittivity ($\varepsilon = \varepsilon_r \varepsilon_0$)
ε_0	(A s/V m)	permittivity in vacuum
ε_r	(-)	dielectric constant
η_a	(-)	doping activation efficiency
η_D	(-)	doping efficiency

η_F	(-)	normalized Fermi energy
e	(C)	elementary charge
e_n (e_p)	(s^{-1})	electron (hole) emission probability
e_n^* (e_p^*)	$(cm^{-3} s^{-1})$	electron (hole) emission rate
E	(J)	energy
E_a	(eV)	activation energy
E_a	(eV)	acceptor ionization energy
E_d	(eV)	donor ionization energy
E_{dd}	(eV)	deep donor ionization energy
E_g	(eV)	bandgap energy
E_{kin}	(J)	kinetic energy
E_{pot}	(J)	potential energy
E_{sd}	(eV)	shallow donor ionization energy
E_t	(eV)	trap ionization energy
E_A	(eV)	acceptor energy
E_{BE}	(eV)	energy of bound exciton
E_C	(eV)	energy of conduction band edge
E_D	(eV)	donor energy
E_{DD}	(eV)	deep donor energy
E_F	(eV)	Fermi energy
E_{FE}	(eV)	energy of free exciton
E_{SD}	(eV)	shallow donor energy
E_{Ryd}	(eV)	Rydberg energy
E_{Ryd}^*	(eV)	effective Rydberg energy
E_T	(eV)	trap energy
E_V	(eV)	energy of valence band edge
E_0	(eV)	energy of quantized ground-state subband
$\mathcal{E}$	(V/cm)	electric field
ϕ	(V)	potential
ϕ_B	(V)	barrier height
ϕ	(°)	cylindrical or spherical coordinate (azimuthal angle)
f	(Hz)	frequency
f_B	(-)	Boltzmann distribution function
f_F	(-)	Fermi distribution function
f_M	(-)	Maxwell distribution function
f_{MB}	(-)	Maxwell–Boltzmann distribution function
F	(N)	force
$F_{1/2}$	(-)	Fermi–Dirac integral of order ½
g	(-)	state degeneracy
g_0	(-)	ground-state degeneracy
G	(J)	Gibb's free energy
h	(Js)	Planck's constant
$\hbar$	(Js)	Planck's constant divided by 2π
H	(J)	enthalpy
H	(J)	hamiltonian operator $(-(\hbar^2/2m^*)\Delta + eV)$

I		intensity
j	(A/cm^2)	current density
j	(cm^{-2}s^{-1})	particle flow density
k	(J/K)	Boltzmann's constant
k	(cm^{-1})	wave vector ($2\pi/\lambda$)
K	(-)	equilibrium constant
K_c	(-)	compositional distribution coefficient
K_t	(-)	thermodynamic distribution coefficient
λ	(cm)	wavelength
λ	(cm)	transition region thickness
l	(cm)	length
l	(-)	orbital angular momentum quantum number
l_h	(cm)	hop distance
L_z	(cm)	thickness of quantum well
L_D	(cm)	diffusion length
μ	(cm^2/V s)	mobility
μ_0	(V s/A m)	permeability in vacuum
m	(kg)	mass
m_0	(kg)	electron mass
m^*	(kg)	effective mass
m_e^*	(kg)	electron effective mass
m_h^*	(kg)	hole effective mass
m_{hh}^*	(kg)	heavy hole effective mass
m_l	(-)	azimuthal quantum number
m_l^*	(kg)	effective mass for motion along a longitudinal direction
m_{lh}^*	(kg)	light hole effective mass
m_t^*	(kg)	effective mass for motion along a transverse direction
m_x^*	(kg)	effective mass for motion along the x-direction
m_y^*	(kg)	effective mass for motion along the y-direction
m_z^*	(kg)	effective mass for motion along the z-direction
M	(kg)	mass of lattice atoms
M_c	(-)	number of equivalent minima in the conduction band
∇	(cm^{-1})	nabla operator, ($\partial/\partial x$, $\partial/\partial y$, $\partial/\partial z$)
n	(cm^{-3})	electron concentration
n	(-)	principal quantum number
n_i	(cm^{-3})	intrinsic electron concentration
n_{2DEG}	(cm^{-2})	density of two-dimensional electron gas
n^{2D}	(cm^{-2})	two-dimensional electron concentration
n^{1D}	(cm^{-1})	one-dimensional electron concentration
N_c	(cm^{-3})	effective density of states at conduction band edge
N_c^{2D}	(cm^{-2})	two-dimensional effective density of states at conduction band
N_c^{1D}	(cm^{-1})	one-dimensional effective density of states at conduction band
N_{crit}	(cm^{-3})	critical Mott concentration
N_v	(cm^{-3})	effective density of states at valence band edge
N_A	(cm^{-3})	acceptor concentration

N_{Avo}	(mol^{-1})	Avogadro's number
N_{CV}	(cm^{-3})	CV concentration
N_D	(cm^{-3})	donor concentration
N_{DD}	(cm^{-3})	deep donor concentration
N_{SD}	(cm^{-3})	shallow donor concentration
N^0	(cm^{-3})	concentration of neutral centers
N^+	(cm^{-3})	concentration of positively charged centers
N^-	(cm^{-3})	concentration of negatively charged centers
N^{2D}	(cm^{-2})	two-dimensional impurity density
N^{1D}	(cm^{-1})	one-dimensional impurity density
ω	(s^{-1})	angular frequency $(2\pi/f)$
Ω	(steradian)	solid angle
ψ	$(cm^{-3/2})$	time-independent three-dimensional wave function
Ψ	$(cm^{-3/2})$	time-dependent three-dimensional wave function
p	(cm^{-3})	hole concentration
p	(kg m/s)	momentum
p		probability
p^{2D}	(cm^{-2})	two-dimensional hole concentration
p^{1D}	(cm^{-1})	one-dimensional hole concentration
P	(N/m^2)	pressure
Q	(C/cm^2)	charge per unit area
ρ_{DOS}	$(cm^{-3}eV^{-1})$	density of states
ρ_{DOS}^{2D}	$(cm^{-2}eV^{-1})$	two-dimensional density of states
ρ_{DOS}^{1D}	$(cm^{-1}eV^{-1})$	one-dimensional density of states
r	(cm)	radius
r	(cm)	polar coordinate
r_s	(cm)	screening radius
r_D	(cm)	Debye screening radius
r_H	(-)	Hall factor
r_{TF}	(cm)	Thomas–Fermi screening radius
R	$(J\,mol^{-1}K^{-1})$	universal gas constant
R	(Ω)	resistance
R_H	(cm^3/C)	Hall coefficient
R_n (R_p)	$(cm^{-3}s^{-1})$	net capture rate of electrons (holes)
R_p	(cm)	projected range
R_{SR}	$(cm^{-3}s^{-1})$	Shockley–Read recombination rate
σ	$(\Omega^{-1}cm^{-1})$	conductivity
σ	(cm^2)	capture cross section
σ		standard deviation of a gaussian distribution
σ		step function
s	(-)	spin quantum number
S	(eV/K)	entropy
θ	(°)	spherical coordinate (polar angle)
τ	(s)	time constant
τ_c	(s)	capture time constant

τ_e	(s)	emission time constant
τ_{inel}	(s)	mean time between inelastic scattering events
τ_n (τ_p)	(s)	minority carrier lifetime
t	(s)	time
t	(cm)	thickness
t_{dep}	(cm)	depletion thickness
t_{spacer}	(cm)	spacer layer thickness
T	(K)	absolute temperature
v	(cm/s)	velocity
v_d	(cm/s)	drift velocity
v_g	(cm/s)	growth rate
$\bar{v}$	(cm/s)	mean velocity
v_{rms}	(cm/s)	root-mean-square velocity
v_x	(cm/s)	velocity along the x-direction
V	(V)	potential
V	(V)	voltage (potential difference)
V	(m^3)	volume
V_{exc}	(cm^3)	excitonic volume
W_D	(cm)	depletion layer width
x	(cm)	cartesian coordinate
x	(-)	composition of alloy semiconductors
y	(cm)	cartesian coordinate
y	(-)	composition of alloy semiconductors
z	(cm)	cartesian coordinate
z_D	(cm)	depth of doped layer
z_p	(cm)	period of superlattice
z_0	(cm)	width of ground-state subband
Z	(-)	charge state of center

Some quantities have no dimension and these are marked by '(-)'. If a quantity can have
different dimensions, then the dimensions are omitted. Infrequently used symbols are
defined in the text.

Introduction

Impurities are the lifeblood of virtually all semiconductor devices. Impurities determine the semiconductive properties of materials and allow us to vary the conductivity from the semi-insulating via the semi-conductive to the semi-metallic range of the conductivity spectrum. The species and concentration of impurities determine the conductivity type and the free carrier concentration of semiconductors. For p-type and n-type impurities the electronic conduction occurs predominantly in the valence and conduction band, respectively. Employment of both conductivity types in one semiconductor makes possible pn-junctions, i.e. rectifying devices. Employment of layered n-p-n-type conductivity allows amplification and switching of electronic signals in transistors. Finally, free carriers must be injected or extracted in all optoelectronic devices. Thus, impurities providing free carriers form the basis of light-emitting diodes, current-injection lasers, photodetectors, and solar cells.

Impurities incorporated into a semiconductor lattice predominantly occupy substitutional lattice sites. In compound semiconductors, the lattice site can be either a cation or an anion site. Which of the two sites is preferred by an impurity depends on a number of factors, including the valence electron correlation between impurity and host, the bond strength between the impurity atom and the surrounding host lattice, and the size of the impurity (strain effects). Examples of impurities incorporated into a III–V semiconductor lattice are schematically shown in Fig. 1. Such compound semiconductors have a zincblende structure, in which cations and anions are bonded with tetrahedral symmetry. Intentional impurities are selected from columns adjacent to column III or column V of the Periodic Table, i.e. from column II, IV, or VI. Group-II impurities predominantly occupy group-III sites due to their valence electron configuration correlation. Substitution of the group-III element by a group-II impurity provides a 7 valence

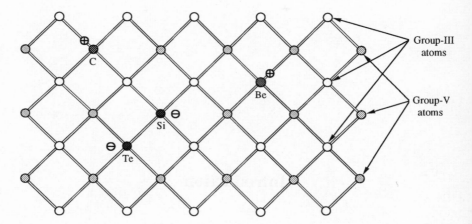

Fig. 1. Schematic illustration of the zincblende lattice consisting of group-III and group-V elements. The lattice is doped with donors (Si, Te) and acceptors (Be, C) which have an excess electron and a deficit electron, respectively.

electron configuration which is close to the 8 electron noble gas electron shell. Substitution of a group-V element by a group-II impurity would result in a 5 valence electron configuration, i.e. far from the 8 electron noble-gas configuration. For analogous reasons, group-VI impurities occupy group-V sites. Group-IV impurities can occupy both cation as well as anion sites. Thus, substitutional group-IV impurities have either one electron less or one electron more than the lattice atom they replaced, i.e. they can have an *excess electron* or a *deficit electron*.

It is a fundamental property of semiconductors that they have, in the intrinsic state, a completely filled valence band and an empty conduction band. Valence band states are formed from atomic states of p-orbital symmetry, while conduction band states derive themselves from atomic states of s-orbital symmetry. We first consider an impurity with an excess electron. If such an impurity is ionized in an otherwise pure semiconductor, the empty continuum state of lowest energy is a conduction band state. In the process of an electronic excitation, such impurities can *donate* their electron to the conduction band and are therefore called *donors*. We next consider an impurity with a deficit electron (hole). In an otherwise pure semiconductor, the hole state of such an impurity can be filled only by an electron from a filled band, e.g. the valence band. That is, such impurities can *accept* electrons from the valence band and are therefore called *acceptors*. Both types of impurities are schematically shown in Fig. 1.

Fig. 2. Schematic illustration of a hydrogenic wave function of a donor. The effective Bohr radius is much larger than the period of the lattice.

The energy required to ionize a donor or acceptor is called the activation energy of the impurity. As we will see in Chap. 1, the activation energy is mainly determined by the coulombic interaction between the impurity atom and the charge carrier. This interaction is relatively weak due to dielectric screening and due to the light effective mass of the free carriers. The weak binding of charged carriers to an impurity has two notable consequences. First, the wave function of the bound state of the impurity is much larger than the unit cell of the lattice. This situation is shown in Fig. 2, which shows a schematic electronic wave function of s-orbital symmetry which extends over many lattice periods. In III–V semiconductors, the effective Bohr radius of donors can exceed 100 Å, i.e. the bound state wave function extends over more than 10 000 unit cells. Second, the activation energy of such shallow impurities is smaller than or comparable to the thermal energy kT at room temperature. This very much simplifies the treatment of the activation of shallow impurities. At room temperature, all shallow impurities can be considered as activated, i.e. the free carrier concentration coincides, in the absence of compensation effects, with the impurity concentration.

In practice, the characteristics of impurities are more complex than the simple coulombic substitutional center described above. An impurity can occupy sites other than substitutional sites, e.g. interstitial sites. Moreover, an impurity may form complexes with lattice atoms, other impurities, and native defects. Electrically, an impurity can be inactive, or compensating, which affects the efficiency of the dopant. The saturation of the electrical activity at high doping concentrations, and the amphoteric nature of some impurity species determine the properties of doped semiconductors as well. Furthermore, the characteristics of doped semiconductors change quite drastically at high doping concentrations. Effects associated with high doping concentrations are the Mott transition, the Burstein–Moss shift, band tail formation, impurity band formation, and bandgap shrinkage. These effects can change the characteristics of doped semiconductors drastically and therefore need to be taken into account.

Doping in compound semiconductor microstructures and heterostructures requires an unprecedented control of the spatial distribution of dopants. To illustrate the necessity of highly controlled doping profiles, we consider a hetero-bipolar transistor as an example. The band diagram of a hetero-bipolar transistor is shown in Fig. 3(a). The structure consists of a wide-gap n-type emitter, a narrow-gap p-type base and an n-type collector. One of the key advantages of a npn hetero-bipolar transistor is the reduction of the hole current from the base to the emitter. The hole current is reduced as compared to homojunction transistors due to the band discontinuity in the valence band. Under forward bias conditions of the emitter–base junction, the band edges are flat as shown in Fig. 3(a). Then, holes must overcome the energy discontinuity denoted as ΔE in Fig. 3(a). We next consider a hetero-bipolar transistor whose emitter–base pn-junction is displaced with respect to the heterojunction, as shown in Fig. 3(b). In this case, the effective energy barrier for the injection of holes into the emitter is reduced, which increases the magnitude of the hole current from the base into the emitter. The calculation of the barrier height for hole injection shows that even a minor displacement of the pn-junction by 20–50 Å has a deleterious effect. Accurate control of the spatial distribution of dopants is not only required in hetero-bipolar transistors but in many other structures, for example in selectively doped heterostructures.

The unprecedented control of the spatial distribution of impurities caused much interest in well-defined distributions including δ-function-like doping distributions. Impurities can redistribute during epitaxial growth as well as during post-growth or post-implantation processing. Such redistribution effects include diffusion, surface segregation, and surface migration. The assessment of the properties of impurities requires the characterization of semiconductor

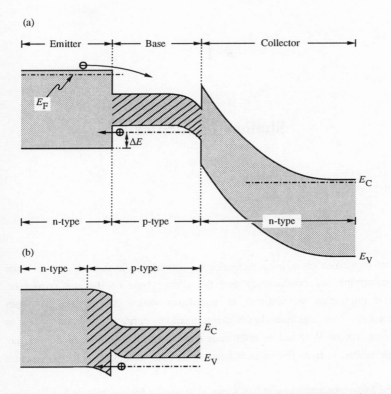

Fig. 3. Schematic band diagram of a hetero-bipolar transistor consisting of wide-gap emitter, base, and wide-gap collector. The locations of the emitter–base pn-junction and of the heterojunction coincide and are displaced in (a) and (b), respectively.

materials by means of electronic, optical, and structural measurement techniques. The techniques frequently reveal that III–V semiconductors contain shallow levels but also a wide variety of deep levels which can be due to impurity atoms, native defects, and complexes.

During the last decade, doping of III–V semiconductors has made tremendous progress. During this time, doping has emerged as an entire discipline of III–V semiconductor research and development. This discipline has many facets which are of great importance for further development and for production of photonic and electronic components and integrated microsystems. This book is dedicated to the many aspects of doping and may provide further insight into this discipline and also serve as a reference for those working with compound semiconductors.

1

Shallow impurities

Shallow impurities are of great technological importance in semiconductors since they determine the conductivity and the carrier type of the semiconductor. Shallow impurities are defined as impurities which are ionized at room temperature. This condition limits the ionization energy of such impurities to values << 100 meV. Shallow impurities can be either acceptors or donors, i.e. 'accept' electrons from the valence band or 'donate' electrons to the conduction band.

The hydrogen atom model can serve as the basis for the calculation of many properties of shallow impurities such as ionization energy and state wave functions. In this chapter, the hydrogen atom is analyzed in terms of Bohr's semi-classical model and in terms of a quantum mechanical approach. The hydrogen atom model is then applied to shallow impurities. Properties such as ionization energies, wave functions, central cell correction terms, and screening of impurity potentials by free carriers are summarized. Effects associated with high impurity or high free carrier concentrations are also discussed including the Mott transition, the Burstein–Moss shift, band tails, impurity bands, and bandgap shrinkage.

1.1 Hydrogenic impurities

Impurities in semiconductors can be incorporated on substitutional sites, interstitial sites, or as impurity complexes. Here, we restrict ourselves to *substitutional, shallow* impurities. Examples for such impurities are Be, Zn, Si, and Sn. These impurities are shallow, i.e. their ionization energy is comparable to the thermal energy kT at room temperature. As a consequence, shallow impurities are fully ionized at room temperature. The hydrogen atom model has proven to predict accurately many properties of shallow impurities. This section summarizes the properties of the semi-classical hydrogen atom, the quantum mechanical hydrogen atom, hydrogenic donors, and hydrogenic acceptors. Finally, central-cell corrections to the hydrogenic model are discussed.

1.1.1 Coulomb potentials and Bohr's hydrogen atom model

The electrostatic potential of a point charge is called the Coulomb potential or the $1/r$ potential. The Coulomb potential of a positive point charge $(+e)$ in vacuum located at $r = 0$ is obtained from Poisson's equation and is given in spherical coordinates by

$$V(r) = \frac{e}{4\pi\varepsilon_0 r} \tag{1.1a}$$

where e is the elementary charge and ε_0 is the permittivity of vacuum. Analogously, the Coulomb potential of a positively charged impurity located at $r = 0$ in a semiconductor with the dielectric constant $\varepsilon_r = \varepsilon/\varepsilon_0$ is given by

$$V(r) = \frac{e}{4\pi\varepsilon r} \tag{1.1b}$$

where ε is the permittivity of the semiconductor.

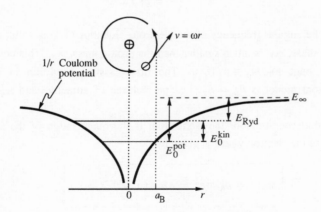

Fig. 1.1. Illustration of Bohr's hydrogen atom model. The electron orbits the positive proton with a $1/r$ Coulomb potential at a radius r_B. The ionization energy of the atom is the Rydberg energy $E_{Ryd} = E_\infty - |E_0^{pot}| + |E_0^{kin}|$.

The hydrogen atom model developed by Bohr is based on (i) classical mechanics of an electron in the Coulomb potential of a positive point charge and on (ii) the quantization of the electron angular momentum. The Bohr model predicts many of the physical properties of the hydrogen atom most notably the emission spectra of the atom. The model is a fascinating example of the simplicity and the power of quantum mechanics. For the classical motion of an electron in the Coulomb potential, the potential energy is given by

$$E_{pot} = \frac{-e^2}{4\pi\varepsilon_0 r}. \tag{1.2}$$

For the hydrogen atom, the permittivity is that of vacuum since a vacuum is assumed between the proton and the electron. The schematic Coulomb potential and an electron orbiting the proton at a distance r are shown in Fig. 1.1. The *attractive* electrostatic Coulomb force F_c on the proton and the electron towards each other is given by

$$F_c = e\mathcal{E}_c = \frac{1}{4\pi\varepsilon_0} \frac{e^2}{r^2} \tag{1.3}$$

where $\mathcal{E}_c = dE_{pot}/dr$ is the Coulomb field. The *repulsive* centrifugal force F_z on the orbiting electron with mass m_0 and velocity v is given by

$$F_z = \frac{m_0 v^2}{r} = m_0 r\omega^2 \tag{1.4}$$

where ω is the angular frequency of the electron. Equation (1.4) is valid, only if the electron mass, m_0, is much smaller than the proton mass, m_p. This condition is fulfilled, since $m_0/m_p \cong 1/1840$. The first classical condition in Bohr's hydrogen atom model is $F_c = F_z$, i.e. the balance of attractive and repulsive forces.

The second condition for the Bohr atom is the quantization of the angular momentum of the electron which is given by

$$\boxed{m_0 vr = m_0 \omega r^2 = n\hbar \quad (n = 1,2,3\cdots)} \tag{1.5}$$

The validity of the angular momentum quantization can be visualized by recalling the wave character of the electron. An electron wave around the positive proton is shown in Fig. 1.2. The electron wave is stable, only if the wave is interfering constructively with itself, i.e. when the length of the electron orbit equals integer multiples of the electron wavelength, λ,

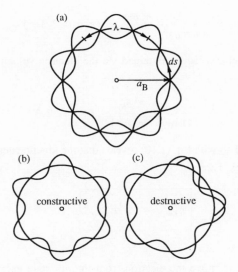

Fig. 1.2. Quantization of the electron orbit of an atom in terms of integer multiples of the electron de Broglie wavelength. Only if $2\pi r_B = n\lambda$, does constructive interference of the electron wave occur.

$$2\pi r = n\lambda \quad (n = 1,2,3\cdots) \ . \tag{1.6}$$

The reader can easily verify that Eqs. (1.5) and (1.6) are identical by recalling that the kinetic energy of a particle is given by $E = \hbar^2 k^2 / 2m_0$ where $k = 2\pi/\lambda$ and m_0 are the electron wave vector and mass, respectively.

Elimination of force and velocity from Eqs. (1.3)–(1.5) yields the radii of the allowed electron orbits in the hydrogen atom

$$a_{B,n} = \frac{4\pi\varepsilon_0 \, n^2 \hbar^2}{m_0 e^2} \quad (n = 1,2,3\cdots) \tag{1.7}$$

The radius of the ground state orbit ($n=1$) is given by

$$a_B = 0.53 \ \text{Å} \tag{1.8}$$

which is called the **Bohr radius** of the hydrogen atom.

Insertion of the Bohr radii into Eq. (1.2) yields the potential energy of the electron

$$E_{\text{pot},n} = \frac{-1}{(4\pi\varepsilon_0)^2} \frac{e^4 m_0}{n^2 \hbar^2} \quad (n = 1,2,3\cdots) \ . \tag{1.9}$$

Furthermore, the kinetic energy is obtained via the electron velocity of Eq. (1.5) according to

$$E_{\text{kin},n} = \frac{1}{2} \frac{1}{(4\pi\varepsilon_0)^2} \frac{e^4 m_0}{n^2 \hbar^2} \quad (n = 1,2,3,\cdots) \ . \tag{1.10}$$

Comparison of Eq. (1.9) with Eq. (1.10) reveals that the kinetic energy is just half of the potential energy, i.e.

$$E_{\text{kin},n} = \tfrac{1}{2} \, |E_{\text{pot},n}| \tag{1.11}$$

The energy required to move the electron from the nth state energy, E_n, to the vacuum level at infinite distance from the proton, i.e. $E_\infty = E(r \rightarrow \infty)$, is given by

$$E_{\text{Ryd},n} = \frac{1}{2} \frac{1}{(4\pi\varepsilon_0)^2} \frac{e^4 m_0}{n^2 \hbar^2} \quad (n = 1,2,3\cdots) \tag{1.12a}$$

which, for $n = 1$, is called the **Rydberg energy**. This energy is required to ionize a hydrogen atom. For $n = 1$ the Rydberg energy is given by

$$E_{\text{Ryd}} = 13.6 \text{ eV} \ . \tag{1.12b}$$

In the classical orbital motion of the electron around the proton, the ratio of the electron velocity and the velocity of light c can be calculated from the Bohr model. The ratio is obtained as

$$\alpha = \frac{v}{c} = \frac{e^2}{4\pi\varepsilon_0 \hbar c} \cong \frac{1}{137} \tag{1.13}$$

which is called the **Sommerfeld fine structure constant**. Evaluation of Eq. (1.13) yields that the electron orbits the proton with a velocity of approximately 2200 km/s.

The magnetic field generated by the circular current of the orbiting electron can be calculated from Bohr's model using the Maxwell equations. It is given by

$$\mu_B = \frac{1}{2} \mu_0 \hbar \frac{e}{m_0} \tag{1.14}$$

and is called the ***Bohr magneton***.

The above calculation demonstrates that the relatively simple Bohr model, i.e. classical mechanics and angular momentum quantization, provides many physical quantities of the hydrogen atom. The calculated state energies of the hydrogen atom were found to agree with hydrogen emission spectra. The Bohr model and its prediction of the electron energies was one of the first successes of the quantum theory. Further refinement of the model is obtained by considering not only circular orbits but also elliptical orbits. On such an elliptical orbit the velocity of the electron is a function of the position, i.e. the velocity is not constant as in the circular orbit. (The position dependence of the velocity is analogous to the planetary motion around the sun).

Using the momentum $p = \hbar k$ and $k = 2\pi/\lambda$, the angular momentum quantization condition, which for circular motion is given by Eq. (1.6), can be written for any orbit as

$$\boxed{\frac{1}{2\pi} \int_0^{2\pi} p_\phi \, d\phi = n_\phi \hbar \quad (n_\phi = 1,2,3,\cdots)} \tag{1.15}$$

where $p_\phi = m\omega_\phi r^2$ is the position-dependent (that is angle-dependent) angular momentum. For a circular orbit, the angular momentum is a constant and Eq. (1.15) reduces to Eq. (1.5).

The condition of classical mechanics for motion on an elliptical orbit and the angular momentum quantization condition lead to the total energy of the electron. The total energy is given by

$$E_{n_\phi} = \frac{1}{2} \frac{1}{(4\pi\varepsilon_0)^2} \frac{e^4 m_0}{n_\phi^2 \hbar^2} \quad (n_\phi = 1,2,3) \tag{1.16}$$

which is identical to Eq. (1.12). Thus, elliptical orbits for the electron exist and have the same energy as electrons on circular orbits. The total energy of a particle on an elliptical orbit in a $1/r$ potential can be calculated by classical mechanics and depends only on the main axis a of the ellipsis. The main axis is then given by

(a)

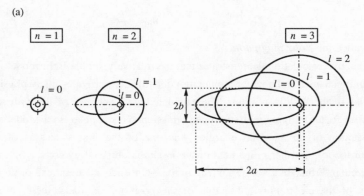

(b)

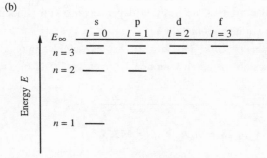

Fig. 1.3. (a) Allowed orbits of an electron in a hydrogen atom. The energies of different electron states are identical for electron orbits with the same main axis of the ellipses. (b) Energies of the electron in the hydrogen atom for different principal and angular momentum quantum numbers.

$$a = \frac{4\pi\hbar^2 \, \varepsilon_0}{m_0 e^2} \, n_\phi^2 . \tag{1.17}$$

The angular momentum of the particle on the elliptical orbit is given by

$$p_\phi^2 = \frac{a(1 - \varepsilon^2) \, e^2 m_0}{4\pi\varepsilon_0} \tag{1.18a}$$

where ε is the eccentricity of the ellipse. Since the angular momentum is quantized according to $p_\phi = n_\phi \hbar$, one obtains with Eq. (1.18a)

$$n_\phi^2 \hbar^2 = \frac{a(1 - \varepsilon^2) e^2 m_0}{4\pi\varepsilon_0} . \tag{1.18b}$$

Inserting the main axis, a (see Eq. (1.17)) into Eq. (1.18b) yields

$$\frac{b}{a} = \frac{n_\phi}{n} \qquad (1.19)$$

where a and b are the axes of the ellipsis. The angular momentum quantum number n_ϕ can assume values of 1, 2, 3, $\cdots$, n, which represents a family of ellipses. If the angular quantum number coincides with the principal quantum number, i.e. $n_\phi = n$, the previously calculated circular orbit is obtained. The ellipses for the $n = 1, 2$ and 3 states are shown in Fig. 1.3(a).

If the angular momentum quantum number is formally introduced as $l = n_\phi - 1$, then l can assume values of

$$l = 0, 1, 2, \cdots, n-1. \qquad (1.20)$$

The value of $l = 0$ corresponds to the ellipsis with the largest eccentricity. The value of $l = n-1$ represents the circular orbit.

Each orbit of an electron around the proton of a hydrogen atom is fully determined by the principle quantum number, n, and the angular quantum number, l. The orbits of the electron for different quantum numbers and the corresponding energies are shown in Fig. 1.3. Frequently, the $l = 0, 1, 2$ and 3 orbitals are historically denoted as the s, p, d and f orbitals (White, 1934). If several quantum states have the same energy the states are called *degenerate states*. For example, the two states determined by $n = 2, l = 0$ and $n = 2, l = 1$ are degenerate. In addition to the principal quantum number and the angular momentum number, the quantum number m describes the quantization of the azimuthal angular momentum in units of $\hbar$. For a discussion of the azimuthal quantum number, we refer to the literature (see, for example, Bohm, 1951).

Upon any perturbation of the hydrogen atom, different electron orbits respond in different ways to the perturbation. Therefore, degenerate electron states will split and become non-degenerate upon a suitable perturbation. The perturbation of the hydrogen atom can be achieved, for example, by an electric field (Stark effect) or a magnetic field (Zeeman effect).

Even though the Bohr model explains many characteristics of the hydrogen atom it is limited in its applicability. For example, if the principles of the Bohr model are applied to the helium atom, incorrect results are obtained for the energy levels in that atom. In addition, the deterministic Bohr hydrogen model violates the quantum mechanical uncertainty principle. That is, momentum and position of the electron are exactly determined at all times in Bohr's model, which contradicts $\Delta x \Delta p \cong \hbar$. Nevertheless, due to its simplicity and clarity, the Bohr model has not lost its attractiveness.

The Bohr model was refined in 1925 by inclusion of the electron spin. The spin of an electron is also called its intrinsic angular momentum and can be

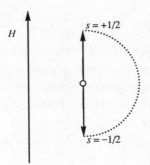

Fig. 1.4. Alignement of the spin of an electron in an external magnetic field H. The spin is either parallel (spin 'up') or antiparallel (spin 'down') to the magnetic field, as represented by the spin quantum number $s = \pm 1/2$. The difference in angular momentum between the two orientations is $\hbar$.

visualized as the rotation of an electron around its own symmetry axis. Goudsmit and Uhlenbeck postulated the spin when conducting Zeeman effect experiments on hydrogen. Due to angular momentum quantization the difference between intrinsic angular momenta is $\hbar$. The intrinsic angular momentum is then given by

$$p_s = s\hbar \quad \text{with} \quad s = \pm\tfrac{1}{2} \tag{1.21}$$

where s is the spin quantum number which can assume values of $s = \pm 1/2$. If an electron with spin is subjected to an external magnetic field, the spin-axis will orientate (align) itself in a parallel or antiparallel manner, as shown in Fig. 1.4. All other orientations are of transient nature since the repulsive and attractive magnetic forces of the spin and external field tend to reorientate the spin to the parallel or antiparallel orientation.

1.1.2 Wave functions of the hydrogen atom

The Bohr model predicts the energy levels of the hydrogen atom with amazing accuracy. However, the wave functions ψ_{nlm} and the probability distributions $\psi_{nlm}\,\psi_{nlm}^{*}$ cannot be obtained from the Bohr model. The Schrödinger equation must be solved in order to obtain the exact solutions. Although the exact solution of the hydrogen atom model goes beyond the scope of this book, the results of the exact hydrogen atom are summarized in the following. For further study, the reader is referred to textbooks on quantum mechanics (Bohm, 1951; Sherwin, 1959; Davydov, 1965; Borowitz, 1967; Saxon, 1968; Merzbacher, 1970).

The wave functions of the hydrogen atom can be obtained by solving the Schrödinger equation. For a particle with mass m and charge e in a potential V, the Schrödinger equation is given by

$$-\frac{\hbar^2}{2m}\,\Delta\Psi + eV\Psi = -\frac{\hbar}{i}\,\frac{\partial}{\partial t}\,\Psi \qquad (1.22)$$

where $\Psi = \Psi(x, y, z, t) = \psi(x, y, z)\,e^{i\omega t}$ is the time-dependent wave function and $\Delta = (\partial^2/\partial x^2 + \partial^2/\partial y^2 + \partial^2/\partial z^2)$ is the delta operator. Not being interested in the time dependence of the solution, we use the kinetic energy operator $E = -(\hbar/i)(\partial/\partial t)$ to obtain the time-independent Schrödinger equation

$$\boxed{-\frac{\hbar^2}{2m}\,\Delta\psi + (eV-E)\psi = 0} \qquad (1.23)$$

where $\psi = \psi(x,y,z)$ is the time-independent wave function. The Coulomb potential is a spherically symmetric potential and can be expressed solely as a function of the radius r, i.e. $V = V(r)$ (see Eq. (1.1)). It is useful to convert the Δ-operator into spherical coordinates (r, θ, ϕ), i.e.

$$\Delta = \frac{\partial^2}{\partial x^2} + \frac{\partial^2}{\partial y^2} + \frac{\partial^2}{\partial z^2}$$

$$\qquad (1.24)$$

$$= \frac{1}{r^2}\,\frac{\partial}{\partial r}\left[r^2\,\frac{\partial}{\partial r}\right] + \frac{1}{r^2\sin\theta}\,\frac{\partial}{\partial\theta}\left[\sin\theta\,\frac{\partial}{\partial\theta}\right] + \frac{1}{r^2\sin^2\theta}\,\frac{\partial^2}{\partial\phi^2}$$

where r, θ and ϕ are the radius, the polar angle, and azimuthal angle, respectively. The Schrödinger equation is a separable linear differential equation and can be solved by employing the product method. The wave functions can then be written as

$$\psi(r,\theta,\phi) = R(r)\,\Theta(\theta)\,\Phi(\phi)\,. \qquad (1.25)$$

Since $\psi\psi^*$ is the quantum mechanical probability density of the particle, the wave function ψ must satisfy the condition

$$\int_{-\infty}^{\infty}\int_{-\infty}^{\infty}\int_{-\infty}^{\infty} \psi\psi^*\,\mathrm{d}x\,\mathrm{d}y\,\mathrm{d}z = \int_{0}^{\infty}\int_{0}^{2\pi}\int_{0}^{\pi} \psi\psi^*\,r^2\,\sin\theta\,\mathrm{d}\theta\,\mathrm{d}\phi\,\mathrm{d}r = 1\,. \qquad (1.26)$$

The solution of the Schrödinger equation in spherical coordinates for the Coulomb potential is a set of orthogonal functions which are usually classified by the four quantum numbers n, l, m_l, and s. The quantum numbers are

n = the principal quantum number

l = the orbital angular momentum number

m_l = the azimuthal quantum number

s = the spin quantum number

The quantum number can assume values of

$$n = 1,2,3,\ldots \tag{1.27a}$$

$$l = 0,1,2,\ldots,n-2,\, n-1 \tag{1.27b}$$

$$m_l = -l,-l+1,\ldots,l-1,+l \tag{1.27c}$$

$$s = -\tfrac{1}{2},\, +\tfrac{1}{2}\,. \tag{1.27d}$$

Two electrons with different spin can occupy an orbit defined by the three quantum numbers n, l, and m_l (Pauli principle).

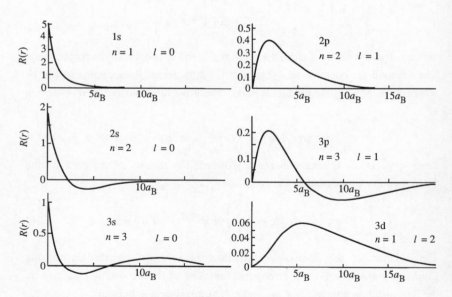

Fig. 1.5. Calculated radial parts of the wave functions for some elementary states of the hydrogen atom.

The wave functions corresponding to the three quantum numbers n, l, and m_l are designated as ψ_{nlm_l}. Correspondingly, the radial part of the wave functions are denoted as R_{nlm_l} (see (Eq. 1.25)). The wave functions for some of the lowest states of the hydrogen atom are (Sherwin, 1959)

$$\psi_{100} = \frac{1}{\sqrt{\pi}} \left[\frac{1}{a_B} \right]^{3/2} e^{-r/a_B} \tag{1.28a}$$

$$\psi_{200} = \frac{1}{4\sqrt{2\pi}} \left[\frac{1}{a_B} \right]^{3/2} \left[2 - \frac{r}{a_B} \right] e^{-r/2a_B} \tag{1.28b}$$

$$\psi_{210} = \frac{1}{4\sqrt{2\pi}} \left[\frac{1}{a_B} \right]^{3/2} (\cos\theta) \frac{r}{a_B} e^{-r/2a_B} \tag{1.28c}$$

$$\psi_{211} = \frac{1}{4\sqrt{2\pi}} \left[\frac{1}{a_B} \right]^{3/2} \frac{e^{i\phi}}{\sqrt{2}} (\sin\theta) \frac{r}{a_B} e^{-r/2a_B} \tag{1.28d}$$

$$\psi_{21-1} = \frac{1}{4\sqrt{2\pi}} \left[\frac{1}{a_B} \right]^{3/2} \frac{e^{-i\phi}}{\sqrt{2}} (\sin\theta) \frac{r}{a_B} e^{-r/2a_B} . \tag{1.28e}$$

Further solutions of the hydrogen atom state can be found in the literature (Bohm, 1951). The radial parts R_{nl0} ($m_l = 0$) for some of the lowest hydrogen atom states are shown in Fig. 1.5. States of 's-type' (i.e. $l = 0$) symmetry have a maximum of the wave function at $r = 0$. States of 'p-type' (i.e. $l = 1$) symmetry have a node at $r = 0$.

The probability of finding an electron at radius r can be obtained by integration over all angles θ and ϕ

$$p(r)\,dr = \int_0^{2\pi} \int_0^{\pi} \psi\psi^* \, r^2 \sin\theta \, d\theta \, d\phi \, dr . \tag{1.29}$$

For wave functions of s-type symmetry, ψ does not depend on θ and ϕ. Eq. (1.29) then yields

$$p(r) = \psi\psi^* 4\pi r^2 . \tag{1.30}$$

Using $\psi = \psi_{100}$ according to Eq. (1.28) one obtains

$$p(r) = 4 \frac{r^2}{a_B^3} e^{-2r/a_B} . \tag{1.31}$$

The probability $p(r)$ has a maximum at $r = a_B$. Thus, the classical Bohr radius is the radius of maximum probability in the quantum mechanical picture.

1.1.3 Hydrogenic donors

The hydrogen atom model can be applied to shallow donors in III–V semiconductors. Properties predicted by the hydrogen atom model agree amazingly well with experimentally determined properties of shallow donors. Such donors are called *effective-mass-like donors, hydrogen-like donors*, or briefly *hydrogenic donors*. Ionization energy, wave functions, and effective Bohr radius are well predicted for such donors. The similarity of the hydrogen atom and donor impurities originates in the $1/r$ coulombic potential of both entities.

Two modifications are required in order to apply Bohr's hydrogen atom model to shallow donors. These corrections are related to the effective mass of carriers and of the dielectric constant of the semiconductor (Bethe, 1942). First, the effective mass of electrons in semiconductors, m_e^*, differs from the electron mass quite significantly. The dispersion relation of a semiconductor with a spherically symmetric band structure is given by

$$E = \frac{\hbar^2 k^2}{2m_e^*} .$$

(1.32)

Thus, the dispersion relation $E(k)$ allows one to determine the effective mass according to

$$m_e^* = \hbar^2 \left[\frac{d^2 E}{dk^2} \right]^{-1} .$$

(1.33)

The dispersion relation of Eq. (1.32) differs from the dispersion relation of a free electron just by the magnitude of the electron mass. In order to apply the hydrogen atom model, the electron mass must be replaced by the effective electron mass. The second correction arises from the dielectric properties of semiconductors. The Coulomb potential of a positive point-charge in a semiconductor located at $r = 0$ is given by

$$V(r) = \frac{e}{4\pi\varepsilon r}$$

(1.34)

which differs from Eq. (1.1a) by the static dielectric constant $\varepsilon_r = \varepsilon/\varepsilon_0$. How does the potential change in the presence of an electron orbiting the impurity charge? The polarization of the valence electrons is then more complicated and cannot be taken into account by the substitution of ε for ε_0. To answer the question we first make the simplifying assumption that the electronic charge can be described by a diffuse electron cloud with a spatial extent much larger than the lattice constant. In the limit of an infinitely large electron cloud, the potential of the positive impurity is correctly described by Eq. (1.34). The situation changes, however, if the electron were to orbit the impurity atom with a radius comparable to the lattice constant. In this case the polarization of the lattice depends on the donor as well as on the electron charge. The true potential is not given by the dielectrically screened potential of Eq. (1.34). For a finite size of the electron orbit, the polarization of lattice atoms is overestimated by Eq. (1.34). A smaller dielectric constant $\varepsilon_r^* < \varepsilon_r$ can be used to account for the reduction in polarization. It should therefore be noted that Eq. (1.34) assumes a Coulomb potential screened by the dielectric properties of the semiconductor (i.e. by polarization of tightly bound valence electrons and nuclei of the lattice) and that the equation can only be used if the electron can be described by a diffuse electron cloud with a large spatial extent. The true potential $V(r)$ which arises from the positive donor ion, the electron bound to the donor ion, and the polarization of the surrounding semiconductor is rather complicated and cannot be expressed in terms of a simple $1/r$ potential (Kohn, 1957a, 1957b). Employing the approximate $1/r$ potential, the Schrödinger equation is given by

$$-\frac{\hbar^2}{2m_e^*} \Delta\psi + \frac{e}{4\pi\varepsilon r} \psi = E\psi \qquad (1.35)$$

which is called the effective-mass equation for a hydrogenic impurity.

Using the effective-mass and the dielectric constant corrections, the following properties of hydrogenic impurities can be derived. The *effective Bohr radius* is obtained from Eq. (1.7) and is given by

$$a_{B,n}^* = \frac{4\pi\varepsilon n^2 \hbar^2}{m_e^* e^2} \qquad (n=1,2,3,\cdots) . \qquad (1.36)$$

The radius of the donor ground state ($n = 1$) is then given by

$$a_B^* = \frac{4\pi\varepsilon\hbar^2}{m_e^* e^2} = \frac{\varepsilon_r}{m_e^*/m_0} a_B = \frac{\varepsilon_r}{m_e^*/m_0} 0.53 \text{ Å} \qquad (1.37)$$

The effective Bohr radius is also called the *donor Bohr radius*. As an example, we consider a hydrogenic donor in GaAs with $\varepsilon_r = 13.1$ and $m_e^* = 0.067 \, m_0$. Insertion of these values into Eq. (1.37) yields $a_B^* = 103$ Å which is the effective Bohr radius of donors in GaAs.

The *effective Rydberg energy* is obtained by applying the effective-mass and the dielectric constant corrections to Eq. (1.12).

$$E_{\text{Ryd},n}^* = \frac{1}{2} \frac{1}{(4\pi\varepsilon)^2} \frac{e^4 m_e^*}{n^2 \hbar^2} \quad (n=1,2,3, \dots) \ . \tag{1.38}$$

The **donor ionization energy** is required for a transition from $n=1$ to $n\rightarrow\infty$ and is given by

$$E_{\text{d}} = \frac{e^4 m_e^*}{2\,(4\pi\varepsilon\hbar)^2} = \frac{m_e^*/m_0}{\varepsilon_r^2}\,E_{\text{Ryd}} = \frac{m_e^*/m_0}{\varepsilon_r^2}\,13.6\ \text{eV} \tag{1.39}$$

The donor ionization energy is occasionally also referred to as donor Rydberg energy. As an example, we consider a hydrogenic donor in GaAs and obtain $E_{\text{d}} = 5.3$ meV which is in agreement with experimental results.

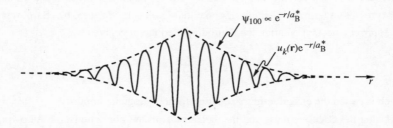

Fig. 1.6. Schematic illustration of a hydrogenic donor wave function with the quantum numbers $n = 1$ and $l = m_l = 0$. The wave function is the product of the lattice-periodic Bloch function $u_k(\mathbf{r})$ and the envelope function ψ_{100}.

Finally, the wave functions of hydrogenic donors can be obtained from Eq. (1.28) by substituting the effective Bohr radius for the Bohr radius. The ground-state envelope wave function is then obtained as

$$\psi_{100}(r) = \frac{1}{\sqrt{\pi}} \left[\frac{1}{a_{\text{B}}^*}\right]^{3/2} e^{-r/a_{\text{B}}^*} \ . \tag{1.40}$$

It should be noted that Eq. (1.40) describes the donor *envelope* function rather than the donor wave function. The actual donor ground-state wave function is given by (Kohn, 1957a)

$$\psi_{\text{d},100}(\mathbf{r}) = \psi_{100}(r)\,u_k(\mathbf{r}) \tag{1.41}$$

where $u_k(\mathbf{r})$ is the lattice-periodic factor of the well-known Bloch function of conduction band electrons. The function $u_k(\mathbf{r})$ has translational symmetry with respect to the semiconductor lattice constant. The ground-state wave function according to Eq. (1.41) is schematically shown in Fig. 1.6. The dashed curve represents the impurity envelope function of Eq. (1.40).

It should be noted that the use of Bohr's hydrogen atom model for shallow impurities is not self-sufficient. The *ab initio* assumption of a 'large' electron cloud and the substitution of effective electron mass and dielectric constant cannot be justified solely on the basis of the hydrogenic model. Even though the hydrogenic model yields a relatively large electron orbit, this result does not justify the initial assumptions. However, more rigorous calculations (Kohn 1957a, 1957b; Madelung, 1978; Altarelli and Bassani, 1982) indeed demonstrate that the electron distribution has a spatial extent much larger than the lattice constant. The substitution of the electron mass, m_0, by the effective mass, m_e^*, is therefore justified since the electron orbit around the donor extends over many lattice constants. The effective mass of electrons in semiconductors is a direct consequence of the periodic potential of the lattice. Thus, electrons bound to donors are subject to the periodic potential, since the effective Bohr radius is much larger than the lattice constant, $a_B^* \gg a_0$. If, in contrast, electrons were tightly bound ($a_B^* \cong a_0$) the effective-mass correction could not be applied. Similar arguments apply to the substitution of the permittivity of the semiconductor, ε, for the permittivity of vacuum, ε_0. The lattice atoms are polarized by the Coulomb field which results in its reduction as compared to the field without polarization. The effect of the polarization is taken into account via the dielectric constant. Since the effective Bohr radius extends over many lattice constants, the use of the dielectrically screened Coulomb potential is justified. Despite the simplicity of the hydrogen atom model for shallow donors, the model yields quite accurate results.

The ***degeneracy*** of the donor ground state is a quantity required for the occupancy probability of the donor state (see Chap. 3). The ground state has the quantum numbers $n = 0$, $l = 0$, $m_l = 0$ and $s = \pm\frac{1}{2}$. Thus, since the donor ground state can be occupied by an electron with spin $+\frac{1}{2}$ or $-\frac{1}{2}$, the ground state degeneracy is $g = 2$.

1.1.4 Hydrogenic acceptors

The application of the hydrogenic model to acceptors in III–V semiconductors is complicated by their degenerate valence band structure. For hydrogenic donors, the effective electron mass was substituted for the electron mass. The substitution was possible, since the conduction band was assumed to be parabolic, isotropic, and non-degenerate (as for most III–V semiconductors). Such a simple substitution is not possible for acceptors, since the valence band structure of III–V semiconductors is much more complicated than the conduction band structure. The electronic band structure of several III–V semiconductors with zincblende structure was calculated by Chelikowsky and Cohen (1976). The band structure near the center of the Brillouin zone is schematically shown in Fig. 1.7. The highest point of the valence band is located at $k = 0$. Without spin-orbit coupling this point would be sixfold degenerate with three dispersion relations and twofold spin degeneracy (Kohn, 1957a). The simplest way to understand this degeneracy is to consider the tight binding limit, in which the wave functions corresponding to the highest point go over into atomic 3p functions. The spin-orbit coupling lifts the degeneracy partially and leads to the situation shown in Fig. 1.7. The top of the valence band remains fourfold degenerate at $k = 0$. The corresponding dispersion relations are called the heavy hole (hh) and light hole (lh) dispersion relations. The top of the twofold (spin) degenerate split-off (so) band is at $k = 0$ at an energy E_{so} below the valence band maximum where E_{so} is the spin-orbit coupling energy. The top of the valence band corresponds to atomic $j = 3/2$ states (Kohn, 1957a), where j is the total angular momentum (obit + spin), i.e. $j = l+s = l + \frac{1}{2} = 3/2$. The *inner quantum number j* is formally not necessary since it can be expressed by l and s. The introduction of j by Sommerfeld (1920) has historic reasons (Finkelnburg, 1958). The top of the valence band has Γ_8 symmetry. The split-off band corresponds to atomic $j = l-s = 1 - \frac{1}{2} = \frac{1}{2}$ states which have Γ_7 symmetry.

For strong spin-orbit coupling, the split-off band is far removed from the top of the valence band, i.e. $\Delta E_{so} \gg E_a$ where E_a is the acceptor binding energy. In this case, both the heavy hole and light hole band must be taken into account and the Hamiltonian is, therefore, a 4×4 matrix (Kohn, 1957a). In the limit of weak spin-orbit coupling, i.e. $\Delta E_{so} \cong E_a$, all three valence bands must be taken into account and the Hamiltonian is a 6×6 matrix.

Kohn (1957a) used a 6×6 Hamiltonian matrix to calculate acceptor energies in cubic semiconductors. The author used variational envelope wave functions for acceptors of the form

$$\psi_i(r) = A_i e^{-r/r_i} \tag{1.42}$$

where r_i is a variational parameter. Subsequently, Baldareschi and Lipari (1973) developed a now widely accepted model for shallow acceptor states in cubic semiconductors with degenerate valence bands. In their approach, the Hamiltonian is written as the sum of a spherical term and a cubic correction, thus pointing out the relevance of the spherical symmetry in the acceptor problem and the strong similarity to the case of atoms with spin-orbit interaction. Neglecting the cubic term, Hamiltonians with radial symmetry were obtained. Variational

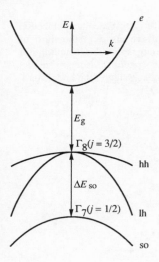

Fig. 1.7. Schematic illustration of the electron, heavy hole (hh), light hole (lh) and split-off (so) dispersion relations near the center of the Brillouin zone ($k = 0$).

wave functions were used to calculate acceptor ionization energies.

In the limit of strong spin-orbit interaction, that is for spin-orbit splittings (ΔE_{so}) much larger than the acceptor energy (E_a), Baldareschi and Lipari (1973) calculated the effective Bohr radius. For an effective hole mass m_h^*, they obtained the effective Bohr radius

$$a_B^* = \frac{4\pi\varepsilon\hbar^2\gamma_1}{e^2 m_h^*} \, , \qquad (1.43)$$

the effective Rydberg energy

$$E_{Ryd}^* = \frac{e^4 m_h^*}{2(4\pi\varepsilon\hbar)^2\gamma_1} \, , \qquad (1.44)$$

and the acceptor ionization energy

$$E_a = E_{Ryd}^* \, f(\mu) \, , \qquad (1.45)$$

with

$$\mu = \frac{6\gamma_3 + 4\gamma_2}{5\gamma_1} \, . \qquad (1.46)$$

The parameters γ_1, γ_2, and γ_3 are the so-called Luttinger parameters which describe the hole dispersion relation near the center of the Brillouin zone (Luttinger, 1956). The function $f(\mu)$ relates the acceptor energy with the

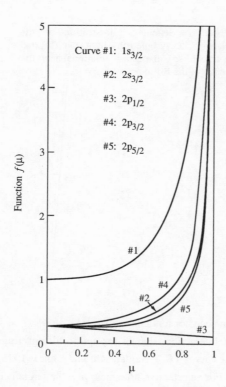

Curve #1: $1s_{3/2}$

#2: $2s_{3/2}$

#3: $2p_{1/2}$

#4: $2p_{3/2}$

#5: $2p_{5/2}$

Fig. 1.8. Calculated function $f(\mu)$ versus μ in the limit of strong spin-orbit coupling for the spherical acceptor model (after Baldareschi and Lipari, 1973).

effective Rydberg energy (see Eq. (1.45)). The function $f(\mu)$ is shown in Fig. 1.8. For the Luttinger parameters of III–V semiconductors, μ assumes values of $\mu \cong 0.6-0.9$ and the function $f(\mu)$ assumes values of $f(\mu) \cong 1.5-4$ for the ground state energy of acceptors. Baldareschi and Lipari (1973) obtained the numerical values of $f(\mu)$ using a variational approach and using spherical trial functions (Kohn, 1957a) for the acceptor wave functions (see Eq. (1.42)). The Luttinger parameters, the variable μ, the effective Rydberg energy, and the ground state ($1s_{3/2}$) and excited state energies of hydrogenic acceptors in several III–V semiconductors are shown in Table 1.1 (Luttinger, 1956; Baldareschi and Lipari, 1973). The calculated and experimental acceptor energies agree reasonably well.

In subsequent work, Baldareschi and Lipari (1974) investigated the contribution of cubic symmetry terms of the Hamiltonian to the spherical model for acceptor states. The effects of the cubic symmetry were studied using perturbation theory which allowed the authors to reproduce all the details of acceptor spectra. The quantitative changes in acceptor energy caused by the cubic term are small, i.e. less than 1 meV for the III–V semiconductors listed in Table 1.1.

Table 1.1. *Values for the Luttinger parameters* $(\gamma, \gamma_2, \gamma_3)$, *the parameter* μ, *the effective Rydberg energy, and ground and excited state energies (in meV) for acceptors in III–V semiconductors.*

	γ_1	γ_2	γ_3	μ	E_{Ryd}^{*} (meV)	$E_{1s_{3/2}}$	$E_{2s_{3/2}}$	$E_{2p_{1/2}}$	$E_{2p_{3/2}}$	$E_{2p_{5/2}}$
AlSb	4.15	1.01	1.75	0.701	22.8	42.4	12.4	3.3	17.5	10.5
GaP	4.20	0.98	1.66	0.661	28.0	47.5	13.7	4.2	19.1	11.7
GaAs	7.65	2.41	3.28	0.767	11.3	25.6	7.6	1.6	11.1	6.5
GaSb	11.80	4.03	5.26	0.808	4.7	12.5	3.8	0.65	5.6	3.2
InP	6.28	2.08	2.76	0.792	14.1	35.2	10.5	2.0	15.5	8.9
InAs	19.67	8.37	9.29	0.907	3.2	16.6	5.1	0.4	7.9	4.4
InSb	35.08	15.64	16.91	0.935	1.2	8.6	2.7	0.2	4.2	2.3

The **degeneracy** of the acceptor ground states in III–V semiconductors is $g = 4$. Typical acceptor energies are much smaller than the spin-orbit splitting energy, i.e. $E_a \ll \Delta E_{so}$. The top of the valence band is fourfold degenerate due to heavy and light hole dispersion and due to twofold spin degeneracy. Since the acceptor wave functions are composed of valence band wave functions near the top of the band (Kohn, 1957a), the acceptor degeneracy is $g = 4$ as well.

1.1.5 Central cell corrections

The ionization energy of hydrogenic donors and acceptors as calculated from effective-mass theory does not depend on the chemical nature of the impurity atom. On the other hand, experimental values of the ionization energy do depend on the chemical nature especially for acceptors in III–V semiconductors. The difference in ionization energy between chemically different impurities is attributed to **central cell potentials**. The central cell potential is assumed to be due to the chemical characteristics (e.g. electronegativity) of the impurity atom and thus the potential leads to a correction of the hydrogenic ionization energy. This correction is frequently referred to as **chemical shift** (Pantelides, 1975).

The total impurity potential is the sum of the Coulomb potential and the central cell potential V_{cc} and can be written as

$$V(r) = \frac{e}{4\pi\varepsilon r} + V_{cc}(r) \ . \tag{1.47}$$

The central cell potential is a short range potential and has a spatial extent of no more than the unit cell (central cell) of the host semiconductor. *Donor* wave functions are usually quite delocalized in III–V semiconductors. Therefore, the central cell corrections play a minor role for donors and their ionization energy is well described by the hydrogen model. In contrast, the *acceptor* Bohr radius is usually much smaller resulting in a significant central cell correction. This difference between donor and acceptor states is indeed observed experimentally, for example in GaAs. Several model potentials have been used for the central potential including a constant potential extending over the unit cell (Abarenkov and Heine, 1965), and δ-function-like potentials. Various models for central cell potentials were reviewed by Pantelides (1978), Stoneham (1975, 1986), and Altarelli and Bassani (1982). Finally, it is worthwhile to note that *isoelectronic* impurities lack the Coulomb term in Eq. (1.47). For isoelectronic impurities, the central cell potential is the only potential that can bind electrons (Thomas and Hopfield, 1966).

A simple model explaining the chemical shift of impurity ionization energies was proposed by Phillips (1970a, 1970b). The model is based on the local strain around the impurity atom. The strain is caused by the mismatch of the valence bonds of the impurity with valence bonds of the host lattice. Using this model, the chemical trend in donor ionization energies of Te, S, and Se in GaP were qualitatively explained (Phillips, 1970b). Phillips (1973) developed a second model in which the chemical shift in ionization energy is based on the difference in electronegativity, ΔX, between the impurity atom and the host lattice. The author showed that a large difference in electronegativity between impurity and the atom replaced by the impurity results in a large chemical shift. The chemical shift was assumed to be proportional to the heat of formation, i.e.

$$\Delta E \propto (\Delta X)^2 \tag{1.48}$$

where ΔE is the difference between the calculated effective mass impurity energy and the actual impurity energy (i.e. ΔE is the chemical shift). Using Eq. (1.48), differences in chemical shifts of impurities occupying cation and anion sites in GaP were explained.

Note that chemical shifts are expected to be larger for impurity states with s-type symmetry as compared to states with p-type symmetry. The central cell potential is spatially restricted to the atomic vicinity of the impurity atom. In this region the amplitude of s-type wave functions is large while p-type wave functions have a node. Thus, perturbation theory predicts greater corrections for s-type symmetry states as compared to p-symmetry states.

1.1.6 Impurities associated with subsidiary minima

The conduction band structure of III–V semiconductors consists of three local minima, which occur at the L, Γ, and X-point of the Brillouin zone. The Γ minimum is located at the center of the Brillouin zone at $k=0$, while the L and X minima occur at finite wave vectors. Donors, which by their very nature are associated with the conduction band, can form donor levels with all local minima of the conduction band. Impurity states associated with subsidiary minima of the conduction band were first analyzed by Bassani et al. (1969). The theoretical study revealed that the impurity states are generally formed from Bloch functions of many Brillouin zones and their respective contributions depend on the particular band structure and on the strength and nature of the impurity potential. The existence of several conduction band minima with large effective masses can increase the number of bound states as compared to the single valley hydrogenic model. It was further shown that resonant states in the continuum of one local minima can be produced by impurity states associated with another minima. This situation is shown in Fig. 1.9 which depicts two donor levels associated with two conduction band minima and a resonant state in the continuum of the energetically lower minimum (Altarelli and Bassani, 1982).

Donor levels associated with subsidiary conduction band minima were experimentally observed in III–V semiconductors. Adler (1969) used hydrostatic pressure to study the effect of donors on the electron transfer in n-type GaAs. Onton et al. (1972) directly observed a subsidiary conduction band minimum and its associated donor levels in InP by optical absorption measurements.

1.2 Screening of impurity potentials

Variations of the electrostatic potential are reduced in magnitude by the spatial redistribution of free carriers. Variations in the band-edge potential can occur due to local doping concentration changes or local compositional changes of a semiconductor. An example of a potential fluctuation is shown in Fig. 1.10. At some location, an excess positive donor charge causes a dip in the band-edge potential. The potential dip attracts electrons and results in a locally higher concentration of electrons at the potential dip. The potential generated by the negative charge of the excess electrons reduces the original fluctuation and smooths the potential, i.e. free carriers *screen* the potential fluctuation. This is

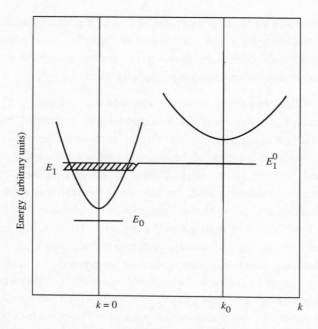

Fig. 1.9. Schematic representation of a bound state E_0 and a resonant state E_1 associated with the absolute minimum and a subsidiary minimum of the band structure, respectively. The resonant state E_1 (originating from the unperturbed state E_1^0) has a finite width due to its degeneracy with band states (after Altarelli and Bassani, 1982).

what screening of potential fluctuations by free carriers is all about. As an example of a potential fluctuation, consider the Coulomb potential of an impurity. As we recall, the impurity potential has a $1/r$ dependence. In the presence of free carriers the Coulomb potential is screened. The resulting potential is called the **screened Coulomb potential** which does *not* have the $1/r$ dependence of the unscreened Coulomb potential.

The potential variation is modeled in terms of the spatially non-uniform electrostatic potential $V(\mathbf{r})$, where $\mathbf{r} = (x,y,z)$ is the spatial coordinate. To find the energy levels in the perturbed potential $V(\mathbf{r})$, Schrödinger's and Poisson's equation must be solved simultaneously in order to find the free carrier distribution and the (screened) potential.

This procedure to calculate the screened potential is quite elaborate and it is usually possible to circumvent it, if the potential $V(\mathbf{r})$ is relatively smooth, i.e. it changes only little over the length of the electron wavelength. The electrons then

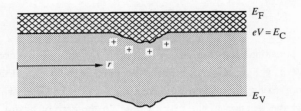

Fig. 1.10. Perturbation of the conduction band edge of a semiconductor by excessive positive charges. Electrons accumulate in the potential minimum and screen (i.e. compensate) the excess charge.

'see' only the potential at their own location. The total energy of the electron then follows the classical sum of potential and kinetic energy, i.e.

$$E = eV(\mathbf{r}) + \hbar^2 k^2 / 2m_{\mathrm{e}}^* . \qquad (1.49)$$

In other words, we assume that the spatial dimensions of the potential perturbation are so large that size quantization of the carrier system does not need to be considered.

The electron concentration in the potential perturbation shown in Fig. 1.10 is next calculated using classical quantization, i.e. using the semiclassical density of states. The conduction band edge sufficiently far away from the perturbation has a potential V_0 and an electron concentration of n_0. The electron concentration at the potential perturbation is given by (see Chap. 3)

$$n(\mathbf{r}) = \int_{eV(\mathbf{r})}^{\infty} f(E) \, \rho_{\mathrm{DOS}}(E) \, \mathrm{d}E \qquad (1.50)$$

where $f(E)$ and $\rho_{\mathrm{DOS}}(E)$ are the Fermi–Dirac distribution and the density of states, respectively. In the following, we differentiate between a degenerate and a non-degenerate electron gas. The results obtained for the two cases are, as will be seen, quite different. For a *non-degenerate* electron gas, Boltzmann statistics is employed and the local electron concentration is obtained as (see Chap. 3)

$$n(\mathbf{r}) = \frac{1}{\sqrt{2}} \left[\frac{m_{\mathrm{e}}^* kT}{\pi \hbar^2} \right]^{3/2} \exp \left[\frac{E_{\mathrm{F}} - eV(\mathbf{r})}{kT} \right] \qquad (1.51)$$

where E_{F} is the Fermi level. For a *degenerate* electron gas, Fermi–Dirac statistics must be employed and the electron concentration in the limit of *extreme degeneracy* $(E_{\mathrm{F}} - eV(\mathbf{r}) \gg kT)$ is given by (see Chap. 3)

$$n(\mathbf{r}) = \frac{(2m_e^*)^{3/2}}{3\pi^2 \, \hbar^3} \, (E_F - eV(\mathbf{r}))^{3/2} \,. \tag{1.52}$$

If the potential perturbation is a 'dip' in the conduction band then $E_F - eV(\mathbf{r}) > E_F - eV_0$ and electrons accumulate in the dip. If the perturbation is a 'bump' in the conduction band then $E_F - eV(\mathbf{r}) < E_F - eV_0$ and electrons deplete at the location of the perturbation. In the case of a 'dip', the potential generated by the excess electrons reduces the magnitude of the dip, i.e. smooths the potential. The resulting potential is obtained from Poisson's equation, which relates the charge density and the potential $V(\mathbf{r})$ according to

$$\nabla^2 V(\mathbf{r}) = -\frac{e}{\varepsilon} \, [\xi(\mathbf{r}) - n(\mathbf{r}) - n_0] \tag{1.53}$$

where $e\xi(\mathbf{r})$ is the concentration of fixed (positive) charge, $\nabla = (\partial/\partial x, \partial/\partial y, \partial/\partial z)$ is the Nabla operator, and $\nabla^2 = \Delta = (\partial^2/\partial x^2 + \partial^2/\partial y^2 + \partial^2/\partial z^2)$. The concentration $\xi(\mathbf{r})$ is homogeneous except at the location of interest. The average concentration of electrons is n_0, while $n(\mathbf{r})$ is the deficiency or excess of charge which depends on $\mathbf{r}$. Equation (1.53) is called a quasi-classical equation of the Thomas–Fermi type. Since $n(\mathbf{r})$ depends in a non-linear manner on $V(\mathbf{r})$, the differential equation is non-linear. It is the basic equation of non-linear screening theory. Unfortunately, the equation has no general solution. However, in the limit of small variations $n(\mathbf{r})$, it is possible to linearize the differential equation. Suppose the potential fluctuations of $V(\mathbf{r})$ are small as compared to $E_F - eV(\mathbf{r})$. Then Eqs. (1.51) and (1.52) can be linearized, i.e.

$$n(\mathbf{r}) = n_0 + \left. \frac{dn}{dV(\mathbf{r})} \right|_{n \, = \, n_0} V(\mathbf{r}) \tag{1.54}$$

which greatly simplifies screening theory. The linearization of $n(\mathbf{r})$, i.e. our restriction to *small* potential fluctuations is the basis of **linear screening theory**. The linearization of Eq. (1.54) allows one to write the Poisson equation (Eq. (1.53)) as

$$\nabla^2 V(\mathbf{r}) = \frac{V(\mathbf{r})}{r_s^2} - \frac{e\xi(\mathbf{r})}{\varepsilon} \tag{1.55a}$$

where

$$r_s = \left[-\left(\frac{dn}{dV} \right)_{V=0} \frac{e}{\varepsilon} \right]^{-1/2} \tag{1.55b}$$

is called the screening radius. The screening radius is usually referred to as the **Debye screening radius** and the **Thomas–Fermi screening radius** for non-degenerate and degenerate electron systems, respectively. Using Eq. (1.55b) and non-degenerate and degenerate statistics for an isotropic and parabolic conduction band (see Chap. 3), the screening radii are obtained as:

Debye screening radius

$$r_D = (\varepsilon kT/e^2 n)^{1/2} \tag{1.56}$$

Thomas–Fermi screening radius

$$r_{TF} = \pi^{2/3} \sqrt{\frac{\varepsilon \hbar^2}{e^2 m^* (3n)^{1/3}}} \tag{1.57}$$

Note the different functional dependences of the Debye and the Thomas–Fermi screening radius on temperature and free carrier concentration. While r_D depends on temperature, r_{TF} is temperature-independent. Furthermore, the Thomas–Fermi screening radius depends very weakly on the electron concentration, i.e. $n^{-1/6}$. Now the analytic solution of the screened potential can be obtained by integration of Eq. (1.55a) and the solution is given by

$$V(\mathbf{r}) = \int_r K(\mathbf{r} - \mathbf{r}') \, \xi(\mathbf{r}') \, d\mathbf{r}' \tag{1.58}$$

with $K(r) = (e/4\pi\varepsilon r) \, e^{-r/r_s}$.

We now calculate the screened potential of a single ionized impurity. The charge distribution of such a single impurity located at the origin of the coordinate system is $\xi(\mathbf{r}) = \delta(\mathbf{r})$. Insertion into Eq. (1.58) yields the **screened Coulomb potential**

$$V(r) = \frac{e}{4\pi\varepsilon r} \, e^{-r/r_s} \tag{1.59}$$

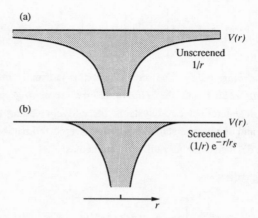

Fig. 1.11. (a) Unscreened and (b) screened Coulomb potentials. The unscreened and screened Coulomb potentials coincide for $r \to 0$ but strongly deviate for large radii.

where the screening radius, r_s, can be either the Debye or the Thomas–Fermi radius depending on the degeneracy of the screening electron gas. The screened Coulomb potential (Debye and Hückel, 1923) is also called the **Yukawa potential** in analogy to a potential in meson theory. The screened Coulomb potential is modified as compared to the unscreened Coulomb potential by the factor e^{-r/r_s}. Note that for $r \to 0$ the screened and the unscreened Coulomb potential become identical. For $r \to \infty$ the screened and unscreened Coulomb potential are strongly diverging due to the e^{-r/r_s} factor. The screened and the unscreened Coulomb potential are illustrated in Fig. 1.11 for an impurity located at $r = 0$. The unscreened potential is a long range potential while the screened one has a shorter spatial range.

There are several major results which are inferred from linear screening theory which are summarized in the following. *First*, the screening of potential perturbations leads to an exponential decay of the perturbing potential with distance, i.e. e^{-r/r_s}, where r_s is the screening radius. *Second*, in the case of several potential perturbations, each perturbation is screened separately. According to Eq. (1.58), a potential fluctuation can be expressed as a superposition of (unscreened) Coulomb potentials. The screened potential is then given by the superposition of the *screened Coulomb potentials. The applicability of the superposition principle is a consequence of the linearity* of the screening theory. *Third*, the screening radius r_s does not depend on the *magnitude* of the potential perturbation to be screened. Instead the screening radius depends on the properties of the screening electron (hole) gas only. The independence of the screening radius on the magnitude of the potential fluctuation is again due to the linearization of screening theory.

The linear screening theory becomes invalid for very large potential fluctuations. The linearization of screening theory given in Eq. (1.54) is based on the assumption of *small* fluctuations. As an example of a large potential perturbation, consider a metal–semiconductor junction (Schottky barrier). Such junctions can induce perturbations of e.g. 1 eV (Schottky barrier height). For such large potential perturbations ***non-linear screening theory*** must be employed to obtain realistic results. Linear screening theory cannot be applied to Schottky barriers. The dependence of the depletion region thickness (i.e. screening length) on the magnitude of the Schottky barrier height of a metal–semiconductor interface is a result of non-linear screening theory.

The nature of screening changes drastically, if the screening carriers are confined to a two-dimensional plane. This situation is referred to as ***two-dimensional screening***. Ando et al. (1982) considered a Coulomb charge located at the cylindrical coordinates $r = 0$ and $z = z_0$ which is screened by electrons confined to the plane $z = 0$. The screened Coulomb potential in the electron plane ($z = 0$) is given by

$$V(r, z = 0) = \int_0^\infty q\, A(q)\, J_0(qr)\, \mathrm{d}q \qquad (1.60)$$

where J_0 is the Bessel function of zero order. The constant $A(q)$ is

$$A(q) = \frac{e}{4\pi\varepsilon} \frac{e^{qz_0}}{q + 1/r_s^{2\mathrm{D}}} \qquad (1.61)$$

where $r_s^{2\mathrm{D}}$ is the two-dimensional screening radius

$$\boxed{r_s^{2\mathrm{D}} = \frac{2\varepsilon}{e^2} \frac{\pi\hbar^2}{m^*}} \qquad (1.62)$$

Note that the two-dimensional screening radius depends neither on temperature, nor on the sheet charge density. This characteristic is a result of the two-dimensional density of states which is constant and does not depend on energy (see Chap. 3). For large values of r, where $r/r_s^{2\mathrm{D}} \gg 0$, the asymptotic form of the potential seen by the electrons is (Stern, 1967)

$$V(r, z = 0) \cong \frac{e(1 + z_0/r_s^{2D})}{4\pi\varepsilon r^3/(r_s^{2D})^2} . \qquad (1.63)$$

This inverse-cube dependence of the potential on distance is much weaker than the exponential decay found in the three-dimensional case, and is one of the principal qualitative differences between two-dimensional and three-dimensional screening (Ando et al., 1982).

1.3 High doping effects

At high doping concentrations, many characteristics of semiconductors change. The changes are due to either the high concentration of impurities or due to the high free carrier concentration accompanying the high doping concentrations. Among the characteristics changed at high doping concentrations are the impurity ionization energy, the fundamental absorption edge, the density of states in the vicinity of the band edges, and the energy of the fundamental gap. Effects causing these changes are the Mott transition, the Burstein–Moss shift, band tailing effects, and bandgap renormalization.

1.3.1 The Mott transition

The Mott transition refers to a *insulator-to-metal* transition occurring in semiconductors at high doping concentrations. Consider an n-type semiconductor with low doping concentration. At low temperatures $(T \rightarrow 0)$, shallow impurities are neutral, i.e. electrons occupy the ground state of the donor impurities. All continuum states in the conduction band are unoccupied. In this case, the semiconductor has the properties of an *insulator*. As the doping concentration increases, the Coulomb potentials of impurities overlap as schematically shown in Fig. 1.12. As a result of the overlapping impurity potentials, electrons can transfer more easily from one donor to another donor. Electrons transfer from one donor state to a state of an adjacent donor by either *tunneling* or by *thermal emission* over the barrier. The probability of both processes increases with decreasing donor separation. In other words, the activation energy for electron transport is reduced. In the extreme case, the activation energy approaches zero, i.e. the conductivity remains finite even for $T \rightarrow 0$. The semiconductor then has *metal*-like properties.

Screening of impurity potentials is a second contribution to the reduction of the impurity ionization energy. Impurity potentials are effectively screened at high free carrier concentrations. Screened potentials are less capable of binding electrons. Thus, the effective ionization energy decreases due to screening.

The insulator-to-metal transition occurs at the impurity concentration at which the distance between impurities becomes comparable to the Bohr radius. If

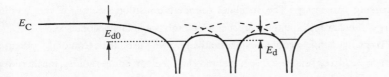

Fig 1.12. Conduction band edge with three donor potentials. For high donor concentrations the Coulomb potentials overlap and the ionization energy E_{d0} is reduced to E_d.

donors with concentration N were to occupy sites of a simple cubic lattice, their separation would be $N^{-1/3}$. The Mott transition would then occur at a concentration

$$2a_B^* = N_{crit}^{-1/3} \tag{1.64a}$$

where a_B^* is the effective Bohr radius and N_{crit} is called the critical concentration. However, Eq. (1.64a) does not give the correct result, because impurities are distributed *randomly* in semiconductors. Using a poissonian distribution of impurities, one can show that the Bohr orbital of an impurity is likely to overlap with the orbitals of one, two, or three neighboring impurities if

$$2a_B^* = \frac{3}{2\pi} N_{crit}^{-1/3} . \tag{1.64b}$$

Rearrangement of the equation yields the *Mott criterion*

$$\boxed{a_B^* \, N_{crit}^{1/3} \cong 0.24} \tag{1.64c}$$

As an example, we calculate the critical concentration of donors in GaAs with an effective Bohr radius of $r_B^* = 103$ Å. Equation (1.64c) yields a critical density of $N_{crit} = 1.2 \times 10^{16}$ cm^{-3}.

The Mott criterion can be also obtained by a fundamentally different approach, i.e. by considering the binding of electrons to screened Coulomb potentials (see, for example, Mott, 1990). At low concentrations, electrons are bound to the (essentially) unscreened Coulomb potentials of the impurities. At higher free carrier concentrations, screening becomes relevant and Coulomb potentials must be replaced by screened Coulomb (Yukawa) potentials. The binding energy of screened Coulomb potentials is smaller than the binding energy of Coulomb potentials. Furthermore, the binding energy decreases with increasing screening. The insulator-to-metal transition occurs, if the binding energy of electrons bound to screened Coulomb potentials becomes zero.

To solve this problem quantitatively, we use the Coulomb potential $V(r) = (e/4\pi\varepsilon r)\exp(-r/r_{TF})$, where r_{TF} is the screening radius. Furthermore, we use the variational wave function $\psi(r) = c\exp(-r/\alpha)$, where c is a normalization constant and α is the variational parameter (Flügge, 1971). The binding energy, E, and the spatial extent of the wave function, α, can be obtained by the variational method. From the condition $E = 0$ one obtains the variational parameter $\alpha = 2a_B^*$ and $r_{TF}/a_B^* = 1$. Using Eq. (1.57) for r_{TF} and Eq. (1.37) for a_B^* yields

$$a_B^* \, N_{crit}^{1/3} = \frac{1}{4}\left[\frac{\pi}{3}\right]^{1/3} \cong 0.25 \tag{1.65}$$

which is similar to Eq. (1.64c).

There is yet another approach to calculate the Mott density. The calculation is based on the reduction of the bandgap energy due to many-body effects. Haug and Schmitt-Rink (1984) showed that the energy level of an exciton merges with the conduction band due to the lowering of the conduction band edge at high concentrations (bandgap renormalization, see Sect. 1.3.5). The authors further showed that the energy of the exciton is remarkably constant with carrier concentration due to the charge neutrality of the exciton. The critical concentration, estimated from the bandgap reduction due to many-body effects (see Sect. 1.3.5), agrees well with the Mott criterion of Eq. (1.65). Even though the result of Haug and Schmitt-Rink (1984) was obtained for excitons, it also applies to neutral donors which can be thought of as excitons with an infinitely heavy hole mass.

The insulator-to-metal transition does not occur abruptly at the critical concentration. Instead the transition is gradually evolving with increasing impurity concentration. Quantitatively, the gradual nature of the Mott transition can be expressed in terms of a continuously changing impurity activation energy. Experimental donor activation energies have been described by the equation (Debye and Conwell, 1954)

$$E_d = E_{d0}[1 - (N_D/N_{crit})^{1/3}] \tag{1.66}$$

where E_{d0} is the donor activation energy for $N_D \ll N_{crit}$. The reduction in donor ionization energy is thus proportional to the distance between the donor

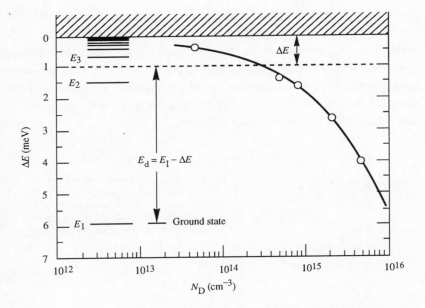

Fig. 1.13. Hydrogenic donor levels in GaAs, E_n, and donor ionization energy E_d as a function of doping concentration (Stillman et al., 1982).

atoms. The ionization energy of donors in GaAs as a function of the donor concentration is shown in Fig. 1.13 (Stillman et al., 1982). The effective ionization energy is measured from the impurity level to the quasi-continuum. The effective ionization energy approaches zero at the critical density which is $N_{crit} \cong 10^{16}$ cm^{-3} for donors in GaAs.

1.3.2 The Burstein–Moss shift

The shift of the absorption edge to higher energies occurring at high doping concentrations is referred to as the **Burstein–Moss effect** or **shift** (Burstein, 1954; Moss, 1961). This shift is also called **band filling** or **phase space filling**. The up-shift of the absorption edge is related to band filling which is schematically shown in Fig. 1.14 for n-type doping. The conduction band becomes significantly filled at high doping concentrations due to the finite density of states. Due to band filling, absorption transitions cannot occur from the top of the valence band to the bottom of the conduction band. As a result, the fundamental edge of absorption transitions shifts from $E_C - E_V = E_g$ for undoped semiconductors, to

$E_F - E_V > E_g$ in heavily doped n-type semiconductors. This shift was first observed in n-type InSb (Moss, 1961) and has since been used to make semiconductors transparent in the near-band-edge region (Verie, 1967; Dapkus et al., 1969, Deppe et al., 1990). It should be pointed out that the Burstein–Moss shift competes with the formation of impurity bands (Sect. 1.3.3) and band tails (Sect. 1.3.4). As will be seen, the Burstein–Moss shift dominates in semiconductors with light effective carrier mass.

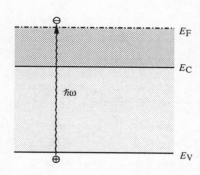

Fig. 1.14. Schematic illustration of conduction band filling due to heavy n-type doping. The absorption edge occurs at an energy E_F–E_C which can be significantly larger than the bandgap energy E_C–E_V.

Quantitatively, the Burstein–Moss shift can be calculated from the filling of the conduction or valence band in n-type and p-type semiconductors, respectively. Assuming a degenerately doped n-type semiconductor, the band filling for a single-valley, isotropic, and parabolic band is in the limit of extreme degeneracy given by (see Chap. 3)

$$E_F - E_C = \frac{\hbar^2}{2m_e^*}(3\pi^2 n)^{2/3} \tag{1.67}$$

where n is the free carrier density and m_e^* is the effective carrier mass. The absorption edge occurs then at the energy

$$E = E_g + (E_F - E_C) . \tag{1.68}$$

Note that the Burstein–Moss shift is inversely proportional to the effective mass. This explains the fact that the Burstein–Moss effect is more prominent in semiconductors with light carrier masses. For example, the shift clearly manifests itself in n-type GaAs ($m_e^* = 0.067\, m_0$) but not in p-type GaAs ($m_{hh}^* = 0.45\, m_0$) as illustrated in Figs. 1.15 and 1.22.

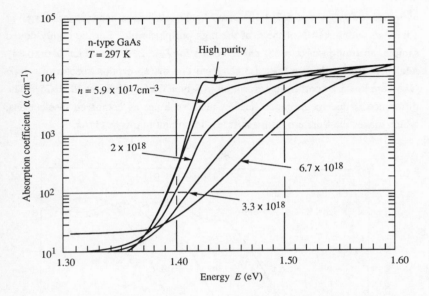

Fig. 1.15. Absorption coefficient in n-type GaAs at $T = 297$ K for different doping concentrations. The absorption edge shifts to higher energies with increasing doping concentration (after Casey et al., 1975).

For the sake of completeness, we consider the Burstein–Moss effect in *two-dimensional systems*. In such systems, free carriers are confined to a thin sheet. The confinement can be achieved in terms of quantum well structures or at the heterojunctions of two semiconductors. The band filling of a semiconductor with a single, parabolic subband is, in the high doping limit, given by (see Chap. 3)

$$E_F - E_0 = \frac{\pi \hbar^2}{m_e^*} n^{2D} \tag{1.69}$$

where E_0 is the bottom of the subband and n^{2D} is the two-dimensional carrier concentration per cm^2. The energy given by Eq. (1.69) represents the shift with respect to the absorption edge in an undoped two-dimensional semiconductor.

The Burstein–Moss shift is illustrated in Fig. 1.15 for n-type GaAs (Casey et al., 1975). The graph shows the absorption coefficient as a function of photon energy for n-type doping concentrations up to 6.7×10^{18} cm^{-3}. As the carrier concentration exceeds 6×10^{17} cm^{-3}, the Burstein–Moss shift due to the filling of the conduction band begins to have a significant effect. This shift of the absorption coefficient to higher energies is readily shown in Fig. 1.15 for samples with $5.9 \times 10^{17} < n < 6.7 \times 10^{18}$ cm^{-3}. These curves tend to converge at 1.6 eV to within $\pm 10\%$ of the α of the high-purity sample. For the highly doped sample, absorption of $\alpha = 20$ cm^{-1} occurs for $E < 1.38$ eV, which is probably related to band tails (see Sect. 1.3.4). Note that the absorption spectra of n-type GaAs are in stark contrast to absorption in p-type GaAs (Casey et al., 1975). The difference is due to the much heavier mass of holes as compared to electrons, which makes the Burstein–Moss shift less important in p-type GaAs.

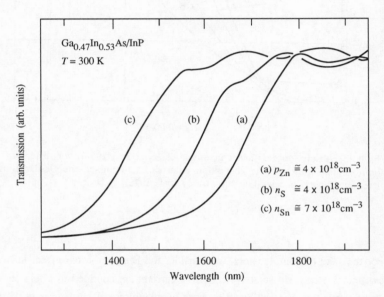

Fig. 1.16. Optical transmission versus wavelength for p- and n-type $Ga_{0.47}In_{0.53}As$ epitaxial layers grown on InP. The layers are doped with Zn, S, and Sn. Near-band-edge absorption decreases as the n-type doping level increases (after Deppe et al., 1990).

Transmission spectra of a $Ga_{0.47}In_{0.53}As$ lattice matched to InP are shown in Fig. 1.16 for various doping concentrations (Deppe et al., 1990). For a p-type Zn doping level of $N_{Zn} \cong 4 \times 10^{18}$ cm^{-3}, no appreciable shift of the absorption edge is observed as compared to undoped bulk $Ga_{0.47}In_{0.53}As$ $(\lambda^{300K}_{GaInAs} \cong 1.65$ μm). However, a significant shift is observed for n-type $Ga_{0.47}In_{0.53}As$ at a doping level of 4×10^{18} cm^{-3} and an even larger shift at a doping level of 7×10^{18} cm^{-3}. At the highest n-type doping level, the semiconductor is virtually transparent at a wavelength of 1.55 μm. The highly n-type doped $Ga_{0.47}In_{0.53}As$ was used for quarter-wave reflectors operating at

1.55 μm, which require optical transparency at that wavelength (Deppe et al., 1990).

1.3.3 Impurity bands

At impurity concentrations well below the critical Mott concentration, impurities can be considered as isolated, non-interacting entities. As the concentration increases but is still well below the Mott concentration, impurities begin to *interact*. Carrier transport at low temperatures occurs via thermally assisted tunneling between impurity states. This transport process is called *hopping conduction*. At still higher impurity concentrations but below the critical Mott concentration, overlapping impurity states form an *impurity band*. At low temperatures, carriers can propagate within the impurity band without entering the conduction band. This transport process is known as *impurity band conduction*. This section summarizes the elementary characteristics of impurity-assisted conduction mechanisms for concentrations below the Mott concentration. Extensive reviews of the topic were given by Mott (1987, 1990) and Shklovskii and Efros (1984).

Consider a semiconductor with a donor density well below the critical Mott transition density. Upon cooling the semiconductor to low temperature, the conductivity is expected to decrease as free electrons freeze out onto localized donor states. For $kT \ll E_d$ the conductivity of an n-type semiconductor is expected to become vanishingly small. Experimentally, zero conductivity is not observed in semiconductors containing a net concentration of shallow impurities. Instead, the temperature dependence of the conductivity is less drastic than expected from free carrier freeze-out. The conductivity in this regime is not given by electrons excited to the conduction band but rather by electrons hopping from neutral donors to ionized donors. The conductivity is referred to as *hopping conductivity* and can be described by the general hopping conductivity formula

$$\sigma_{hop} = (\alpha/T^{\beta}) \exp\left[-\frac{E_{hop}}{kT}\right]^{\gamma} \tag{1.70}$$

where α and β are constants, E_{hop} is the thermal activation energy for the hopping process and γ determines the functional dependence of the exponential factor. For simplicity, the factor γ is frequently assumed to be unity. Austin and Mott (1969) showed for gaussian localization, in which transport occurs via tunneling to remote but energetically similar donors, that the value of $\gamma = \frac{1}{4}$. Efros and Shklovskii (1975) showed that $\gamma = \frac{1}{2}$, if Coulomb interaction between adjacent donors determines the hopping transport between adjacent donors. In this case the thermal activation energy for the hopping process is given by

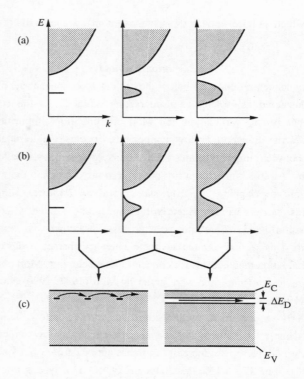

Fig. 1.17. Donor impurity level/band at low, medium and high doping concentrations for (a) an ordered impurity distribution and (b) a random impurity distribution. (c) Band diagram illustrating hopping conduction and impurity band conduction.

$$E_{\text{hop}} \cong \alpha \frac{e^2}{4\pi\varepsilon} \left[\frac{4\pi}{3} N_{\text{D}} \right]^{1/3} \tag{1.71}$$

where the factor α has a numerical value of about 0.60. The qualitative physical explanation of the hopping conduction process is as follows. Consider a lightly compensated n-type semiconductor at low temperatures. Most of the donors are neutral, some donors ionized due to the slight compensation. When the donor impurities are closely spaced, their energy levels split. Electrons can tunnel from a donor state to an empty state of an adjacent donor. A so-called *Coulomb gap* develops between filled donor states and empty donor states. The Coulomb gap is caused by the long-range Coulomb interaction of localized electrons (Pollak and Knotek, 1974; Efros and Shklovskii, 1975), and occurs at the Fermi level. The tunneling from filled donor states to adjacent empty donor states therefore requires a small thermal activation energy given by Eq. (1.71). The activation energy can be interpreted as the energy from the Fermi level to the energy of the maximum of the density of empty state distribution (Böer, 1990). Typically the activation energy is much smaller than the donor ionization energy.

The regime of hopping conduction is schematically illustrated in Fig. 1.17 which shows the dispersion relations and donor impurity states at different doping densities. At low doping density, states of adjacent impurities do not interact ($N \ll N_{crit}$) and impurity ground states are discrete and energetically well-defined levels. As the doping concentration increases, states of adjacent impurities interact, split, and finally form an impurity band ($N \leq N_{crit}$). At even higher doping concentrations, the impurity band widens and merges with the continuum band ($N \geq N_{crit}$).

It is instructive to consider the effect of *ordered* and *random* impurity distribution on the formation of an impurity band, which is shown in Fig. 1.17(a) and (b), respectively. In the case of an *ordered* impurity distribution, the potentials of the impurities are strictly periodic, similar to the periodic potential assumed in the Kronig–Penney model. As a result, the impurity band has a well-defined width and well-defined edges, as shown in Fig. 1.17(a). However, the case of *random* impurity distribution in semiconductors is a more realistic assumption (Shockley, 1961). For a random impurity distribution, the impurity band does not have well-defined band edges but the impurity states will tail into the forbidden gap, as shown in Fig. 17(b). Tail states of impurities occurring at high doping density are discussed in Sect. 1.3.4. Figure 1.17(c) schematically shows hopping conduction and impurity band conduction.

Impurity bands are formed at sufficiently high doping concentrations. The impurity band is formed by the quasi-periodic Coulomb potentials of the impurities. Since the impurity potentials are not strictly periodic, a periodic band calculation (e.g. Kronig–Penney) cannot be used. The hydrogenic potentials of donors can be used to form a Hubbard band (Hubbard, 1963). In the Hubbard model, the center of the impurity band coincides with the energy level of the unperturbed impurity state. If the donors forming the impurity band are partially compensated by acceptors, the band is only partially filled and impurity band conduction can occur within the impurity band, i.e. electrons need not occupy conduction band states for carrier transport (Adler, 1982). The impurity band width, ΔE_D, is approximately given by the overlap integral between donors separated by the average distance $N_D^{-1/3}$. The band width is approximately equal to the interaction energy

$$\Delta E_D \cong \frac{e^2}{4\pi\varepsilon \, N_D^{-1/3}} . \tag{1.72}$$

As an example, we consider GaAs with a donor concentration of 1×10^{16} cm^{-3} and $\varepsilon_r = 13.1$ and obtain $\Delta E_D \cong 2.4$ meV.

The mobility associated with impurity band conduction are very low, typically <1 cm^2/V s. The low mobility is due to the heavy dispersion mass. In order to qualitatively relate the low impurity band mobility with the narrow width of the impurity band, let us recall some of the basic results of one-dimensional band theory. The dispersion mass of a parabolic band is given by

$$m^* = \frac{\hbar^2}{d^2 E/dk^2} . \tag{1.73}$$

In the simple Kronig–Penney model, the dispersion (i.e. effective) mass is given by

$$m^* = 2\hbar^2/z_p^2 \Delta E_D \tag{1.74}$$

where z_p is the period of the periodic potential and ΔE_D is the width of the band. Although the result of the one-dimensional Kronig–Penney model cannot be rigorously applied to an impurity band, the model provides a qualitative understanding of the functional dependences. It is therefore reasonable to conclude that a heavy effective mass is associated with transport of carriers in a narrow impurity band. Since the mobility and the effective mass are related according to the Drude model by

$$\mu = \frac{e\tau}{m^*} , \tag{1.75}$$

a low mobility results for carrier transport in a partially filled impurity band.

1.3.4 Band tails

The random distribution of charged impurities results in potential fluctuations of the band edges. In an *undoped* semiconductor such potential fluctuations are absent and the band edges are well defined. In a highly doped semiconductor the potential fluctuations cause the band edges to vary spatially. This situation is schematically shown in Fig. 1.18. States with energy below the unperturbed conduction band edge or above the unperturbed valence band edge are called *tail states*. The tail states significantly change the density of states in the vicinity of the band edge.

The magnitude of band-edge energy fluctuations caused by the random distribution of charged donors and acceptors was first calculated by Kane (1963). For an ionized donor concentration of N_D and an ionized acceptor concentration

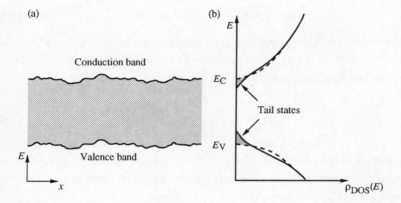

Fig. 1.18. (a) Spatially fluctuating band edges caused by random distribution of impurities. (b) Resulting densities of states in the conduction and valence band with tail states extending into the forbidden gap. The dashed lines show the parabolic densities of states in undoped semiconductors.

of N_A, the root-mean-square (energy) fluctuation of the band edges is given by (Kane, 1963; Morgan, 1965)

$$\sigma_E = \frac{e^2}{4\pi\varepsilon} \, [2\pi(N_D + N_A)r_s]^{1/2} \qquad (1.76)$$

where r_s is the screening radius given by (see Eq. 1.55b)

$$r_s = \left[-\left(\frac{dn}{dV} \right)_{V=0} \frac{e}{\varepsilon} \right]^{-1/2}. \qquad (1.77)$$

Note that the square-root dependence of the energy fluctuation, σ_E, on the doping concentration, $N_D + N_A$, is a result of the Poisson distribution assumed for impurities. The Poisson distribution can be replaced by a gaussian distribution, if the energy fluctuations are small. Small fluctuations occur if the random variation of the number of donor atoms within a spherical volume defined by the screening radius is much smaller than the average number of donors within the sphere (Kane, 1963), i.e.

$$\sqrt{(N_D + N_A) \, r_s^3} \ll (N_D + N_A) \, r_s^3 \, . \tag{1.78}$$

The fluctuation of the potential, e.g. the conduction band edge potential, can then be expressed in terms of a gaussian distribution. The probability for the conduction band edge energy to occur at an energy E_C, is given by

$$p(E_C) = \frac{1}{\sqrt{2\pi} \, \sigma_E} \, e^{-\frac{1}{2} \left[\frac{E_C}{\sigma_E} \right]^2} \tag{1.79}$$

where we defined the mean energy of the conduction band edge as $E_C = 0$ and σ_E is given by Eq. (1.76). The unperturbed density of states in the conduction band is given by (see Chap. 3)

$$\rho_{DOS}(E) = \frac{1}{2\pi^2} \left[\frac{2m^*}{\hbar^2} \right]^{3/2} \sqrt{E - E_C} \, . \tag{1.80}$$

The density of states in a perturbed potential is obtained by a summation over all locations in space, i.e. by an integral over the probability distribution of the band edge

$$\rho_{DOS, Kane} = \frac{1}{2\pi^2} \left[\frac{2m^*}{\hbar^2} \right]^{3/2} \int_{-\infty}^{E} \sqrt{E - E_C} \, \frac{1}{\sqrt{2\pi} \, \sigma_E} \, e^{-\frac{1}{2} \left[\frac{E_C}{\sigma_E} \right]^2} \, dE_C \tag{1.81}$$

where $\rho_{DOS, Kane}$ is the density of states according to Kane (1963), or briefly the Kane function. The density of states given by Eq. (1.81) is valid for the conduction band. The corresponding density of states of the valence band can be obtained by replacing E_C by $-E_V$, E by $-E$, and the electron effective mass by the heavy-hole effective mass.

The parabolic density of states, the gaussian probability distribution of the band edge, and the Kane function are shown in Fig. 1.19. The density of states shown in Fig. 1.19(c) tails into the gap. Tail states have energies lower than the average band-edge energy. It is important to note that the density of states of the Kane function does not change the *bandgap* energy, i.e. the spatially *averaged* position of the conduction band edge does not change. The tailing of the band into the forbidden gap changes the density of state drastically in the vicinity of the band edge. However, the density of states is practically unchanged for energies $E - E_C \gg \sigma_E$, as can be easily inferred from Eq. (1.81). The Kane function is of great practical use for the simple, quantitative description of band edges. For example, Casey and Stern (1976) used the Kane function to calculate absorption and spontaneous emission in doped GaAs.

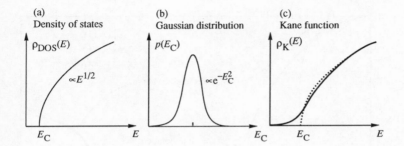

Fig. 1.19. (a) Density of states of a parabolic, spherical valley. (b) Gaussian probability distribution. (c) The Kane function describes the density of states of a parabolic band including tail states at $E < E_C$ which arise from potential fluctuations caused by randomly distributed impurities.

The density of states according to Kane (Eq. (1.81)) overestimates the extent of band tailing since tunneling of carriers through the potential barriers is not taken into account. In addition, the quantization of carriers in the potential minima is not taken into account in the Kane model. However, in the limit of carriers with a large effective mass, the Kane function applies, since quantum effects can be neglected for such heavy carriers.

Halperin and Lax (1966, 1967) and Lax and Halperin (1966) calculated the density of states in band tails taking into account (i) the quantization of carriers in the potential minima as well as (ii) tunneling of carriers through potential barriers. The importance of carrier quantization can be estimated by comparing the magnitude of the potential fluctuations with the quantization energy of carriers. The former is given by Eq. (1.76) for poissonian (random) impurity distributions. The latter can be estimated from

$$E_Q \cong \frac{\hbar^2}{2m^* r_s^2} \tag{1.82}$$

where r_s is the screening radius. The density of states calculated by Halperin and Lax (1966) is shown in Fig. 1.20 along with the density of states of the unperturbed bands and the density of states of the Kane model for GaAs with doping concentrations of $N_A = 1.1 \times 10^{19}$ cm^{-3} and $N_D = 9 \times 10^{18}$ cm^{-3} (Casey and Stern, 1976). The density of states calculated according to the Kane model shown in Fig. 1.20 was used to interpolate between the unperturbed band and the Halperin–Lax result. The parameter σ_E (see Eq. (1.81)) was adjusted in order to obtain a good fit between the two functions.

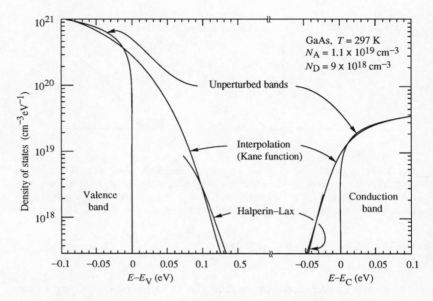

Fig. 1.20. Densities of states in the conduction and valence band versus energy for GaAs with a net acceptor concentration of 2×10^{18} cm^{-3}. The curves show the densities of states in the unperturbed bands and the densities of states in the band tails calculated by Halperin and Lax. The curves which join them are interpolated Kane functions (after Casey and Stern, 1976).

1.3.5 Bandgap narrowing

At high doping concentrations the bandgap energy of semiconductors decreases. The magnitude of the reduction increases with doping concentration and is usually referred to as *bandgap narrowing, bandgap shrinkage*, or *bandgap renormalization*. There are many reasons for the reduction of the band gap which have been reviewed by Abram et al. (1978). The most important reasons for bandgap narrowing are many-body effects of free carriers which lower the electron energies as compared to a non-interacting carrier system. Many-body effects describe the interaction of free carriers. The interaction becomes important at small carrier-to-carrier distances, i.e. at high free carrier concentrations. Electrons can interact with each other either by their long-range Coulomb potential or via their spin. We first consider coulombic electron–electron interactions. Consider an electron which is added to a highly doped, neutral semiconductor. When the electron is added to the semiconductor, other electrons in the vicinity of the added electron spatially redistribute in order to reduce the long-range coulombic interaction energy. The energy of the electron added to the semiconductor is *reduced* by the redistribution of neighboring electrons. Many-body effects are thus related to screening, i.e. the spatial redistribution of carriers in the presence of potential perturbations. Carriers can also interact via their spin (Mahan, 1990). Due to the Fermion nature of electrons, each volume element in phase space can be occupied by at most two electrons with opposite spin (Pauli principle). Electrons with like spin have a repulsive interaction while electrons with opposite spin have an attractive interaction. If electrons were distributed uniformly throughout the crystal, the attractive and repulsive energies would exactly cancel each other. However, due to the interaction, electrons with like spin tend to stay away from each other and electrons with opposite spin tend to stay closer. As a result, the interaction energy reduces the total energy of the electron system.

Many-body interactions can occur between free carriers and between free carriers and ionized impurities. Such interactions are called carrier–carrier and carrier–impurity interactions. Thus, bandgap narrowing occurs in highly doped semiconductors as well as in undoped, but highly excited semiconductors with high free carrier concentrations. We summarize the many-body interactions in n- and p-type semiconductors as follows:

i. *Electron–electron interactions.* The repulsive and attractive interactions of electrons with like and opposite spin and the long-range coulombic interactions lead to a net attractive term. As a consequence, the conduction band edge is lowered (i.e. lowering of gap energy).

ii. *Electron–donor interactions.* The interaction of electrons and ionized donors is attractive and leads to another lowering of the conduction band edge.

iii. *Hole–hole interactions.* The spin interaction energy and the coulombic interaction result in a net attractive energy and an increase in the valence band edge (i.e. lowering of gap energy).

iv. *Hole–acceptor interactions.* The interaction between holes and donors is attractive and leads to an increase of the valence band edge (i.e. decrease in gap energy).

v. *Electron-hole interactions.* Highly excited semiconductors have a large concentration of electrons and holes. Interaction effects lead to a reduction of the energy gap in highly excited semiconductors.

Next we calculate the magnitude of bandgap renormalization due to electron–electron interaction effects and follow the calculation of Haug and Schmitt-Rink (1985). At room temperature, the magnitude of bandgap renormalization can be approximated by the classical self energy of an electron interacting with an electron gas, i.e. interacting with its own polarization field. Using the classical analogy, we define the self energy of an electron interacting with an electron gas as

$$\Delta E_g \cong e \lim_{r \to 0} [V_s(r) - V(r)] \tag{1.83}$$

where $V(r)$ is the Coulomb potential and $V_s(r)$ is the screened Coulomb potential of an electron in an electron gas. Equation (1.83) thus represents the change in electrostatic energy of an electron before and after the electron gas has spatially redistributed itself to reduce the Coulomb interaction energy. In other words, ΔE_g describes the energy that the electrons gain by avoiding each other. Assuming that the screened potential is of the Yukawa form, we obtain

$$\Delta E_g \cong -\frac{e^2}{4\pi\varepsilon r_s} \tag{1.84}$$

The screening radius, r_s, is given by the Debye and the Thomas–Fermi radii in non-degenerate and degenerate semiconductors, respectively. Insertion of the screening radii into Eq. (1.84) yields the bandgap renormalization energies

$$\Delta E_g \quad -\frac{e^3 \sqrt{n}}{4\pi\varepsilon^{3/2}\sqrt{kT}} \quad \text{(Debye)} \tag{1.85a}$$

$$\Delta E_g = -\frac{e^3 \sqrt{m^*}(3n)^{1/3}}{4\,\pi^{5/3}\,\varepsilon^{3/2}\,\hbar} \quad \text{(Thomas–Fermi)} \tag{1.86b}$$

The momentum-dependence of the renormalization energy is relatively small (Wolf, 1962; Haug and Schmitt-Rink, 1985). If the momentum-dependence is neglected, bandgap renormalization produces a *rigid downward movement* of the conduction band dispersion (or, in p-type semiconductors, a rigid upward movement of the valence band). Finally, bandgap renormalization also occurs in undoped but highly excited semiconductors. Brinkman and Rice (1973) examined the effect of electron–hole interactions in highly excited semiconductors and showed that bandgap renormalization occurs due to electron–hole interactions. Phenomenological expressions for the carrier–carrier interaction and the carrier–impurity interaction were given by Mahan (1980) and Landsberg et al. (1985). Assuming that the interaction energies are proportional to the carrier–carrier distance, the change in energy gap follows a 1/3 power of the doping concentration, that is

$$\Delta E_g \propto (N_D)^{1/3} \qquad (1.86)$$

where full activation of the donors was assumed, i.e. $n = N_D$. Similar considerations are valid for p-type semiconductors. In this case carrier-carrier and carrier–acceptor interactions must be considered. Note that the phenomenological dependence of ΔE_g on n given in Eq. (1.86), i.e. $\Delta E_g \propto n^{1/3}$, is weaker than the Debye result, $\Delta E_g \propto n^{1/2}$ (see Eq. (1.85a)), but stronger than the Thomas–Fermi result, $\Delta E_g \propto n^{1/6}$ (see Eq. (1.85b)).

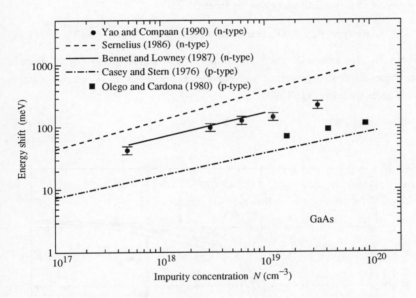

Fig. 1.21. Bandgap narrowing for n-type and p-type GaAs as a function of impurity concentration (after Yao and Compaan, 1990).

Experimental and theoretical results of bandgap narrowing as a function of doping concentration are shown in Fig. 1.21 for n-type and p-type doping of GaAs (Yao and Compaan, 1990). The graph includes experimental data for n-type (Yao and Compaan, 1990), and p-type GaAs (Olego and Cardona, 1980), as well as theoretical data for n-type (Sernelius, 1986; Bennet and Lowney, 1987) and p-type GaAs (Casey and Stern, 1976). The data of Fig. 1.21 shows that bandgap narrowing can assume quite large values, such as 200 meV, at high n-type doping concentrations. Furthermore, the change in bandgap energy follows the $N^{1/3}$ dependence predicted by Eq. (1.86). The magnitudes of bandgap narrowing can be expressed by the following simple formulas:

$$\text{n-type GaAs:} \quad \Delta E_g(\text{meV}) \cong -6.6 \times 10^{-5} \; \sqrt[3]{N_D(\text{cm}^{-3})} \quad (1.87)$$

$$\text{p-type GaAs:} \quad \Delta E_g(\text{meV}) \cong -2.4 \times 10^{-5} \; \sqrt[3]{N_A(\text{cm}^{-3})} \; . \quad (1.88)$$

These equations represent a fit to the experimental data shown in Fig. 1.21. As expected, the bandgap narrowing for n-type GaAs is larger as compared to p-type GaAs (Yao and Compaan, 1990). Bandgap narrowing was studied to a smaller extent in other III–V semiconductors. An approximate formula for bandgap narrowing in n-type InP was given by Böer (1990)

$$\text{n-type InP:} \quad \Delta E_g(\text{meV}) \cong -2.25 \times 10^{-5} \; \sqrt[3]{N_D(\text{cm}^{-3})} \quad (1.89)$$

where it is assumed that all donors are active ($n = N_D$). Further phenomenological expressions and parameters for other III–V semiconductors were given by Jain et al. (1990).

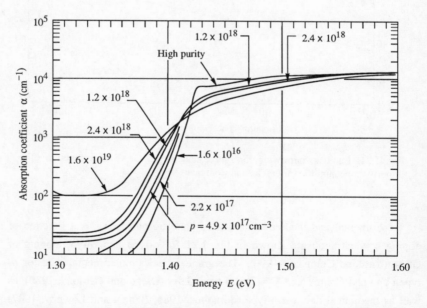

Fig. 1.22. Absorption coefficient for p-type GaAs at 297 K for different doping concentrations (after Casey and Stern, 1976).

The Burstein–Moss shift and bandgap narrowing are two phenomena which cause the Fermi level of highly doped semiconductors to change in opposite directions. While bandgap narrowing causes the Fermi level of an n-type semiconductor to decrease, the Burstein–Moss shift is due to an increase of the Fermi level. Absorption measurements allow one to measure $E_F - E_V$ and $E_C - E_F$ for n-type and p-type semiconductors, respectively. In n-type GaAs, the Burstein–Moss shift prevails, resulting in a blue-shift of the absorption edge (see Sect. 1.3.2). In p-type GaAs, bandgap narrowing prevails and results in a red-shift of the absorption edge. The absorption coefficient of p-type GaAs is shown in Fig. 1.22 for different p-type doping levels (Casey and Stern, 1976). High-purity GaAs exhibits a rapidly decreasing absorption coefficient at the fundamental gap. However, as the doping concentration increases, the absorption below the gap energy increases as well. Furthermore, the absorption coefficient decreases less rapidly as compared to the high-purity GaAs. Both characteristics indicate that bandgap narrowing and the formation of tail states dominate the near-band-edge optical absorption rather than the Burstein–Moss shift. The strongly different absorption characteristics of p-type (see Fig. 1.22) and n-type GaAs (see Fig. 1.15) are due to the heavier hole mass as compared to the effective electron mass. The Burstein–Moss shift is inversely proportional to the carrier effective mass, which results in less band filling in p-type GaAs.

2

Phenomenology of deep levels

Deep levels are states within the forbidden gap of a semiconductor that are far removed from either conduction or valence band. Many of the deep levels are closer to the center of the gap than to either of the band edges. The name *midgap center* is frequently used for such deep levels. Due to their ability to capture free electrons and holes, deep levels are also called *traps* or *deep traps*. Deep levels caused by substitutional, non-hydrogenic impurities are referred to as *deep impurities*. Chromium in GaAs and Fe in InP are examples of such deep impurities. Finally, deep levels can be due to point defect centers. Examples of such *deep centers* are anti-site defects and interstitials.

Deep levels can be caused not only by point defects but also by spatially extended defects. Examples for extended defects are dislocations such as threading dislocations, misfit dislocations, or screw dislocations. Another example of extended 'defects' are semiconductor surfaces. Electronic states at semiconductor surfaces are called *surface states* or *Bardeen states* after Bardeen (1947) who demonstrated the influence of such states on the surface potential. In many semiconductors, Bardeen states are located in the vicinity of the center of the gap (e.g. GaAs and InP) and have properties similar to deep levels occurring in the bulk.

Deep centers can be classified according to their charge state. Centers with a neutral and positively charged state are called *donor-like states*. Centers with a neutral and negatively charged state are called *acceptor-like states*. However, deep levels can have many levels within the forbidden gap as well as within the valence or conduction band. Examples for such multileveled deep centers are Ga and As vacancies in GaAs (Baraff and Schlüter, 1985).

It is not the purpose of this chapter to discuss exhaustively experimental and theoretical characteristics of deep levels since deep levels are only indirectly related to the topic of doping. A near-complete discussion of the properties of deep levels can be found in the literature (Queisser, 1971; Jaros, 1982; Pantelides, 1986). Nevertheless, deep impurities can cause the properties of doped semiconductors to deteriorate strongly and are therefore of importance for the understanding of such properties. First, deep levels can compensate shallow dopants and, as a consequence, reduce the free carrier concentration. A concomitant effect is a reduction in carrier mobility, since the concentration of charged levels increases while the free carrier concentration decreases. Second, deep centers are recombination centers and decrease the radiative efficiency as well as the minority carrier lifetime. Recombination via deep traps (Shockley–Read recombination) can drastically degrade the performance of minority carrier devices such as light-emitting diodes, lasers, and bipolar transistors. Due to this characteristic, deep levels are dubbed *lifetime killers* and *luminescence killers*. To summarize, the negative consequences of traps are (i) reduced luminescence efficiency, (ii) shorter minority carrier lifetime, (iii) a shorter diffusion length of minority carriers, (iv) compensation of intentional shallow impurities, and (v) reduced drift mobility.

This chapter contains a phenomenological description of deep levels and the physical consequences of deep levels. The reader is referred to Chap. 9 for experimental characteristics of specific deep centers in III–V semiconductors.

2.1 General characteristics

The energy level or energy levels of deep traps are far removed from the band edges. Consider a donor-like deep level with the ground-state energy E_T. Such a donor is considered to be *deep*, if the condition

$$E_t \gg kT \quad (T = 300 \text{ K}) \tag{2.1}$$

is fulfilled, where $E_t = E_C - E_T$ is the activation energy of the deep level (see Fig. 2.1a). An analogous definition can be made for acceptor-like deep levels. As a consequence of Eq. (2.1), donor- and acceptor-like deep centers are not ionized at room temperature. That is, donor-like and acceptor-like deep centers are neutral in n-type and p-type semiconductors, respectively. Note, however, that donor-like deep centers are ionized in p-type semiconductors and the free hole concentration is reduced due to the presence of donor-like traps.

The confining potential of a deep level is a short-range potential rather than the long-range Coulomb potential of shallow impurities. The short-range potential of deep impurities typically extends over one or a few atomic radii

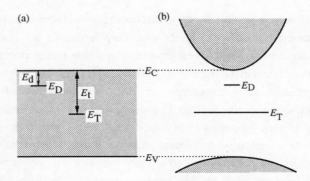

Fig. 2.1. Illustration of a deep level and a shallow level in (a) real space and in (b) momentum space.

(Madelung, 1978). The spatial extent of the impurity wave function decreases as the ionization energy of the center increases. According to the WKB approximation, the impurity wave function decay length, L, and the trap ionization energy are related by $L \propto \exp{-\sqrt{E_t^{-1}}}$. Electrons bound to deep levels are typically confined to the volume of one or a few unit cells. This characteristic of deep centers is in marked contrast to shallow impurities. Hydrogenic shallow impurities (see Chap. 1) confine carriers to a sphere defined by the Bohr radius which can be on the order of 100 $\overset{\circ}{\text{A}}$, i.e. a sphere comprising thousands of unit cells. Due to this striking difference, the properties of deep levels are mainly determined by the chemical nature of the center, while the chemical nature is of minor importance for shallow impurities with a large effective Bohr radius.

The different degrees of spatial confinement of deep and shallow impurities can be represented as different degrees of uncertainty in momentum space. Figure 2.1(b) shows the dispersion relations as well as impurity levels which are represented as horizontal lines of various lengths. For shallow impurities the horizontal line has a width of approximately $1/a_B^*$ where a_B^* is the effective Bohr radius

$$a_B^* = \frac{\varepsilon}{\varepsilon_0} \frac{m_0}{m_e^*} a_B \,, \tag{2.2}$$

and a_B is the Bohr radius of the hydrogen atom. The width of the line is a result of the position–momentum uncertainty relation $\Delta x \Delta p \cong h$ which, using $p = \hbar k$, can be written as $\Delta x \Delta k \cong 2\pi$. Here Δx represents the position uncertainty of an electron bound to the center. For deep impurities, the position uncertainty Δx is much smaller as compared to hydrogenic impurities, which results in a larger momentum uncertainty. Thus, the wave function of a deep impurity state requires contributions from a large range of k values, i.e. the deep impurity wave function contains contributions from a large range of the Brillouin zone (Seeger, 1982). Transitions between the conduction or valence band and the deep level are therefore possible for a large range of carrier momenta rather than only for $k \cong 0$ as in the case of shallow levels.

Another consequence of the strong localization of electrons on deep levels is the breakdown of effective mass theory for deep levels. The effective mass of carriers is a result of the periodic lattice potential. In effective mass theory, the periodic crystal potential is taken into account by the employment of the effective mass. Due to the strong localization to only one or a few lattice constants, the carrier bound to the deep center is *not* subject to a periodic potential. Thus, the basis for the effective mass concept is not given, which complicates the theoretical treatment of deep levels.

Due to the breakdown of effective mass theory, the crystal potential must be taken into account explicitly in addition to the potential of the deep impurity. Madelung (1978) pointed out that in the limit of a strong deep impurity potential, the crystal potential can be considered as a perturbation of the impurity potential. In this case, the theoretical treatment of deep centers is simplified. The energy levels of the deep impurity would then depend mostly on the nature of the center rather than on the host semiconductor.

The chemical origin of deep levels is frequently unknown. This is because deep levels can have many origins, e.g. impurities, point defects, or extended defects. Due to the uncertainty of their origin, deep levels are frequently labeled by their activation energy, e.g. the '0.8 eV level' in GaAs. Martin et al. (1977) introduced a systematic nomenclature for deep levels in n-type GaAs. Subsequently, Mitonneau et al. (1977) provided a systematic nomenclature for deep levels in p-type GaAs. An Arrhenius plot of the emission rates and the activation energy of the deep levels are shown in Figs. 2.2 and 2.3 for n-type and p-type GaAs, respectively. The first two letters in the trap label refer to the type of trap (E = electron trap, H = hole trap) and to the institution of publication (e.g. B = Bell Laboratories). The following number is a 'serial' number of the trap. Some of the different labels are most likely a trap with the same chemical origin (e.g. EB5 = EL4). Further discussion on specific deep levels in III–V semiconductors is referred to Chap. 9.

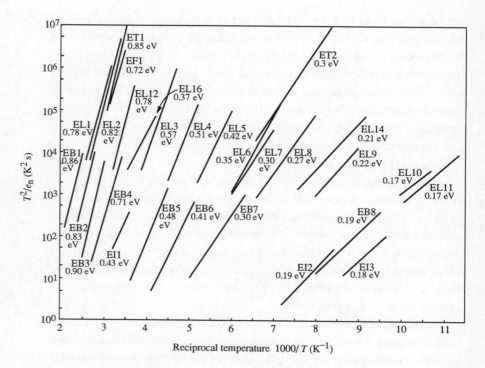

Fig. 2.2. Arrhenius plots and activation energies of deep electron traps in GaAs.
The levels are labeled as ET, EF, EL, EB, and EI (after Martin et al., 1977).

Two important classes of deep substitutional impurity atoms are impurities
with the tetrahedral sp^3 bond configuration and impurities with d-orbital
participation in the bonding to host atoms. The sp^3 bond configuration is a
common configuration and arises from the hybridization of spherically symmetric
s orbitals and three p orbitals. The hybridized electron configuration is called
sp^3-type and consists of four orbitals directed to the corners of a tetrahedron.
Thus, sp^3-bonded impurities match the tetrahedral symmetry of atoms in the
zincblende lattice of III–V semiconductors. Theoretical models for sp^3 bonding
were reviewed by Lannoo and Bourgoin (1981) and Bourgoin and Lannoo (1983).
A particularly simple model for sp^3-bonded deep impurities was developed by
Hjalmarson et al. (1980). In this model, the impurity has an electrically inactive
filled 'hyperdeep' level which forms the bond between the impurity and the host.
The antibonding state of this level is predicted to be in the forbidden gap and
forms the electrically active deep level. Based on this model, Vogl (1981)

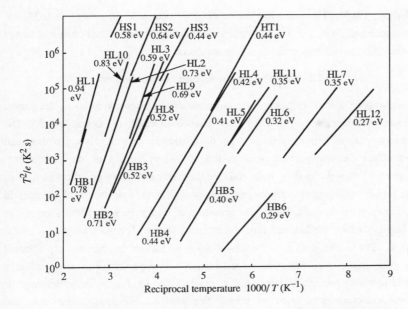

Fig. 2.3. Arrhenius plot and activation energies of deep hole traps in GaAs. The levels are labeled as HS, HT, HL, and HB (after Mitonneau et al., 1977).

predicted chemical trends for sp^3-bonded substitutional impurities in covalent semiconductors. In this approach, the trends were predicted based solely on the energy bands of the undisturbed host semiconductor and on the impurities' atomic structure. The response of the host electrons and the lattice to the perturbation caused by the impurity were not taken into account. Major chemical trends in deep trap energies were well explained by the model.

Transition metals are a second important class of deep impurities in III–V semiconductors. Transition metals with the 3d orbitals partially filled include Sc, Ti, V, Cr, Mn, Fe, Co, and Ni. For these impurities, d-orbital electrons participate in the bonding to the host semiconductor. Chromium and Fe are of great technological importance, since they are used to render III–V semiconductors semi-insulating (see Chap. 9). In addition, transition metals are frequently found as background impurities in III–V semiconductors. The oxidation states and charge states of transition metal impurities were categorized by Zunger (1986). Transition metal impurities are unique since internal optical transitions can occur between d orbitals. These transitions are typically found in the infra-red range. For example, internal d-to-d transitions occur in Fe^{2+} impurities in InP:Fe

(Bishop, 1986). The main luminescence peak occurs at an energy of 0.351 eV corresponding to $\lambda \cong 3.5$ μm. Vogl (1985) studied systematical chemical trends in transition metal impurities in semiconductors.

2.2 Capture, emission, and recombination

The dynamic process of capture and emission of electrons and holes by deep traps allows considerable insight in the phenomenology of deep levels. The consequences of capture and emission of carriers by deep levels are widespread and affect the transport as well as the optical properties of semiconductors. Capture of carriers by deep traps reduces the free carrier concentration as well as the mobility. Transient effects can be expected after rapid potential changes in semiconductor devices due to the presence of traps. Finally, traps shorten the minority carrier lifetime and thus the minority carrier diffusion length L_D (since $L_D \cong \sqrt{D\tau}$). The optical properties of semiconductors can be strongly affected by deep traps. Non-radiative recombination via traps competes with radiative band-to-band recombination. As a result, the luminescence efficiency of semiconductors is reduced by traps. The analysis of capture, emission, and recombination via deep traps allows one to quantitatively evaluate the deleterious effects alluded to.

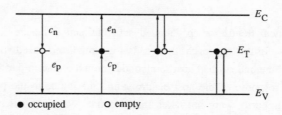

Fig. 2.4. Schematic illustration of emission and capture of electrons and holes. Traps that interact predominantly with the conduction band are called electron traps. Hole traps interact predominantly with the valence band.

The capture and emission process of a deep trap is schematically shown in Fig. 2.4. An empty level can either capture an electron from the conduction band or emit a hole into the valence band (i.e. bind an electron from the valence band). A filled trap can either emit an electron to the conduction band or capture a hole from the valence band (i.e. release an electron into the valence band). Traps that interact predominantly with the conduction band are called electron traps. The following inequalities allow one to differentiate between electron and hole traps

(Sah, 1967)

$$c_n \gg e_p \quad \text{and} \quad e_n \gg c_p \quad \text{for electron traps ,} \tag{2.3}$$

$$e_p \gg c_n \quad \text{and} \quad c_p \gg e_n \quad \text{for hole traps .} \tag{2.4}$$

where c_n and c_p are the electron and hole capture rates and e_n and e_p are the electron and hole emission rates, respectively. Since the capture rates depend on the free carrier concentration, the same centers can act as electron traps and hole traps in n-type and p-type semiconductors, respectively.

The capture of carriers by a trap can be described in terms of a *capture cross section*, σ, which has the dimension of an area. If a free carrier traverses the area of the size of the capture cross section, the carrier will be captured by the trap. Assuming an n-type semiconductor with electron concentration of n and an average carrier velocity v, the *capture probability* of one deep trap, i.e. the probability of *one* trap capturing a free carrier per unit time is given by

$$\boxed{c_n = nv\sigma} \tag{2.5}$$

The velocity of carriers can either be the *thermal* velocity (for a non-degenerate distribution) or the *Fermi* velocity (for a degenerate carrier density). The mean time that an initially unoccupied deep trap remains unoccupied is

$$\tau_c = 1/c_n . \tag{2.6}$$

The total *capture rate* is determined by the concentration of traps. If the concentration of unoccupied traps is N_t^0, then the capture rate is given by

$$\boxed{c_n^* = N_t^0 c_n = N_t^0 \, nv\sigma} \tag{2.7}$$

Thus the capture rate gives the *number of carriers captured per unit volume and per unit time*. The reader is cautioned that the capture *probability* and the capture *rate* are frequently not distinguished in the literature, even though they are different quantities.

Under thermal equilibrium conditions, the emission probability and the capture probability are identical. The equality of emission and capture probabilities is a result of the *principle of detailed balance* (Blakemore, 1987). This principle states *for systems in equilibrium that the rate (and probability) of a process and its inverse are equal and balance in detail irrespective of other processes*. The equilibrium occupancy of traps and of the conduction band is given by the Boltzman distribution. If we consider electrons of concentration n in the conduction band and occupied traps of concentration N_t^-, then the Boltzmann distribution yields

$$n/N_t^- = (N_c/N_t^0) \, g_0 \, \exp\left[-\frac{E_C - E_T}{kT}\right] \tag{2.8}$$

where g_0 is the ground-state degeneracy factor of the trap and N_t^0 is the concentration of empty traps. Assuming that all traps are initially filled with electrons, i.e. $N_t^- = N_t$, with the principle of detailed balance, and with Eq. (2.5), one obtains the *emission probability* per trap and per unit time according to

$$e_n = v\sigma N_c \, g_0 \, \exp\left[-\frac{E_C - E_T}{kT}\right] \tag{2.9}$$

The mean time that a trap is occupied by an electron is given by the inversion of the emission probability, i.e.

$$\tau_e = 1/e_n \tag{2.10}$$

Accordingly, the *emission rate*, that is the number of electrons emitted per unit volume in the unit time interval is given by

$$e_n^* = N_t^- e_n = N_t^- \, v\sigma N_c \, g_0 \, \exp\left[-\frac{E_C - E_T}{kT}\right] \tag{2.11}$$

which should not be confused with the emission probability.

The knowledge of the emission and capture rates allows us to formulate the *rate equations* for transient emission and transient capture of electrons. Consider first the simple case of an n-type semiconductor with a concentration of N_t^- filled traps. Neglecting any capture processes, the rate equation for electron emission is given by

$$\frac{dN_t^-}{dt} = -N_t^- \, e_n \,. \tag{2.12a}$$

Assuming that all traps are initially filled with electrons [i.e. $N_t^- (t=0) = N_t$] the differential equation has the solution

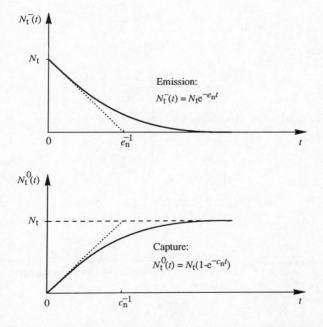

Fig. 2.5. Transient emission of electrons from traps and transient capture of electrons by traps. At $t = 0$ all traps are assumed to be filled and empty for emission and capture, respectively.

$$N_t^-(t) = N_t e^{-e_n t} . \tag{2.12b}$$

Consider next the opposite case in which electron capture dominates and any emission processes can be neglected. The rate equation is then given by

$$\frac{dN_t^0}{dt} = N_t^0 c_n . \tag{2.13a}$$

Assuming that all traps are initially unoccupied [i.e. $N_t^0(t=0) = N_t$] the differential equation has the solution

$$N_t^0(t) = N_t(1 - e^{-c_n t}) . \tag{2.13b}$$

The emission and capture transients of Eqs. (2.12b) and (2.13b) are shown in Fig. 2.5. If both processes, i.e. emission and capture, are considered, the rate equation is

$$\frac{dN_t^-}{dt} = -N_t^- \, e_n + N_t^0 \, c_n \qquad (2.14a)$$

where the identity

$$\frac{dN_t^-}{dt} = -\frac{dN_t^0}{dt} \qquad (2.14b)$$

was used. Under steady state conditions, $dN_t^- / dt = 0$, and one obtains

$$\boxed{\frac{N_t^-}{N_t^0} = \frac{c_n}{e_n}} \qquad (2.15)$$

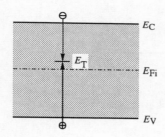

Fig. 2.6. Recombination of electrons and holes via a Shockley–Read recombination center.

In semiconductor structures with an appreciable number of minority carriers, electron–hole recombination can occur via deep traps. The electron–hole recombination via a deep level is schematically shown in Fig. 2.6. As a result, the minority carrier lifetime as well as the luminescence efficiency is reduced. Properties of minority carrier devices such as bipolar transistors, light-emitting diodes, and lasers are strongly affected by traps. Recombination via deep levels was first analyzed by Shockley and Read (1952) and is therefore frequently referred to as *Shockley–Read recombination*.

For the calculation of the Schockley–Read recombination (Shockley and Read, 1952; Hall 1951, 1952) we assume an equilibrium electron and hole concentration of n_0 and p_0, respectively. We first consider the capture of electrons and subsequently the capture of holes. We assume that the trap has a concentration of N_t and the charge states of 0 and -1 when unoccupied and occupied by an electron, respectively. We further assume that the trap has a single level at an energy E_t below the conduction band edge. The capture rate of electrons is given by Eq. (2.7) according to

$$c_n^* = N_t^0 \, n v_n \, \sigma_n$$

$$= N_t(1 - f_t) n v_n \, \sigma_n \tag{2.16}$$

where N_t^0 is the concentration of unoccupied traps. The Fermi–Dirac factor f_t is the probability that the trap is filled, i.e. $(1 - f_t)$ is the probability that the trap is empty. The Fermi–Dirac distribution (see Chap 3) is given by

$$f_t = \left[1 + e^{(E_T - E_F)/kT}\right]^{-1}. \tag{2.17}$$

The emission rate of electrons into the conduction band is given by Eq. (2.11) according to

$$e_n^* = N_t^- \, v_n \sigma_n \, N_c \, g_0 \, e^{-E_t/kT}$$

$$= N_t f_t \, v_n \, \sigma_n N_c \, e^{-E_t/kT} \tag{2.18}$$

where the ground-state degeneracy g_0 is assumed to be unity, and $E_t = E_C - E_T$. Using the Boltzmann distribution (see Chap. 3) for the occupancy of the conduction band

$$n = N_c \, e^{(E_F - E_C)/kT}, \tag{2.19}$$

Eq. (2.18) can be rewritten as

$$e_n^* = N_t f_t \, v_n \sigma_n \, n_1 \tag{2.20}$$

where n_1 is the free electron concentration if the Fermi level coincides with the trap level. Under equilibrium conditions, the emission rate and the capture rate coincide, that is

$$c_n^* = e_n^*,$$

$$N_t(1 - f_t) n v_n \sigma_n = N_t \, f_t n_1 v_n \sigma_n . \tag{2.21}$$

However, if conditions depart from equilibrium, then $c_n^* \neq e_n^*$ and a **net capture rate**, R_n, can be defined as

$$R_n = c_n^* - e_n^* = N_t v_n \sigma_n [(1 - f_t) n - f_t n_1] . \tag{2.22}$$

The net capture rate of holes, R_p, is obtained in an analogous way and is given by

$$R_p = N_t \, v_p \sigma_p [f_t p - (1 - f_t) p_1] \tag{2.23}$$

where $p_1 = n_i^2/n_1$, σ_p is the capture cross section of the trap for holes and v_p is either the thermal or the Fermi velocity of holes. The trap alternatively captures electrons and holes. In a steady state process the net hole capture rate equals the net electron capture rate, that is $R = R_n = R_p$. This condition gives us the following equation for f_t

$$f_t = \frac{n v_n \sigma_n + p_1 v_p \sigma_p}{v_n \sigma_n (n + n_1) + v_p \sigma_p (p + p_1)} . \qquad (2.24)$$

With $R = R_n = R_p$ and the insertion of Eq. (2.24) into Eq. (2.22) or (2.23) one obtains, using $n_1 p_1 = n_i^2$, the net recombination rate

$$R_{SR} = R = \frac{N_t v_n v_p \sigma_n \sigma_p (np - n_i^2)}{v_n \sigma_n (n + n_1) + v_p \sigma_p (p + p_1)} \qquad (2.25)$$

which gives us the net non-radiative **Schockley–Read recombination rate** in semiconductors. Note that for equilibrium conditions $np = n_i^2$ and the Shockley–Read rate equals zero. The Shockley–Read rate has the dimension cm^{-3}/s. The non-equilibrium carrier concentrations is conveniently written in terms of $n = n_0 + \Delta n$ and $p = p_0 + \Delta p$ with $\Delta n = \Delta p$ due to neutrality. Using this notation, the Shockley–Read recombination rate is

$$\boxed{R_{SR} = \frac{n_0 \Delta p + p_0 \Delta p + \Delta p^2}{(N_t v_p \sigma_p)^{-1} (n_0 + n_1 + \Delta p) + (N_t v_n \sigma_n)^{-1} (p_0 + p_1 + \Delta p)}} \qquad (2.26)$$

The lifetime of excess carriers is given by the equation $R_{SR} = \Delta p / \tau$. Consequently the lifetime is given by

$$\frac{1}{\tau} = \frac{n_0 + p_0 + \Delta p}{(N_t v_p \sigma_p)^{-1} (n_0 + n_1 + \Delta p) + (N_t v_n \sigma_n)^{-1} (p_0 + p_1 + \Delta p)} . \qquad (2.27)$$

We now differentiate between majority carriers and minority carriers and assume that electrons are in the majority, i.e. $n_0 \gg p_0$ and $n_0 \gg n_1$. If we further assume a small deviation from equilibrium i.e. $\Delta p \ll n$, then

$$\boxed{\frac{1}{\tau} = \frac{1}{\tau_{p_0}} = N_t v_p \sigma_p} \qquad (2.28)$$

If holes are the majority carriers, the lifetime is obtained in an analogous way.

$$\frac{1}{\tau} = \frac{1}{\tau_{n_0}} = N_t v_n \sigma_n . \qquad (2.29)$$

The results show that the Shockley–Read recombination rate is limited by the rate of capture of minority carriers. This result suggests itself since the capture of majority carriers is a much more likely event than the capture of minority carriers. Equation (2.27) can then be written as

$$\frac{1}{\tau} = \frac{n_0 + p_0 + \Delta p}{\tau_{p_0}(n_0 + n_1 + \Delta n) + \tau_{n_0}(p_0 + p_1 + \Delta p)} . \tag{2.30}$$

For small deviations from equilibrium, i.e. $\Delta p \ll n_0$, the equation simplifies to

$$\tau = \tau_{p_0} \frac{n_0 + n_1}{n_0 + p_0} + \tau_{n_0} \frac{p_0 + p_1 + \Delta p}{n_0 + p_0} \cong \tau_{p_0} \frac{n_0 + n_1}{n_0 + p_0} . \tag{2.31}$$

Inspection of the equation reveals that the lifetime does not change for small deviations from equilibrium in an extrinsic semiconductor.

For further insight we assume that the trap captures electrons and holes at the same rate, i.e. $v_n \sigma_n = v_p \sigma_p$ and $\tau_{p_0} = \tau_{n_0}$. One obtains from Eq. (2.31)

$$\tau = \tau_{p_0} \left[1 + \frac{n_1 + p_1}{n_0 + p_0} \right] . \tag{2.32}$$

For the special case of intrinsic material, i.e. $n_0 = p_0 = n_i$, the equation simplifies to

$$\tau_i = \tau_{p_0} \left[1 + \frac{n_1 + p_1}{2 n_i} \right] = \tau_{p_0} \left[1 + \cosh \left[\frac{E_T - E_{F_i}}{kT} \right] \right] \tag{2.33}$$

where E_{F_i} is the intrinsic Fermi level which is typically close to the middle of the gap. Inspection of the equation reveals that the lifetime is minimized, if $E_T - E_{F_i}$ is small, that is if the trap level is close to midgap. For such midgap levels the lifetime is given by $\tau = 2\tau_{p_0}$. This result demonstrates that deep levels are effective recombination centers if they are in the middle of the gap.

2.3 Effects relating to experimental properties

Several properties of deep levels directly relate to characteristics that can be observed experimentally. For example, if the capture of an electron from the conduction band onto the trap level is an optical transition, then the transition can be detected by luminescence measurements. Another example is the optical excitation of traps. What is the optical ionization threshold for traps? How does the threshold depend on the atomic structure and configuration of the trap? Other properties relate to transport properties of traps. The ionization energy of traps is known to depend on the strength of an electric field applied to the semiconductor. What is the functional dependence of the trap ionization energy on the electric field? These questions can be answered using phenomenological models of traps. The following section summarizes several effects and characteristics of deep levels that are relevant to experiments. The characteristics include (i) recombination characteristics, (ii) configurational relaxation of the trap and the surrounding lattice, and (iii) thermal ionization of traps in the presence of an electric field (Poole–Frenkel effect).

2.3.1 Recombination mechanisms

For shallow impurities, thermal ionization and recombination is primarily assisted by a single phonon. That is, the ionization and neutralization of an impurity is associated by the absorption and the emission of a phonon. For deep levels, a recombination process involving one phonon is not possible. Energies of acoustical and optical phonons in III–V semiconductors are typically limited to energies < 50 meV. Thus, the conservation of energy cannot be satisfied by the emission of a single phonon for deep level energies that are removed by > 50 meV from one of the band edges. Free carriers cannot recombine with deep levels unless there is a way of disposing of the carrier's energy. The recombination process of deep levels is therefore fundamentally different from the recombination process of a shallow level and has been the subject of considerable interest. Since multiphonon recombination has a very small probability (Lax, 1959), other mechanisms are required to explain the relatively large capture cross sections of deep traps which were observed experimentally.

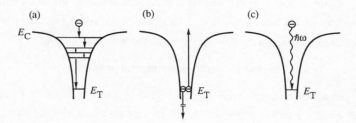

Fig. 2.7. Capture of an electron in the conduction band by a deep trap (a) via a cascade process, (b) assisted by an Auger process, and (c) by a radiative transition.

There are four processes by which recombination of free carriers with deep levels can occur. These processes are *optical* recombination, *cascade* recombination, *Auger* recombination, and *multiphonon* recombination. The former three of the mechanisms are schematically shown in Fig. 2.7. The mechanisms will be discussed in the following.

In the *cascade mechanism* for non-radiative recombination with a deep level, electrons cascade down excited states of the deep level via single phonon emission (Lax, 1959, 1960). To illustrate the cascade mechanism, consider a donor-like deep trap that is unoccupied and positively charged. Electrons in the conduction band are attracted by the long-range Coulomb potential of the deep trap. The deep level generally has a ground state and excited states associated with the short-range potential and a hydrogenic series of levels associated with the long-range Coulomb potential. In the initial stages of the capture process, the electron will be captured by the long-range potential, i.e. occupy states of the hydrogenic series. Within the hydrogenic series of states, the electron can jump to lower states by the emission of a single phonon. Since the probability of one-phonon transitions is substantial, such transitions enable the electron to escape from the conduction band (Jaros, 1982). The electron descends down the ladder of excited states, losing its energy by the successive emission of single phonons until the separation between the excited states becomes too large. Although it is obviously impossible to bring the electron down to the ground state without a multiphonon transition (or without the emission of a photon), it is sufficient that the electrons are trapped in the excited states and become unavailable for conduction. Lax argued that if a center is to behave as an efficient trap at a given temperature, its ionization energy of the hydrogenic series must be substantially larger than kT so that there are enough excited states from which the electron cannot be thermally ejected back into the conduction band. The motion up and down among the excited states is so rapid compared with the rate at which the electron disappears from the conduction band that the corresponding delay may be neglected and be accounted for by expressing the fraction of electrons that 'stick'. The sticking probability $p(U)$ simply depends upon the probability than an electron with energy U will emit a phonon of required energy. To obtain the capture cross section, one must calculate the total probability, i.e. (Jaros, 1982)

$$\sigma(E) = \int \sigma(E,U)p(U)\,dU \qquad (2.34)$$

where E is the kinetic energy of the incoming (free) electron and $\sigma(E,U)$ is the cross section for the corresponding collision event. Lax argued that Eq. (2.34) leads to realistic values for the calculated capture cross section. He pointed out that multiphonon transitions to a deep level gives capture cross sections that are

10 orders of magnitude too small. In III–V semiconductors the energy of the hydrogenic donor series is $\leq kT$. This enables carriers to jump to the hydrogenic ground state by emission of a single phonon. Even though the cascade capture is less effective if the hydrogenic ground state is shallow, the cascade process is more likely than multiphonon capture (Lax, 1992). Capture cross sections of centers capturing carriers by the cascade process are typically on the order of the Bohr radius. For an effective Bohr radius of 100 $\overset{\circ}{A}$, the capture cross section is on the order of $\sigma = a_B^{*2} \pi \cong 10^{-12} \text{ cm}^2$. The cascade process does not work equally well for *neutral* centers since neutral centers do not have the attractive long-range Coulomb potential. Lax (1959) argued that the large cross sections found for neutral centers can be explained on the same basis, the attractive potential in this case being provided by the large polarizability of the neutral center. This problem was further discussed by Jaros and Brand (1976). They concluded that neutral centers may indeed have several energy levels within the forbidden gap. Cascade recombination processes can occur via such energy levels. However, the authors also concluded that some neutral centers may have one single energy level only. In this case, a cascade recombination process cannot occur.

The second non-radiative recombination mechanism is the *Auger recombination mechanism* which is schematically shown in Fig. 2.7(b). Sheinkman (1965) proposed this mechanism for multiply charged deep centers. As an example, consider a double negatively charged center (see Fig. 2.7b). The total energy can be preserved if one of the electrons transfers to the valence band and the other is lifted to the conduction band. Thus, neither phonons nor photons are required for the recombination process. Similarly, the level can be doubly charged by capturing an electron from the conduction band and another carrier simultaneously from the valence band (this situation can be obtained from Fig. 2.7(b) by reversing the direction of the arrows). Auger band-to-band recombination was analyzed by Landsberg (1970). The probability of such Auger band-to-band processes increases with the square of the free carrier concentration, since two electrons are required for the process. However, deep level Auger processes involving pairs of electrons bound to the same defect do not exhibit this characteristic concentration dependence. Riddoch and Jaros (1980) showed that the capture cross section for Auger effects is very sensitive to the relationship between the impurity wave function and the relevant distorted Bloch states participating in the transition. The authors showed that the Auger capture cross section is approximately $10^{-16} - 10^{-18} \text{ cm}^2$, which is comparable to the capture cross section of a multiphonon capture.

Radiative recombination of a conduction band electron with a deep level is shown in Fig. 2.7(c). Transitions between deep levels and continuum states are characterized in terms of an optical capture cross section σ_c^{opt}, and a photoionization cross section, σ^{opt}. The optical photoionization cross section is defined as the probability that an absorption transition takes place between a deep level state and the continuum states for one incoming photon per unit area per second (Jaros, 1982). The optical capture cross section is defined as the probability that a photon is emitted due to the transition involving a continuum level and the deep level for one incoming electron per unit area and per unit time. Under thermal equilibrium conditions, the photon capture and photoionization rates balance in detail. The theoretical framework of radiative deep level transitions is based on the well-known theory of radiative band-to-band transitions developed by van Roosbroeck and Shockley (1954). This approach was modified by Solar and Burstein (1955) and by Blakemore (1967) to radiative extrinsic processes. There are only few reports on optical transitions involving deep levels. Banks et al. (1980) reported photoionization cross sections of deep level transitions in GaAs and GaP. Line shape analysis of the deep level transitions was also performed.

Multiphonon recombination processes usually have a very small capture cross section, e.g. $< 10^{-20}$ cm^2 (Lax, 1959, 1960). The small capture cross section is due to the fact that several phonons are required for a single recombination process. However, if the deep center is neutral and does not have excited states, then cascade recombination is less likely to occur. Multiphonon recombination may then be the dominating recombination mechanism. Henry and Lang (1977) argued that multiphonon emission during recombination processes with deep centers in GaAs and GaP is the prevailing process. The authors further showed that large lattice relaxation (see Sect. 2.3.2) is required for multiphonon recombination. This result is in agreement with Kovarskii (1962), who pointed out that multiphonon processes are likely in centers with large lattice relaxation.

2.3.2 Large lattice relaxation

The model of large lattice relaxation of deep centers can satisfactorily explain a host of experimental characteristics frequently observed with deep centers. In the model of lattice relaxation, the generalizing assumption is made that the atomic configuration of the center and of the surrounding host lattice depends on the charge state of the deep center. For example, the position, bonding structure, and electron configuration of the center can change with the charge state of the center. Based on this assumption, many features of deep levels can be explained including (i) a thermally activated capture cross section, (ii) a thermally activated emission energy that exceeds the thermal ionization energy, and (iii) a optical ionization energy that is much larger than the thermal activation energy of the center.

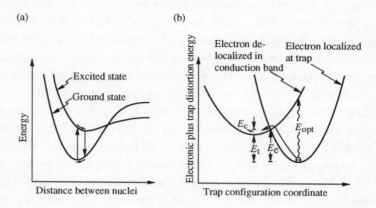

(a)   (b)

Fig. 2.8. Schematic illustration of the configuration coordinate diagram for (a) a molecule exhibiting a Franck–Condon shift (small relaxation) and for (b) a deep center with large lattice relaxation. Thermally activated capture with energy E_c is required for the case of large lattice relaxation, while thermal activation is not required for the electron transition to the ground state in the case of small lattice relaxation.

To appreciate the effect of large lattice relaxation, it is useful to recall the Franck–Condon principle (Davydov, 1976). The Franck–Condon principle applies to optical transitions between two electronic states of a molecule. The total energy of the molecule is shown in Fig. 2.8(a) for two different electronic configurations as a function of the distance between the nuclei. In the first configuration, the electron of interest occupies the ground state. In the second configuration, the electron is in an excited state. Other electrons are not considered here. The distance between the nuclei assumes a value to minimize the total energy of the molecule. For the dependences shown in Fig. 2.8(a), the distance between the nuclei is smaller if the electron occupies the ground state orbital. The two atoms are more loosely bound, i.e. the distance between the radii is larger, if the electron occupies the excited state. Franck argued that a change in the *electronic* configuration can occur much more rapidly than a change in the *atomic* configuration, due to the much smaller masses of electrons as compared to those of the nuclei (Davydov, 1976). Therefore, the transitions, indicated by arrows in Fig. 2.8(a), can only be 'vertical', i.e. without a change in the atomic configuration. As a result, the emission energy is always smaller than the absorption energy. The difference in energy is referred to as the ***Franck–Condon shift*** or ***Franck–Condon energy***. The excess energy, i.e. the difference between absorption and emission energy, usually results in vibrational energy of the molecule (Finkelnburg, 1958).

The Franck–Condon principle can be applied to deep centers in semiconductors by replacing the distance between nuclei by a ***generalized configuration coordinate***. Movement along the configuration coordinate implies the continuous change of the atomic configuration of the deep trap as well as of its surrounding lattice. A change in the configuration coordinate can represent the displacement of an impurity out of its equilibrium position, the breakage of an electronic bond, the distortion of the lattice, or a similar effect. It is further assumed that the configuration of the center is changed upon changing the charge state of the center. In the following, we consider a trap with two charge states, namely with an electron localized at the center (charged) and with the electron delocalized in the conduction band. Furthermore, we assume that the elastic energy changes quadratically upon changing the configuration coordinate (Hooke's law). Such a situation is illustrated in Fig. 2.8(b) which shows the total energy, i.e. electronic energy, plus trap distortion energy as a function of the configuration coordinate.

It is useful to differentiate between different magnitudes of lattice relaxation. Lattice relaxation is termed *small*, if vertical absorption or emission transitions can occur from the minimum of each parabola to the other parabola. Such *small lattice relaxation* is shown in Fig. 2.8(a). Lattice relaxation is called *large*, if vertical absorption or emission transitions are not possible from the minimum of each parabola to the other parabola. Such a *large lattice relaxation* is illustrated in the configuration coordinate diagram shown in Fig. 2.8(b). The figure shows an absorption transition from the lower to the upper parabola. However, a vertical transition from the minimum of the upper parabola to the lower parabola is not possible. As a consequence, systems with large lattice relaxation exhibit metastability, i.e. cannot return to the ground state via a vertical transition.

Recombination of an electron with the deep level exhibiting large lattice relaxation cannot occur via a *vertical* transition but requires the thermal excitation of the center to the crossover point of the two parabola. Once excited to the crossover point, the center can relax to its ground state. The capture cross section of the deep center can be written as (Henry and Lang, 1977)

$$\sigma = \sigma^{\infty} e^{-E_c/kT} \qquad (2.35)$$

where E_c is the thermal activation energy for capture and σ^{∞} is the capture cross section for $T \rightarrow \infty$. The thermally activated capture cross section of Eq. (2.35) has a fundamentally different functional dependence on temperature as compared to ordinary deep levels with a capture cross section independent of temperature. A second possibility for the relaxation of the deep trap is optical excitation. 'Vertical' excitation from the parabola representing the empty deep center to the parabola representing the (filled) center results in a finite probability that the trap relaxes to its filled ground state. The photon energy required for this transition exceeds the thermal capture energy, E_c, for the configuration coordinate diagram shown in Fig. 2.8(b). Large lattice relaxation of deep centers can also account for a large difference between the thermal ionization energy and the transient emission energy of deep levels. As illustrated in Fig. 2.8(b), the emission of an electron from the deep level into the conduction band requires an energy of E_e. This energy is relevant in transient emission experiments such as deep level transient spectroscopy (Lang and Logan, 1977; Lang et al., 1979). However, under (non-transient) thermal equilibrium conditions, the distribution of electrons between the deep level ground state and conduction band states is given by the energy difference between the two states, E_t. This thermal ionization energy is relevant in Hall measurements carried out at a given temperature. Note that the activation energy for transient emission is the sum of the thermal activation energy and the capture activation energy, i.e.

$$E_e = E_t + E_c \qquad (2.36)$$

as shown in Fig. 2.8(b).

The optical excitation energy of a electron occupying the deep level can be much larger than the thermal excitation energy for levels exhibiting large lattice relaxation. As with the Franck–Condon effect in molecules, it is reasonable to assume that changes in the electronic configuration of the center occur much more rapidly than changes in the structural (atomic) configuration. Therefore, the optical excitation of an electron to the conduction band must be a 'vertical' transition. The optical excitation energy, E_{opt}, can therefore be much larger than the thermal excitation energy, as shown in the example of Fig. 2.8(b). Much larger optical excitation energies have indeed been observed in centers exhibiting large lattice relaxation (Lang et al., 1979).

Large lattice relaxation can also explain the ***persistent photoconductivity*** effect which is observed at low temperatures in several III–V semiconductors (see e.g. Nelson, 1977). After photoexcitation of the deep centers at low temperatures, the conductivity increases. Most of the increase in photoconductivity persists even after the exciting illumination is turned off (Schubert and Ploog, 1984). The effect is referred to as persistent photoconductivity or persistent photocurrent. To explain the phenomenon of persistent photoconductivity, we refer to Fig. 2.8(b). After photoexcitation, the electron occupies conduction band states and contributes to the conductivity of the semiconductor. Recombination with the deep center requires that the thermal barrier of energy E_c be overcome. For $E_c \gg kT$, recombination can be very slow and can last for more than 24 hours, since carriers cannot overcome the thermal capture barrier. The persistent photoconductivity effect can be quenched by heating the sample to higher temperatures at which carriers have sufficient thermal energy to overcome the capture barrier.

Centers with large lattice relaxation are frequently characterized by the **Huang–Rhys parameters** S (Huang and Rhys, 1950). This parameter describes the shape of the parabola in configuration space (see Fig. 2.8). The parameter is given by $S = E_e/\hbar\omega$, where E_e is the energy required for electron emission from the ground state of the center, and $\hbar\omega$ is the vibrational energy of the center. For $S \gg 1$, i.e. $E_e \gg \hbar\omega$, a thermally activated emission process is expected.

We finally note that deep centers exhibiting lattice relaxation can have different symmetries for the two different configurations, i.e. in their occupied and empty states. Therefore, experimental procedures sensitive to the symmetry of the center can contribute to the structural identification of the two configurations (for examples see Chap. 9).

2.3.3 The Poole–Frenkel effect

The emission characteristics of deep centers depend on the electric field applied to the semiconductor. For electric fields with a magnitude of $\sim 10^5$ V/cm, the change in emission characteristics can be quite significant (see, for example, Lang, 1974a, 1974b; Vincent et al., 1979). The change in emission characteristics is attributed to the Poole–Frenkel effect, which implies a lowering of the effective emission barrier due to an electric field (Frenkel, 1938a and 1938b). The effect is also named after Poole (1914) who observed that the conductivity of a dielectric increases exponentially with the electric field in the pre-breakdown regime (see also Bethe and Salpeter, 1957).

The schematic potential of a deep impurity is shown schematically in Fig. 2.9 without (dash-dotted line) and with (solid line) a superimposed electric field. The emission of an electron from the deep level can be either thermionic emission or thermally assisted tunneling as indicated by the two arrows (a) and (b), respectively. Below, we first consider the effect of thermionic emission and

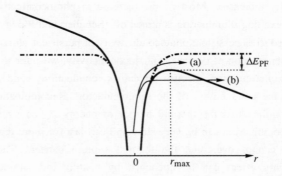

Fig. 2.9. Deep center potential with (solid line) and without (dash-dotted line) an electric field. Electron emission to the conduction band in the presence of an electric field can occur (a) via thermal emission (Poole–Frenkel emission) or (b) via thermally assisted tunneling.

subsequently the effect of tunneling through the barrier.

The potential of a deep level can be separated into a short-range potential and a long-range Coulomb potential. Since the short-range potential extends over only a few unit cells, the effect of an electric field on the short-range potential is small. The long-range Coulomb potential and the superimposed electric field are given by

$$V(r) = \frac{e}{4\pi\varepsilon r} + \mathcal{E}\,r \qquad (2.37)$$

where $\mathcal{E}$ is the magnitude of the electric field and $r = 0$ is the location of the impurity core. The potential has an extremum $(dV/dr = 0)$ at $r_{max} = (e/4\pi\varepsilon\mathcal{E})^{1/2}$. The lowering of the emission barrier is then obtained by inserting r_{max} into Eq. (2.37), i.e.

$$\Delta E_{PF} = eV(r_{max}) = e\sqrt{\frac{e\mathcal{E}}{\pi\varepsilon}} \qquad (2.38)$$

The emission probability for a deep center in an electric field is obtained from Eq. (2.9) by including the Poole–Frenkel energy, i.e.

$$e_n^{PF} = v\sigma N_c \, e^{-(E_t - \Delta E_{PF})/kT} \tag{2.39}$$

where $E_t = E_C - E_T$, and the ground-state degeneracy of the trap is assumed to be unity. Equation (2.39) is based on purely thermionic emission and tunneling through the barrier is neglected. Furthermore, Eq. (2.39) is a one-dimensional approximation. The three-dimensional case was developed by Hartke (1968) and by Milnes (1973).

In addition to thermal emission, tunneling is an alternative mechanism for electron emission from a deep level. This mechanism is indicated by the arrow (b) in Fig. 2.9. As a simple approximation, the tunnel barrier can be considered to be triangular. Tunneling through triangular barriers is known as Fowler–Nordheim tunneling which has the classical tunneling probability of

$$p(E) = e^{-K} \tag{2.40a}$$

with

$$K = \frac{4}{3} \frac{\sqrt{2m^*}}{e\hbar\mathcal{E}} E_t . \tag{2.40b}$$

The emission probability due to tunneling is then given by (Korol, 1977)

$$e_n^{tunnel} = \gamma \frac{E_t}{eK} e^{-K} . \tag{2.41}$$

The prefactor γ depends on the short-range potential of the deep level. Korol (1977) obtained a value of $\gamma = e/3\hbar$ for the model potential used. Subsequently, thermally assisted tunneling was considered by Vincent et al. (1979) and Pons and Makram-Ebeid (1979). At high temperatures, the emission probability can be approximated by

$$e_n \cong e_{n_0} \, e^{-2SkT/\hbar\omega} \tag{2.42}$$

where the Huang–Rhys parameter S represents the strength of the electron–phonon coupling.

3

Semiconductor statistics

The concentration of neutral impurities, ionized impurities, and free carriers in a doped semiconductor depends on a large number of parameters such as the impurity atom concentration, the free carrier mass, the bandgap energy, and the dielectric constant. The interdependences of the free majority and minority carrier concentration, the impurity concentration, impurity ionization energy as well as some other constants and materials parameters are given by *semiconductor statistics*. Semiconductor statistics describes the probabilities that a set of electronic states are either vacant or populated.

Electronic states include localized impurity states as well as delocalized conduction and valence band states. In the simplest case, an impurity has a single state with no degeneracy ($g_0 = 1$). However, an impurity may have a degenerate ground state ($g_0 > 1$) as well as excited levels which may need to be considered. The states in the bands and their dependence on energy are described by the *density of states*. In semiconductor heterostructures, the free motion of carriers is restricted to two, one, or zero spatial dimensions. In order to apply semiconductor statistics to such systems of reduced dimensions, the density of states in quantum wells (two dimensions), quantum wires (one dimension), and quantum dots (zero dimensions), must be known. The density of states in such systems will also be calculated in this chapter.

Classical Maxwell–Boltzmann statistics and quantum mechanical Fermi–Dirac statistics are introduced to calculate the occupancy of states. Special attention is given to analytic approximations of the Fermi–Dirac integral and to its approximate solutions in the non-degenerate and the highly degenerate regime. In addition, some numerical approximations to the Fermi–Dirac integral are summarized.

The *activation energy* of impurities will be frequently used in this chapter. It is useful to recall the interdependence of free energy, internal energy, enthalpy, entropy, and activation energy. To do so, consider the electronic ionization of an impurity, for example a donor

$$D^0 \leftrightarrow D^+ + e \ . \tag{3.1}$$

The effective work necessary to accomplish the ionization process at a constant temperature and pressure equals to the change of *Gibbs free energy* of the system (Kittel and Kroemer, 1980; Reif, 1965). In thermodynamics, Gibbs free energy G is defined as

$$G = H - TS \ , \tag{3.2}$$

$$H = E + PV \tag{3.3}$$

where H is the reaction enthalpy, S the entropy, E the internal energy, and PV the product of pressure and volume of the system. The change in Gibbs free energy occurring during the donor ionization process of Eq. (3.1) at a constant temperature T is then given by

$$\Delta G = \Delta H - T\Delta S \ , \tag{3.4}$$

$$\Delta H = \Delta E + P\Delta V \tag{3.5}$$

where constant temperature and constant pressure is assumed. Gibbs free energy is the proper energy to be used in a Boltzmann factor or Fermi function (see Sect. 3.3). The change in volume of the system occurring during chemical reactions can be quite significant. However, the change in volume during the electronic reaction of Eq. (3.1) is very small since the valence electron configuration does not change. The change in volume can therefore be neglected. In this chapter, the change in entropy as well as the mechanical work $(P\Delta V)$ are neglected. In this case, $\Delta G \cong \Delta H \cong \Delta E$. The energy required for the ionization reaction of Eq. (3.1) is the difference in internal energy, i.e. the difference in energy of states occupied by the electron before and after the ionization process. The change in free energy for donors can then be written as $\Delta G \cong \Delta H \cong \Delta E = E_C - E_D = E_d$, that is, the ionization energy equals the donor level energy relative to the bottom of the conduction band. The enthalpy and the entropy of ionization of centers in semiconductors were further considered by Thurmond (1975) and by Van Vechten and Thurmond (1976a, 1976b). The authors made simple estimates of the entropy of ionization of coulombic, isoelectronic, and vacancy-type defects in semiconductors by considering the effect of localized and free-carrier charge distributions upon the lattice modes. The empirical values of these entropies are observed as the temperature variation of the corresponding ionization levels (i.e. the term $T\Delta S$ in Eq. (3.4)). The change in entropy during the ionization reaction of Au-related levels in Si was considered by Lang et al. (1980), who differentiated between the entropy change due to electronic degeneracy and due to atomic vibrational changes. The authors showed that the change in entropy can be a small fraction (10%) of the ionization enthalpy.

3.1 Density of continuum states

Carriers occupy either localized impurity states or delocalized continuum states in the conduction band or valence band. In the simplest case, each impurity has a single, non-degenerate state. Thus, the density of impurity states equals the concentration of impurities. The energy of the impurity states is the same for all impurities (of the same species) as long as the impurities are sufficiently far apart and do not couple. The density of continuum states is more complicated and will be calculated in the following sections. Several cases will be considered including (i) a spherical, single-valley band, (ii) an anisotropic band, (iii) a band with multiple valleys, and (iv) the density of states in a semiconductor with reduced degrees of freedom such as quantum wells, quantum wires, and quantum boxes. Finally the *effective* density of states will be calculated.

3.1.1 Single-valley, spherical, and parabolic band

The simplest band structure of a semiconductor consists of a single valley with an isotropic (i.e. spherical), parabolic dispersion relation. This situation is closely approximated by, for example, the conduction band of GaAs. The electronic density of states is defined as the number of electron states per unit volume and per unit energy. The finiteness of the density of states is a result of the *Pauli principle*, which states that only two electrons of opposite spin can occupy one volume element in phase space. The *phase space* is defined as a six-dimensional space composed of real space and momentum space. We now define a 'volume' element in phase space to consist of a range of positions and momenta of a particle, such that the position and momentum of the particle are distinguishable from the positions and momenta of other particles. In order to be distinguishable, the range of positions and momenta must be equal or exceed the range given by the *uncertainty relation*. The volume element in phase space is then given by

$$\Delta x \Delta y \Delta z \Delta p_x \Delta p_y \Delta p_z = (2\pi\hbar)^3 . \qquad (3.6)$$

The 'volume' element in phase space is $(2\pi\hbar)^3$. For systems with only one degree of freedom, Eq. (3.6) reduces to the one-dimensional Heisenberg uncertainty principle $\Delta x \Delta p_x = 2\pi\hbar$. The Pauli principle states that two electrons of opposite spin occupy a 'volume' of $(2\pi\hbar)^3$ in phase space. Using the de Broglie relation ($p = \hbar k$) the 'volume' of phase space can be written as

$$\Delta x \Delta y \Delta z \Delta k_x \Delta k_y \Delta k_z = (2\pi)^3 . \tag{3.7}$$

The *density of states* per unit energy and per unit volume, which is denoted by $\rho_{DOS}(E)$, allows us to determine the total number of states per unit volume in an energy band with energies E_1 and E_2 according to

$$N = \int_{E_1}^{E_2} \rho_{DOS}(E) \, dE . \tag{3.8}$$

Note that N is the total number of states per unit volume, and $\rho_{DOS}(E)$ is the density of states per unit energy *and* per unit volume. To obtain the density of states per unit energy dE, we have to determine how much unit-volumes of k-space is contained in the energy interval E and $E + dE$, since we already know that one unit volume of k-space can contain two electrons of opposite spin.

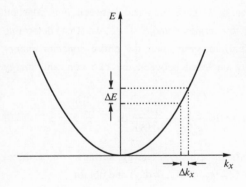

Fig. 3.1. Parabolic dispersion relation with a k-space interval Δk_x and a corresponding energy interval $\Delta E = (\partial E/\partial k_x)\Delta k_x$.

In order to obtain the volume of k-space included between two energies, the *dispersion relation* will be employed. A one-dimensional, parabolic dispersion relation $E = E(k_x)$ is shown in Fig. 3.1. For a given dE one can easily determine the corresponding length in k-space, as illustrated in Fig. 3.1. The k-space length associated with an energy interval dE is simply given by the slope of the dispersion relation. While the one-dimensional dispersion relations can be illustrated easily, the three-dimensional dispersion relation cannot be illustrated in three-dimensional space. To circumvent this difficulty, *surfaces of constant*

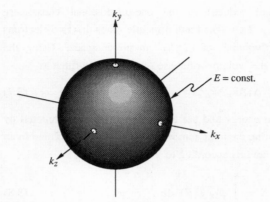

Fig. 3.2. Constant energy surface for a single valley, isotropic band.

energy in k-space are frequently used to illustrate a three-dimensional dispersion relation. As an example, the constant energy surface in k-space is illustrated in Fig. 3.2 for a spherical, single-valley band. A large separation of the constant energy surfaces, i.e. a large Δk for a given ΔE, indicates a weakly curved dispersion and a large effective mass.

In order to obtain the volume of k-space enclosed between two constant energy surfaces, which correspond to energies E and $E + dE$, we (first) determine dk associated with dE and (second) integrate over the entire constant energy surface. The 'volume' of k-space enclosed between the two constant energy surface is thus given by

$$V_{k\text{-space}}(E) = dE \int_{\text{Surface}} \frac{\partial k}{\partial E(k)} \, ds \tag{3.9}$$

where ds is an area element of the constant energy surface. In a three-dimensional k-space we use $\nabla_k = (\partial/\partial k_x, \partial/dk_y, \partial/dk_z)$ and obtain

$$V_{k\text{-space}}(E) = dE \int_{\text{Surface}} \frac{ds}{\nabla_k E(k)} \, . \tag{3.10}$$

Since an electron requires a volume of $4\pi^3$ in phase space, the number of states per unit volume is given by

$$N(E) = \frac{1}{4\pi^3} \, dE \int_{\text{Surface}} \frac{ds}{\nabla_k E(k)} \, . \tag{3.11}$$

Finally, we obtain the density of states per unit energy and unit volume according to

$$\rho_{DOS}(E) = \frac{1}{4\pi^3} \int_{Surface} \frac{ds}{\nabla_k E(k)} . \tag{3.12}$$

In this equation, the surface element ds is always perpendicular to the vector $\nabla_k E(k)$. Note that the surface element ds is in k-space and that ds has the dimension m^{-2}.

Next we apply the expression for the density of states to *isotropic parabolic* dispersion relations of a three-dimensional semiconductor. In this case the surface of constant energy is a sphere of area $4\pi k^2$ and the parabolic dispersion is $E = \hbar^2 k^2 / 2m^* + E_{pot}$ where k is the wave vector. Insertion of the dispersion in Eq. (3.12) yields the density of states in a semiconductor with a single-valley, isotropic, and parabolic band

$$\rho_{DOS}^{3D}(E) = \frac{1}{2\pi^2} \left[\frac{2m^*}{\hbar^2} \right]^{3/2} \sqrt{E - E_{pot}} \tag{3.13}$$

where E_{pot} is a potential energy such as the conduction band edge or the valence band edge energy, E_C or E_V, respectively.

3.1.2 Single-valley, anisotropic, parabolic band

In an anisotropic single-valley band, the dispersion relation depends on the spatial direction. Such an anisotropic dispersion is found in III–V semiconductors in which the L- or X- point of the Brillouin zone is the lowest minimum, for example in GaP or AlAs. The surface of constant energy is then no longer a sphere, but an ellipsoid, as shown in Fig. 3.3. The three main axes of the ellipsoid may have different lengths, and thus the three dispersion relations are curved differently. If the main axes of the ellipsoid align with a cartesian coordinate system, the dispersion relation is

$$E = \frac{\hbar^2 k_x^2}{2m_x^*} + \frac{\hbar^2 k_y^2}{2m_y^*} + \frac{\hbar^2 k_z^2}{2m_z^*} . \tag{3.14}$$

The vector $\nabla_k E$ is given by $\nabla_k E = (\hbar^2 k_x / m_x^*, \hbar^2 k_y / m_y^*, \hbar^2 k_z / m_z^*)$. Since the vector $\nabla_k E$ is perpendicular on the surface element, the *absolute* values of ds and $\nabla_k E$ can be taken for the integration. Integration of Eq. (3.12) with the dispersion relation of Eq. (3.14) yields the density of states in an anisotropic semiconductor with parabolic dispersion relations, i.e.

$$\rho_{DOS}(E) = \frac{\sqrt{2}}{\pi^2 \hbar^3} \sqrt{m_x^* m_y^* m_z^*} \sqrt{E - E_{pot}} . \tag{3.15}$$

If the main axes of the constant-energy ellipsoid do not align with the k_x, k_y, and k_z axes of the coordinate system then m_x^*, m_y^*, and m_z^* can be formally replaced by m_1^*, m_2^*, and m_3^*.

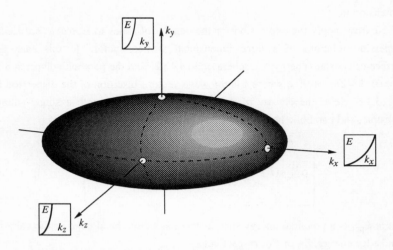

Fig. 3.3. Ellipsoidal constant energy surface with a weakly curved dispersion relation along the k_x axis and strongly curved dispersion along the k_y and k_z axis.

Frequently, the constant energy surfaces are rotational ellipsoids, that is, two of the main axes of the ellipsoid are identical. The axes are then denoted as the transversal and the longitudinal axes for the short and long axes, respectively. Such a rotational ellipsoid is schematically shown in Fig. 3.3. A relatively light mass is associated with the (short) transversal axis, while a relatively heavy mass is associated with the (long) longitudinal axis. If the masses are denoted as m_t^* and m_l^* for the transversal and the longitudinal mass, respectively, Eq. (3.15) can be modified according to

$$\rho_{\text{DOS}}(E) = \frac{\sqrt{2}}{\pi^2 \hbar^3} \sqrt{m_l^* m_t^{*2}} \sqrt{E - E_{\text{pot}}} \, . \tag{3.16}$$

The anisotropic masses m_x^*, m_y^*, m_z^*, m_l^*, and m_t^* are frequently used to define a *density-of-states effective mass*. This mass is given by

$$m^*_{\text{DOS}} = (m^*_x m^*_y m^*_z)^{1/3} \qquad (3.17\text{a})$$

$$m^*_{\text{DOS}} = (m^2_t m_l)^{1/3} \qquad (3.17\text{b})$$

The density of states is then given by

$$\rho_{\text{DOS}}(E) = \frac{\sqrt{2}}{\pi^2 \hbar^3} (m^*_{\text{DOS}})^{3/2} \sqrt{E - E_{\text{pot}}} . \qquad (3.18)$$

Note that for isotropic semiconductors the effective mass coincides with the density-of-states effective mass.

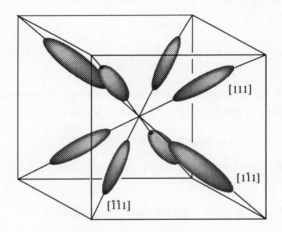

Fig. 3.4. Constant energy surface for the *L*-point of the Brillouin zone. The band structure consists of eight equivalent rotational ellipsoids.

3.1.3 Multiple valleys

At several points of the Brillouin zone, several equivalent minima occur. For example, eight equivalent minima occur at the *L*-point as schematically shown in Fig. 3.4. Each of the valleys can accommodate carriers, since the minima occur at different k_x, k_y, and k_z values, i.e. the Pauli principle is not violated. The density of states is thus obtained by multiplication with the number of equivalent minima, that is

$$\rho_{DOS}(E) = \frac{M_c \sqrt{2}}{\pi^2 \hbar^3} \sqrt{m_1^* m_2^* m_3^*} \sqrt{E - E_{pot}} \tag{3.19}$$

where M_c is the number of equivalent minima and m_1^*, m_2^*, and m_3^* are the effective masses for motion along the three main axes of the ellipsoid.

3.1.4 Reduced spatial dimensions

Semiconductor heterostructure allows one to change the band energies in a controlled way and confine charge carriers to two (2D), one (1D), or zero (0D) spatial dimensions. Due to the confinement of carriers, the dispersion relation along the confinement direction is changed. The change in dispersion relation results in a change in the density of states.

Confinement of a carrier in one spatial dimension, e.g. the z-direction results in the formation of quantum states for motion along this direction. Consider the ground state in a quantum well of width L_z with infinitely high walls. The ground-state energy is obtained from the solution of Schrödinger's equation and is given by

$$E_0 = \frac{\hbar^2}{2m^*} \left[\frac{\pi}{L_z} \right]^2 . \tag{3.20}$$

The particle in the quantum well can assume a range of momenta in the z-direction; the range is given by the uncertainty principle, i.e.

$$\Delta k_z = \Delta p_z / \hbar = 2\pi / L_z . \tag{3.21}$$

The dispersion relation for motion along the confinement (z-) direction is thus given by

$$E = E_0 \quad \text{for the entire range of } k_z . \tag{3.22}$$

The dispersion is flat, i.e. constant for all values of k_z. The z-component of the vector $\nabla_k E$ (see Eq. (3.12)) is therefore zero and need not be considered.

We next consider the x- and y- direction and recall that the Schrödinger equation is separable for the three spatial dimensions. Thus, the kinetic energy in the xy-plane is given by

$$E = \frac{\hbar^2}{2m^*} (k_x^2 + k_y^2) \tag{3.23}$$

for a parabolic dispersion.

The surface of constant energy for the dispersion relation given by Eq. (3.23) is shown in Fig. 3.5, and is a circle around $k_x = k_y = 0$. The density of states of such a 2D electron system is obtained by similar considerations as for the 3D

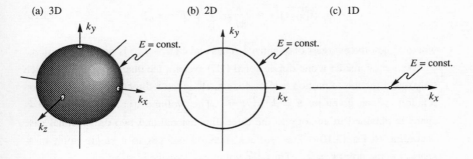

Fig. 3.5. Constant energy surfaces of a (a) 3-dimensional, (b) 2-dimensional, and (c) 1-dimensional system. The surfaces are a sphere, a circle, and a point for 3D, 2D, and 1D systems, respectively.

case. The reduced phase space now consists only of the xy-plane and the k_x and k_y coordinates. Correspondingly, the two-dimensional density of states is the number of states per *unit-area* and unit-energy. The volume of k-space between the circles of constant energy is given by Eq. (3.10). The equation is evaluated most conveniently in polar coordinates in which $k_r = (k_x^2 + k_y^2)^{1/2}$ is the radial component of the k-vector. The surface integral reduces to a line integral and the total length of the circular line is $2\pi k_r$. The volume of k-space then obtained is

$$V_{k-\text{space}}^{2D}(E) = dE \int\limits_{\text{Surface}} \frac{ds}{\nabla_k E(k)} = \frac{2\pi m^*}{\hbar^2} \, . \qquad (3.24)$$

Since two (one) electrons of opposite spin require a volume element of $(2\pi)^2$ in phase space, the density of states of a 2D electron system is given by

$$\boxed{\rho_{\text{DOS}}^{2D}(E) = \frac{m^*}{\pi \hbar^2} \quad \text{for } E \geq E_0} \qquad (3.25)$$

where E_0 is the ground state of the quantum well system. For energies $E \geq E_0$, the 2D density of states is a constant and does not depend on energy. If the 2D semiconductor has more than one quantum state, each quantum state has a state density of Eq. (3.25). The total density of states can be written as

$$\rho_{DOS}^{2D}(E) = \frac{m^*}{\pi \hbar^2} \sum_n \sigma(E - E_n) \qquad (3.26)$$

where E_n are the energies of quantized states and $\sigma(E - E_n)$ is the step function.

We next consider a one-dimensional (1D) system, the quantum wire, in which only one direction of motion is allowed, e.g. along the x-direction. The dispersion relation is then given by $E = \hbar^2 k_x^2 / 2m^*$. The 'volume' (i.e. length-unit) in k-space is obtained in analogy to the three-dimensional and two-dimensional case according to Eq. (3.10). The 'surface' integral reduces to a single point in k-space, i.e. the point $k = k_x$. Thus, the volume of k-space is given by

$$V_{k-\text{space}}^{1D}(E) = \int_{\text{Surface}} \frac{\delta(k_x - k_{x0})\, ds}{\nabla_k E(k_x)}$$

$$= \sqrt{\frac{m^*}{2\hbar^2(E - E_0)}} \quad \text{for } E \geq E_0 . \qquad (3.27)$$

The volume in phase space of two electrons with opposite spin is given by 2π and thus the 1D density of states is given by

$$\rho_{DOS}^{1D}(E) = \frac{1}{\pi \hbar} \sqrt{\frac{m^*}{2(E - E_0)}} \quad \text{for } E \geq E_0 \qquad (3.28)$$

Note that the density of states in a 3-, 2- and 1-dimensional system has a functional dependence on energy according to $E^{1/2}$, E^0, and $E^{-1/2}$, respectively. For more than one quantized state, the 1D density of states is given by

$$\rho_{DOS}^{1D}(E) = \frac{1}{\pi \hbar} \sum_n \sqrt{\frac{m^*}{2\sigma(E - E_n)}} \qquad (3.29)$$

where E_n are the energies of the quantized states of the wire.

Finally, we consider the density of states in a zero-dimensional (0D) system, the quantum box. No free motion is possible in such a quantum box, since the electron is confined in all three spatial dimensions. Consequently, there is no k-space available which could be filled up with electrons. Each quantum state of a 0D system can therefore be occupied by only two electrons. The density of states is therefore described by a δ-function.

Degrees of freedom

Density of states

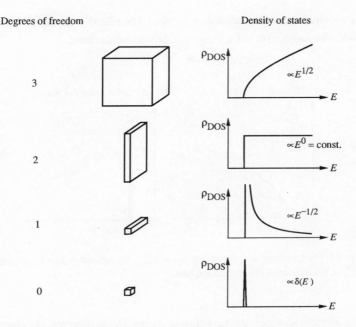

Fig. 3.6. Electronic density of states of semiconductors with 3, 2, 1, and 0 degrees of freedom for electron propagation. Systems with 2, 1, and 0 degrees of freedom are referred to as quantum wells, quantum wires, and quantum boxes, respectively.

$$\rho_{DOS}^{0D}(E) = 2\delta(E - E_0) \tag{3.30}$$

For more than one quantum state, the density of states is given by

$$\rho_{DOS}^{0D}(E) = \sum_n 2\delta(E - E_n) . \tag{3.31}$$

The densities of states for one quantized level for a 3D, 2D, 1D, and 0D electron system are schematically illustrated in Fig. 3.6.

3.1.5 Effective density of states

The *effective density of states* is introduced in order to simplify the calculation of the population of the conduction and valence band. The basic simplification made is that all band states are assumed to be located directly at the band edge. This situation is illustrated in Fig. 3.7 for the conduction band. The 3D density of states has square-root dependence on energy. The effective density of states is δ-function-like and occurs at the bottom of the conduction band.

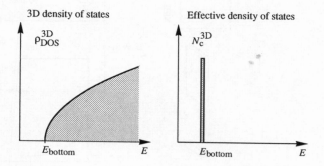

Fig. 3.7. Energy dependent density of states, ρ_{DOS}^{3D}, and *effective* density of states, N_c^{3D}, at the bottom of the conduction band.

An electronic state can be either occupied by an electron or unoccupied. Quantum mechanics allows us to attribute to the state a probability of occupation. The total electron concentration in a band is then obtained by integration over the product of state density and the probability that the state is occupied, that is

$$n = \int_{E_{bottom}}^{E_{top}} \rho_{DOS}(E) f(E) \; dE \qquad (3.32)$$

where $f(E)$ is the (dimensionless) probability that a state of energy E is populated (see Sect. 3.2). The limits of the integration are the bottom and the top energy of the band, since the electron concentration in the entire band is of interest.

As will be seen in the next section, the probability of occupation, $f(E)$, is given by the Maxwell–Boltzmann distribution (the Maxwell–Boltzmann distribution applies only to non-degenerate situations which will be defined in the next section). The Maxwell–Boltzmann distribution, also frequently referred to as the Boltzmann distribution, is given by

$$f_B(E) = \exp \left[- \frac{E - E_F}{kT} \right] \qquad (3.33)$$

where E_F is the Fermi energy (for a definition of the Fermi energy the reader is again referred to the next section). Using Eq. (3.32), the electron concentration can be determined by evaluating the integral.

The *effective* density of states at the bottom of the conduction band is now defined as the density of states which yields, with the Boltzmann distribution, the *same* electron concentration as the true density of states, that is

$$n = \int_{E_{\text{bottom}}}^{E_{\text{top}}} \rho_{\text{DOS}}(E)\, f_{\text{B}}(E)\, dE = N_c f_{\text{B}}(E = E_C) \tag{3.34}$$

where N_c is the effective density of states at the bottom of the conduction band and E_C is the energy of the bottom of this band. Strictly speaking, the effective density of states has no physical meaning but is simply a mathematical tool to facilitate calculations. For completeness, Eqs. (3.32) and (3.34) are now given explicitly using the Boltzmann distribution and the density of states of an isotropic three-dimensional semiconductor:

$$n = \int_{E_C}^{\infty} \frac{1}{2\pi^2} \left[\frac{2m^*}{\hbar^2} \right]^{3/2} \sqrt{E - E_C}\; e^{-(E - E_F)/kT}\, dE\ , \tag{3.35}$$

$$n = N_c\, e^{-(E_C - E_F)/kT}\ . \tag{3.36}$$

The upper limit of the integration can be taken to be infinity without loss of accuracy due to the strongly converging Boltzmann factor. Evaluation of the integral in Eq. (3.35) and comparison with Eq. (3.36) yields the effective density of states

$$\boxed{N_c = \frac{1}{\sqrt{2}} \left[\frac{m^* kT}{\pi \hbar^2} \right]^{3/2}} \tag{3.37}$$

Note that the effective density of states given by Eq. (3.37) applies to one minimum in the conduction band. If there are a number of M_c equivalent minima in the conduction band, the corresponding density of states must be multiplied by M_c. Furthermore, if the band structure is anisotropic, the effective mass m^* must be replaced by the density-of-states effective mass m^*_{DOS}. For a degenerate valence band with heavy and light holes, the effective density of states is the sum of both effective state densities, that is

$$N_v = \frac{1}{\sqrt{2}} \left[\frac{m^*_{\text{hh}} kT}{\pi \hbar^2} \right]^{3/2} + \frac{1}{\sqrt{2}} \left[\frac{m^*_{\text{lh}} kT}{\pi \hbar^2} \right]^{3/2}\ . \tag{3.38}$$

The effective density of states in a two-dimensional system (i.e. a system with two degrees of freedom) is obtained by the identical procedure as the three-dimensional effective density of states. The equations analogue to Eqs. (3.35) and (3.36) then read

$$n^{2D} = \int_{E_C}^{\infty} \frac{m^*}{\pi \hbar^2} \, e^{-(E - E_F)/kT} \, dE \; , \tag{3.39}$$

$$n^{2D} = N_c^{2D} \, e^{-(E_C - E_F)/kT} \tag{3.40}$$

where N_c^{2D} is the two-dimensional effective density of states. The carrier concentration n^{2D} represents the number of electrons per unit-area and is also referred to as the 2D density. Evaluation of the integral yields

$$\boxed{N_c^{2D} = \frac{m^*}{\pi \hbar^2} \, kT} \tag{3.41}$$

Finally, the effective density of states of a one-dimensional (1D) system is obtained in a similar way. The 1D density, i.e. the number of carriers per unit length is given by

$$n^{1D} = \int_{E_C}^{\infty} \frac{1}{\pi \hbar} \sqrt{\frac{m^*}{2(E - E_C)}} \, e^{-(E - E_F)/kT} \, dE \; , \tag{3.42}$$

$$n^{1D} = N_c^{1D} \, e^{-(E_C - E_F)/kT} \; . \tag{3.43}$$

The one-dimensional effective density of states is obtained as

$$\boxed{N_c^{1D} = \sqrt{\frac{m^* kT}{2\pi \hbar^2}}} \tag{3.44}$$

The evaluation of a zero-dimensional density of states does not yield a simplification of the carrier-density calculation, since the zero-dimensional density of states is δ-function like. Table 3.1 summarizes the dispersion relation, the density of states, and the effective density of states of semiconductors with various degrees of freedom.

Table 3.1. *Density of states for semiconductor with 3, 2, 1, and 0 degrees of freedom for propagation of electrons. The dispersion relations are assumed to be parabolic. The formulas can be applied to anisotropic semiconductors if the effective mass m^* is replaced by the density-of-states effective mass m_{DOS}^*. If the semiconductor has a number of M_c equivalent minima, the corresponding density of states must be multiplied by M_c. The bottom of the band is denoted as E_C and $\sigma(E)$ is the step-function.*

Degrees of freedom	Dispersion (kinetic energy)	Density of states	Effective density of states
3 (bulk)	$E = \dfrac{\hbar^2}{2m^*}(k_x^2 + k_y^2 + k_z^2)$	$\rho_{DOS}^{3D} = \dfrac{1}{2\pi^2}\left(\dfrac{2m^*}{\hbar^2}\right)^{3/2}\sqrt{E - E_C}$	$N_c^{3D} = \dfrac{1}{\sqrt{2}}\left(\dfrac{m^* kT}{\pi\hbar^2}\right)^{3/2}$
2 (slab)	$E = \dfrac{\hbar^2}{2m^*}(k_x^2 + k_y^2)$	$\rho_{DOS}^{2D} = \dfrac{m^*}{\pi\hbar^2}\,\sigma(E - E_C)$	$N_c^{2D} = \dfrac{m^*}{\pi\hbar^2}\,kT$
1 (wire)	$E = \dfrac{\hbar^2}{2m^*}k_x^2$	$\rho_{DOS}^{1D} = \dfrac{1}{\pi\hbar}\sqrt{\dfrac{m^*}{2(E - E_C)}}$	$N_c^{1D} = \sqrt{\dfrac{m^* kT}{2\pi\hbar^2}}$
0 (box)	—	$\rho_{DOS}^{0D} = 2\delta(E - E_C)$	$N_c^{0D} = 2$

3.2 Classical and quantum statistics

Typical densities of free carriers in semiconductors range from 10^{15} cm^{-3} to 10^{20} cm^{-3}. It is impossible to describe the energies or velocities of those carriers individually. An alternative to the individual characterization of particles is the *statistical* description of a carrier system. The statistical description uses *probabilities* of velocities or energies rather than knowing these quantities for all individual carriers. Thus, the statistical treatment represents a simplification. The derivation of the energy distribution function treats the carrier system as an *ideal gas*, for example a gas of oxygen molecules. The ideal gas is assumed to have only *elastic* collisions between atoms or molecules. Furthermore, the energy of the gas molecules is assumed to be purely translational *kinetic*. Since these properties are applied to the electron or hole system, those systems are frequently referred to as *electron-gases* or *hole-gases*.

Semiconductor statistics includes both classical statistics and quantum statistics. Classical or Maxwell–Boltzmann statistics is derived on the basis of purely classical physics arguments. In contrast, quantum statistics takes into account two results of quantum mechanics, namely (i) the Pauli exclusion principle which limits the number of electrons occupying a state of energy E and (ii) the finiteness of the number of states in an energy interval E and $E + dE$. The finiteness of states is a result of the Schrödinger equation. In this section, the basic concepts of classical statistics and quantum statistics are derived. The fundamentals of ideal gases and statistical distributions are summarized as well since they are the basis of semiconductor statistics.

3.2.1 Probability and distribution functions

Consider a large number N of free classical particles such as atoms, molecules or electrons which are kept at a constant temperature T, and which interact only weakly with one another. The energy of a single particle consists of *kinetic energy* due to translatory motion and an internal energy for example due to rotations, vibrations, or orbital motions of the particle. In the following we consider particles with only kinetic energy due to translatory motion. The particles of the system can assume an energy E, where E can be either a discrete or a continuous variable. If N_i particles out of N particles have an energy between E_i and $E_i + dE$, the probability of any particle having any energy within the interval E_i and $E_i + dE$ is given by

$$f(E_i) \; dE = \frac{N_i}{N} \tag{3.45}$$

where $f(E)$ is the **energy distribution function** of a particle system. In statistics, $f(E)$ is frequently called the *probability density function*. The total number of particles is given by

$$\sum_i N_i = N \tag{3.46}$$

where the sum is over all possible energy intervals. Thus, the integral over the energy distribution function is

$$\int_0^\infty f(E) \, dE = \sum_i \frac{N_i}{N} = 1 \,. \tag{3.47}$$

In other words, the probability of any particle having an energy between zero and infinity is unity. Distribution functions which obey

$$\int_0^\infty f(E) \, dE = 1 \tag{3.48}$$

are called **normalized** distribution functions.

The **average energy** or **mean energy** $\overline{E}$ of a single particle is obtained by calculating the total energy and dividing by the number of particles, that is

$$\overline{E} = \frac{1}{N} \sum_i N_i E = \int_0^\infty E f(E) \, dE \,. \tag{3.49}$$

In addition to energy distribution functions, velocity distribution functions are valuable. Since only the kinetic translatory motion (no rotational motion) is considered, the velocity and energy are related by

$$E = \tfrac{1}{2} mv^2 \,. \tag{3.50a}$$

The average velocity and the average energy are related by

$$\overline{E} = \tfrac{1}{2} m\overline{v^2} \tag{3.50b}$$

where $\overline{v^2}$ is called the *mean square velocity*. The *root-mean-square* (rms) velocity is defined as

$$v_{\text{rms}} = \sqrt{\overline{v^2}} \tag{3.50c}$$

and is the velocity corresponding to the average energy

$$\bar{E} = \frac{1}{2} m v_{\text{rms}}^2 .$$
$$\tag{3.50d}$$

In analogy to the energy distribution we assume that N_i particles have a velocity within the interval v_i and $v_i + dv$. Thus,

$$f(v)\,dv = \frac{N_i}{N}$$
$$\tag{3.51}$$

where $f(v)$ is the normalized velocity distribution. Knowing $f(v)$, the following relations allow one to calculate the mean velocity, the mean square velocity, and the root-mean-square velocity

$$\bar{v} = \int_0^\infty v f(v)\,dv ,$$
$$\tag{3.52a}$$

$$\overline{v^2} = \int_0^\infty v^2 f(v)\,dv ,$$
$$\tag{3.52b}$$

$$v_{\text{rms}} = \sqrt{\overline{v^2}} = \left[\int_0^\infty v^2 f(v)\,dv \right]^{1/2} .$$
$$\tag{3.52c}$$

Up to now we have considered the velocity as a scalar. A more specific description of the velocity distribution is obtained by considering each component of the velocity $v = (v_x, v_y, v_z)$. If N_i particles out of N particles have a velocity in the 'volume' element $v_x + dv_x$, $v_y + dv_y$, and $v_z + dv_z$, the distribution function is given by

$$f(v_x, v_y, v_z)\,dv_x\,dv_y\,dv_z = \frac{N_i}{N} .$$
$$\tag{3.53}$$

Since $\sum_i N_i = N$, the velocity distribution function is normalized, i.e.

$$\int_{-\infty}^{\infty} \int_{-\infty}^{\infty} \int_{-\infty}^{\infty} f(v_x, v_y, v_z)\,dv_x\,dv_y\,dv_z = 1 .$$
$$\tag{3.54}$$

The average of a specific propagation direction, for example v_x is evaluated in analogy to Eqs. (3.52a–c). One obtains

$$\bar{v}_x = \int\limits_{-\infty}^{\infty} \int\limits_{-\infty}^{\infty} \int\limits_{-\infty}^{\infty} v_x \, f(v_x, v_y, v_z) \, dv_x \, dv_y \, dv_z \; , \tag{3.55a}$$

$$\overline{v_x^2} = \int\limits_{-\infty}^{\infty} \int\limits_{-\infty}^{\infty} \int\limits_{-\infty}^{\infty} v_x^2 \, f(v_x, v_y, v_z) \, dv_x \, dv_y \, dv_z \; , \tag{3.55b}$$

$$v_{x,\text{rms}} = \sqrt{\overline{v_x^2}} = \left[\int\limits_{-\infty}^{\infty} \int\limits_{-\infty}^{\infty} \int\limits_{-\infty}^{\infty} v_x^2 \, f(v_x, v_y, v_z) \, dv_x \, dv_y \, dv_z \right]^{1/2} . \tag{3.55c}$$

In a closed system the mean velocities are zero, that is $\bar{v}_x = \bar{v}_y = \bar{v}_z = 0$. However, the mean square velocities are, just as the energy, not equal to zero.

3.2.2 Ideal gases of atoms and electrons

The basis of classical semiconductor statistics is ideal gas theory. It is therefore necessary to make a small excursion into this theory. The individual particles in such ideal gases are assumed to interact weakly, that is collisions between atoms or molecules are a relatively seldom event. It is further assumed that there is no interaction between the particles of the gas (such as electrostatic interaction), unless the particles collide. The collisions are assumed to be (i) *elastic* (i.e. total energy and momentum of the two particles involved in a collision are preserved) and (ii) of very short duration.

Ideal gases follow the universal gas equation (see e.g. Kittel and Kroemer, 1980)

$$PV = RT \tag{3.56}$$

where P is the pressure, V the volume of the gas, T its temperature, and R is the universal gas constant. This constant is independent of the species of the gas particles and has a value of $R = 8.314 \, \text{J K}^{-1} \, \text{mol}^{-1}$.

Next, the pressure P and the kinetic energy of an individual particle of the gas will be calculated. For the calculation it is assumed that the gas is confined to a cube of volume V, as shown in Fig. 3.8. The quantity of the gas is assumed to be 1 mole, that is the number of atoms or molecules is given by Avogadro's number, $N_{\text{Avo}} = 6.023 \times 10^{23}$ particles per mole. Each side of the cube is assumed to have an area $A = V^{2/3}$. If a particle of mass m and momentum mv_x (along the x-direction) is elastically reflected from the wall, it provides a momentum $2mv_x$ to reverse the particle momentum. If the duration of the collision with the wall is dt, then the force acting on the wall during the time dt is given by

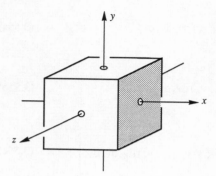

Fig. 3.8. Cubic volume confining one mole ($N_{Avo} = 6.023 \times 10^{23}$ atoms/mole) of an ideal gas. The pressure of the ideal gas exerted on one side of the cube (shaded area) is given by Eq. (3.59).

$$F = \frac{dp}{dt} \qquad (3.57a)$$

where the momentum change is $dp = 2mv_x$. The pressure P on the wall during the collision with one particle is given by

$$dP = \frac{F}{A} = \frac{1}{A}\frac{dp}{dt} \qquad (3.57b)$$

where A is the area of the cube's walls. Next we calculate the total pressure P experienced by the wall if a number of N_{Avo} particles are within the volume V. For this purpose we first determine the number of collisions with the wall during the time dt. If the particles have a velocity v_x, then the number of particles hitting the wall during dt is $(N_{Avo}/V) A v_x dt$. The fraction of particles having a velocity v_x is obtained from the velocity distribution function and is given by $f(v_x, v_y, v_z) dv_x dv_y dv_z$. Consequently, the total pressure is obtained by integration over all positive velocities in the x-direction

$$P = \int_{-\infty}^{\infty} \int_{-\infty}^{\infty} \int_{0}^{\infty} \frac{N_{Avo}}{V} A v_x dt \, f(v_x, v_y, v_z) \, dv_x dv_y dv_z \, \frac{2mv_x}{A \, dt} \, . \qquad (3.58a)$$

Since the velocity distribution is symmetric with respect to positive and negative x-direction, the integration can be expanded from $-\infty$ to $+\infty$

$$P = \frac{N_{Avo}}{V} m \int_{-\infty}^{\infty} \int_{-\infty}^{\infty} \int_{-\infty}^{\infty} v_x^2 f(v_x, v_y, v_z) \, dv_x dv_y dv_z = \frac{N_{Avo}}{V} m \, \overline{v_x^2} \, . \qquad (3.58b)$$

Since the velocity distribution is isotropic, the mean square velocity is given by

$$\overline{v^2} = \overline{v_x^2} + \overline{v_y^2} + \overline{v_z^2} \quad \text{or} \quad \overline{v_x^2} = \frac{1}{3}\,\overline{v^2}\,. \tag{3.58c}$$

The pressure on the wall is then given by

$$P = \frac{1}{3}\,\overline{v^2}\,\frac{N_{\text{Avo}}}{V}\,m\,. \tag{3.59}$$

Using the universal gas equation, Eq. (3.56), one obtains

$$RT = \frac{2}{3}\,N_{\text{Avo}}\frac{1}{2}m\overline{v^2}\,. \tag{3.60}$$

The average kinetic energy of one mole of the ideal gas can then be written as

$$\overline{E} = \overline{E_{\text{kin}}} = \frac{3}{2}\,RT\,. \tag{3.61}$$

The average kinetic energy of one single particle is obtained by division by the number of particles, i.e.

$$\boxed{\overline{E} = \overline{E_{\text{kin}}} = \frac{3}{2}\,kT} \tag{3.62}$$

where $k = R/N_{\text{Avo}}$ is the Boltzmann constant. The preceding calculation has been carried out for a three-dimensional space. In a one-dimensional space (one degree of freedom), the average velocity is $\overline{v^2} = \overline{v_x^2}$ and the resulting kinetic energy is given by

$$\overline{E_{\text{kin}}} = \tfrac{1}{2}kT \text{ (per degree of freedom)}\,. \tag{3.63}$$

Thus the kinetic energy of an atom or molecule is given by $\tfrac{1}{2}kT$. Equation (3.63) is called the **equipartition law**, which states that each 'degree of freedom' contributes $\tfrac{1}{2}kT$ to the total kinetic energy.

Next we will focus on the energetic distribution of electrons. The properties which have been derived in this section for atomic or molecular gases will be applied to free electrons of effective mass m^* in a crystal. To do so, the interaction between the electrons and the lattice must be negligible and electron–electron collisions must be a relatively seldom event. Under these circumstances we can treat the electron system as a classical ideal gas.

3.2.3 Maxwell velocity distribution

The Maxwell velocity distribution describes the distribution of velocities of the particles of an ideal gas. It will be shown that the Maxwell velocity distribution is of the form

$$f_M(v) = A \exp\left[-\frac{\frac{1}{2}mv^2}{kT}\right] \tag{3.64a}$$

where $\frac{1}{2}mv^2$ is the kinetic energy of the particles. If the energy of the particles is purely kinetic, the Maxwell distribution can be written as

$$f_M(E) = A \exp\left[-\frac{E}{kT}\right]. \tag{3.64b}$$

The proof of the Maxwell distribution of Eq. (3.64) is conveniently done in two steps. In the first step, the exponential factor is demonstrated, i.e. $f_M(E) = A \exp(-\alpha E)$. In the second step it is shown that $\alpha = 1/kT$.

In the theory of ideal gases it is assumed that collisions between particles are elastic. The total energy of two electrons before and after a collision remains the same, that is

$$E_1 + E_2 = E_1' + E_2' \tag{3.65}$$

where E_1 and E_2 are the electron energies before the collision and E_1' and E_2' are the energies after the collision. The probability of a collision of an electron with energy E_1 and of an electron with energy E_2 is proportional to the probability that there is an electron of energy E_1 and a second electron with energy E_2. If the probability of such a collision is p, then

$$p = Bf_M(E_1)f_M(E_2) \tag{3.66}$$

where B is a constant. The same consideration is valid for particles with energies E_1' and E_2'. Thus, the probability that two electrons with energies E_1' and E_2' collide is given by

$$p' = B f_M(E_1')f_M(E_2') . \tag{3.67}$$

If the change in energy before and after the collision is ΔE, then $\Delta E = E_1' - E_1$ and $\Delta E = E_2 - E_2'$. Furthermore, if the electron gas is in equilibrium, then $p = p'$ and one obtains

$$f_M(E_1)f_M(E_2) = f_M(E_1 + \Delta E)f_M(E_2 - \Delta E) . \tag{3.68}$$

Only the exponential function satisfies this condition, that is

$$f_M(E) = A \exp(-\alpha E) \tag{3.69}$$

where α is a positive yet undetermined constant. The exponent is chosen negative to assure that the occupation probability decreases with higher energies. It will become obvious that α is a universal constant and applies to all carrier systems such as electron-, heavy- or light-hole systems.

Next, the constant α will be determined. It will be shown that $\alpha = 1/kT$ using the results of the ideal gas theory. The energy of an electron in an ideal gas is given by

$$E = \tfrac{1}{2} mv^2 = \tfrac{1}{2}m(v_x^2 + v_y^2 + v_z^2) \ . \tag{3.70}$$

The exponential energy distribution of Eq. (3.69) and the normalization condition of Eq. (3.54) yield the normalized velocity distribution

$$f(v_x, v_y, v_z) = \left[\frac{m\alpha}{2\pi}\right]^{3/2} \exp\left[-\tfrac{1}{2}\, m\alpha(v_x^2 + v_y^2 + v_z^2)\right] \ . \tag{3.71}$$

The average energy of an electron is obtained by (first) calculating the mean square velocities, $\overline{v_x^2}, \overline{v_y^2}, \overline{v_z^2}$ from the distribution and (second) using Eq. (3.70) to calculate E from the mean square velocities. One obtains

$$E = (3/2)\, \alpha^{-1} \ . \tag{3.72}$$

We now use the result from classic gas theory which states according to Eq. (3.62) that the kinetic energy equals $E = (3/2)\, kT$. Comparison with Eq. (3.72) yields

$$\alpha = (kT)^{-1} \tag{3.73}$$

which concludes the proof of the Maxwell distribution of Eq. (3.64).

Having determined the value of α, the explicit form of the normalized *maxwellian velocity distribution* in cartesian coordinates is

$$\boxed{\,f_M(v_x, v_y, v_z) = \left[\frac{m}{2\pi kT}\right]^{3/2} \exp\left[-\frac{\tfrac{1}{2}m(v_x^2 + v_y^2 + v_z^2)}{kT}\right]\,} \tag{3.74a}$$

Due to the spherical symmetry of the maxwellian velocity distribution, it is useful to express the distribution in spherical coordinates. For the coordinate transformation we note that $f_M(v_x, v_y, v_z)\, dv_x dv_y dv_z = f_M(v)\, dv$, and that a volume element $dv_x dv_y dv_z$ is given by $4\pi v^2\, dv$ in spherical coordinates. The maxwellian velocity distribution in spherical coordinates is then given by

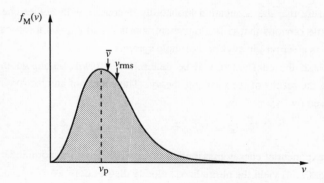

Fig. 3.9. Schematic maxwellian velocity distribution $f_M(v)$ of an ideal elecron gas. The velocity with the highest probability, v_p, is lower than the mean velocity, $\bar{v}$, and the root-mean-square velocity, v_{rms}.

$$f_M(v) = \left[\frac{m}{2\pi kT}\right]^{3/2} (4\pi v^2) \exp\left[-\frac{\frac{1}{2}mv^2}{kT}\right].$$ (3.74b)

The maxwellian velocity distribution is shown in Fig. 3.9. The peak of the distribution, that is the most likely velocity, is $v_p = (2kT/m)^{1/2}$. The mean velocity is given by $\bar{v} = (8kT)/(\pi m)^{1/2}$. The root-mean-square velocity can only be obtained by numerical integration.

3.2.4 The Boltzmann factor

The maxwellian velocity distribution can be changed to an energy distribution by using the substitution $E = \frac{1}{2}mv^2$. Noting that the energy interval and the velocity interval are related by $dE = mv\,dv$ and that the number of electrons in the velocity interval, $f_M(v)\,dv$, is the same as the number of electrons in the energy interval, $f_{MB}(E)\,dE$, then the energy distribution is given by

$$f_{MB}(E) = \frac{2}{\sqrt{\pi}} \frac{\sqrt{E}}{(kT)^{3/2}} e^{-E/kT}$$ (3.75)

which is the *Maxwell–Boltzmann distribution*.

For large energies, the exponential term in the Maxwell–Boltzmann distribution essentially determines the energy dependence. Therefore, the high-energy approximation of the Maxwell–Boltzmann distribution is

$$f_B(E) = A\,e^{-E/kT} \tag{3.76}$$

which is the **Boltzmann distribution**. The exponential factor of the distribution, $\exp(-E/kT)$, is called the **Boltzmann factor** or **Boltzmann tail**. The Boltzmann distribution does not take into account the quantum mechanical properties of an electron gas. The applicability of the distribution is therefore limited to the classical regime, i.e. for $E \gg kT$.

3.2.5 The Fermi–Dirac distribution

In contrast to classical Boltzmann statistics, the quantum mechanical characteristics of an electron gas are taken into account in Fermi–Dirac statistics. The quantum properties which are explicitly taken into account are

1. The *wave character* of electrons. Due to the wave character of electrons the Schrödinger equation has only a *finite number of solutions* in the energy interval E and $E + dE$.

2. The *Pauli principle* which states that an eigenstate can be occupied by only two electrons of opposite spin.

Since the Pauli principle strongly restricts the number of carriers per energy level, higher states are populated even at zero temperature. This situation is illustrated in Fig. 3.10, where two electron distributions are illustrated at zero temperature. The distribution in Fig. 3.10(a) does not take into account the Pauli principle while that in Fig. 3.10(b) does.

The first restriction imposed by quantum mechanics is the *finiteness of states* within an energy interval E and $E + dE$. The finiteness of states played a role in the derivation of the density of states, which was determined in Sect. 3.1. The density of states in an isotropic semiconductor was shown to be

$$\rho_{DOS}(E) = \frac{1}{2\pi^2}\left[\frac{2m^*}{\hbar^2}\right]^{3/2}\sqrt{E} \tag{3.77}$$

where E is the kinetic energy. Note that for the derivation of the density of states the Pauli principle has been taken into account. Therefore, the states given by Eq. (3.77) can be occupied only by *one* electron. Since the number of states per velocity-interval will be of interest, Eq. (3.77) is modified using $E = \frac{1}{2}mv^2$ and $dE = mv\,dv$. Note that the number of states per energy interval dE is the same as

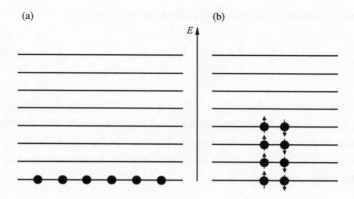

Fig. 3.10. Distribution of electrons at zero temperature among discrete energy levels (a) without Pauli principle and (b) with Pauli principle and spin taken into account. Spin 'up' and 'down' is illustrated by arrows.

the number of states per velocity interval dv, i.e. $\rho_{DOS}(E) dE = \rho_{DOS}(v) dv$. The number of states per velocity interval (and per unit volume) is then given by

$$\rho_{DOS}(v) = \frac{m^3}{\pi^2 \hbar^3} v^2 \tag{3.78}$$

for an isotropic semiconductor.

The *Fermi–Dirac distribution*, also called the Fermi distribution, gives the probability that a state of energy E is occupied. Since the Pauli principle has been taken into account in the density of states given by Eq. (3.77), each state can be occupied by at most one electron. The Fermi distribution is given by

$$f_F(E) = \left[1 + \exp\left(\frac{E - E_F}{kT} \right) \right]^{-1} \tag{3.79}$$

where E_F is called the Fermi energy. At $E = E_F$ the Fermi distribution has a value of ½. For small energies the Fermi distribution approaches 1; thus low-energy states are very likely to be populated by electrons. For high energies the Fermi distribution decreases exponentially; states of high energy are less likely to the populated. Particles which follow a Fermi distribution are called *fermions*. Electrons and holes in semiconductors are such fermions. A system of particles which obey *Fermi* statistics are called a *Fermi gas*. Electrons and holes constitute such Fermi gases.

An approximate formula for the Fermi distribution can be obtained for high energies. One obtains for $E \gg E_F$

$$f_F(E) \cong \exp\left[-\frac{E - E_F}{kT}\right] = f_B(E) . \qquad (3.80)$$

This distribution coincides with the Boltzmann distribution. Thus the (quantum-mechanical) Fermi distribution and the (classical) Boltzmann distribution coincide for high energies, i.e. in the classical regime.

Next we prove the Fermi distribution of Eq. (3.79) by considering a collision between two electrons. For simplification we assume that one of the electrons has such a high energy that it belongs to the classical regime of semiconductor statistics. Quantum statistics applies to the other low-energy electron. During the collision of the two electrons, the energy is conserved

$$E_1 + E_2 = E_1' + E_2' \qquad (3.81)$$

where, as before (Eq. (3.65)), E_1 and E_2 are electron energies before the collision and E_1' and E_2' are the energies after the collision.

The probability for the transition $(E_1, E_2) \rightarrow (E_1', E_2')$ is given by

$$p = f_F(E_1)f_B(E_2)[1 - f_F(E_1')][1 - f_B(E_2')] \qquad (3.82a)$$

where it is assumed that E_2 and E_2' are relatively large energies and the corresponding electron can be properly described by the Boltzmann distribution. The terms $[1 - f_F(E_1')]$ and $[1 - f_B(E_2')]$ describe the probability that the states of energies E_1' and E_2' are empty, and are available for the electron after the collision. Further simplification is obtained by considering that E_2' is large and therefore $[1 - f_B(E_2')] \cong 1$. Equation (3.82a) then simplifies to

$$p = f_F(E_1)f_B(E_2)[1 - f_F(E_1')] . \qquad (3.82b)$$

The same considerations are valid for the transition $(E_1', E_2') \rightarrow (E_1, E_2)$. The probability of this transition is given by

$$p' = f_F(E_1')f_B(E_2')[1 - f_F(E_1)] . \qquad (3.83)$$

Under equilibrium conditions both transition probabilities are the same, i.e. $p = p'$. Equating Eqs. (3.82b) and (3.83), inserting the Boltzmann distribution for $f_B(E)$, and dividing by $f_F(E_1)f_F(E_1')f_B(E_2)$ yields

$$\frac{1}{f_F(E_1')} - 1 = \left[\frac{1}{f_F(E_1)} - 1\right] \exp\left[\frac{E_2 - E_2'}{kT}\right] \qquad (3.84)$$

which must hold for all E_1 and E_1'. This condition requires that

$$\frac{1}{f_F(E)} - 1 = A \exp \frac{E}{kT} \qquad (3.85)$$

where A is a constant. If the value of the constant is taken to be $A = \exp(-E_F/kT)$ one obtains the **Fermi–Dirac distribution**

$$f_F(E) = \left[1 + \exp \left(\frac{E - E_F}{kT} \right) \right]^{-1} \qquad (3.86)$$

which proves Eq. (3.79).

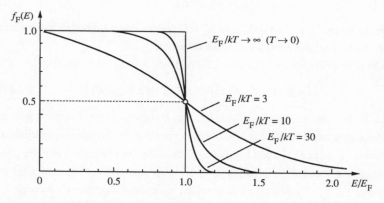

Fig. 3.11. Fermi–Dirac distribution as a function of E/E_F for different temperatures.

The Fermi–Dirac distribution is shown for different temperatures in Fig. 3.11. At the energy $E = E_F$ the probability of a state being populated has always a value of ½ independent of temperature. At higher temperatures, states of higher energies become populated. Note that the Fermi–Dirac distribution is symmetric with respect to E_F, that is

$$f_F(E_F + \Delta E) = 1 - f_F(E_F - \Delta E) \qquad (3.87)$$

where ΔE is any energy measured with respect to the Fermi energy.

The Fermi–Dirac velocity distribution of the particles in a Fermi gas is obtained by multiplication of Eq. (3.78) with Eq. (3.79)

$$g(v) = \rho_{DOS}(v) f_F(v)$$

$$= \frac{m^3 v^2}{\pi^2 \hbar^3} \left[1 + \exp\left[\frac{\frac{1}{2}mv^2 - E_F}{kT} \right] \right]^{-1} \tag{3.88}$$

where we have used the fact that the energy of the Fermi gas is purely kinetic, i.e. $E = \frac{1}{2}mv^2$. Note that $g(v)$ is the number of carriers per velocity interval v and $v + dv$ and per unit volume. If the velocity v is expressed in terms of its components, then the spherical volume element, $4\pi v^2 dv$, is modified to a volume element in rectangular coordinates, $dv_x dv_y dv_z$. Thus, using $g(v) dv = g(v_x, v_y, v_z) dv_x dv_y dv_z$, one obtains

$$g(v_x, v_z, v_z) = \frac{m^3}{\pi^2 \hbar^3} \frac{1}{4\pi} \left\{ 1 + \exp\left[\frac{\frac{1}{2}m(v_x^2 + v_y^2 + v_z^2) - E_F}{kT} \right] \right\}^{-1} \tag{3.89}$$

which is the Fermi velocity distribution (per unit volume) in cartesian coordinates.

The Fermi distribution of energies of an ideal gas is obtained by multiplication of Eq. (3.77) with Eq. (3.79) and is given by

$$g(E) = \frac{1}{2\pi^2} \left[2\frac{m}{\hbar^2} \right]^{3/2} \sqrt{E} \left[1 + \exp\left[\frac{E - E_F}{kT} \right] \right]^{-1} \tag{3.90}$$

when $g(E)$ is the number of particles in the energy interval E and $E + dE$ and per unit volume.

3.2.6 The Fermi–Dirac integral of order $j = +\frac{1}{2}$

The Fermi–Dirac integral of order $j = +\frac{1}{2}$ allows one to calculate the free carrier concentration in a three-dimensional (3D) semiconductor. The free carrier concentration in one band, e.g. the conduction band, of a semiconductor is obtained from the product of density of states and the state occupation probability, i.e.

$$n = \int_{E_C}^{E_{top}} \rho_{DOS}(E) f_F(E) dE . \tag{3.91}$$

Integration over all conduction band states is required to obtain the total concentration. The upper limit of integration is the top of the conduction band and can be extended to infinity. This extension of $E_C^{top} \rightarrow \infty$ can be done without

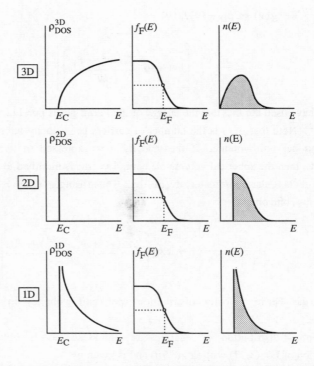

Fig. 3.12. Density of states (ρ_{DOS}), Fermi–Dirac distribution function (f_F) and carrier concentration (n) as a function of energy for a 3D, 2D, and 1D system. The shaded areas represent the total carrier concentration in the conduction band.

losing accuracy, since $f_F(E)$ converges strongly at high energies. The two functions, $\rho_{DOS}(E)$, $f_F(E)$, and their product are schematically shown in Fig. 3.12 for a semiconductor with three, two, and one, spatial degrees of freedom. The concentration per unit energy $n(E)$ is the product of state density and distribution function.

Equation (3.91) is evaluated by inserting the explicit expressions for the state density (Eq. (3.13)) and the Fermi–Dirac distribution (Eq. (3.79)). One obtains

$$ n = \frac{1}{2\pi^2} \left[\frac{2m^*kT}{\hbar^2} \right]^{3/2} \int\limits_0^\infty \frac{\eta^{1/2}}{1 + \exp(\eta - \eta_F)} \, d\eta \qquad (3.92) $$

where $\eta = E/kT$ and $\eta_F = -(E_C - E_F)/kT$ is the reduced Fermi energy. η_F is positive when E_F is inside the conduction band. The equation can be written in a more convenient way by using the Fermi–Dirac integral of order j, which is defined by (Sommerfeld, 1928; Sommerfeld and Frank, 1931)[†]

$$F_j(\eta_F) = \frac{1}{\Gamma(j+1)} \int_0^\infty \frac{\eta^j}{1 + \exp(\eta - \eta_F)} \, d\eta \qquad (3.93)$$

where $\Gamma(j+1)$ is the Gamma-function. With $j = ½$ and $\Gamma(3/2) = \sqrt{\pi}/2$ one obtains

$$n = N_c F_{½}(\eta_F) \qquad (3.94)$$

where N_c is the effective state density at the bottom of the conduction band (see Table 3.1). The Fermi–Dirac integral $F_{½}$ is shown in Fig. 3.13 along with several approximations which will be discussed later. For $j = ½$ the Fermi–Dirac integral is

$$F_{½}(\eta_F) = \frac{1}{\Gamma(3/2)} \int_0^\infty \frac{\eta^{½}}{1 + \exp(\eta - \eta_F)} \, d\eta \ . \qquad (3.95)$$

The evaluation of the integral cannot be done analytically. Even though the numerical calculation of the Fermi–Dirac integral is straightforward, it proves frequently convenient to use *approximate* analytic solutions of $F_{½}(\eta_F)$.

For analytic approximations of the Fermi-Dirac integral $n/N_c = F_{½}(\eta_F)$, the inverse function is frequently used, that is the reduced Fermi energy η_F is expressed as a function of n/N_c. A number of analytic approximations developed prior to 1982 have been reviewed by Blakemore (Blakemore, 1982). To classify various approximations, we differentiate between non-degeneracy and degeneracy. In the *non-degenerate* regime, the Fermi energy is below the bottom of the conduction band, $E_F \ll E_C$. In the *degenerate* regime the Fermi energy is at or above the bottom of the conduction band.

(1) **Extreme non-degeneracy (3D)**
     In the case of extreme non-degeneracy (i.e. $E_C - E_F \gg kT$ or $F_{½} \ll 1$

[†] Sommerfeld's original definition of the Fermi–Dirac integral omitted the term of the Gamma–function, i.e. $F_j^s(\eta_F) = \int_0^\infty \eta^j/[1 + \exp(\eta - \eta_F)] \, d\eta$. The modern definition of the Fermi–Dirac integral of Eq. (3.93) has the following advantages: (i) Unlike F_j^s, the functions F_j exist for negative orders of j, e.g. $j = -½, -1, -3/2$ etc. (ii) In the non-degenerate limit in which $\eta_F \ll 0$, all members of the $F_j(\eta_F)$ family reduce to $F_j(\eta_F) \to \exp\eta_F$ for all j. (iii) The *derivative* of the Fermi–Dirac integral of integer order j can be expressed as a Fermi–Dirac integral of order $(j-1)$, i.e. $(\partial/\partial\eta_F)F_j(\eta_F) = F_{j-1}(\eta_F)$.

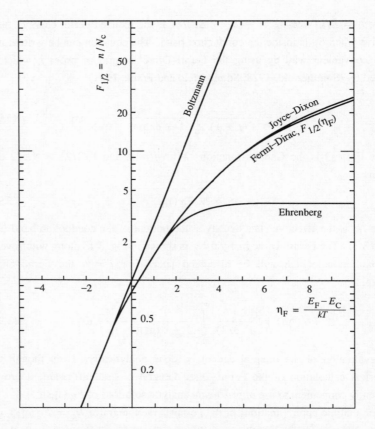

Fig. 3.13. Fermi–Dirac integral of order 1/2 as a function of the reduced Fermi energy η_F. Also shown are the Boltzmann distribution, the Joyce–Dixon approximation and the Ehrenberg approximation.

or $n \ll N_c$) the Fermi–Dirac distribution approaches the Boltzmann distribution. One obtains

$$\eta_F = -\frac{E_C - E_F}{kT} \cong \ln\,(n/N_c) \tag{3.96a}$$

which is shown in Fig. 3.14. This approximation is good when the Fermi

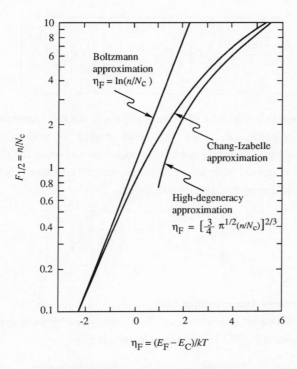

Fig. 3.14. Approximations for the Fermi–Dirac integral $F_{1/2}(\eta_F)$. Shown are the low-density approximation (Boltzman distribution), the high-degeneracy approximation and the full-range Chang–Izabelle approximation.

energy is $2kT$ or more below the bottom of the conduction band. Rearrangement of the equation yields the carrier concentration as a function of the Fermi energy in the non-degenerate limit

$$n = N_c \exp\left[-\frac{E_C - E_F}{kT}\right] \qquad (3.96b)$$

(2) **Extreme degeneracy (3D)**
In the case of extreme degeneracy (i.e. $(E_F - E_C) \gg kT$ or $F_{1/2} \gg 1$ or $n \gg N_c$) the Fermi–Dirac integral reduces to

$$\eta_F = -\frac{E_C - E_F}{kT} \cong \left[\frac{3}{2} \Gamma \left(\frac{3}{2} \right) \frac{n}{N_c} \right]^{2/3} \cong \left[\frac{3}{4} \sqrt{\pi} \frac{n}{N_c} \right]^{2/3} \quad (3.97a)$$

which is shown in Fig. 3.14. The range of validity for this approximation is $E_F - E_C > 10 \, kT$, i.e. when the Fermi energy is well within the conduction band. Rearrangement of the equation and using Eq. (3.37) for the effective density of states yields the carrier concentration as a function of the Fermi energy in the degenerate limit

$$n = \frac{1}{3\pi^2} \left[\frac{2m^* (E_F - E_C)}{\hbar^2} \right]^{3/2} \quad (3.97b)$$

(3)   *The Ehrenberg Approximation (3D)*

This approximation (Ehrenberg, 1950) was developed for weak degeneracy and is shown in Fig. 3.13. The approximation is given by

$$\eta_F = -\frac{E_C - E_F}{kT} \cong \ln \frac{n}{N_c} - \ln \left(1 - \frac{1}{4} \frac{n}{N_c} \right). \quad (3.98)$$

For small n the second logarithm term approaches zero, that is the Boltzmann distribution is recovered. The range of validity of the approximation is limited to $E_F - E_C \leq 2kT$, i.e. to weak degeneracy.

(4)   *The Joyce–Dixon approximation (3D)*

An approximation valid for a wider range of degeneracy was developed by Joyce and Dixon (Joyce and Dixon, 1977; Joyce, 1978). This approximation expresses the reduced Fermi energy as a sum of the Boltzmann term and a polynomial, i.e.

$$\eta_F = -\frac{E_C - E_F}{kT} \cong \ln \frac{n}{N_c} + \sum_{m=1}^{4} A_m \left[\frac{n}{N_c} \right]^m \quad (3.99a)$$

where the first four coefficients A_m are given by

$$A_1 = \frac{\sqrt{2}}{4} = 3.535\ 53 \times 10^{-1} \ ,$$

$$A_2 = -4.950\ 09 \times 10^{-3} \ , \tag{3.99b}$$

$$A_3 = 1.483\ 86 \times 10^{-4} \ ,$$

$$A_4 = -4.425\ 63 \times 10^{-6} \ .$$

The Joyce–Dixon approximation given here is shown in Fig. 3.13 and can be used for degeneracies of $E_F - E_C \le 8kT$. Inclusion of higher terms in the power series ($m > 4$) allows one to extend the Joyce-Dixon approximation to higher degrees of degeneracy.

(5) **The Chang–Izabelle Approximation (3D)**

The Chang–Izabelle approximation (Chang and Izabelle, 1989) is a full-range approximation which is valid for non-degenerate as well as degenerate semiconductors. The approximation is motivated by the fact that low-density and high-density approximations are available (see Eqs. (3.96) and (3.97)) which are the exact solutions in the two extremes. The Chang–Izabelle approximation represents the construction of a function, which approaches the low-density solution and the high-density solution of the Fermi–Dirac integral as shown in Fig. 3.14. The reduced Fermi energy is then given by

$$\eta_F = \frac{E_C - E_F}{kT} \cong \ln \frac{n}{N_c + n} + \left[\frac{3}{2}\Gamma\left(\frac{3}{2}\right) \right]^{2/3} \frac{n/N_c}{(A + n/N_c)^{1/3}} \tag{3.100a}$$

where

$$n/N_c = F_{\frac{1}{2}} \ , \tag{3.100b}$$

$$\frac{3}{2}\Gamma\left(\frac{3}{2}\right) = \frac{3}{2}\frac{\sqrt{\pi}}{2} = 1.329\ 34 \ , \tag{3.100c}$$

$$A = \frac{[(3/2)\,\Gamma(3/2)]^2\,(n_0/N_c)^3}{[\ln(1 + N_c/n_0)]^3} \ , \tag{3.100d}$$

$$\frac{n_0}{N_c} = \frac{n(E_F = E_C)}{N_c} = F_{\frac{1}{2}}(\eta_F = 0) = 0.765\ 15 \ . \qquad (3.100e)$$

One can easily verify that Eq. (3.100a) recovers the low-density approximation and the high-density approximation for $n \ll N_c$ and $n \gg N_c$, respectively. Furthermore, the approximation yields an exact solution for $\eta_F = 0$, i.e. when the Fermi energy touches the bottom of the conduction band. The largest relative error of η_F is 1% in the Chang–Izabelle approximation. Chang and Izabelle (1989) showed that the relative error can be further reduced by a weighting function and a polynomial function. Using these functions, the maximum relative error is reduced to 0.033%.

(6) **The Nilsson approximation (3D)**
    The Nilsson approximation (Nilsson, 1973) is valid for the entire range of Fermi energies. It is given by

$$\eta_F = -\frac{\ln(n/N_c)}{n/N_c - 1} + \left[\frac{3}{4}\sqrt{\pi}\,\frac{n}{N_c}\right]^{2/3} + \frac{(3/2)\sqrt{\pi}\,(n/N_c)}{[3 + (3/4)\sqrt{\pi}\,(n/N_c)]^2} \ .$$

$$(3.101)$$

The maximum relative error of the approximation is 1.1%.

3.2.7 The Fermi–Dirac integral of order $j = 0$

The Fermi–Dirac integral of order $j = 0$ allows one to calculate the free carrier density in a two-dimensional (2D) semiconductor. For semiconductor structures with only two degrees of spatial freedom, the Fermi–Dirac integral is obtained from Eq. (3.91) by insertion of the two-dimensional density of states. One obtains for the 2D carrier density

$$n^{2D} = \int_{E_C}^{E_{top}} \rho_{DOS}^{2D}(E)\, f_F(E)\, dE = \frac{m^* kT}{\pi\hbar^2} \int_0^\infty [1 + \exp(\eta - \eta_F)]^{-1}\, d\eta \quad (3.102a)$$

where $\eta = E/kT$ and $\eta_F = (E_F - E_C)/kT$ are reduced energies. The integral can be written as the Fermi–Dirac integral of zero ($j = 0$) order

$$F_{j=0}(\eta_F) = \frac{1}{\Gamma(1)} \int_0^\infty [1 + \exp(\eta - \eta_F)]^{-1}\, d\eta \qquad (3.102b)$$

where $\Gamma(1) = 1$ is the Gamma-function. The two-dimensional carrier density can be written by using the effective density of states of a 2D system given by

Eq. (3.41). One obtains

$$n^{2D} = N_c^{2D} F_0(\eta_F)$$ (3.103)

which is formally similar to the corresponding equation in three dimensions (Eq. (3.94)). The Fermi–Dirac integral of zero order $(j = 0)$ can be solved analytically. Using the integral formula

$$\int \frac{dx}{1 + e^x} = -\ln(1 + e^{-x}) ,$$ (3.104)

one obtains

$$F_0(\eta_F) = \ln(1 + e^{\eta_F}) .$$ (3.105)

Thus, the two-dimensional carrier density depends on the reduced Fermi energy according to

$$\frac{n^{2D}}{N_c^{2D}} = \ln(1 + e^{\eta_F}) .$$ (3.106)

Rearrangement of the equation yields the Fermi energy as a function of the carrier density

$$\eta_F = -\frac{E_C - E_F}{kT} = \ln\left[\exp\left[\frac{n^{2D}}{N_c^{2D}}\right] - 1\right]$$ (3.107)

(1)   *Extreme non-degeneracy (2D)*
   Approximation for the low-density regime $(n^{2D} \ll N_c^{2D})$ and the high-density regime $(n^{2D} \gg N_c^{2D})$ can be easily obtained from Eqs. (3.106) and (3.107). In the low-density regime one obtains

$$\eta_F = -\frac{E_C - E_F}{kT} = \ln\frac{n^{2D}}{N_c^{2D}}$$ (3.108a)

which is valid if the Fermi energy is much below the bottom of the conduction subband. Rearrangement of the equation yields the two-dimensional carrier density as a function of the Fermi energy in the non-degenerate limit

$$n^{2D} = N_c^{2D} \exp \left[-\frac{E_C - E_F}{kT} \right].$$ (3.108b)

(2)   **Extreme degeneracy (2D)**

In the high-density regime one obtains

$$\eta_F = -\frac{E_C - E_F}{kT} = \frac{n^{2D}}{N_c^{2D}}$$ (3.109a)

Rearrangement of the equation and insertion of the explicit expression for N_c^{2D} given in Eq. (3.41) yields

$$n^{2D} = \frac{m^*}{\pi \hbar^2} (E_F - E_C) \quad \text{(for high degeneracy)}$$ (3.109b)

that is the Fermi energy and the two-dimensional density follow a linear relation for two-dimensional structures.

3.2.8 The Fermi–Dirac integral of order $j = -\frac{1}{2}$

The Fermi-Dirac integral of order $j = -\frac{1}{2}$ allows one to calculate the free carrier density (per unit length) in a one-dimensional (1D) semiconductor. In semiconductor structures with only one degree of spacial freedom, the Fermi–Dirac integral is obtained from Eq. (3.91) by insertion of the 1D density of states of Eq. (3.28). One obtains

$$n^{1D} = \int_{E_c}^{E_{top}} \rho_{DOS}^{1D}(E) \, f_F(E) \, dE$$

$$= \frac{kT}{\pi \hbar} \sqrt{\frac{m^*}{2kT}} \int_0^\infty \frac{\eta^{-\frac{1}{2}}}{1 + \exp(\eta - \eta_F)} \, d\eta$$ (3.110)

where $\eta = E/kT$ and $\eta_F = -(E_C - E_F)/kT$ are reduced energies. The integral can be written as the Fermi–Dirac integral of order $j = -\frac{1}{2}$

$$F_{-\frac{1}{2}}(\eta_F) = \frac{1}{\Gamma(\frac{1}{2})} \int_0^\infty \frac{\eta^{-\frac{1}{2}}}{1 + \exp(\eta - \eta_F)} \, d\eta$$ (3.111)

where $\Gamma(\frac{1}{2}) = \sqrt{\pi}$ is the Gamma-function. Using the effective density of states of a one-dimensional system (see Eq. (3.44)), the one-dimensional density can be written as

$$n^{1D} = N_c^{1D} \, F_{-\frac{1}{2}}(\eta_F) \tag{3.112}$$

which is similar to the corresponding equations in three (Eq. (3.94)) and two (Eq. (3.103)) dimensions. The Fermi–Dirac integral of order $j = -\frac{1}{2}$ can only be obtained by numerical integration or by approximate solutions which will be discussed in the following sections. The $j = -\frac{1}{2}$ Fermi–Dirac integral has asymptotic solutions for the regimes of non-degeneracy and high degeneracy.

(1) *Extreme non-degeneracy (1D)*
In the regime of extreme non-degeneracy ($n^{1D} \ll N_c^{1D}$) the Fermi–Dirac integral of order $j = -\frac{1}{2}$ approaches the Boltzmann distribution. One obtains

$$\eta_F = -\frac{E_C - E_F}{kT} = \ln \frac{n^{1D}}{N_c^{1D}} \tag{3.113a}$$

This approximation is good for $E_F - E_C \leq 2kT$, i.e. when the Fermi energy is at least $2kT$ below the bottom of the conduction subband. Rearrangement of the equation yields the one-dimensional carrier density as a function of the Fermi energy in the non-degenerate limit

$$n^{1D} = N_c^{1D} \exp\left[-\frac{E_C - E_F}{kT}\right] . \tag{3.113b}$$

(2) *Extreme degeneracy (1D)*
In the case of extreme degeneracy ($n^{1D} \gg N_C^{1D}$) the Fermi–Dirac integral of order $j = -\frac{1}{2}$ reduces to

$$\eta_F = -\frac{E_C - E_F}{kT} \cong \frac{1}{4} \, [\Gamma(\tfrac{1}{2})]^2 \left[\frac{n^{1D}}{N_c^{1D}}\right]^2 \cong \frac{\pi}{4} \left[\frac{n^{1D}}{N_c^{1D}}\right]^2 \tag{3.114}$$

The range of validity of the approximation is $E_F - E_C > 10 \, kT$, i.e. when the Fermi energy is well within the conduction band.

3.3 Carrier concentrations

The free carrier concentration in semiconductors depends on a number of parameters such as the doping concentration, impurity activation energy, temperature, and other parameters. Given the results of the previous sections on the density of states and the distribution functions, the carrier concentration can now be calculated. In the calculation intrinsic, extrinsic, and compensated semiconductors will be considered.

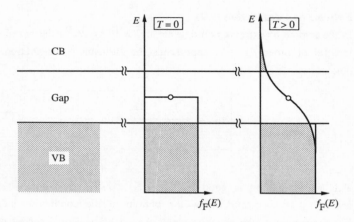

Fig. 3.15. Carrier distributions in the conduction band (CB) and valence band (VB) and Fermi distribution function in an intrinsic semiconductor at $T = 0$ K and at finite temperatures $T > 0$ K.

3.3.1 Intrinsic semiconductors

The carrier concentration of pure, undoped semiconductors is determined by thermal excitation of electrons from the valence band to states in the conduction band. An intrinsic semiconductor has a filled valence band and an empty conduction band at zero temperature. This property is the very definition of semiconductors. The band diagram along with the $T = 0$ K Fermi distribution function is shown in Fig. 3.15.

As the temperature increases, a small fraction of electrons in the valence band is excited into the conduction band. Thus the number of holes (unoccupied states) p in the valence band coincides with the number of electrons n in the conduction band. Semiconductors for which $n = p$ are called *intrinsic*. The condition that the concentration of electrons coincides with the concentration of holes requires

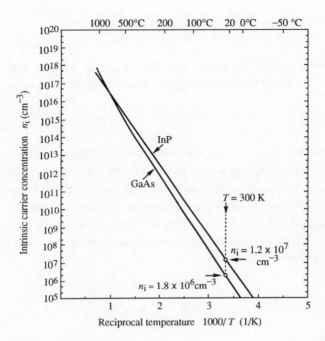

Fig. 3.16. Intrinsic carrier concentrations of GaAs and InP as a function of temperature. The slopes of the curves are proportional to the gap energy (after Thurmond, 1975; Laufer et al., 1980).

that the Fermi energy be within the forbidden gap. The position of the Fermi energy in the gap is visualized in Fig. 3.15. The electron and hole concentrations are given by

$$n = p \,, \tag{3.115a}$$

$$N_c F_{1/2}(\eta_F) = N_v F_{1/2}(\eta_F) \,. \tag{3.115b}$$

Since the Fermi energy is within the forbidden gap, i.e. many values of kT below the conduction band and many values of kT above the valence band, simpler Boltzmann statistics can be used instead of Fermi–Dirac statistics. Equation (3.115b) then simplifies to

$$N_c \exp \left[- \frac{E_C - E_F}{kT} \right] = N_v \exp \left[- \frac{E_F - E_V}{kT} \right] . \qquad (3.116)$$

Rearrangement of the equation and the definition of the gap energy $E_g = E_C - E_V$ yields for the Fermi energy of an *intrinsic* semiconductor

$$E_F = E_V + \frac{1}{2} E_g + \frac{kT}{2} \ln \frac{N_v}{N_c} . \qquad (3.117)$$

In this equation, $E_V + \frac{1}{2} E_g$ represents the mid-gap energy. Since the logarithmic function changes weakly with N_v / N_c, the Fermi energy of an intrinsic semiconductor is approximately at mid-gap. The temperature dependence of the intrinsic Fermi energy is weak due to the (weak) logarithmic dependence of the Fermi energy on the temperature. Using Boltzmann statistics the Fermi energy allows us to determine the **intrinsic carrier concentration**, n_i, of electrons and holes in an undoped semiconductor.

$$n_i = \sqrt{N_v N_c} \exp \left[- \frac{E_g}{2kT} \right] \qquad (3.118)$$

According to this equation the intrinsic carrier concentration increases exponentially with temperature. In addition, the effective density of states have the comparatively weak temperature dependence of $N_{c,v} \propto T^{3/2}$. The intrinsic carrier concentration is of special importance. Calculating the product of electron and hole concentration for *any* (non-degenerate) Fermi level using Boltzmann statistics yields

$$np = n_i^2 = N_v N_c \exp \left[- \frac{E_g}{kT} \right] . \qquad (3.119)$$

Thus the product np is a constant at a given temperature and, since the result does not depend on the Fermi level, is independent of the doping concentration. The intrinsic carrier concentrations of GaAs and InP are shown as a function of temperature in Fig. 3.16.

3.3.2 Extrinsic semiconductors (single donor species)
Substitutional donors and acceptors have an excess or a deficit electron in their outer electron shell, respectively, as compared to the replaced lattice atom.

Donors have one excess electron which can be *donated* to the conduction band. **Acceptors** have one less electron than the replaced lattice atom and can *accept* an electron from the filled valence band of the semiconductor, thereby creating a *hole*. The detailed physical properties of shallow impurities were considered in Chap. 1. Here we consider donors and acceptors being represented by an energy state close to the conduction band edge (donor) or close to the valence band edge (acceptor), as shown in Fig. 3.17. In the following, the free carrier concentration as a function of temperature is investigated in a semiconductor with donor impurities of concentration N_D of one chemical species.

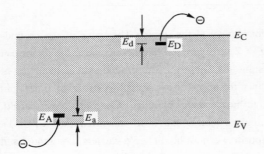

Fig. 3.17. Energy levels of acceptors and donors in the semiconductor band diagram. A donor (acceptor) level at energy E_D (E_A) has an ionization energy E_d (E_a).

The charge state of donors is *neutral* when occupied by an electron and positively charged if the electron is excited to the conduction band. The total concentration of donors is the sum of neutral donor concentration and ionized donor concentration, i.e.

$$N_D = N_D^0 + N_D^+ \ . \tag{3.120}$$

The energy of the donor impurity state is denoted as E_D. The donor energy is frequently given with respect to the conduction band edge, that is

$$E_d = E_C - E_D \ . \tag{3.121}$$

The probability of occupation of an acceptor or donor follows Fermi–Dirac statistics. Consequently, the concentration of neutral donors, i.e. donors occupied by an electron is

$$N_D^0 = N_D \, f_F(E_D) \tag{3.122}$$

where $f(E_D)$ is the value of the Fermi–Dirac distribution (Eq. (3.79)) at the energy of the donor. With $N_D^+ = N_D - N_D^0$ one obtains the concentration of ionized donors

$$N_D^+ = N_D[1 - f_F(E_D)] = N_D \left[1 - \cfrac{1}{1 + \cfrac{1}{g} \exp\left[\cfrac{E_D - E_F}{kT} \right]} \right]$$

$$= N_D \left[1 + g \exp\left[\frac{E_F - E_D}{kT} \right] \right]^{-1} \qquad (3.123)$$

where g is the ground-state degeneracy of the donor. The value of the ground-state degeneracy in GaAs is $g = 2$ for hydrogen-like donors since the donor can donate one electron of either spin (see Chap. 1). The ground-state degeneracy of acceptors in GaAs is $g = 4$, since the acceptor can accept electrons of either spin from the heavy-hole and the light-hole valence band (see Chap. 1). Note that Eq. (3.123) is limited to concentrations below the Mott transition (see Chap. 1). Above the Mott transition, impurities cannot bind charge carriers, i.e. donors and acceptors cannot be in the neutral charge state.

If a semiconductor has one carrier type dominating due to doping, the other carrier type has an extremely small equilibrium concentration. If, for example, GaAs is doped with $N_D = 10^{17}$ cm^{-3} donors and $n \cong 10^{17}$ cm^{-3}, the hole-concentration inferred from Eq. (3.118) at 300 K is $p = n_i^2/n = 3.2 \times 10^{-3}$ cm^{-3}. Thus, there are approximately 3 holes in 1000 cm^3 of this n-type semiconductor. The very small concentration of the *minority carrier* allows us to completely neglect minority carriers in many semiconductor structures. Such semiconductor devices are called *majority carrier* devices.

Charge neutrality is maintained in a doped semiconductor and has to be taken into account in addition to Fermi–Dirac statistics. Since minority carriers can be neglected, the free carrier concentration coincides with the ionized dopant concentration. If we restrict ourselves to n-type semiconductors, then

$$n = N_D^+ . \qquad (3.124)$$

We now consider the semiconductor at low temperatures, when most electrons occupy donor states. Then Boltzmann statistics can be used for the occupation of conduction band states according to

$$n = N_c \exp \left[-\frac{E_C - E_F}{kT} \right] . \qquad (3.125)$$

If Fermi–Dirac statistics are used for the occupation of the donor level according to Eq. (3.123), one obtains a quadratic equation for the free carrier concentration

$$n^2 - \frac{1}{g} N_D N_c e^{-E_d/kT} + \frac{1}{g} n N_c e^{-E_d/kT} = 0 . \qquad (3.126)$$

At low temperatures the free carrier concentration, n, is much smaller than the donor concentration, N_D. Thus, the third term of the quadratic equation is much smaller than the second term. The free carrier concentration is given by

$$n \cong \left[\frac{1}{g} N_D N_c \right]^{1/2} \exp \left[-\frac{E_d}{2kT} \right] \qquad (3.127)$$

where the ground state degeneracy for donors is $g = 2$.

Equation (3.127) was first obtained by de Boer and van Geel (1935) by the method described here. The formulas can also be obtained by minimizing the free energy change due to thermal excitation of electrons from donor states to conduction band states (Mott and Gurney, 1940). At higher temperatures all donors become ionized. The carrier concentration is then constant $n = N_D^+ = N_D$ and independent of temperature. This temperature regime is called the *saturation regime*.

As the temperature is increased even further, the *intrinsic* carrier concentration n_i increases and at sufficiently high temperatures assumes values comparable or higher than the dopant concentration. For most technologically useful semiconductors, the crossover from the saturation to the *intrinsic regime* occurs at temperatures much higher than room temperature. The three temperature regimes (i) thermal ionization regime (ii) saturation regime and (iii) intrinsic regime are shown schematically in Fig. 3.18 along with the associated activation energies.

The thermal ionization energy of a donor can be obtained from the slope of n versus reciprocal temperature. Rearrangement of Eq. (3.127) yields

$$E_d = -2k \frac{d(\ln n)}{d(1/T)} \qquad (3.128)$$

which allows one to determine E_d directly from the temperature dependent carrier concentration. The change in carrier concentration with increasing temperatures

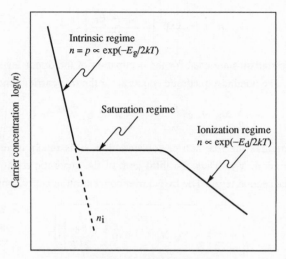

Reciprocal temperature $1/T$

Fig. 3.18. Carrier concentration as a function of reciprocal temperature for an uncompensated n-type semiconductor. The donor is assumed to form one level at energy E_d below the conduction band edge. Three regimes, namely (i) the ionization regime, (ii) the saturation regime, and (iii) the intrinsic regime and their corresponding activation energies are indicated (after Smith, 1986).

also implies a continuously changing Fermi level in the semiconductor. Consider an n-type semiconductor. At low temperatures the donor levels are filled, while the conduction band is empty. Thus, the Fermi level must be slightly above the donor level. As the temperature increases, the Fermi distribution becomes smeared out, as conduction band states become filled and donor states become unoccupied. Simultaneously the Fermi level moves deeper into the forbidden gap. At still higher temperatures, the Fermi level approaches the (near) mid-gap level and the semiconductor becomes intrinsic.

3.3.3 Extrinsic semiconductors (two donor species)

In the following, the free carrier concentration as a function of temperature is investigated in a semiconductor with two different species of donor impurities. It is assumed that the two donors form two different energy levels in the gap of the semiconductor. The two types of donor levels can originate from two different chemical species (e.g. Sn and Te donors in GaAs).

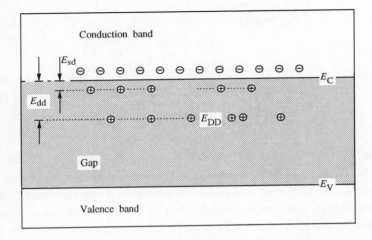

Fig. 3.19. Band diagram of a semiconductor with two species of donors namely a shallow donor with energy E_{SD} (ionization energy E_{sd}) and a deep donor with energy E_{DD} (ionization energy E_{dd}).

If both donor types have very similar thermal ionization energies, there is no necessity to differentiate between the two types of donors. It is therefore assumed that the two types of donors have markedly different ionization energies. In particular, we assume that one of the donors is relatively shallow and the other one is relatively deep. The band diagram of a semiconductor with two types of donors with energies E_{SD} and E_{DD} is shown in Fig. 3.19. If the shallow and deep donor concentrations are given by N_{SD} and N_{DD}, respectively, then the free carrier concentration is given by

$$n = N_{SD}^+ + N_{DD}^+ . \tag{3.129}$$

The carrier concentration in the conduction band is given by the Boltzmann distribution

$$n = N_c \exp\left[-\frac{E_C - E_F}{kT}\right] \tag{3.130}$$

while Fermi–Dirac statistics is assumed for the donor levels (see Eq. (3.123)).

$$N_{\text{SD}}^{+} = N_{\text{SD}} - \cfrac{N_{\text{SD}}}{1 + \cfrac{1}{g} \exp\left[\cfrac{E_{\text{SD}} - E_{\text{F}}}{kT}\right]} \, , \qquad (3.131\text{a})$$

$$N_{\text{DD}}^{+} = N_{\text{DD}} - \cfrac{N_{\text{DD}}}{1 + \cfrac{1}{g} \exp\left[\cfrac{E_{\text{DD}} - E_{\text{F}}}{kT}\right]} \qquad (3.131\text{b})$$

where E_{SD} and E_{DD} are the energies of the donor states. We assume that $E_{\text{sd}} \ll E_{\text{dd}}$ and $E_{\text{dd}} \ll E_g$, where $E_{\text{sd}} = E_{\text{C}} - E_{\text{SD}}$ and $E_{\text{dd}} = E_{\text{C}} - E_{\text{DD}}$.

All donors are neutral at very low temperatures. As temperature increases, shallow donors will donate their electrons to the conduction band, until all shallow donors are ionized. As the temperature is further increased, the deep donors start to become ionized until all deep donors are ionized. At even higher temperatures the intrinsic carrier concentration exceeds the dopant concentration and the semiconductor becomes intrinsic. In the following, the ionization regimes of the shallow and the deep donor are investigated.

At low temperatures when both types of donors are neutral, the Fermi energy is higher than the shallow donor energy. Then the energy difference between the Fermi energy and the deep donor energy is relatively large and according to Eq. (3.131b), the deep donor can be considered as neutral, i.e. $N_{\text{DD}}^{+} = 0$. The carrier concentration is then given by

$$n = N_{\text{SD}}^{+} = N_{\text{SD}} - \cfrac{N_{\text{SD}}}{1 + \cfrac{1}{g} \exp\left[\cfrac{E_{\text{SD}} - E_{\text{F}}}{kT}\right]} \, . \qquad (3.132)$$

Using Boltzmann statistics for the conduction band one obtains the quadratic equation

$$n^2 - \frac{1}{g} N_{\text{SD}} N_{\text{c}} \exp(-E_{\text{sd}}/kT) + \frac{1}{g} n N_{\text{c}} \exp(-E_{\text{sd}}/kT) = 0 \quad (3.133)$$

which is identical to the single donor equation Eq. (3.126). Thus, the low temperature solution is

$$n \cong \left[\frac{1}{2} N_{SD} N_c \right]^{1/2} \exp \left[- \frac{E_{sd}}{2kT} \right] \qquad (3.134)$$

where the ground-state degeneracy is assumed to be $g = 2$. Ionization of the shallow donor continues until all shallow donors are ionized, i.e. $n = N_{SD}^+ = N_{SD}$.

As the temperature is increased further, deep donors become ionized. Using Boltmann statistics for the conduction band (Eq. (3.130)) and Fermi–Dirac statistics for the deep donor (Eq. 131b) one obtains the quadratic equation

$$(N_{DD}^+)^2 + N_{DD}^+ \left[N_{SD}^+ + \frac{1}{g} N_c \exp(-E_{dd}/kT) \right] - \frac{1}{g} N_{DD} N_c \exp(-E_{dd}/kT) = 0.$$

$$(3.135)$$

For $N_{SD}^+ \gg (g^{-1}) N_c \exp(-E_{dd}/kT)$ one obtains

$$(N_{DD}^+)^2 + N_{SD}^+ N_{DD}^+ \cong \frac{1}{g} N_{DD} N_c \exp(-E_{dd}/kT) . \qquad (3.136a)$$

Since the free carrier concentration is the sum of ionized deep and shallow donor concentration ($n = N_{SD}^+ + N_{DD}^+$) the equation can be written as

$$n(n - N_{SD}^+) \cong \frac{1}{2} N_{DD} N_c \exp(-E_{dd}/kT) \qquad (3.136b)$$

where the donor ground-state degeneracy is assumed to be $g = 2$.

At the elevated temperatures considered here, the shallow donor is ionized ($N_{SD}^+ = N_{SD}$) and therefore the slope of the carrier density with respect to temperature follows the proportionality

$$n^2 - n N_{SD} \propto \exp(-E_{dd}/kT) \qquad (3.136c)$$

Note that this equation is significantly different from the simple relation $n \propto \exp(-E_d/kT)$, which would lead to incorrect results if applied to a semiconductor with two types of donors. Equation (3.136c) was applied to shallow and deep Si donors in $Al_x Ga_{1-x} As$ (Schubert and Ploog, 1984).

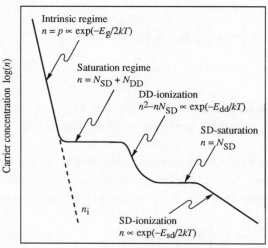

Fig. 3.20. Carrier concentration as a function of reciprocal temperature for a semiconductor with two different dopant species, namely a shallow donor (SD) and a deep donor (DD). At low temperatures the shallow donor becomes ionized and saturates when $n = N_{SD}$. At higher temperatures the deep donor becomes ionized and saturates ($n = N_{SD} + N_{DD}$). At even higher temperatures the semiconductor becomes intrinsic ($n = n_i$). The activation energies of each regime are indicated.

The ionization of the deep donor continues until shallow and deep donors are ionized, which corresponds to the saturation regime. At even higher temperatures the intrinsic carrier concentration increases above the dopant concentration and the semiconductor becomes intrinsic. The carrier concentration is shown in Fig. 3.20 for a semiconductor containing two different donor species as a function of temperature. The different saturation and ionization regimes along with their activation energies are indicated in the figure. Note that the different ionization regimes discussed above may not be as clearly distinguishable if the difference between E_{sd} and E_{dd} is small.

3.3.4 Compensated semiconductors

A partially compensated semiconductor contains dopant atoms of one type (n- or p-type) and, in addition, a smaller number of dopants of the other type. The band diagram of a partially compensated n-type semiconductor with a small number of acceptors is shown in Fig. 3.21. The free carrier concentration is given by

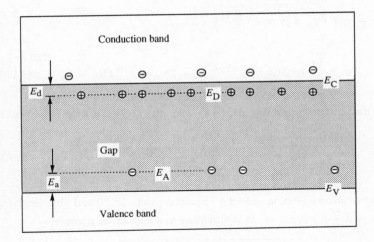

Fig. 3.21. Band diagram of a lightly compensated n-type semiconductor.

$$n = N_D^+ - N_A^- .$$ (3.137)

Electrons from donors prefer to occupy lower-energy acceptor states at all temperatures. Thus, even for low temperatures $(T \to 0 \text{ K})$ *all* acceptors and *some* donors are ionized. Fermi–Dirac statistics for the occupation of the donor state (Eq. (3.122)) and Eq. (3.137) yields

$$n + N_A^- = N_D^+ = N_D - \frac{N_D}{1 + \dfrac{1}{g} \exp\left[\dfrac{E_D - E_F}{kT}\right]} .$$ (3.138)

Using Boltzmann statistics for the population of the conduction band (Eq. (3.96b)) allows one to eliminate the Fermi energy. One obtains the quadratic equation

$$n^2 + n \left[N_A + \frac{1}{g} N_c \, e^{-E_d/kT}\right] - \frac{1}{g} N_c \, e^{-E_d/kT} (N_D - N_A) = 0$$ (3.139)

where acceptors are assumed to be ionized at all temperatures, i.e. $N_A = N_A^-$. The solution of the quadratic equation is, for $g = 2$:

$$n = -\tfrac{1}{2} \left[N_A + \tfrac{1}{2} N_c \, e^{-E_d/kT} \right]$$

$$+ \tfrac{1}{2} \left[\left(N_A + \tfrac{1}{2} N_c \, e^{-E_d/kT} \right)^2 + 2N_c \, e^{-E_d/kT} (N_D - N_A) \right]^{1/2} \quad (3.140)$$

At low temperatures when $\tfrac{1}{2} N_c \exp(-E_d/kT) \ll N_A$ the equation simplifies to

$$n = -\tfrac{1}{2} N_A + \tfrac{1}{2} N_A \left[1 + 2N_c \, e^{-E_d/kT} \left(\frac{N_D - N_A}{N_A^2} \right) \right]^{1/2} . \quad (3.141)$$

An approximate solution of the equation can be found by applying $\sqrt{1 + x} \cong 1 + \tfrac{1}{2}x$ (for $x \ll 1$) to the square root term of the equation.

$$n \cong \tfrac{1}{2} N_c \, \frac{N_D - N_A}{N_A} \, e^{-E_d/kT} \qquad\qquad (3.142)$$

Note that the temperature dependence of the carrier concentration as a function of temperature has a different slope as compared with the uncompensated semiconductor (see Eq. (3.127)). The slopes are different by a factor of two for the compensated and uncompensated case.

As the temperature is further increased, $N_D \gg \tfrac{1}{2} N_c \exp(-E_d/kT) \gg N_A$ and Eq. (3.140) simplifies to

$$n \cong \left[\tfrac{1}{2} N_D N_c \right]^{1/2} \exp \left[-\frac{E_d}{2kT} \right] \qquad\qquad (3.143)$$

which is identical to the uncompensated case given by Eq. (3.127).

Even further increase of the temperature results in fully ionized donors. The free carrier concentration is then given by $n = N_D^+ - N_A = N_D - N_A$. The two different slopes for the carrier concentration vs. temperature have indeed been observed experimentally (Morin, 1959).

3.4 Law of mass action

The law of mass action is frequently applied to chemical reactions and allows one to calculate the concentration of reagents as a function of temperature. Consider the two starting reagents A and B and the products C and D. The chemical reaction between the reagents can be written as

$$A + B \overset{\Delta G}{\leftrightarrow} C + D \tag{3.144}$$

where ΔG is the change in Gibbs free energy of the reaction. We refer to the introduction of this chapter and assume that $\Delta G \cong \Delta H \cong \Delta E$, i.e. the reaction enthalpy is a good approximation of the Gibbs free energy. ΔE is defined to be positive and negative for *endothermic* and *exothermic* reactions, respectively. Under thermal equilibrium conditions, the reaction of Eq. (3.144) follows the *principle of detailed balance* which states that at a given temperature the reaction rate of a process and the reaction rate of the reverse process balance in detail independent of other processes. Therefore, the *concentrations* of the reagents, [A], [B], [C], and [D] are *constant*. The *law of mass action* states that the ratio of the products of the concentrations of the final and initial reagents of the reaction is a constant which depends only on (i) temperature and (ii) the energy of the reaction, i.e.

$$\boxed{\frac{[C][D]}{[A][B]} = K = c \exp\left(-\frac{\Delta E}{kT}\right)} \tag{3.145}$$

where K and c are constants. At a given temperature the ratio of the concentrations is a constant. For low temperatures ($T \to 0$) and positive ΔE the right-hand side of Eq. (3.145) approaches zero and consequently $[C][D] \to 0$. Thus, the equilibrium of the endothermic chemical reaction is strongly shifted to the left-hand side in Eq. (3.144) at low temperatures.

The law of mass action is also applicable to *electronic reactions* in semiconductors. Such electronic processes are, for example, the ionization process of an impurity, the band-to-band excitation of an electron–hole pair, and other electronic equilibrium reactions. The application of the law of mass action to such processes will be discussed below.

Note that the law of mass action does not tell us anything about reaction *rates*. Such rates depend on the activation energy, which is illustrated by a potential hill in Fig. 3.22. Low activation energies result in fast reaction rates. However, the ratio of the concentrations of the reagents under equilibrium conditions (see Eq. (3.145)) depends only on ΔE. Note further that the law of mass action applies to *diluted* chemical solutions. Application of the law of mass action to electronic processes in semiconductors is therefore limited to *non-degenerate* carrier

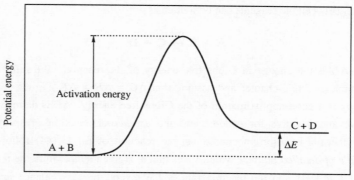

Fig. 3.22. Schematic potential energy diagram of a chemical reaction with the initial reagents A and B and the final reagents C and D. The equilibrium constant is determined by ΔE. The reaction rate is determined by the activation energy.

concentrations (Madelung, 1970). In such non-degenerate semiconductors, the (effective) density of states (in either the conduction or the valence band) is much larger than the free carrier density. Thus, the free carrier concentration is not limited by the finite density of states.

3.4.1 Intrinsic semiconductors

In an undoped semiconductor, free carriers are solely generated by excitation of an electron from the (filled) valence band to the (empty) conduction band. Excitation of an electron from the top of the valence band to the bottom of the conduction band requires the gap energy, E_g. (For a discussion on how the excitation energy relates to free energy, entropy, and enthalpy, we refer to the introduction of this chapter). The band-to-band excitation process can be expressed as

$$\text{VE} \overset{E_g}{\leftrightarrow} e + h \tag{3.146}$$

where VE is a valence electron, and h and e are a free hole in the valence band and a free electron in the conduction band, respectively. The concentration of h and e are p and n, respectively. The concentration of valence electrons is very large and is not significantly changed by the process of Eq. (3.146). It is therefore reasonable to assume that [VE] = const. The law of mass action is then given in analogy to Eqs. (3.144) and (3.145) by

$$np = c \; e^{-E_g/kT} \tag{3.147a}$$

where c is a constant. This result has the same functional dependence as Eq. (3.119), which was obtained from semiconductor statistics and the charge neutrality condition. The as yet unknown constant c is determined by comparison with Eq. (3.119) to be $c = N_c N_v$. With the definition of the intrinsic carrier concentration $n_i = (np)^{1/2}$ one obtains

$$\boxed{np = n_i^2 = N_c N_v e^{-E_g/kT}} \tag{3.147b}$$

The exponential dependence of the intrinsic carrier concentration has thus been derived using a second independent method in terms of the law of mass action.

3.4.2 Extrinsic semiconductors (single donor species)

For the calculation of the temperature dependence of extrinsic semiconductors we consider first an n-type semiconductor with donor energy E_d. The ionization process can be written as

$$D^0 \overset{E_d}{\leftrightarrow} D^+ + e \tag{3.148}$$

where D^0 and D^+ are neutral and ionized donors, respectively. If the concentration of D^0, D^+ and e are given by N_D^0, N_D^+ and n, respectively, then the law of mass action is

$$\frac{n \, N_D^+}{N_D^0} = c \, e^{-E_d/kT} \; . \tag{3.149a}$$

The equation can be simplified by considering $n \ll N_D$. Recalling that $n = N_D^+$ and with $N_D^0 = N_D - N_D^+ \cong N_D$ one obtains

$$n^2 \cong c \, N_D \; e^{-E_d/kT} \tag{3.149b}$$

which is the same functional dependence as Eq. (3.127). Comparison of the two equations yields the value of the constant $c = \tfrac{1}{2} N_c$. The electron concentration is then given by

$$n \cong (\tfrac{1}{2} N_D N_c)^{1/2} e^{-E_d/2kT} \qquad (3.149c)$$

Note that the factor of 2 arises from the twofold spin-degeneracy of the donor.

3.4.3 Extrinsic semiconductors (two donor species)

Consider a semiconductor with two types of dopants with different thermal ionization energies. For simplicity we assume a shallow and a deep donor with ionization energies E_{sd} and E_{dd}, respectively. At low temperature, the deep donor is assumed to be neutral and the reaction is given by

$$SD^0 \overset{E_{sd}}{\leftrightarrow} SD^+ + e \qquad (3.150)$$

where SD^0 and SD^+ refer to the neutral and ionized shallow donor, respectively. As shown in the preceding section, this reaction leads to the concentration

$$n \cong (\tfrac{1}{2} N_{SD} N_c) \exp(-E_{sd}/2kT) . \qquad (3.151)$$

As the temperature is further increased all shallow donors are ionized, i.e. $N_{SD}^+ = N_{SD}$ and deep donors start to become ionized. The corresponding reaction is

$$DD^0 \overset{E_{dd}}{\leftrightarrow} DD^+ + e \qquad (3.152)$$

where DD^0 and DD^+ are the neutral and ionized donor, respectively. Application of the law of mass action yields

$$\frac{N_{DD}^+ n}{N_{DD}^0} = c \exp(-E_{dd}/kT) . \qquad (3.153a)$$

This equation can be simplified by considering the temperature range in which only a few deep donors are ionized, i.e. $N_{DD}^+ \ll N_{DD}$ and $N_{DD} \cong N_{DD}^0$. Since the free electron density is the sum of shallow and ionized deep donor concentrations, i.e. $n = N_{SD} + N_{DD}^+$ one obtains

$$n(n - N_{SD}) \cong c N_{DD} \exp(-E_{dd}/kT) \qquad (3.153b)$$

which has the same functional dependence as Eq. (3.136b). A comparison of the two equation allows one to determine the constant $c = \tfrac{1}{2} N_c$. The electron concentration is then given by

$$n(n - N_{SD}) \cong \tfrac{1}{2} N_{DD} N_c \exp(-E_{dd}/kT) \ . \tag{3.153c}$$

3.4.4 Compensated semiconductors

In n-type semiconductors, donors are the dominant dopant species; however some residual acceptors are always present as well. Even at very low temperatures, some donors are ionized, because of the availability of lower energy acceptor states. For $T \rightarrow 0$ the free carrier concentration tends to zero, $n \rightarrow 0$, but $N_D^+ = N_A^- = N_A$. As the temperature is increased, additional donors become ionized. The corresponding process is

$$D^0 \overset{E_d}{\leftrightarrow} D^+ + e \tag{3.154}$$

where E_d is the thermal ionization energy of the donor. If the concentrations of D^0, D^+ and e are N_D^0, N_D^+, and n, respectively, then the law of mass action is

$$\frac{n \, N_D^+}{N_D^0} = c \, e^{-E_d/kT} \ . \tag{3.155a}$$

The equations can be simplified by considering low temperatures; then $n \rightarrow 0$, $N_D^+ \cong N_A$, and $N_D^0 \cong N_D - N_A$ which yields

$$n \cong c \, \frac{N_D - N_A}{N_A} \, e^{-E_d/kT} \tag{3.155b}$$

which has the same functional dependence as Eq. (3.143) which in turn was obtained by statistical considerations. The constant c is obtained by comparison with Eq. (3.143) and has a value of $c = \tfrac{1}{2} N_c$. The law of mass action for a partially compensated semiconductor then is given by

$$\boxed{n \cong \tfrac{1}{2} N_c \, \frac{N_D - N_A}{N_A} \, e^{-E_d/kT}} \tag{3.155c}$$

Note that the slope of n versus reciprocal temperature is described by the activation energy E_d.

If the temperature is increased further and $N_A \ll n \ll N_D$ then $n = N_D^+ - N_A \cong N_D^+$ and $N_D^0 = N_D - N_D^+ \cong N_D$. Insertion of these conditions into Eq. (3.155a) yields

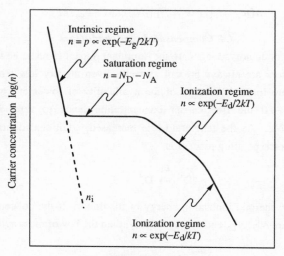

Fig. 3.23. Carrier concentration as a function of reciprocal temperature for a partially compensated n-type semiconductor. The ionization regime is characterized by two different activation energies (E_d and $E_d/2$).

$$n \cong \left[\tfrac{1}{2}\, N_c\, N_D\right]^{1/2} e^{-E_d/2kT} . \tag{3.156}$$

Note that the slope of n versus reciprocal temperature is described by the activation energy $\tfrac{1}{2}E_d$, which is a factor of two smaller than the activation energy of the compensated semiconductor of Eq. (3.155c). The carrier concentration versus temperature of a compensated semiconductor is schematically illustrated in Fig. 3.23, where the different regimes are indicated.

4

Growth technologies

The two growth technologies molecular-beam epitaxy (MBE) and vapor-phase epitaxy (VPE) are frequently employed for epitaxial growth, i.e. the crystalline growth of a thin layer on a crystalline substrate. High-quality epitaxial semiconductors with well controlled doping, composition, and thickness can be grown by MBE and VPE. Furthermore, the two growth methods MBE and VPE are capable of growing atomically abrupt doping and compositional profiles. This chapter covers the basic concepts of MBE, VPE, and related growth technologies.

4.1 Molecular-beam epitaxy

Molecular-beam epitaxy (MBE) is a powerful epitaxial growth technique whose historical origin dates back to the 1950s when Günther (1958) developed the 'three temperature method'. For the epitaxial growth of a compound semiconductor, e.g. GaAs, by coevaporation, the temperatures of the effusion cells (Ga and As) and the GaAs substrate must conform with the three temperature rule. Günther showed that the substrate temperature must be *between* the two effusion cell temperatures for stoichiometric growth. In the experiments of Günther, the temperature of the group-V oven (As, Sb) was much lower than the temperature of group-III (In) oven. The substrate temperature was kept at a temperature between the two oven temperatures. Stoichiometric, polycrystalline growth of InAs and InSb was demonstrated on glass substrates using the three temperature method. The choice of the group-V to group-III flux ratio is of crucial importance. The group-V element flux is selected to be larger than the group-III flux. Excess group-V atoms on the substrate reevaporate from the substrate (due to its higher temperature) except for the group-V atoms that have reacted with group-III atoms on the substrate. Stoichiometric growth of compound semiconductors thus requires excess group-V pressure.

The first epitaxial growth of III–V semiconductors was achieved by Davey and Pankey (1968). They used single-crystal GaAs substrates and a simple glass tube as a growth chamber.

The technology of molecular-beam epitaxy, as it is known today, is mainly due to Cho and coworkers (Cho, 1971; Cho and Casey, 1974; Cho and Arthur, 1975; Cho, 1985). Among the innovations introduced by Cho are a large area liquid N_2 cryoshroud encircling effusion cells and substrate, improved beam flux controllability, separate effusion cells for the group-III and group-V elements, and increased cell capacity. He demonstrated n-type and p-type conductivity in GaAs using Si, Sn, and Mg as doping elements. He made the first use of RHEED in an MBE system. Furthermore, he demonstrated the suitability of high-quality epitaxial layers for microwave and optoelectronic devices. In the 1980s the technology of MBE was further advanced by a large number of improvements including a load-lock sample introductory chamber, rotatable substrate holder, and better vacuum conditions.

The epitaxial growth of GaAs was studied in detail by several groups. Arthur (1967, 1968) investigated the kinetic behavior of Ga and As_2 beams impinging onto GaAs surfaces. The atomic and molecular beams, were obtained by evaporation of GaAs under ultra-high vacuum (UHV) conditions. The chopping of the atomic and molecular beams made possible the study of the growth kinetics. The growth kinetics of GaAs growth were studied in further detail by Foxon et al. (1974) and by Foxon and Joyce (1975, 1977) using a modulated molecular beam and reflection high-energy electron diffraction (RHEED). The thermal accommodation coefficients, surface lifetimes, sticking coefficients, desorption energies, and the reaction order were determined in their studies. The present understanding of GaAs growth from Ga and As_4/As_2 beams emerged mainly from their experiments.

The basic components of an MBE system which are inside the ultra-high vacuum chamber of the growth apparatus are schematically shown in Fig. 4.1. The constituents of the semiconductor are evaporated from effusion cells which are equipped with mechanical shutters. The shutters allow one to initiate and to terminate either the epitaxial growth or the doping. The substrate holder is rotated during growth for better uniformity. Several instruments are included within typical MBE systems which allow *in situ* analysis. These include a reflection high-energy electron diffraction (RHEED) electron gun with fluorescent screen, a residual gas analyzer (RGA), an Auger electron spectrometer (AES), and a mass spectrometer. In the following, the essential components of a MBE system and the materials used for the system will be discussed.

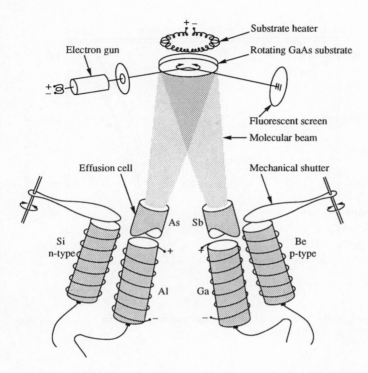

Fig. 4.1. Sketch showing the essential components of a molecular-beam epitaxy (MBE) system consisting of resistively heated Knudsen cells with shutters and the substrate which is rotating for better uniformity. A reflection high-energy electron diffraction (RHEED) gun and fluorescent screen are used to determine the reconstruction of the semiconductor surface. The components shown are enclosed in ultra-high vacuum (UHV). A cryoshroud (not shown) filled with liquid nitrogen encircles the growth area and the effusion cells.

4.1.1 Materials

The materials used in an MBE system need to fulfill several requirements due to the large temperature differences and the ultra-high vacuum conditions within the MBE apparatus. The temperatures range from $T = 77$ K for the cryoshroud filled with liquid N_2 to temperatures exceeding 1400 °C occurring in high temperature effusion cells. Furthermore, the low pressures of 10^{-10} Torr require that the materials do not outgass even at high temperatures. Thus, the requirements for materials used in the MBE apparatus are (i) low vapor pressure at high temperatures, (ii) low outgassing at high temperatures, (iii) small thermal expansion coefficient, and (iv) good mechanical properties.

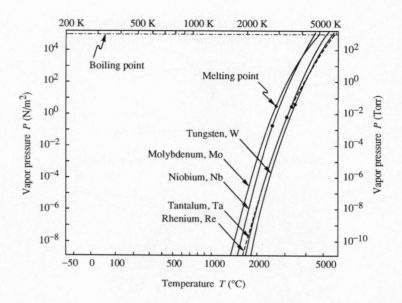

Fig. 4.2. Vapor pressure data of metals frequently used for high-temperature components of MBE systems.

The vapor pressure curves of metals with very low vapour pressure are shown in Fig. 4.2. Tungsten (W) is the metal with the lowest vapor pressure. Other metals with slightly higher vapor pressure are tantalum (Ta), molybdenum (Mo), rhenium (Re), and niobium (Nb). These metals are frequently used for the high temperature components of the MBE system due to the requirement of low vapor pressure. Tungsten is frequently used for heating wires; Ta for heat shields and heater ribbons; Mo for the substrate mounting block; Re and W in thermocouples ($W_{0.95} Re_{0.05} / W_{0.74} Re_{0.26}$). Stainless steel is used for parts not heated to very high temperatures but which require high mechanical stability. For example, stainless steel is used for the vacuum chamber walls and for the cryoshroud filled with liquid N_2.

Dielectric materials used in an MBE chamber are required to have low vapor pressure, low outgassing, small thermal expansion, and good mechanical properties. Boron nitride (BN), graphite (C), and quartz (SiO_2) fulfill these requirements. Very pure BN can be made by pyrolysis. Such pyrolytic BN is used for the crucibles of effusion cells and for the tubes of cracker cells. Graphite is used for some crucibles of effusion cells.

4.1.2 Effusion cells

A critical part of an MBE system are the effusion cells which are used to evaporate the constituent elements of the semiconductor as well as the dopants. The effusion cells are of Knudsen-type geometry. In this geometry, collisions between evaporated atoms (or molecules) are much less frequent than collisions between atoms (or molecules) and the crucible wall. As a result, the flux distribution from a cylindrical effusion cell with Knudsen-type geometry is jet-like and can be expressed by a cosine function (Knudsen, 1909). The distribution of atoms or molecules arriving at the substrate is cosine-like as well.

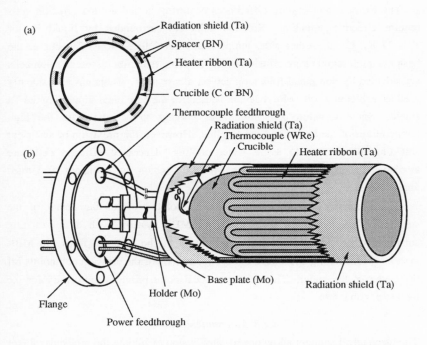

Fig. 4.3. (a) Top-view and (b) side-view of a Knudsen-type effusion cell with a cylindrical crucible. The crucible is heated by low-induction Ta ribbons and heat-shielded by Ta foil. A thermocouple is used to measure the temperature of the crucible. The effusion cell is mounted on a conflat flange with feedthroughs for the heater and the thermocouple.

A sketch of an effusion cell mounted on a conflat flange is shown in Fig. 4.3. The essential components of an effusion cell are the crucible, heating wire or ribbon, heat shield, and thermocouple. The thermocouple touches the bottom of the crucible for good thermal contact and temperature accuracy. The crucible is typically made out of BN or C. Cylindrically shaped as well as conically shaped crucibles are employed. Conically shaped crucibles have better flux homogeneity as compared to cylindrically shaped ones.

The heater of the effusion cell typically consists of W wire or Ta ribbon. The heater filament is mechanically kept in place by BN spacers. It is desirable for the heater filament to have a low inductance. The low-inductance wiring makes possible the employment of alternating current (ac) supplies for the heater and is also aimed to reduce magnetic fields which would perturb the beam of the RHEED electron gun. Therefore, a meander-type rather than the spiral-type filament pattern is used in effusion cells.

The extreme temperature differences occurring in and around effusion cells require effective shielding. Effusion cells are surrounded by liquid N_2 of $T = 77$ K. On the other side, temperatures exceeding 1400 °C occur at the heater of high-temperature effusion cells. To avoid radiation losses, effusion cells are shielded by thin metal foils mounted on a base plate. Tantalum is frequently used as a shielding foil. Molybdenum is used for the base plate supporting the Ta shield. Some effusion cells use water cooling to accommodate the high temperatures of the heater and the crucible. However, the provision of sufficient shielding allows one to avoid water cooling. Uncooled effusion cells are available for temperatures up to 1400 °C, which is sufficient for most applications.

The symmetry axis of an effusion cell points towards the center of the substrate. This geometry is used for all MBE systems with a non-rotating substrate to provide best thickness uniformity of the epitaxial films. In MBE systems with rotating substrates, the symmetry axis of the effusion cell points off the center of the wafer. The off-center focal point provides uniformity for the largest possible area.

4.1.3 Mechanical shutter

The mechanical shutters allow one to rapidly stop or initiate the molecular fluxes of effusion cells. The shutters can thus be used to stop and initiate growth and doping of the semiconductor. The time required to open and to close a shutter must be much shorter than the time required to grow a monolayer of the semiconductor. Such fast shutters are required to grow atomically abrupt interfaces. Since typical growth rates are 1 monolayer/s $\cong$ 2.8 Å/s, opening and closing times of 0.1 s are desirable.

Several shutter constructions are employed in MBE systems including rotating shutters and sliding shutters. An example of a sliding shutter is shown in Fig. 4.4. The shutter is magnetically coupled from the UHV side to the

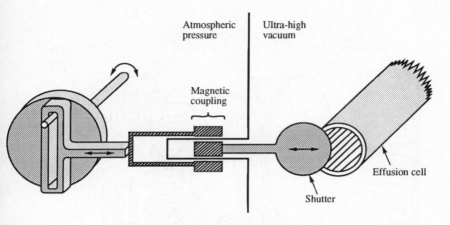

Fig. 4.4. Schematic illustration of a mechanical shutter with translational movement. The coupling between the UHV and the atmospheric pressure environment is provided by magnets. The shutter has 'soft stops', i.e. the translational velocity of the shutter continuously decreases to zero velocity at the end points of the shutter movement.

atmospheric pressure side. The magnets on the atmospheric pressure side move back and forth by means of a rod with an elongated opening. A disk driven by a motor provides the translational motion of the actuator rod. The shutter design shown in Fig. 4.4 has proven to be exceptionally reliable and can be used for thousands of opening and closing cycles. A particular advantage of the shutter is its soft start and stop. That is, the velocity of the actuator rod goes continuously to zero at its starting and at its stopping point. This 'soft' action is due to the conversion of the rotational movement of the disk into the translational movement of the shutter. Thus, the shutter velocity is a sinusoidal function of time. The two end points of the shutter movement are the zeros of the sinusoidal function.

4.1.4 Substrate holder and temperature

The substrate holder provides a mechanical means of holding and transporting the semiconductor substrate in and out of the MBE chamber. The mounting of the semiconductor substrate is either done by soldering with liquid indium or by indium-free clamping of the substrate onto the holder. The material used for substrate holders is Mo which is characterized by a low vapor pressure and little outgassing.

(a)   (b)   (c)

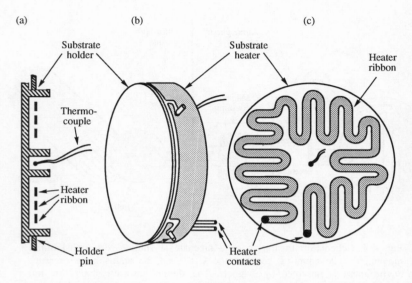

Fig. 4.5. Schematic illustration of a substrate holder (a,b) and a substrate heater (b, c) of an MBE system. The thermocouple is surrounded by the substrate holder to provide good thermal contact. The thermocouple does not 'see' the heating ribbon which is low-inductance meander-shaped in order not to perturb the beam of the RHEED electron gun.

An example of a substrate holder is shown in Fig. 4.5, which also shows the heater ribbon and the thermocouple. The heater ribbon is arranged in a low-inductive pattern. Such a pattern reduces magnetic fields which deflect the beam of the RHEED electron gun. Either tantalum (Ta) heater ribbons or tungsten (W) heater wire are used. However, to date, Ta heater ribbons are more popular since they are less brittle than W wire. Thermocouples are used to measure the substrate temperature.

The calibration and measurement of the substrate temperature can be done using several methods. At the congruent sublimation temperature of the substrate (e.g. 640 °C for GaAs) the surface reconstruction changes from a 2×4 reconstruction to a 3×4 reconstruction. The reconstruction can be monitored by means of the RHEED pattern. The congruent sublimation temperature can then be used as a calibration point for the thermocouple measurement. Another method of measuring the substrate temperature makes use of the InSb melting point at 525 °C. The solid-to-liquid transition can be visually observed through a viewport. The congruent sublimation temperature and the InSb melting temperature can then be used as calibration temperatures for the thermocouple.

Another convenient method of measuring substrate temperatures makes use of a pyrometer which analyzes the radiant spectrum of the substrate holder. Comparison with the known black-body radiation spectrum allows one to estimate the substrate temperature. The pyrometer method can be used if the emissivity of the substrate is known. Furthermore, the windows of the MBE apparatus must be clean and uncoated to provide the true radiant spectrum of the substrate.

Another method of measuring the GaAs substrate temperature was proposed by Hellman and Harris (1987). The GaAs fundamental absorption edge shifts by approximately 50 meV during heating from 300K to typical MBE growth temperatures and can be measured by using the infra-red light emitted by the substrate heater filament. The light transmitted through the GaAs substrate can be measured and the fundamental absorption edge can be determined. The authors claimed an accuracy of ± 10 °C and a precision of ± 2 °C of this method.

4.1.5 Pumps

The ultra-high vacuum of an MBE system require high-efficiency pumping systems. A number of different pumping systems are used in MBE systems which are all based on different physical principles. *Ion getter pumps* have an *ionization stage* in which residual gas atoms are ionized. The ionized atoms (or molecules) are then accelerated toward a metal foil (Ta) by means of a magnetic and electric field. The gas atoms impinge with high impact onto the metal foil where they are adsorbed. *Cryopumps* consist of a cold stage which is cooled to approximately 10 K by means of a closed cycle He refrigeration system. At the low temperature of 10 K most gases (e.g. O_2, N_2, CO_2) condense at the cold stage. The residual pressures achievable with cryopumps are very low and can be in the 10^{-11} Torr range. *Titanium sublimation pumps* consist of a Ti coil that is periodically heated for several seconds. The metal evaporates from the coil and is deposited on the inside walls of the MBE chamber. The freshly deposited Ti layer is very reactive and getters gas molecules from the vacuum. The Ti getter pump is frequently used during bake out times when the evaporating Ti does not interfere with the growing crystal. *Turbo pumps* are turbines with high rotation speed. Turbo pumps are capable of high pumping volumes which are required in gas-source MBE and chemical-beam epitaxy systems. *Liquid N_2 sorption pumps* are used to rough-pump UHV systems after venting. The sorption pumps are activated by cooling to liquid N_2 temperature, where the pumping action is due to the highly absorptive molecular sieve in the pump. Sorption pumps are preferred over mechanical roughing pumps due to their cleanliness.

4.2 Gas-source molecular-beam epitaxy

The solid sources of group-III and group-V elements of a molecular-beam epitaxy (MBE) growth system can be replaced by gaseous sources. These sources are gaseous compounds containing the group-III, group-V, or doping elements. Compounds used are hydrides (e.g. AsH_3, PH_3) for group-V elements and organometallics [e.g. $Ga(CH_3)_3$, $Al(C_2H_6)_3$] for group-III elements. In this section we restrict our discussion to the replacement of group-V elements with gas sources, i.e. *hydride* sources. The group-III elements are supplied by solid sources as in conventional MBE. For brevity we refer to hydride gas-source MBE as GSMBE.

The motivation for the usage of hydrides for the group-V elements is due to the difficulty of conventional MBE to grow compounds containing phosphorus. Phosphorus-containing compounds such as $Ga_x In_{1-x} P_y As_{1-y}$ lattice matched to InP substrates are of paramount importance for long-wavelength ($\lambda = 1.3 -$ 1.6 μm) optoelectronic components such as light-emitting and light-detecting devices. However, the evaporation of red phosphorus from a solid source results in tetramers (P_4) in the gas phase. Phosphorous tetramers have an extremely low sticking coefficient ($< 10^{-3}$) on InP substrates (Tsang, 1985b). Therefore, phosphorus is hardly incorporated during growth of $Ga_x In_{1-x} P_y As_{1-y}$. Furthermore, it is difficult to control the composition 'y' of the quaternary compound due to the low sticking coefficient of P_4. To date it is well accepted that high quality $Ga_x In_{1-x} P_y As_{1-y}$ epitaxial films with good compositional control cannot be grown by solid-source MBE.

The group-V solid sources of As and P were replaced with arsine (AsH_3) and phosphine (PH_3) by Panish (1980). The replacement was intended to achieve a well-controlled flux of phosphorus and phosphorus plus arsenic for the MBE growth of heterostructures of $Ga_x In_{1-x} P_y As_{1-y}$ and InP on InP substrates. A schematic illustration of a GSMBE system is shown in Fig. 4.6, which has solid, elemental effusion cells for the use of group-III elements and dopants. The hydrides arsine and phosphine are thermally cracked into atoms and dimers (i.e. P, P_2, As, and As_2) and enter the MBE chamber through a single port. Many other components of the GSMBE system such as substrate holder, cryoshroud etc. are identical or similar to components in conventional MBE systems.

The flux of the hydrides into the hydride cracker cells is controlled outside the high-vacuum growth chamber. The flow of hydrides can be controlled by either high precision *flow* or high precision *pressure* controllers. Ventilation lines are frequently employed to avoid flow transients. Flow transients typically occur as a gas flow is turned on or off until a steady-state flow rate is achieved. Such flow

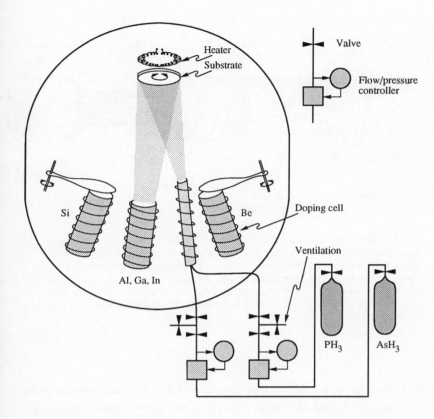

Fig. 4.6. Schematic illustration of a gas-source molecular-beam epitaxy (GSMBE) system. Group-V elements are provided as hydrides (AsH_3 and PH_3) which are thermally cracked upon entering the high vacuum chamber. Group-III elements and doping elements are evaporated from effusion cells.

transients can be avoided by using ventilation lines. The gas flow is not switched on or off but rather directed into the MBE chamber or to the ventilation line using three valves as illustrated in Fig. 4.6.

It is desirable to thermally crack the hydrides into their constituent elements using a cracker cell, since uncracked hydrides result in insufficient incorporation of group-V elements into the III–V film. The use of uncracked hydrides relies on the reaction and decomposition at the heated substrate surface which was shown to be insufficiently small (Panish, 1980; Veuhoff et al., 1981). Thus, hydrides need to be cracked upon entering the MBE chamber.

A schematic illustration of a hydride cracker cell is shown in Fig. 4.7. The

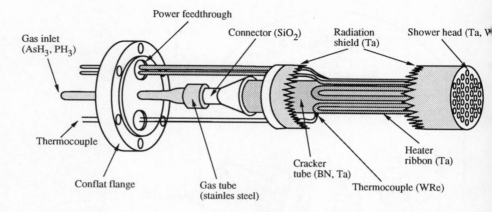

Fig. 4.7. Schematic illustration of a cracker cell used to thermally decompose
arsine and phosphine. The cracker tube may be made of a material (e.g. Ta) with
cathalytic characteristics to faciliate cracking. The cracker tube is heated by a Ta
ribbon and is monitored by a thermocouple. A 'shower head' is provided at the
exit of the cracker cell to provide a homogeneous flux of group-V elements. The
cracker cell is mounted on a conflat flange with feedthroughs for the heater and
the thermocouple.

cell is mounted on a conflat flange for compatibility with conventional MBE
vacuum ports. Thermal cracking is achieved in the cracker tube which is heated
by a Ta heater ribbon to temperatures of 1000 °C. The cracker tube can be made
from several materials including Ta and pyrolytic BN. Tantalum was shown to
have catalytic properties for the cracking process and therefore has a higher
cracking efficiency at a given temperature as compared to the inert BN (Chow and
Chaoi, 1983; Panish et al., 1985). Another method of thermally cracking hydrides
was demonstrated by Calawa (1981) who used heated Ta wire for thermal
decomposition. As illustrated in Fig. 4.7, the temperature of the cracker tube is
monitored by a thermocouple. A Ta foil surrounds the cracker tube as a radiation
shield. The exit end of the cracker cell is equipped with a 'shower head' in order
to diffuse and homogenize the group-V element beam. The 'shower head'
consists of many small holes through which the group-V elements exit the cracker
cell.

All group-V elements enter the MBE chamber through a single port. The
single port provides good homogeneity if two group-V elements are used
simultaneously. That is, the growth of $Ga_x In_{1-x} P_y As_{1-y}$ is expected to have
good P/As homogeneity and little variation of the composition 'y' across the

wafer.

The species created by cracking of hydrides are As and P atoms, As_2 and P_2 dimers, as well as As_4 and P_4 tetramers. Although As and P atoms may be obtained at very high cracker temperatures ($\gg 1000\,°C$) and low hydride pressures, the predominant species obtained by cracking are dimers. These dimers are desirable due to their large sticking coefficient on the substrate. A cracker temperature of or in excess of $1000\,°C$ insures that dimer species dominate (Panish and Temkin, 1989).

The range of operating pressures in GSMBE is higher compared to conventional solid-source MBE. The higher pressure is mainly due to hydrogen which is generated by the cracking of arsine and phosphine. The typical operating pressure during GSMBE is in the 10^{-5} Torr range. Appropriate pumping systems for this pressure range are high-capacity cryopumps, turbo molecular pumps, and diffusion pumps. Diffusion pumps are considered unacceptable for solid-source MBE due to possible contamination of the UHV environment of MBE. In contrast, diffusion pumps are occasionally employed in GSMBE systems with no drastic degradation of the material quality.

High-quality epitaxial materials have been grown by GSMBE including $Al_x Ga_{1-x} As/GaAs$, $Ga_{0.47} In_{0.53} As/InP$, $Al_{0.48} In_{0.52} As/InP$, and $Ga_x In_{1-x} P_y As_{1-y}/InP$. A review on quantum well structures, superlattice *pin* photodetectors, and bipolar transistors grown by GSMBE was given by Panish (1987). He demonstrated $Ga_{0.47} In_{0.55} As$ and $Ga_x In_{1-x} P_y As_{1-y}$ quantum wells of thickness 5 Å and near atomic monolayer abruptness. Excitonic enhancement of the photoluminescence and absorption spectra were also demonstrated. In the $Al_x Ga_{1-x} As/GaAs$ material system, high-quality selectively doped heterostructures were demonstrated (Schubert et al., 1987). Thus, it is well established that GSMBE is capable of producing device quality epitaxial material in the GaAs as well as in the InP material systems. Even though most of the highest purity material and the highest electron mobilities were reported for growth by MBE, the epitaxial layers grown by GSMBE are of near comparable quality.

A drawback of GSMBE with hydrides is the toxicity of arsine and phosphine. Arsine is an extremely toxic compound. The usage of hydrides necessitates continuous gas analysis and monitoring of the growth facility.

4.3 Chemical-beam epitaxy

The growth of III–V semiconductors by chemical-beam epitaxy (CBE) employs gas sources for both the group-III as well as the group-V elements. Ultra-high vacuum chambers as used for solid-source MBE and hydride GSMBE are also employed for CBE. The motivation for chemical beam epitaxy is to combine the advantages of MBE and organometallic vapor-phase epitaxy (Sect. 4.4).

The sources used for group-V elements are the hydrides of arsenic and phosphorus which are thermally cracked upon entering the reactor chamber. The reagents providing the group-V elements are thus the same as in gas-source MBE. The precursors used for group-III sources are organometallics, i.e. the group-III element is attached to three alkyl molecules. The alkyl groups most frequently used are derived from methane (CH_4) and ethane (C_2H_6). The corresponding organometallic molecules are trimethylaluminum [$Al(CH_3)_3$, TMAl], trimethylgallium [$Ga(CH_3)_3$, TMGa], trimethylindium [$In(CH_3)_3$, TMIn], triethylaluminum [$Al(C_2H_5)_3$, TEAl], triethylgallium [$Ga(C_2H_5)_3$, TEGa], and triethylindium [$In(C_2H_5)_3$, TEIn]. The employment of organometallic compounds for the group-III precursors has given rise to the name of organometallic molecular-beam epitaxy (OMMBE). Frequently, CBE and OMMBE are used as synonyms.

Organometallic group-III sources and hydride group-V sources in an MBE chamber were first used at the beginning of the 1980s (Veuhoff et al., 1981). The authors reported the use of TMGa and arsine for the growth of GaAs. This initial work was not encouraging due to a high carbon incorporation obtained in GaAs and, as was shown later, especially in $Al_xGa_{1-x}As$. High p-type conductivity was obtained in these materials as a result of the C incorporation. Tsang (1984) reported the CBE growth of GaAs and InP using organometallic group-III compounds and thermally cracked arsine and phosphine for the group-V compounds. Low p-type background doping in GaAs grown by CBE of 1.2×10^{15} cm^{-3} was reported by Horiguchi et al. (1986).

A schematic illustration of a CBE system is shown in Fig. 4.8. The components related to the hydrides are identical to the components of the GSMBE system discussed in the previous section. It will not be further discussed here. Both group-III and group-V chemicals are introduced into the high-vacuum reactor in which the molecular beams are directed to the substrate. Many components of CBE systems not related to gas handling are identical to corresponding components of MBE systems. Among these components are the substrate holder and heater, in-situ analysis instruments, and the UHV recipient.

The gas handling of the organometallic precursors is schematically shown in Fig. 4.8. Although many organometallic compounds, such as TMGa or TMAl, are liquid at room temperature, some (e.g. TMGa) have a sufficiently high vapor pressure to enter the reaction chamber at a high enough rate to enable growth. Others, however, (e.g. TEGa) have a lower vapor pressure that is insufficient to drive the compound into the reaction chamber. In such cases of low pressure,

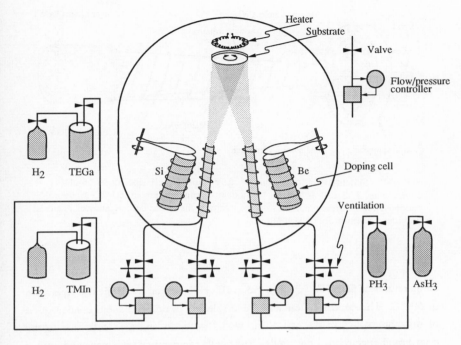

Fig. 4.8. Schematic illustration of a chemical-beam epitaxy (CBE) system which uses organometallic group-III compounds and cracked hydrides as group-V compounds.

organometallic bubblers are used to increase the group-III flux. In such bubblers, an inert gas (e.g. H_2 or N_2) bubbles through the liquid organometallic compound and serves as a carrier gas for the group-III compound. The flow of the carrier gas plus the organometallic compound is regulated by flow or pressure controllers. The flow is then switched between the reaction chamber and a ventilation line, while the flow through the controller is maintained constant. Such a constant flow is provided in order to avoid flux transients (see Sect. 4.2).

The group-III organometallics and their carrier gas are introduced into the high-vacuum recipient through a group-III cell. Figure 4.9 shows a gas introduction cell for organometallics. The two purposes of the cell are to introduce the organometallics into the growth chamber and to provide a homogeneous distribution of the molecular beam inside the reaction chamber. Thermal cracking of the organometallics is *not* provided by the cell, since thermal pyrolysis of the organometallics occurs readily at the heated substrate at

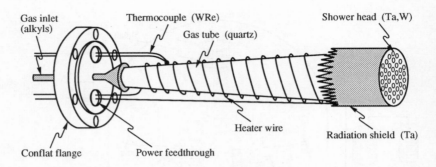

Fig. 4.9. Schematic illustration of a group-III cell for organometallic compounds. The cell is heated to moderate temperatures of 50 °C to avoid condensation of organometallics on the quartz walls of the cell. A 'shower head' provides a uniform flux distribution.

temperatures of 500–700 °C. The group-III cell is heated to a typical temperature of 50 °C. This moderate temperature is selected in order to avoid condensation of organometallics in the cell. This would occur in unheated cells due to the cryoshroud encircling the cell. The cell temperature is measured by a thermocouple. The material used for the gas tube is typically quartz, but the material is not critical since the tube is not heated to high temperatures. The cell is equipped with a 'shower head' at the gas exit in order to provide a homogeneous flux distribution of the organometallics. For heat shielding, the cell is surrounded by a Ta radiation shield which thermally insulates the cell from the liquid N_2 cryoshroud. Finally, the group-III cell is mounted on a conflat flange to provide compatibility with the UHV recipient.

Among the advantages of chemical-beam epitaxy systems are the simplicity and flexibility of source changes without breaking the vacuum of the reaction chamber and good homogeneity over large areas of the substrate. In comparison with OMVPE systems, chemical-beam epitaxy (i) uses less than 1% of AsH_3 and PH_3, (ii) allows in-situ surface analysis instruments to be used, and (iii) has a smaller memory effect. A comprehensive review on the properties of chemical-beam epitaxy was given by Tsang (1989).

The pressure regime of chemical-beam epitaxy is in the 10^{-4} Torr range which is higher than MBE and GSMBE. The higher pressure is due to the unused alkyl radicals as well as to hydrogen. The high gas-load necessitates large-volume pumping systems in order to keep the pressure in the 10^{-4} to 10^{-5} Torr range. Turbo molecular pumps as well as cryopumps are well-suited pumping systems for CBE systems.

High quality $Ga_{0.47}In_{0.53}As/InP$ and quaternary $Ga_xIn_{1-x}P_yAs_{1-y}/InP$ grown by CBE was reported in the literature. Characterization of CBE-grown material includes optical and transport properties of $Ga_{0.47}In_{0.53}As$ (Tsang et al., 1986), optical properties of $Ga_{0.47}In_{0.53}As/InP$ quantum wells (Tsang and Schubert, 1986), and optical properties of quaternary $Ga_xIn_{1-x}P_yAs_{1-y}$ (Tsang et al., 1987). These studies demonstrated the capability of CBE to grow high-quality epitaxial films with low background impurity concentration, high carrier mobility, and intense photoluminescence spectra with narrow linewidths. Furthermore, high-performance electronic devices processed from CBE grown material were reported including heterobipolar transistors (Tsang et al., 1988) and heterostructure field effect transistors (Schubert et al., 1988).

A problem area of CBE is the growth of high-quality $Al_xGa_{1-x}As$ with sufficiently low background impurity concentration. Although GaAs growth with low background concentrations was reported (Pütz et al., 1986; Chiu et al., 1987), the impurity background concentration of $Al_xGa_{1-x}As$ was found to be very high, typically in the $10^{17} - 10^{19}$ cm^{-3} range. Such high background impurity concentrations are unacceptable for device applications. The high C background concentration in $Al_xGa_{1-x}As$ is due to the high reactivity of Al which tends to form strong Al–C bonds (which are stronger than the Ga–C bonds). Carbon is thus incorporated on group-V sites and acts as an acceptor. The employment of chemicals other than TMAl and TEAl have reduced the C background concentration to lower levels (Chiu et al., 1991). However, the $Al_xGa_{1-x}As$ quality grown by CBE is still inferior to MBE grown $Al_xGa_{1-x}As$.

4.4 Organometallic vapor-phase epitaxy

Among the different epitaxial growth techniques, organometallic vapor-phase epitaxy (OMVPE) has proven the ability to grow state of the art heterostructures, quantum wells and superlattices. Low-threshold current-injection lasers and efficient photodetectors were also demonstrated by the OMVPE growth technique. OMVPE is used in manufacturing environments for the production of optoelectronic devices. Multiwafer systems are frequently used in such manufacturing systems. OMVPE is a promising candidate for future high-throughput epitaxial growth systems.

Historically, the development of OMVPE started with the work of Manasevit (1968) and of Manasevit and Simpson (1969). The authors grew single crystal GaAs initially on insulating, lattice mismatched substrates such as Al_2O_3. Subsequently, the authors grew binary (GaAs, GaP) and ternary (GaP_xAs_{1-x}, $GaAs_xSb_{1-x}$) semiconductors on GaAs. The epitaxial growth was achieved by using organometallics and hydrides for the supply of group-III and group-V precursors, respectively. Device-quality materials were grown with OMVPE at the end of the 1970s. Dupuis and Dapkus (1978) reported the first continuous wave operation of $Al_x Ga_{1-x} As/GaAs$ double-heterostructure lasers at room temperature grown by OMVPE. This publication stimulated further work in the field and is considered as one of the milestones in OMVPE research.

GROWTH METHOD		PRESSURE	GAS FLOW
OMVPE	Atmosph. Pressure	760 Torr	Viscous
	Low Pressure	$>10^{-2}$ Torr	Viscous
GSMBE	CBE/MOMBE	$\cong 10^{-5}$ Torr	Molecular
	Hydride GSMBE	$\cong 10^{-6}$ Torr	Molecular
MBE		$\cong 10^{-9}$ Torr	Molecular

Fig. 4.10. Operating pressures and gas flow characteristics for different epitaxial growth techniques.

The gas flow in VPE reactors is fundamentally different from the gas transport in GSMBE and MBE (Tsang, 1989). The gas flow and the operating pressures of the three growth technologies are compared in Fig. 4.10. Growth techniques which operate under high vacuum ($< 10^{-5}$ Torr) and ultra-high vacuum conditions ($< 10^{-7}$ Torr) such as GSMBE and MBE have gas transport in terms of a *molecular beam*. That is, the source materials evaporated from effusion cells or injected into the reaction chamber have the nature of a *beam*. The molecules forming the beam have one preferential direction and furthermore molecules forming the beam do not interact. In contrast, OMVPE operating at pressures $> 10^{-2}$ Torr has a *viscous* gas flow. That is, the molecules forming the gas flow are strongly interacting. The flow pattern follows the laws of gas dynamics which is in clear contrast to the beam nature found in high-vacuum systems.

The viscous flow of gases in OMVPE systems necessitates that the reaction chamber have a smooth shape, i.e. no sharp corners and dead volumes. Such dead volumes are defined as having a zero or a very small gas velocity. A typical

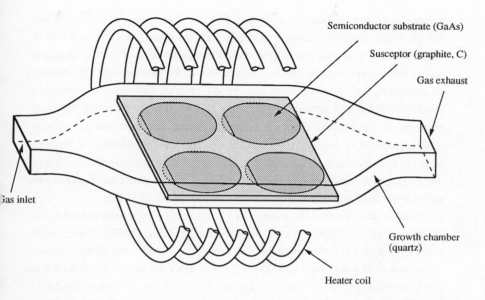

Fig. 4.11. Schematic illustration of the growth chamber of an organometallic vapor-phase epitaxy (OMVPE) system.The wafers are on a conductive susceptor which is heated by an induction coil. The growth chamber has no sharp corners, thus making possible a laminar, non-turbulent gas flow.

reaction chamber of an OMVPE system is shown in Fig. 4.11. The reaction chamber walls are typically made of quartz and have smooth shapes to enable a laminar, non-turbulent gas flow. The reaction chamber shown has a multiwafer susceptor for high-volume throughput. The requirements for the susceptor are (i) low vapor pressure in order to avoid incorporation of the susceptor material into the growing semiconductor and (ii) high electrical and thermal conductivity. Graphite is frequently used as a susceptor material. The desirable high thermal conductivity of the susceptor provides a homogeneous heat distribution across the entire susceptor area.

The susceptor is heated by radio-frequency (rf) induction as indicated in Fig. 4.11. In order to induce currents in the susceptor, it must be conductive. The inductive heating is provided by a coil around the reaction chamber. The coil generates a high-frequency field inside the reaction chamber and generates an induced current in the susceptor which in turn resistively heats the substrate.

A laminar flow of the gases inside the reaction chamber is necessary in order to avoid the *memory effect*. The memory effect manifests itself by a lack of abruptness of compositional and doping profiles. *Dead volumes* inside the gas manifold and inside the reaction chamber gradually release gases over a period of time and result in graded (rather than abrupt) interfaces and in smoothly changing (rather than abrupt) doping profiles. A reduction of the memory effect is obtained by evacuating the reaction chamber, i.e. by low-pressure OMVPE. *Low-pressure OMVPE* operates at a pressure of 10^{-2} Torr to 100 Torr. As a result of the lower pressure, the gas velocity inside the reactor increases. The increased gas velocity is advantageous in several respects. First, as already mentioned, the memory effect is less pronounced. Second, the homogeneity of composition, thickness, and doping across the growing wafer is improved. Third, interfaces between two different semiconductors are more abrupt in low-pressure OMVPE.

Vertical design and *horizontal design* reaction chambers have been realized (Ludowise, 1985). In the vertical design, the gas flow enters the reaction chamber from the top and is deflected at the wafer surface. In horizontal design reactors (as shown in Fig. 4.11) the gas flow is parallel to the surface of the wafer. The wafer is exposed to a more uniform gas stream than in vertical design chambers. To date, horizontal design reactors are the preferred solution due to better homogeneity of doping and thickness of the epitaxial layers. More recently, the horizontal design was further improved by using rapidly rotating substrate holders (see, for example, Tanaka et al., 1987). The rapid rotation of the substrates provides significantly improved homogeneity of the thickness, composition and doping across the wafer.

The chemical starting materials used in OMVPE are organometallics for the group-III elements and hydrides for the group-V elements. The starting chemicals must satisfy several requirements for OMVPE growth. First, the chemicals must thermally decompose at the growth temperature which is typically in the 500–800 °C range for III–V semiconductor growth. If the chemicals are stable and do not decompose at the growth temperature, incorporation at the wafer surface does not occur. Second, the starting chemicals must be either gaseous or have a relatively high vapour pressure at room temperature. Gaseous chemicals are easily transported through tubes from the reservoir to the reaction chamber. Liquid and solid compounds are not suited for such transport.

The vapor pressure of organometallics varies over several orders of magnitude at room temperature. The vapor pressures of the most popular organometallics are shown in Fig. 4.12. As a general tendency, molecules with larger alkyl groups have lower vapor pressures than molecules with small alkyl groups. For example, the vapor pressure of TMGa is higher than the vapor pressure of TEGa. The same trend applies to TMAl, TEAl, TMIn, and TEIn. If the vapor pressure of a

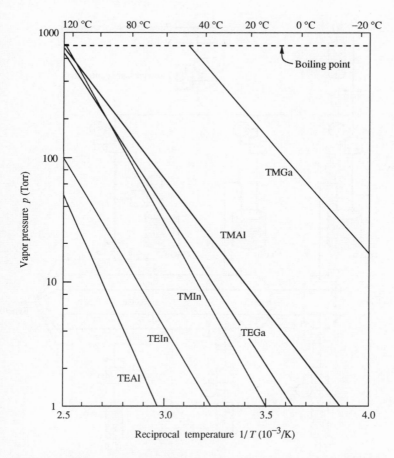

Fig. 4.12. Vapor pressure of organometallic group-III compounds as a function of reciprocal temperature. A carrier gas such as H_2 is needed to transport a compound into the reaction chamber if the vapor pressure of the compound (e.g. TEAl) is very low.

chemical is low, a small flow rate results. The supply rate of such low-pressure chemicals can be enhanced by means of hydrogen which is bubbled through the liquid organometallic compound. Thus, H_2 can serve as a carrier gas for low vapor pressure organometallics.

The starting material containers and the gas manifold system of an OMVPE system is schematically shown in Fig. 4.13. The starting chemicals are either directly fed into the gas manifold or are equipped with an inert gas (H_2) bubbler. Furthermore, valves are used to completely open and close the flow of reagents.

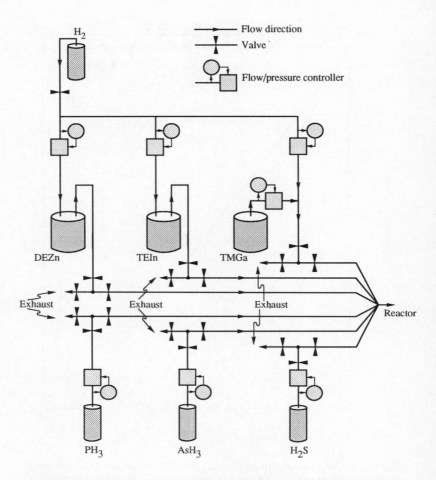

Fig. 4.13. Gas manifold of an organometallic vapor-phase epitaxy (OMVPE) system. The gas flows are switched between the reaction chamber and the ventilation line in order to provide a constant flow through the flow/pressure controllers.

The flow of reagents and of the carrier gas is regulated by either pressure or flow controllers. All chemical reagents are mixed before they enter the growth chamber. The pre-mixing of chemicals results in a homogeneous gas composition which in turn results in a homogeneous composition of the epitaxially grown layers.

The switching of source reagents occurs with the use of exhaust or ventilation valves (see Fig. 4.13). The gas flow is switched from the reaction chamber to the exhaust or vice versa. It is important to note that the gas flow is never shut off, but is instead directed to either the growth chamber or to the exhaust at a *constant* flow rate. The purpose of this procedure is to minimize flow transients. Flow transients usually occur at transitions between different materials during growth and, as a result, the compositions of the growing semiconductor cannot be changed abruptly. The use of exhaust valves reduces the magnitude of such flow transients.

The precursors of the OMVPE technique are thermally cracked at the hot susceptor and substrate surface. As an example, we consider TMGa.

$$(CH_3)_3 Ga + \frac{3}{2} H_2 \leftrightarrow 3 CH_4 + Ga . \qquad (4.1)$$

The TMGa molecule is cracked into alkyl radicals (CH_3) which form stable methane (CH_4) with the excess hydrogen in the growth chamber. The cracking process of the hydrides is straightforward. Considering arsine as an example, the pyrolysis can be written as

$$2 AsH_3 + Energy \leftrightarrow As_2 + 3 H_2 \qquad (4.2a)$$

and

$$4 AsH_3 + Energy \leftrightarrow As_4 + 6 H_2 . \qquad (4.2b)$$

The species of arsenic formed by the pyrolysis are As_2 as well as As_4. The formation of arsenic monomers is unlikely since they are not stable at the pressure and temperature range of interest (Panish and Temkin, 1989).

The combination of the above two chemical reactions yields the complete reaction occurring during OMVPE growth. The complete reaction for the starting reagents TMGa and AsH_3 is thus given by

$$(CH_3)_3 Ga + AsH_3 \leftrightarrow GaAs + 3 CH_4 . \qquad (4.3)$$

The relative fractions of reactions that are pyrolytic (Eq. (4.1) and (4.2)) and of the reactions that are complete (Eq. (4.3)) are not known. However it is clear that pyrolysis does occur and is a necessary condition for the growth of the semiconductor.

A vertical cut through an OMVPE reactor is shown in Fig. 4.14. The reactor shown is of the horizontal type. Pyrolysis (as defined in Eqs. (4.1) and (4.2)) occurs in the vicinity of the susceptor and the semiconductor substrate. A *stagnant gas layer* develops on top of the susceptor, the substrate, and at the reactor walls. This stagnant boundary layer is a gas layer that has a very small or zero velocity. Such low flow velocities are due to mechanical friction of the gas

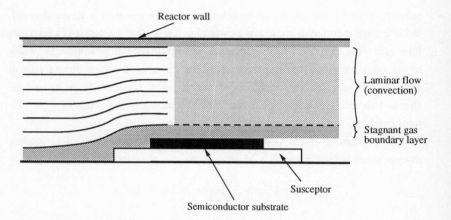

Fig. 4.14. Schematic illustration of the gas flow in a vapor-phase epitaxy (VPE) reaction chamber. Good mixing of gases is achieved in the laminar flow region. A stagnant gas layer develops in the vicinity of the substrate. Gas transport through the stagnant layer occurs through diffusion rather than through convection.

at the substrate surface. The spatial extent of the stagnant layer is limited to the vicinity of the susceptor, substrate, and reactor walls. Laminar gas flow occurs sufficiently far away from the walls and the susceptor. As a result of the stagnant layer, reagents are not supplied directly from the laminar flow region (see Fig. 4.14 for illustration) to the substrate surface by *convection*. Instead, the stagnant layer becomes depleted by the growth reaction and new reagents are supplied to the semiconductor surface by means of *diffusion* of reagents from the laminar flow region to the growing epitaxial layer surface. The occurrence of the stagnant layer and the transport of reagents through this layer by diffusion is one of the major differences between OMVPE and CBE (Tsang, 1989).

One of the drawbacks of OMVPE is the toxicity of its source reagents, especially AsH_3, and the amount needed during growth. Arsine is a highly toxic gas and can only be used with many elaborate and expensive safety precautions. Furthermore, the amount of reagents required for OMVPE is much higher (typically a factor of 10–100, Tsang, 1990) as compared to the CBE growth technique. The use of such large amounts of dangerous chemicals represents a potential safety hazard.

Excellent parameter control is desirable for epitaxial growth. Many parameters of OMVPE growth are interdependent. For example, the substrate temperature influences the epitaxial growth rate. At low growth temperatures, the pyrolysis is incomplete and the growth rate decreases with OMVPE growth. This characteristic of OMVPE is in contrast to MBE, where growth rate and substrate temperature are independent over a large range of growth temperatures. Due to the interdependence of so many parameters in OMVPE, all the parameters must be controlled very accurately in order to achieve small variations of epitaxial layer thickness, composition, and doping. The requirement of accurate parameter control and the interdependence of many parameters makes OMVPE a growth technology which is less flexible than MBE.

On the other hand, OMVPE is very versatile and allows one to grow virtually all III–V semiconductors. Excellent heterostructures, quantum well structures, and superlattice structures have been grown by OMVPE. In addition, electronic as well as optoelectronic devices with impressive performance have been grown by OMVPE. Recently extensive reviews were published on OMVPE (Razeghi, 1989; Stringfellow, 1989; Swaminathan and Macrander, 1991), which provide an overview on the growth, physics, and chemistry of the OMVPE technique as well as device characteristics of OMVPE-grown structures.

4.5 Vapor-phase epitaxy

Conventional vapor phase-epitaxy (VPE) which preceded the now prominent OMVPE, does not employ organometallic compounds for the group-III elements. Instead, conventional VPE uses the chlorides of the group-III elements as source gases. VPE was first demonstrated by Tietjen and Amick (1966). The authors used gaseous arsine and phosphine as group-V precursors and gallium chloride for the growth of GaP_xAs_{1-x}. An alternative to hydrides for the group-V precursors are chlorides (e.g. $AsCl_3$ or PCl_3). VPE is frequently referred to as *hydride VPE* and *chloride VPE* for growth with group-V hydride precursors and group-V chloride precursors, respectively.

The growth of III-V semiconductors by VPE is performed in fused silica reactors consisting of three zones. Such a three zone VPE reactor is shown schematically in Fig. 4.15. The zones of the reactor are the *source zone*, the *baffle zone*, and the *growth* zone. The flow of gases to the reaction chamber is controlled by either flow or pressure regulators. The gas manifold system is typically operated in a growth or ventilation mode, which allows one to maintain a constant flow or pressure. The switching between growth and ventilation mode reduces flow and pressure transients that always occur upon changing the nominal flow or pressure.

The first zone of the reaction chamber is called the *source zone*. The purpose of the source zone is to generate gaseous compounds containing the group-III

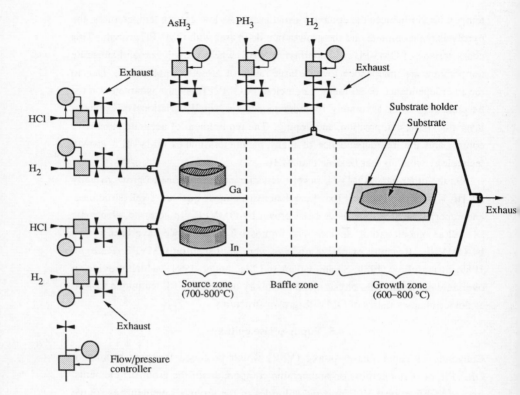

Fig. 4.15. Schematic illustration of a hydride vapor-phase epitaxy (VPE) system. The group-V elements are provided by means of hydrides (AsH$_3$ and PH$_3$). Group-III elements are transported by means of their chlorides using hydrochloric acid as transport reagent.

elements. Chlorides are the most popular inorganic group-III precursors. They are synthesized by hydrogen chloride gas flowing over and reacting with the elemental group-III sources. The chemical reaction for a gallium source is given by

$$HCl + Ga \leftrightarrow GaCl + \tfrac{1}{2} H_2 , \qquad (4.4a)$$

$$HCl + GaCl \leftrightarrow GaCl_2 + \tfrac{1}{2} H_2 , \qquad (4.4b)$$

$$HCl + GaCl_2 \leftrightarrow GaCl_3 + \tfrac{1}{2} H_2 . \qquad (4.4c)$$

All three of the above chlorides, i.e. GaCl, GaCl$_2$, and GaCl$_3$ are present in the gas phase of a VPE reactor. The source zone of the VPE reactor is heated to temperatures of typically 700–800 °C to accelerate the chemical reaction.

The second zone of the reactor is called the *baffle zone* and has the purpose of mixing the group-III with the group-V precursor. Such a mixing is required to obtain epitaxial growth with good homogeneity of composition and uniformity of layer thickness. The baffle zone consists typically of several perforated walls which diffuse the gas flow. Tietjen and Amick (1966) used a so-called *reaction zone* instead of the baffle zone. The reaction zone was heated to a temperature of 850 °C to enhance the chemical reaction between the group-III and group-V constituents.

The third zone of the VPE reactor is the *growth zone*. This zone includes the semiconductor substrate which is placed on top of the substrate holder (substrate holders are termed *susceptors* if heated by radio-frequency induction). The heating of all the three zones of the VPE system is usually provided by a furnace surrounding the reactor. Such reactors are called hot-wall reactors. It is desirable that the furnace has several zones in order to adjust the temperature in each zone independently. Arsine and phosphine are thermally cracked in the hot growth zone of the reactor. The pyrolysis reactions can be written as

$$2\,AsH_3 + Energy \leftrightarrow As_2 + 3H_2 \,, \tag{4.5a}$$

$$4\,AsH_3 + Energy \leftrightarrow As_4 + 6H_2 \,. \tag{4.5b}$$

The corresponding pyrolysis reaction for phosphine is given by

$$2\,PH_3 + Energy \leftrightarrow P_2 + 3H_2 \,, \tag{4.5c}$$

$$4\,PH_3 + Energy \leftrightarrow P_4 + 6H_2 \,. \tag{4.5d}$$

The chemical species of arsenic and phosphorus formed are thus dimers and tetramers. The formation of arsenic and phosphorus monomers is unlikely since the monomers are thermodynamically unstable at the pressure and temperature range of interest (Panish and Temkin, 1989).

For the complete chemical reaction of chloride VPE growth, we consider GaAs as an example. The reaction is given by

$$GaCl + AsH_3 \leftrightarrow GaAs + HCl + H_2 \,. \tag{4.6}$$

Hence, gaseous hydrochloric acid is generated during VPE growth. Since HCl is toxic, it needs to be neutralized by a base before being released into the exhaust system.

The VPE growth reaction (Eq. (4.6)) is an equilibrium reaction which depends strongly on the growth temperature. In other words, the growth rate can be positive (growth) as well as negative (etching). As a result, the VPE growth technique is well suited for growth on previously processed structures, for selective area growth, and for growth on patterned substrates (Johnston et al., 1989). On the other hand, the strong dependence of the growth rate on the substrate temperature can be disadvantageous due to the interdependence of parameters. It is generally desirable for the growth parameters to be independent.

A further disadvantage of chloride/hydride VPE is the difficulty of growing thin quantum well structures. The growth of thin, homogeneous quantum well structures by VPE has not hitherto been demonstrated. This lack is partly due to the high growth rates of VPE of typically 100 Å/s = 36 μm/hr at moderate growth temperatures. Furthermore, polycrystalline growth occurs on the fused silica walls of the reactor in hot-wall reactors used for VPE. The deposition of material on the reactor walls changes the heat-radiation pattern of the VPE reactor and results in changing growth rates, which complicates the growth rate control.

5

Doping with elemental sources

Intentional impurities (dopants) can be introduced during epitaxial crystal growth in terms of either elemental doping sources or chemical compounds. This chapter is devoted to aspects of elemental doping sources and to the characteristics of these dopants. The following chapter discusses non-elemental, gaseous doping precursors. The incorporation of elemental doping impurities is mediated through thermal evaporation of the doping element from an effusion cell. Such elemental solid evaporation sources are commonly used for molecular-beam epitaxy (MBE) and related techniques discussed in Chap. 4.

5.1 Doping techniques and calibration

The incorporation of doping atoms originating from solid elemental doping sources is mediated through thermal evaporation within a vacuum environment. The solid source is heated to a temperature T at which the desired amount of the doping element evaporates. The rate of evaporation, the source and substrate geometry, and the sticking probability of impurity atoms on the crystal surface determine the incorporation rate of the impurities into the growing epitaxial crystal. These parameters are discussed below.

5.1.1 Doping impurity flux

The concentration of doping impurities in an epitaxial layer is usually controlled by the impurity-cell temperature. Consider an epitaxial crystal of GaAs that grows at a rate of 1 monolayer per second, i.e. at a rate of 6×10^{14} GaAs molecules per cm^2 per second. Assume further that doping with concentration of 1×10^{18} cm^{-3} is desired. Using a thickness of one monolayer of 2.83 Å, an impurity atom flux of 2.83×10^{10} cm^{-2} s^{-1} is required (unity sticking of impurity atom on the substrate is assumed). This example shows the simple relationship between the required dopant flux and the doping concentration in the crystal. However, the relationship between the desired dopant *flux* and the effusion cell *temperature* is important as well and is calculated below.

The vapor pressure of a doping source can be calculated from the **ideal gas law** which is given by

$$PV = N_{Avo} \, kT \tag{5.1}$$

where P is the pressure of $N_{Avo} = 6.022 \times 10^{23}$ atoms (or molecules) residing within a volume V at the temperature T. The vapor pressure of a solid can be expressed as $P = P_0 \exp(-H_v/kT)$, where P_0 is a constant and H_v is the latent heat of vaporization. Furthermore, the thermal velocity, v_{th}, of the atoms (molecules) is given by the **equipartition law**, i.e.

$$\frac{1}{2} \, mv_{th}^2 = \frac{3}{2} \, kT \tag{5.2}$$

where m is the mass of one atom (molecule). The average flux of atoms (per cm^2 per second) is then given by

$$F = \frac{N}{V} \, v_{th} = (P/kT)\sqrt{3kT/m} \tag{5.3}$$

which can be used to calculate the required impurity atom flux at the substrate surface. Consider further that the impurity source has an effective area A and is located at a distance r from the substrate. Assuming that the flux is isotropically distributed over a half-sphere, the impurity atom flux at the evaporation source is given by

$$F_{source} = \frac{\frac{1}{2}\pi r^2}{A} \, F = \frac{\frac{1}{2}\pi r^2}{A} \, P\sqrt{3/kTm} \ . \tag{5.4}$$

For a given doping flux F_{source}, the equation gives us one condition for the pressure P and the temperature T. Of course, P and T are also related by the vapour pressure curves (Marton and Marton, 1979) which represent a second condition for P and T.

As an example, consider Si doping of GaAs at a concentration of 5×10^{18} cm^{-3} with a solid Si source with an area of 1 cm^2 located at a distance of 50 cm from the semiconductor substrate. At a growth rate v_g, the required doping atom flux is given by

$$F = N_{Si} v_g \ . \tag{5.5}$$

At a growth rate of 1 monolayer per second ($v_g \cong 2.8$ Å/s) the required doping flux at the growing layer is obtained as $F = 1.4 \times 10^{11}$ cm^{-2}s^{-1}. The flux at the Si source is $F_{source} = \frac{1}{2} F\pi r^2/A = 5.5 \times 10^{14}$ cm^{-2} . Using Eq. (5.4) and the vapor pressure versus temperature curve of Si (Marton and Marton, 1979) allows one to determine the required Si-source vapor pressure and temperature. One obtains, for a Si mass of $m_{Si} = 14 \times 10^{-3}$ kg/$6.022 \times 10^{23} = 2.3 \times 10^{-26}$ kg, a Si vapor pressure of $P_{Si} \cong 1.5 \times 10^{-8}$ Torr and a Si-source temperature of approximately $T_{Si} \cong 1000$ °C.

The above calculation is insufficient in two respects. First, the impurity gas atoms have a maxwellian velocity distribution rather than the average velocity used in the calculation. Second, the impurity atom flux is assumed to be isotropic; this assumption is not justified for Knudsen-type effusion cells. Even for dopant evaporation from filaments an isotropic distribution of the impurity-atom flux is hardly fulfilled. Nevertheless, the calculation provides a fast and simple means of estimating the temperature and pressure of a thermal doping source.

5.1.2 Effusion cells

Doping impurities are frequently evaporated from *effusion cells*. Effusion cells employ filaments for resistively heating crucibles which contain the impurity element. The material used for the crucibles is typically a chemically inert material with a low vapor pressure even at high temperatures. Pyrolytic boron nitride (BN) and graphite (C) are suited for this purpose. In *Knudsen-type* effusion cells (Knudsen, 1909), evaporated atoms or molecules interact weakly with each other and mostly with the walls of the effusion cell. Such effusion cells have a preferred direction of evaporation which is the symmetry axis of the effusion cell. The radiant pattern of the cell is given by a cosine distribution. Such a cosine distribution of the flux is schematically shown in Fig. 5.1(a). In the following the cosine distribution will be derived.

Consider an effusion cell heated to a temperature T. The area of the effusion cell is denoted as ΔA. Atoms (or molecules) evaporate from the source and leave the effusion cell through its opening. The velocity distribution function of the particles is, under equilibrium conditions, given by the maxwellian velocity distribution (see Chap. 3). In cartesian and spherical coordinates, the distribution is given by

(a)

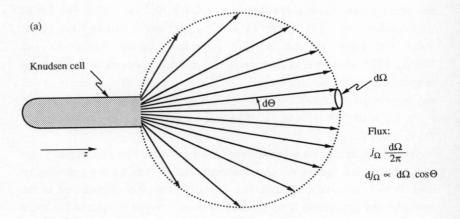

Knudsen cell

Flux:

$$j_\Omega \frac{d\Omega}{2\pi}$$

$$dj_\Omega \propto d\Omega \cos\Theta$$

(b)

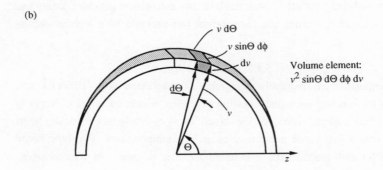

Volume element:
$v^2 \sin\Theta \, d\Theta \, d\phi \, dv$

Fig. 5.1. (a) Schematic illustration of an effusion cell and the cosine-shaped distribution of the effusion flux. (b) Velocity 'volume element' in spherical coordinates (velocity v, polar angle Θ, and azimuthal angle ϕ) used to calculate the spatial flux distribution of a Knudsen cell.

$$f(v_x, v_y, v_z) \, dv_x dv_y dv_z = \left[\frac{m}{2\pi kT} \right]^{3/2} \exp\left[-\frac{\frac{1}{2}m(v_x^2 + v_y^2 + v_z^2)}{kT} \right] dv_x \, dv_y \, dv_z$$

$$\text{(5.6a)}$$

$$f(v) v^2 \sin\theta \, d\theta \, d\phi \, dv = \left[\frac{m}{2\pi kT} \right]^{3/2} \exp\left[-\frac{\frac{1}{2}mv^2}{kT} \right] v^2 \sin\theta \, d\theta \, d\phi \, dv \quad \text{(5.6b)}$$

where m is the mass of one atom (or molecule). It is convenient to transform the velocity distribution into spherical coordinates using the spherical coordinates v,

ϕ, and θ. The 'volume' elements are then given by $dv_x\,dv_y\,dv_z = v^2$ $\sin\theta\,d\theta\,d\phi\,dv$, as illustrated in Fig. 5.1(b). Assuming that the symmetry axis of the effusion cell is the z-axis, all atoms with a positive z-component of the velocity can leave the effusion cell. The number of particles which exit the effusion cell through the area ΔA during the time Δt with a velocity between v and dv is given by

$$dn_{\Delta A}(v) = dn(v)\,\frac{\Delta A\,\Delta t\,v_z}{V} = N\,f(v_x,v_y,v_z)\,dv_x\,dv_y\,dv_z\,\frac{\Delta A\,\Delta t\,v_z}{V}$$

$$= \frac{\Delta A\,\Delta t\,N}{V}\left[\frac{m}{2\pi kT}\right]^{3/2} e^{-mv^2/2kT}\,v\cos\theta\,v^2\sin\theta\,d\theta\,d\phi\,dv$$

$$= \frac{\Delta A\,\Delta t\,N}{V}\left[\frac{m}{2\pi kT}\right]^{3/2} e^{-mv^2/2kT}\,v^3\cos\theta\,d\Omega\,dv \qquad (5.7)$$

where $v_z = v\cos\theta$ is the z-component of the velocity, $d\Omega = \sin\theta\,d\theta\,d\phi$ is the solid angle, N is the total number of gas atoms in the cell, and V is the volume of the cell. Integration over all velocities and division by ΔA and Δt yields the flux of atoms out of the effusion cell into the solid angle $d\Omega$

$$dj_\Omega = \int_0^\infty \frac{N}{V}\left[\frac{m}{2\pi kT}\right]^{3/2} e^{-mv^2/2kT}\,v^3\,dv\cos\theta d\Omega = \frac{1}{4\pi}\,\frac{N}{V}\,\bar{v}\,\cos\theta d\Omega\,. \quad (5.8)$$

where we used $\bar{v} = 4\pi \int_0^\infty vf(v)v^2\,dv$. Defining the total flux out of the effusion cell as $j_{2\pi} = (N/V)\bar{v}/4$, the flux distribution is given by

$$\boxed{dj_\Omega = j_{2\pi}(d\Omega/\pi)\cos\theta} \qquad (5.9)$$

The result shows that the flux of doping atoms out of the effusion is not isotropic. The flux has its preferential direction along the z-direction, i.e. along the symmetry axis of the effusion cell. The angle-dependence of the effusion follows a cosine law. As a check we integrate the flux over the entire half-space and obtain

$$\int_{2\pi} dj_\Omega = j_{2\pi}\int_0^{2\pi} d\phi \int_0^{\pi/2} \frac{1}{\pi}\cos\theta\sin\theta d\theta = j_{2\pi} \qquad (5.10)$$

which is the expected result.

5.1.3 Effusion cell calibration

Doping concentrations in III–V semiconductors need to be controlled over a wide range, for example from 10^{15} to 10^{20} cm^{-3}. Doping control over 5 orders of magnitude requires the *calibration* of the doping process. In a typical calibration process several samples are grown at the same growth rate but different doping effusion cell temperatures. The thicknesses of such calibration epitaxial layers is typically on the order of 1 µm. The total thickness of the epitaxial layer must be much thicker than the width of the surface depletion layer in order to avoid large corrections of free carrier concentrations, e.g. measured by the Hall effect (see below). The surface depletion layer thickness, W_D, depends on the doping concentration according to

$$W_D = \sqrt{\frac{2\varepsilon}{eN_D} \, \phi_B} \qquad (5.11)$$

where ε is the permittivity, e is the electronic charge, and ϕ_B is the barrier height (in volts) of the semiconductor surface.

Typical calibration measurements for Be (p-type) and Si (n-type) doping of MBE-grown GaAs is shown in Figs. 5.2 and 5.3, respectively. The Be effusion cell operates at lower temperatures due to the higher vapor pressure of Be as compared to Si. The growth rate should be constant for all doping cell temperatures. The doping concentration has a quasi-linear relationship if plotted versus reciprocal temperature which reflects the exponential increase of the vapor pressure with temperature.

The doping concentration is frequently measured by the Hall effect. The Hall effect allows one to measure the free electron or hole concentration. The free carrier concentration does not always coincide with the doping concentration. In order to obtain a reliable result (i) all dopants must be electrically active, (ii) no compensation of dopants must occur, and (iii) surface depletion effects must be negligible. Furthermore, the Hall carrier concentration differs from the true free carrier concentration by the so-called Hall factor. The Hall factor causes the measured Hall carrier concentration to deviate from the true free carrier concentration by about 20% depending on the dominant carrier scattering mechanism (Putley, 1960).

5.1.4 Filament doping

Evaporation of doping impurities from filaments was pioneered by Malik (Malik et al., 1988) for C doping of GaAs. The C doping was achieved by resistively heating a graphite filament which was cut in serpentine shape to yield a resistance of 1 Ω and an evaporation area of 1 cm^2. The filament was clamped on both ends

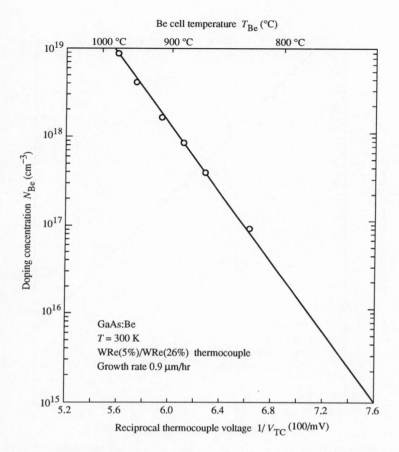

Fig. 5.2. Calibration of Be doping concentration as a function of the effusion cell temperature (thermocouple voltage) for MBE growth. The Be doping concentrations are measured by the Hall effect (courtesy of R. F. Kopf).

using flat tantalum (Ta) tabs attached to Ta rods which in turn were fastened to a dual current vacuum feed through. Malik demonstrated C doping of GaAs up to a concentration of 10^{20} cm^{-3} using C filament doping.

A schematic illustration of a graphite filament is shown in Fig. 5.4. The filament is connected to two electrical conductors which are in turn connected to the contact rods. The entire filament arrangement is S-shaped in order to accommodate the significant thermal expansion of the C filament which may be heated to temperatures as high as 2500 °C.

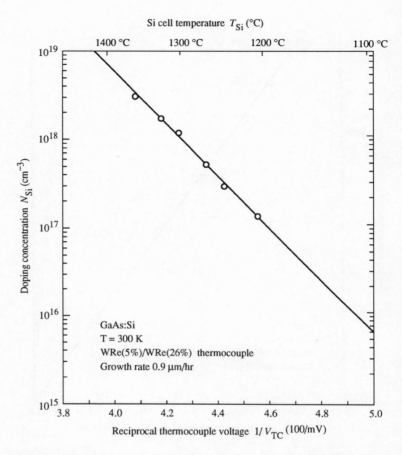

Fig. 5.3. Calibration of Si doping concentration as a function of the effusion cell temperature (thermocouple voltage) for MBE growth. The Si doping concentration is measured by the Hall effect (courtesy of R. F. Kopf).

Doping filaments need to meet several requirements. First, the filament made of the doping element needs to be electrically conductive. Isolating elements such as sulfur (S) are therefore not suited for filament doping. Second, if temperatures >2000 °C are required for evaporation, filament doping is the only choice. Effusion cells are currently available for temperatures as high as 1400 °C (e.g. from EPI Div. Chorus Corp., St. Paul, Minnesota). Graphite, which requires temperatures as high as 2500 °C, cannot be evaporated from effusion cells.

The longevity of filaments is generally shorter compared to effusion cell doping. After many evaporations, the filament becomes thinner and requires an

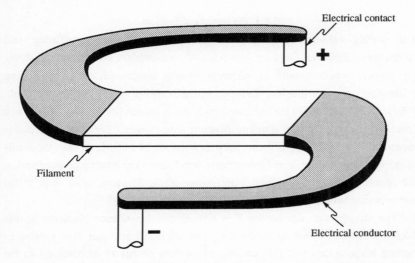

Fig. 5.4. Schematic illustration of a graphite filament for C-doping. The filament is S-shaped in order to accommodate thermal expansion effects.

adjustment of the doping calibration. The purity, homogeneity, and microscopic structure of graphite filaments strongly influence their longevity. Nevertheless, Malik et al. (1988) demonstrated that the relative thickness change of the C filament per typical growth run is relatively small ($< 10^{-3}$). Thus, the graphite filament yields reproducible doping levels over many growth experiments without the need of frequent recalibrations.

Several other groups have used graphite filaments for C doping (Ito and Ishibashi, 1991; Hoke et al. 1991). These reports confirm the feasibility of well-controlled C doping up to doping levels of 10^{20} cm^{-3} using graphite filaments. Hoke et al. (1991) further investigated the influence of the III/V ratio during growth of GaAs and the nature of the As species (As_2 and As_4) on C doping. The authors found no significant change in C doping characteristics for different growth conditions.

Silicon filaments have been employed for n-type doping of III–V semiconductors (Pfeiffer and West, 1987). Single crystal Si wafers of appropriate resistivity cleaved to a stripe-type geometry were used as filament material. The Si filaments have excellent purity, homogeneity and stability resulting in a long lifetime of the filaments.

5.1.5 Filament calibration

The doping calibration of graphite filaments is similar to effusion cell calibrations. Epitaxial films are grown at different graphite filament temperatures (or power) which results in different doping concentrations. The doping calibration curve for graphite filament doping is shown in Fig. 5.5 (Nagle et al., 1991). The figure shows the measured hole concentration (by the Hall effect) as a function of the reciprocal graphite filament temperature. The temperature was measured by a hot-wire pyrometer with an accuracy of ±25 K. Within the limits of the accuracy, the measured temperature was found to be exactly proportional to the fourth root of the electrical input power of the filament, i.e. $T \propto P^{1/4}$ for temperatures between 1600 K and 2400 K.

The temperature dependence $T \propto P^{1/4}$ is the dependence predicted by the *Stefan–Boltzmann law of radiation*. The Stefan–Boltzmann law applies to radiant black bodies and predicts that the radiant energy is proportional to the fourth power of the temperature according to

$$E/V = \pi^2 (kT)^4 / (15 \hbar^3 c^3) \qquad (5.12)$$

where E/V is the radiant heat per unit volume and c is the velocity of light. From the exact $T \propto P^{1/4}$ dependence shown in Fig. 5.5, Nagle et al. (1991) concluded that the heat losses of the filament are dominated by radiation loss. Nagle et al. (1991) used an emissivity factor of 0.8 (Reynolds, 1968) and calculated a radiant area which was very close to the geometrical area of the graphite filament.

5.1.6 Calibration using electrochemical profiling

The doping calibration of epitaxial layers can be achieved efficiently, accurately, and conveniently using electrochemical capacitance–voltage (CV) profiling. This method allows one to calibrate the entire doping range (e.g. 10^{16} cm^{-3} to 10^{19} cm^{-3}) in a *single* growth run. The electrochemical profiling calibration method is therefore less tedious than calibration via the Hall measurements discussed previously.

The electrochemical CV-profiling method consists of alternate capacitance–voltage measurements and electrochemical etching. The CV-technique is a well-known method to measure the net donor ($N_D^+ - N_A^-$) or the net acceptor ($N_A^- - N_D^+$) concentration in semiconductors. The CV-technique using electrochemical etches was proposed by Ambridge and Faktor (1975a, 1975b). The authors demonstrated the use of an electrolyte as a barrier contact (Schottky contact) and also the electrolytical etching of the semiconductor. Both processes are carried out alternately in order to measure a doping *profile*. An

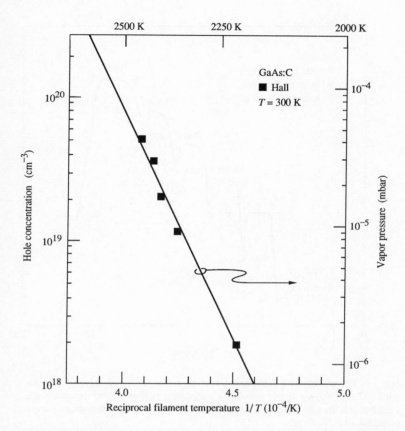

Fig. 5.5. Measured hole concentration at $T = 300$ K in GaAs: C as a function of temperature. The C filament temperature is measured with a pyrometer. The solid line is the calculated vapor pressure of graphite (after Nagle et al., 1991, courtesy of R. J. Malik).

accurate measure of the etching depth is obtained by integrating the etching current and by application of Faraday's law. Recently, the electrochemical CV-technique was reviewed by Blood (1986). This review covers many aspects of the technique including its limitations.

Two electrochemical CV-profiles on p-type and n-type GaAs are shown in Figs. 5.6 and 5.7, respectively. The measurement is carried out at room temperature. Six different effusion cell temperatures are chosen for the calibration which results in six different doping concentrations. The lowest and highest doping concentration are in the 10^{16} cm^{-3} and 10^{18} cm^{-3} range,

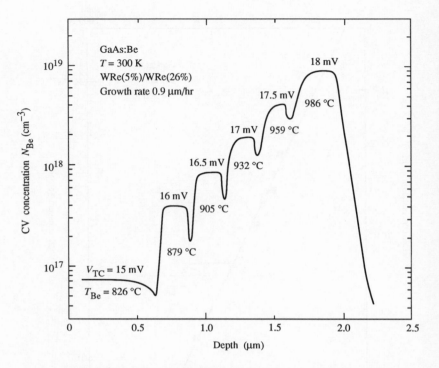

Fig. 5.6. Electrochemical capacitance–voltage depth profile on p-type GaAs:Be with different doping levels grown for calibration purposes. The thermocouple voltages and the corresponding temperatures are indicated for each doping level (courtesy of R. F. Kopf).

respectively. Figures 5.6 and 5.7 demonstrate that the entire doping calibration for one doping cell can be achieved with a single epitaxial growth run.

The accuracy of the CV-measurement is approximately ±5%. Two major factors limit the accuracy of the measurement. First, the dielectric constants of III–V semiconductors are frequently not known very accurately. Second, the contact diameter of the electrolyte Schottky contact may deviate from its nominal value. Since the dielectric constant and the contact diameter enter the calculated concentration, a small systematic error results. The accuracy of the capacitance measurement is usually excellent for state of the art capacitance bridges. Furthermore, the precision (repeatability) of the measurement is excellent, typically much better than 1%. Electrochemical CV-profiling equipment is offered commercially by several companies (e.g. by Polaron Equipment Ltd., Watford, UK).

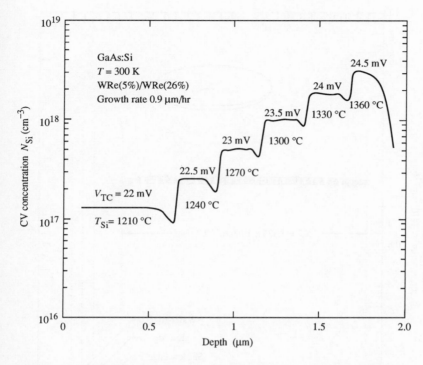

Fig. 5.7. Electrochemical capacitance–voltage depth profile on n-type GaAs:Si with different doping concentrations grown for calibration purposes. The thermocouple voltages and the corresponding temperatures are indicated for each doping level (courtesy of R. F. Kopf).

5.2 Doping homogeneity and uniformity

The production of III–V semiconductor devices from epitaxially grown wafers requires knowledge of the doping homogeneity along the growth axis as well as the uniformity across the wafer. Questions of production yield and manufacturing tolerances are closely related to the homogeneity and uniformity of layer thicknesses and doping. This section summarizes the result of a doping homogeneity study on Si-doped GaAs. The study includes depth-homogeneity, lateral uniformity, and a distribution test. The wafers were grown in a Varian Gen II MBE system with 5 cm rotating In-free substrate mounting.

Doping inhomogeneities are the result of three major causes. First, doping inhomogeneities along the radial direction of a wafer can be the result of non-uniform impurity flux. As shown in the previous section, the flux distribution is

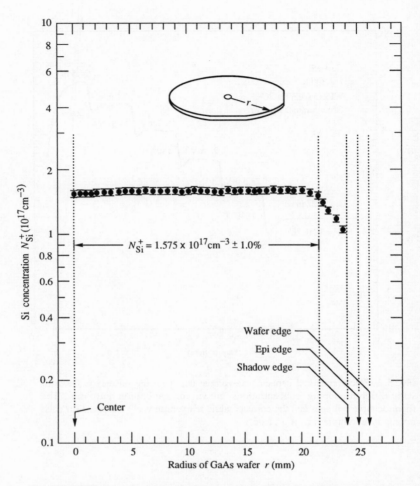

Fig. 5.8. Doping concentration of Si-doped GaAs grown by MBE as a function of distance from the center of the wafer.

cosine-shaped which results in a decrease of the doping concentration at the edges of the wafer. Second, thickness non-uniformities result in doping variation even if the dopant flux is perfectly uniform. Third, variations of the doping concentration along the growth axis can be caused by temporal variations of effusion cell or filament fluxes. If the doping cell (or other growth cells) temperature varies with time, the dopant flux and the doping concentration vary along the growth axis. Such temperature variations may also be caused by temperature changes in the cell environment (e.g. the liquid N_2 cryoshroud).

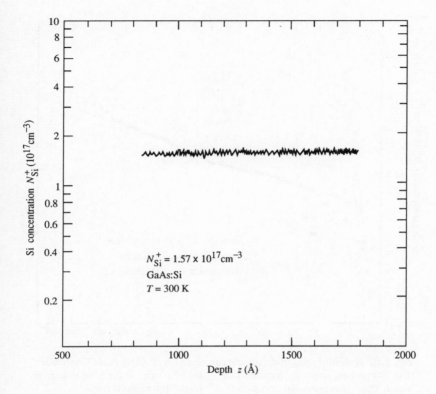

Fig. 5.9. Doping concentration of Si-doped GaAs grown by MBE as a function of depth.

The doping concentration of a 5 cm GaAs:Si wafer is shown in Fig. 5.8 as a function of distance from the center of the wafer. The concentration is measured by the capacitance–voltage technique using 500 μm diameter Ti/Au Schottky contacts. The accuracy of the technique is ±5% limited mostly by the uncertainty of the dielectric constant. The precision of the measurement is typically better than 1%. Since the wafer was rotated during growth, the doping pattern has rotational symmetry. The average doping concentration is 1.57×10^{17} cm^{-3}. The standard deviation of the measured points within a radius < 22 mm is better than 1%. This is a typical result for the MBE system used for this study. Comparable results can be expected for modern MBE systems from other manufactures.

The doping concentration decreases near the edge of the wafer. As indicated in Fig. 5.8, the physical edge of the *substrate* is 26 mm from the center of the

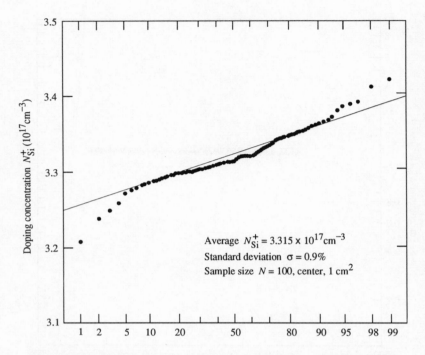

Fig. 5.10. Probability plot of the doping concentration of a Si-doped GaAs wafer. The 100 probing points are located in the center 1 cm^2 area of the 5 cm diameter wafer. The relative standard deviation of N_{Si}^+ for the 100 points is 0.9%.

wafer. The physical edge of the *epitaxial layer* is 25 mm from the center of the wafer. The outer ring with 1 mm width is covered by mounting clamps. The outer 1 mm of the epitaxial layer has rough surface morphology when inspected under an optical microscope and is due to a shadow effect. At the edge of the epilayer the wafer holder used for In-free mounting shadows the epitaxial growth, resulting in non-stoichiometric and polycrystalline growth which results in incomplete doping activation and a general deterioration of the electronic properties. The decrease of the doping concentration at the edge of the wafer is therefore attributed to such edge effects, and cannot be explained by the cosine-shaped flux distribution of the effusion cell.

The doping concentration of Si-doped GaAs as a function of depth is shown in Fig. 5.9. The Si concentration is determined by a C-V measurement using Ti/Au Schottky contacts. The measurement temperature is 300 K. The range of depths accessible to the measurement ranges from 800 Å to 1800 Å at this doping concentration. The doping concentration shown in Fig. 5.9 does not exhibit any depth dependence. The Si concentration is 1.57×10^{17} cm^{-3} for the entire depth range. The constant concentration demonstrates the excellent thermal stability of the doping effusion cell, i.e. the lack of any temperature excursions. It further demonstrates that the growth rate is constant. Any change in growth rate would inevitably result in a change in doping concentration.

The doping concentration of the Si-doped GaAs wafer is shown in a probability plot in Fig. 5.10. The measurement consists of 100 sampling points located in the center 1 cm^2 of a 5 cm GaAs wafer. The average doping concentration is 3.315×10^{17} cm^{-3}, and the lowest and highest concentration are 3.22×10^{17} cm^{-3} and 3.42×10^{17} cm^{-3}, respectively. The standard deviation of the measurement sample is 0.9%. The variation of the doping concentration is attributed to fluctuations in the contact area of the mechanical shadow mask used to define the Ti/Au Schottky contacts. The solid line drawn in Fig. 5.10 represents a fit to the experimental points. If the distribution were a normal (gaussian) distribution, experimental points would follow a straight line. Any non-gaussian distribution would significantly deviate from the linear line in the probability plot. Since the experimental points approximately follow a straight line the distribution of doping concentrations is either gaussian or close to a gaussian distribution.

5.3 Elemental dopants

This section covers the doping elements used in III–V semiconductors and has the sub-sections: 5.3.1 Group-IIa dopants, 5.3.2 Group-IIb dopants, 5.3.3 Group-IV dopants, and 5.3.4 Group-VI dopants. The sections cover the doping impurities used for epitaxial growth, bulk growth, diffusion, and ion implantation. Doping impurities that are introduced as chemical compounds (precursors), which is typical in vapor-phase epitaxy, will be discussed in the next chapter.

The present chapter also covers the specific characteristics of individual impurity elements including amphotericity, saturation, redistribution mechanisms, and incorporation conditions. The properties specific to a certain impurity will be emphasized, for example the autocompensation of Si or the segregation of Sn. However, many questions go beyond the individual characteristics of impurities. For example, *why* does Si autocompensate in GaAs but Be in GaAs does not? *Why* is the diffusion coefficient of Zn in GaAs concentration dependent? *Why* does Be strongly redistribute at concentrations $> 10^{20}$ cm^{-3}? Such questions which go beyond the experimental facts will not be discussed in this chapter. Instead the general characteristics of shallow impurity elements including the underlying physics, chemistry, and material science will be discussed in Chaps. 7

and 8.

5.3.1 Group-IIa dopants

The group-IIa impurities include Be, Mg, Ca, Ba, and Ra. The elements close to the bottom of the IIa column have large atomic radii. They are therefore not expected to incorporate easily in III–V semiconductors. The impurity elements near the top of the IIa column incorporate on the cation sites of III–V semiconductors. The group-IIa impurity atoms have two electrons in their outer shell (oxidation state 2), i.e. one electron less than the cation they substitute for. The group-IIa elements therefore are shallow acceptors in III–V semiconductors.

Beryllium, Be

Summary: Be is a shallow acceptor in III–V semiconductors with high electrical activity. Be diffuses slowly at all except the highest concentrations. Strong redistribution of Be impurities occurs in the very high concentration regime. Be is used in MBE-growth, ion implantation, and as AuBe alloy for p-type ohmic contact metallization.

Be substitutes for cations and is shallow acceptor in all III-V semiconductors. The ionization energy of Be is 28 meV in GaAs and 41 meV in InP as determined by photoluminescence measurements (Ashen et al., 1975; Skromme et al., 1984). However, lower activation energies of 19 meV for Be in GaAs were also reported (Wood, 1985a and 1985b).

Be exhibits near complete activation in III–V semiconductors without compensation effects. The high activation of Be in GaAs is illustrated in Fig. 5.11 which shows the two-dimensional hole density as a function of the Be density (Schubert et al., 1990a). Be impurities were deposited during interruption of epitaxial growth by MBE. A 1000 Å thick GaAs cap layer was grown after impurity deposition on the non-growing surface. The experimental result of Fig. 5.11 demonstrates that the free hole concentration follows the Be impurity concentration in a linear fashion. Furthermore, the experimental free carrier concentration is in agreement with the Be concentration inferred from the deposition rate (which is obtained from an independent secondary ion mass spectrometry calibration) shown by the solid line. Therefore, all Be impurities incorporated into the GaAs crystal are electrically active. The fraction of electrically inactive Be impurities is insignificantly small.

However, the high activation of Be impurities depends on the host material and the incorporation conditions. Experimental results of Hamm et al. (1989) demonstrate that a fraction of Be impurities in $Ga_{0.47}In_{0.53}As$ are inactive especially at high growth temperatures. The lack of activity can be caused by Be

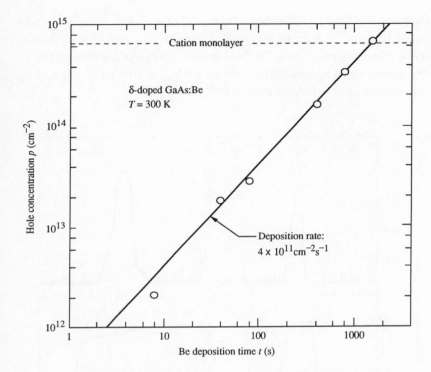

Fig. 5.11. Two-dimensional hole concentration of Be-doped GaAs as a function of the Be deposition time (Be density). The dopants are deposited during the suspension of MBE growth at a temperature of 500 °C. The hole density follows the acceptor density without any indication of saturation effects.

precipitates (Hamm et al., 1989) or by inactive Be atoms on the semiconductor surface. Inactive Be impurities in GaAs were concluded by Duhamel et al. (1981) to be due to Be interstitials. The authors found an increasing density of stacking faults with increasing Be concentration which they attributed to an increasing concentration of Be complexes. A degradation of surface morphology in highly Be doped GaAs was reported Ilegems (1977) which further indicates the dependence of Be incorporation on the growth conditions.

Be in GaAs is a slowly diffusing impurity for all except the highest Be concentrations. Ilegems (1977) and McLevige et al. (1978) found a small diffusion coefficient of $D = 10^{-13} - 10^{-14}$ cm^2/s for $T = 900$ °C and $N_{Be} = 10^{19}$ cm^{-3}. The temperature dependence of the Be diffusion coefficient was determined to be $D = D_0 \exp(-E_a/kT)$ with $E_a = 1.95$ eV and

$D_0 = 2 \times 10^{-5}$ cm^2 (Schubert   et al.   1990a)   for   samples   with $N_{Be} \cong 10^{19}$ cm^{-3}. Both results establish that Be is a very slow diffuser in GaAs. The diffusion of Be in GaP was studied by Ilegems and O'Mara (1972). The authors found a small diffusion length for $N_{Be} < 10^{19}$ cm^{-3} but a strong redistribution of Be at concentrations $> 10^{19}$ cm^{-3}.

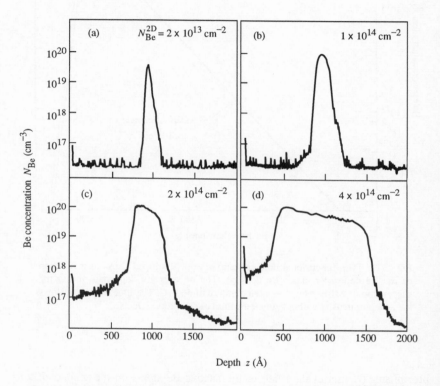

Fig. 5.12. Secondary ion mass spectrometry (SIMS) profiles of Be-doped GaAs with different densities. The dopants are deposited during the suspension of MBE growth at 500 °C. Strong spreading of Be occurs at high densities. The maximum attainable Be concentration is 10^{20} cm^{-3}.

Strong redistribution of Be in GaAs occurs at very high doping concentrations of about 10^{20} cm^{-3}. The occurrence of strong impurity redistribution is illustrated in Fig. 5.12 which shows the SIMS (secondary ion mass spectrometry) profiles of different Be densities in GaAs doped during growth interruption (Schubert et al., 1990a). A 1000 Å thick top layer is grown after the growth interruption. Figure 5.12 illustrates that little Be redistribution occurs at two-dimensional Be densities of $\leq 2 \times 10^{13}$ cm^{-2}. However, for high Be densities, strong redistribution of Be occurs in such a way that peak concentration of Be does not exceed 10^{20} cm^{-3}.

It was proposed by Schubert et al. (1992) that the strong redistribution of Be in GaAs at concentrations $N_{Be} \cong 10^{20}$ cm^{-3} is due to coulombic repulsion of ionized substitutional impurities. The authors showed that high concentration of ionized impurities results in substantial coulombic interaction energies which limit the incorporation of impurities to a maximum concentration which is on the order of $10^{19} - 10^{21}$ cm^{-3} depending on the diffusion coefficient and the growth rate. The authors further showed that the flat top-hat distribution of Be shown in Fig. 5.12 can be fully explained by repulsive coulombic interaction. A further discussion of the impurity concentration limit caused by coulombic interaction is given in Chaps. 7 and 8.

Beryllium is also used as dopant for ohmic contact metallization to p-type semiconductors. A typical contact metal is a AuBe alloy with a composition of 99.1% Au and 0.9% Be. Low contact resistances are obtained by typical alloying cycles of 20 seconds at 380 °C.

Beryllium easily getters oxygen and forms the oxide BeO. Both Be and BeO, but especially BeO, are highly toxic and carcinogenic. Precautions are required for their handling.

Magnesium, Mg

Summary: Mg is a shallow acceptor impurity in III–V semiconductors. The Mg incorporation coefficient is low at typical MBE growth temperatures. Only a small fraction of Mg is incorporated during MBE at high substrate temperatures.

Magnesium occupies cation sites and is a shallow acceptor in III–V semiconductors with an ionization energy of 28 mV in GaAs and 41 meV in InP (Ashen et al., 1975; Skromme et al., 1984). Mg doping of GaAs and Al$_x$G$_{1-x}$As was first reported in 1972 by Cho and Panish. The authors reported a low sticking coefficient of Mg in GaAs grown at 560 °C. However, they found a much larger sticking coefficient for Mg-doped Al$_x$Ga$_{1-x}$As compared to GaAs at the same growth temperature of 560 °C. The authors attributed the increased incorporation of Mg into Al$_x$Ga$_{1-x}$As to the enhanced concentration of oxygen in Al$_x$Ga$_{1-x}$As which forms strong MgO bonds and thus prevents Mg from desorbing. Hole concentrations of $p = 10^{19}$ cm^{-3} in Mg-doped Al$_{0.20}$Ga$_{0.80}$As were reported. These results and their interpretation are somewhat surprising since an impurity-oxygen complex is not expected to act as a simple acceptor.

In subsequent publications Joyce and Foxon (1977) reported a high incorporation coefficient during MBE growth but a low electrical activity of the incorporated Mg impurities. Furthermore, Kirchner et al. (1981) showed that even if Mg is incorporated into GaAs it can be electrically inactive due to the formation of MgO. The authors showed that the formation of MgO can be reduced by adding a nonvolatile oxygen-gettering element (such as Al) to the Ga effusion cell, or by growing $Al_x Ga_{1-x} As$ in order to form AlO complexes rather than electrically inactive MgO complexes. Finally, Wood et al. (1982) carefully analyzed the dependence of Mg incorporation into GaAs as a function of the growth temperature. A low growth temperature of 500 °C results in a Mg impurity incorporation coefficient of 0.3. High growth temperatures of 600 °C reduce the incorporation coefficient by three orders of magnitude to 0.3×10^{-3} probably due to the high vapor pressure of Mg. The reduced incorporation coefficient leads to a reduction of the free hole concentration from the 10^{18} cm^{-3} to the 10^{16} cm^{-3} range which occurs concomitantly with the increase of growth temperature from 500 °C to 600 °C.

The current understanding of Mg impurities in GaAs and $Al_x Ga_{1-x} As$ is summarized as follows. The incorporation coefficient of Mg is close to unity only for temperatures <500 °C and decreases rapidly at higher temperatures. Magnesium has a strong affinity to oxygen and the formation of MgO leads to electrically inactive complexes. These properties restrict the use of Mg to materials grown at low temperatures <500 °C.

Calcium, Ca

Doping of GaAs with Ca revealed no electrical activity (Wood et al., 1982). An effusion cell charged with $Ca_3 As_2$ was used as the Ca doping source in this study.

5.3.2 Group-IIb dopants

The group-IIb impurities include Zn, Cd, and Hg. Mercury has a high vapor pressure and is therefore very volatile. A low incorporation coefficient is therefore expected for Hg doping. Both zinc and cadmium were employed for doping of III–V semiconductors.

Zinc, Zn

Summary: Zn is a shallow acceptor in III–V semiconductors with high electrical activity. Zn diffuses rapidly at high concentrations. Zn doping is used in OMVPE growth, solid-source and vapor-phase diffusion, and in a AuZn alloy for p-type ohmic contact metallization.

Zinc substitutes for cations in III–V semiconductors and is a shallow acceptor. The Zn ionization energy is 31 meV in GaAs and 47 meV in InP as determined by photoluminescence measurements (Asher et al., 1975; White et al., 1973). Zn-acceptors exhibit high electrical activity and show no evidence of autocompensation.

Doping of GaAs with Zn by MBE was first reported by Arthur (1973) who showed that the bonding of Zn to the GaAs surface is weak and that Zn desorbs rapidly from the surface at temperatures of 600 °C. The activation energy for desorption was shown to equal the enthalpy for vaporization of pure Zn. The high vapor pressure of Zn thus makes Zn unsuited for doping during MBE growth.

Zinc is also diffused from a Zn layer into III–V semiconductors. For example, pn-junctions can be formed by diffusing Zn into n-type semiconductors. Evaporated Zn films on the semiconductors serve as a reservoir. It is useful to provide As overpressure during diffusion of Zn into GaAs to reduce As outdiffusion at high temperatures. The As overpressure is usually achieved by an additional, elemental As source enclosed with the semiconductor in an ampoule.

Zinc in GaAs was shown to have a strongly concentration-dependent diffusion coefficient. It was shown that the Zn diffusion profiles cannot be explained by the well-known Fickian (i.e. concentration-independent) diffusion (Tuck and Kadhim, 1972; Kadhim and Tuck, 1972; Tuck, 1988). In order to obtain the concentration-dependent diffusion coefficient Kadhim and Tuck (1972) employed the Boltzmann–Matano method (see for example Tuck, 1988). This method allows one to determine the concentration dependent diffusion coefficient. The Zn diffusion coefficient in GaAs at $T = 1000$ °C as a function of concentration is shown in Fig. 5.13 (Kadhim and Tuck, 1972). The authors found a diffusion coefficient as high as 10^{-7} cm²/s at the highest Zn concentrations. The dependence of the diffusion coefficient is well described by the power law $D \propto N_{Zn}^{1.8}$. An exponent of two (2) in the power law (i.e. $D \propto N_{Zn}^2$) can be explained by the substitutional-interstitial diffusion model discussed in Chap. 8. Zinc is considered to be a model impurity for the substitutional-interstitial diffusion mechanism.

Zinc is a popular dopant in VPE and CBE growth. Diffusion coefficients of Zn in OMVPE-grown GaAs are several orders of magnitudes smaller as compared to the 'infinite' source diffusion experiments (i.e. diffusion from a Zn film). At concentration of 1×10^{18} cm^{-3} the Zn diffusion coefficients reported at 900 °C are $5 \times 10^{-12} - 10 \times 10^{-12}$ cm²/s (Hobson et al., 1989; Enquist et al., 1988). At the lower growth temperature of 625 °C the authors reported a diffusion coefficient as low as 7×10^{-17} cm²/s. Such low diffusion coefficients make feasible the abrupt Zn doping profiles which are required in heterostructure

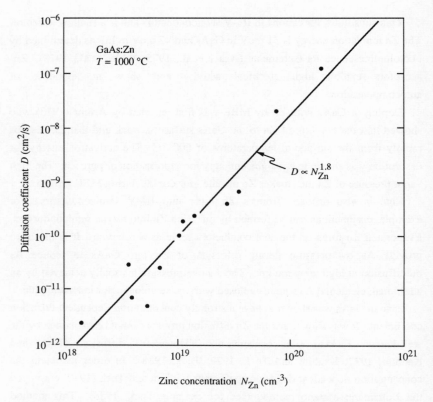

Fig. 5.13. Diffusion coefficient of Zn in GaAs at $T = 1000$ °C as a function of the Zn concentration. The diffusion experiments were performed at the As dissociation pressure (after Kadhim and Tuck, 1972).

devices.

Zinc is also used for ohmic contact formation with p-type III–V semiconductors. A AuZn alloy (95% Au, 5% Zn) is frequently used for contact metallization. A typical alloying cycle for AuZn contacts on GaAs is 20 s at 380 °C.

Cadmium, Cd

Cadmium is a shallow acceptor in GaAs and InP with an activation energy of 35 meV and 56 meV respectively (Ashen et al., 1975; White et al., 1973). It has a vapor pressure approximately 1–2 orders of magnitude higher than Zn. Therefore, Cd is quite volatile which precludes it as a viable doping impurity for MBE growth.

Cadmium has been implanted in $Ga_{0.47}In_{0.53}As$ in order to obtain p-type conductivity (Vescan et al., 1984). The authors attempted to activate the Cd impurities by annealing at a temperature of 800 °C. Despite annealing, only a small fraction of Cd impurities were electrically activated. A further drawback of Cd is its toxicity.

Mercury, Hg

Mercury was reported to form an acceptor level in InP with an ionization energy of 95 meV (Williams et al., 1973). Mercury has a high vapor pressure which makes it very volatile. This unfavorable property precludes Hg as a viable acceptor in III–V semiconductors.

5.3.3 Group-IV dopants

The group-IV impurities include C, Si, Ge, Sn, and Pb. The latter element has a large atomic radius and is therefore not expected to incorporate easily in III–V semiconductors. The remaining group-IV impurities, i.e. C, Si, Ge, and Sn share *four key characteristics* which are summarized as follows. First, all column IV impurities are *amphoteric*, i.e. they can occupy either the group-III or the group-V (cation or anion) site of the zincblende crystal. As a result, group-IV impurities are donors or acceptors for cation site or anion site occupancy, respectively. Second, column-IV impurities *autocompensate*, which is a result of their amphotericity. Consider a column-IV impurity that predominantly occupies the cation site, i.e. is a donor. If some of the impurity atoms occupy anion sites, i.e. acceptor sites, then these acceptors partially compensate the donors which results in a reduction of the free carrier concentration. Since the same species of impurities cause the compensation, the phenomenon is called autocompensation. Third, the electrical activity *saturates* at high impurity concentrations. With increasing doping concentration the tendency of group-IV impurities to autocompensate increases and ultimately leads to the saturation of the free carrier concentration. Fourth, the free carrier concentration depends on *growth temperature* and *V/III flux ratio* during epitaxial growth by MBE, GSMBE and OMVPE. Generally, low growth temperatures result in the highest free carrier concentration. Furthermore, a V/III flux ratio that makes the desired crystal site available (e.g. a high V/III flux ratio for Si doping in order for Si to predominantly occupy group-III sites) results in the highest free carrier concentrations.

What is the materials physics behind these very general properties? There are, of course, several answers and models that explain the specific properties of group-IV impurities. The general models of impurity characteristics will be discussed in Chaps. 7 and 8. In the following, the experimental part of impurity characteristics will be emphasized and the specific properties of group-IV impurities will be illustrated and summarized.

Carbon, C

Summary: Carbon is a shallow acceptor in GaAs and a shallow donor in InAs. It strongly compensates in $Ga_{0.47}In_{0.53}As$. C is a stable impurity and diffuses very slowly. Very high acceptor concentrations $> 10^{19}$ cm^{-3} can be achieved by filament doping during MBE and $> 10^{20}$ cm^{-3} in CBE.

Carbon substitutes for anions in GaAs and forms a shallow acceptor level of 26 meV (Ashen et al., 1975). Although C rarely forms acceptors in InP (Skromme et al., 1984), a C acceptor level of 43 meV in InP has been reported (Hess et al., 1974). The ionization energies were determined by photoluminescence experiments.

Carbon is a very stable impurity in GaAs grown by MBE. Malik et al. (1988) showed for C-filament doping of GaAs that very little diffusion of C impurities occurs at 800 °C. The negligible diffusion found by Malik et al. agrees with earlier reports on C doping by flow-rate modulation epitaxy (Kobayashi et al., 1987). Kobayashi et al. reported a diffusion coefficient of 2×10^{-16} cm^2/s at a temperature of 800 °C and a C concentration of $\cong 10^{18}$ cm^{-3}. Subsequently, the temperature dependence of C diffusion in GaAs was determined and expressed as $D = D_0 \exp(-E_a/kT)$ with $E_a = 1.75$ eV and $D_0 = 5 \times 10^{-8}$ cm^2/s (Schubert, 1990) which yields $D = 3 \times 10^{-16}$ cm^2/s at 800 °C. The diffusion coefficient of C is about two orders of magnitude lower than that of Be in GaAs which further illustrates the stability of C.

Carbon acceptor concentrations of 5×10^{19} cm^{-3} (Malik et al., 1988, Hoke et al., 1991a and 1991b) were achieved in GaAs using C (graphite) filament sources. Even higher concentrations $> 10^{20}$ cm^{-3} were achieved by using alkyl sources as C doping precursors (see Chap. 6).

Carbon incorporation during MBE growth depends on the V/III flux ratio. The measured hole concentration of C-doped GaAs as a function of the V/III flux ratio is shown in Fig. 5.14. (Hoke et al., 1991a and 1991b). The As$_4$/Ga flux ratio was varied between 10 and 34. With increasing flux ratio a decreasing hole concentration is measured. The relative change in the free hole concentration is approximately 20%. The authors also compared C-doped GaAs grown with As$_4$ and As$_2$. They found no difference in carrier concentration and mobility for As$_4$ and As$_2$ grown samples that exceeded their experimental uncertainty of 10%. The experimental results shown in Fig. 5.14 can be understood in terms of increased C autocompensation at high V/III flux ratios. Carbon predominantly

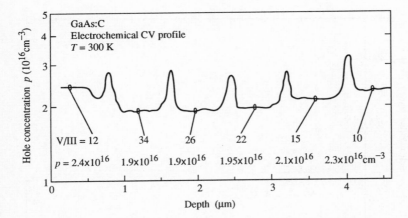

Fig. 5.14. Hole concentration versus depth for epitaxial C-doped GaAs containing six 0.8 μm thick layers grown with different As_4/Ga flux ratios. The layers are delineated by doping spikes (after Hoke et al., 1991a).

occupies group-V sites. However, as the As_4 flux increases, As sites become rapidly occupied and are less likely to be occupied by C. As a result, C is more likely to occupy Ga sites on which C is a donor. The experimental result of Fig. 5.14 can thus be explained by autocompensation of C at high V/III flux ratios. The difference in bond energy of the Ga–C bond (2.65 eV/bond) and the As–C bond (2.47 eV/bond) (Hartley and Patai, 1982) is sufficiently large to make As the preferred C site. However, the difference in bond energy of $\cong 700$ meV (four bonds) is approximately nine times larger than kT at the growth temperature which does not rule out the occupation of the Ga site by a fraction of C atoms. Furthermore, caution is required when Ga–C and As–C bond energies are applied to GaAs:C crystals.

Room-temperature Hall mobilities of C- and Be-doped GaAs and $Al_{0.30}Ga_{0.70}As$ grown by MBE are shown in Fig. 5.15 (Hoke et al., 1991b). The plot shows very similar mobilities for C and Be doping. Furthermore, the hole mobilities agree with values compiled by Blakemore (1982). The hole mobilities in $Al_xGa_{1-x}As$ are lower by a factor of 2–2.5 than the mobilities in GaAs. The reduction in mobility is due to alloy scattering caused by potential fluctuation (due to random cation distribution) in the ternary $Al_xGa_{1-x}As$.

While C is an acceptor in GaAs, it is a donor in InAs. In $Ga_{0.47}In_{0.53}As$ carbon strongly compensates but still shows residual p-type conductivity (Ito and Ishibashi, 1991). It is likely that the conductivity type of C-doped $Ga_{0.47}In_{0.53}As$

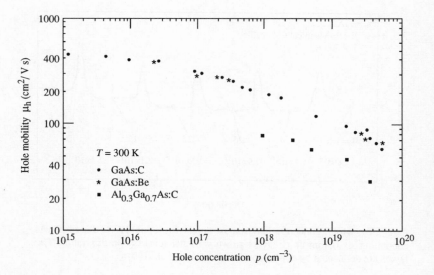

Fig. 5.15. Room temperature Hall mobility as a function of hole concentration for MBE-grown C- and Be-doped GaAs and for C-doped $Al_{0.30}Ga_{0.70}As$ (after Hoke et al., 1991b).

depends strongly on the V/III flux ratio. Ito and Ishibashi showed that under their growth conditions a conversion from p-type to n-type conductivity occurs at an In mole fraction of $x = 0.60$ in $Ga_{1-x}In_xAs$.

High non-radiative recombination rates have been found in MBE-grown highly C-doped GaAs using photoluminescence studies (Calderon et al., 1992). The room-temperature photoluminescence intensity was found to decrease strongly at C doping concentrations exceeding 10^{19} cm^{-3}. The non-radiative rate was larger than the expected Auger recombination rate. The authors therefore concluded that deep centers associated with the high C concentrations are the origin of the high non-radiative recombination.

Silicon, Si

Summary: Si is a shallow donor in all III–V semiconductors. Si auto compensates little at low concentrations, but strongly in the high concentration regime. It diffuses slowly and slightly segregates along with the growing surface during MBE growth. Si is a popular n-type dopant for MBE growth and ion implantation.

Silicon predominantly occupies cation sites in III–V semiconductors and is a hydrogenic shallow donor with 4–6 meV ionization energy in GaAs and InP. A small fraction of Si incorporates as an acceptor. The Si acceptor ionization energy is 35 meV in GaAs (Ashen et al., 1975). Even though Si is predominantly a donor, it can, under specific growth conditions, also be incorporated as an acceptor for example in liquid-phase epitaxial growth (Rupprecht et al., 1966; Spitzer and Panish, 1969).

Silicon is a stable impurity in III–V semiconductors with only little diffusion occurring during GaAs growth at moderate temperatures. The diffusion coefficient was expressed in terms of the equation $D = D_0 \exp(-E_a/kT)$ with an activation energy $E_a = 2.45$ eV and $D_0 = 4 \times 10^{-4}$ cm^2/s for diffusion of atomic Si in GaAs (Schubert et al., 1988). Greiner and Gibbons (1984) investigated the diffusion of Si pairs in GaAs and found a higher diffusion coefficient for pair diffusion. The diffusion coefficient of Si in $Al_x Ga_{1-x} As$ was found to be larger than in GaAs (Lanzilloto et al., 1989). The authors found that Si spreads significantly further in $Al_{0.25} Ga_{0.75} As$ as compared to GaAs. The determination of the Si diffusion coefficient in $Al_{0.30} Ga_{0.70} As$ yielded indeed higher values, namely an activation energy of $E_a = 1.3$ eV and $D_0 = 4 \times 10^{-8}$ cm^2/s (Schubert, et al., 1989).

In addition to the pure diffusive motion of Si in GaAs, Si was also found to redistribute predominantly to the surface during growth by MBE (Beall, et al., 1988). The authors found marked preferential migration towards the surface especially at high growth temperatures. The authors proposed that conventional surface segregation can account for the Si redistribution (Harris et al., 1984). A second model for surface migration is based on the drift of impurities in the surface field (Schubert et al., 1990b). The electric field is the result of the pinning of the Fermi level at the semiconductor surface. The two segregation models will be discussed further in Chap 8.

The free electron concentration of Si-doped III–V semiconducting material exhibits a peculiar saturation in the high concentration regime. The measured free electron concentration of Si-doped GaAs grown by MBE is shown in Fig. 5.16 as a function of reciprocal Si effusion cell temperature (Chai et al., 1981). For effusion cell temperatures ≤ 1250 °C, the free electron concentration does not increase further but saturates at approximately 5×10^{18} cm^{-3}. The exact concentration at which the free carrier concentration saturates depends on the growth conditions. Values of $n = 1.1 \times 10^{19}$ cm^{-3} (Heiblum et al., 1983), 1.6×10^{19} cm^{-3} (Malik et al., 1992), and 1.8×10^{19} cm^{-3} (Ogawa and Baba, 1985) have been reported for the maximum electrically active Si concentration in GaAs. Such high values for the free carrier concentration are obtained under

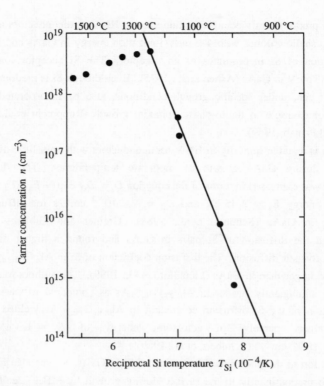

Fig. 5.16. Free electron concentration of Si-doped GaAs as a function of the reciprcal Si effusion cell temperature. For Si cell temperatures > 1250 °C, the free electron concentration saturates (after Chai et al., 1981).

optimized growth conditions, i.e. for low growth temperatures (e.g. 500 °C), high V/III flux ratios (e.g. V/III = 20), or growth with As_2. Under fairly standard growth conditions ($T = 580$ °C, V/III = 4) the onset of free carrier saturation in MBE grown Si-doped GaAs occurs at a Si concentration of $4 \times 10^{18} - 5 \times 10^{18}$ cm^{-3} (Kopf and Schubert, 1990).

The physical origin of the saturation of the free electron concentration in highly Si-doped GaAs was shown to be electrically inactive Si, Si pairs, and Si acceptors, that drastically increase in concentration for nominal doping concentrations above 10^{19} cm^{-3} (Maguire et al., 1987), as demonstrated by infra-red local vibrational mode absorption spectroscopy on highly Si-doped GaAs grown by MBE. Two local vibrational mode spectra are shown in Fig. 5.17 for nominal Si concentrations of 5×10^{18} cm^{-3} and 3×10^{19} cm^{-3}. The spectrum obtained from the sample doped at 5×10^{18} cm^{-3} reveals that ^{28}Si on

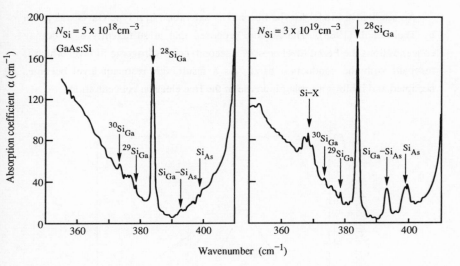

Fig. 5.17. Local vibrational mode (LVM) spectra of heavily Si-doped GaAs with Si concentrations of $5 \times 10^{18} \mathrm{cm}^{-3}$ and $3 \times 10^{19} \mathrm{cm}^{-3}$. At the higher concentration, $\mathrm{Si_{As}}$, $\mathrm{Si_{Ga}-Si_{As}}$ pairs, and Si complexes are detected which compensate for $\mathrm{Si_{Ga}}$ donors, and thus reduce the free carrier concentration (after Maguire et al., 1987).

Ga sites is the dominant species. Furthermore, vibrational resonances of the isotopes $^{29}\mathrm{Si_{Ga}}$ and $^{30}\mathrm{Si_{Ga}}$ are found. Thus Si occupies predominantly substitutional group-III sites at this doping concentration. The concentration of Si acceptors, i.e. Si on As sites, as well as neutral donor–acceptor pairs is comparatively small.

The local vibrational mode spectrum changes markedly at a doping concentration of $N_{\mathrm{Si}} = 3 \times 10^{19}$ cm^{-3}. Three new lines appear which the authors attributed to (i) Si acceptors, i.e. Si on arsenic sites, (ii) neutral Si donor–acceptor pairs, and (iii) a Si-complex (Si–X) possibly being a Ga-vacancy-Si complex. However, an unambiguous identification of the Si–X complex is not provided. The authors assume that all of the three configurations, that is $\mathrm{Si_{As}}$, $\mathrm{Si_{As}} - \mathrm{Si_{Ga}}$, and Si$-$X, contribute to the reduction of the free carrier concentration. The free carrier concentration measured is only 8×10^{17} cm^{-3} at the nominal Si doping concentration of 3×10^{19} cm^{-3}. Maguire et al. concluded that 40% of the Si exists as $\mathrm{Si_{Ga}}$, 19% as $\mathrm{Si_{As}}$, 14% as $\mathrm{Si_{Ga}} - \mathrm{Si_{As}}$, and 27% as Si–X. The work clearly shows that electrically inactive Si causes the saturation in highly Si-doped GaAs.

An alternative model for the free carrier saturation in n-type GaAs was given by Theis et al. (1988). The authors proposed that at sufficiently high doping concentrations the Fermi level crosses a second, non-hydrogenic Si level which is resonant with the conduction band. As a result, this resonant level becomes occupied and inhibits a further increase of the free electron concentration.

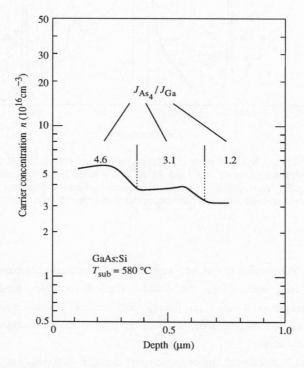

Fig. 5.18. Effect of V/III flux ratio on the free carrier concentration of MBE-grown Si-doped GaAs (after Chai et al., 1981).

Further experiments clarifying the Si incorporation into GaAs were performed by Chai et al. (1981). The authors investigated the dependence of the free carrier concentration of Si-doped GaAs as a function of the MBE growth conditions. Chai et al. varied the As_4/Ga flux ratio and the growth temperature. The experimental free carrier concentration versus depth for different As_4/Ga flux ratios is shown in Fig. 5.18. The flux ratio is varied between V/III = 1.2 − 4.6, i.e. epitaxial growth was under As stabilized conditions. The free carrier concentration   increases   with   increasing   V/III   flux   ratio   from

$n = 3.2 \times 10^{16}$ cm^{-3} (V/III = 1.2) to $n = 5.3 \times 10^{16}$ cm^{-3} (V/III = 4.6). The increase of the free carrier concentration with V/III flux can be understood in terms of the availability of sites and the amphoteric nature of Si in GaAs. Consider the chemical reaction

$$\text{Si}_{\text{Ga}} + \text{V}_{\text{As}} \leftrightarrow \text{Si}_{\text{As}} + \text{V}_{\text{Ga}} \tag{5.13}$$

where V_{As} and V_{Ga} are As and Ga vacancies, respectively. In order to increase the concentration of Si_{Ga} (Si donors) it is, according to the *law of mass action*, advantageous to decrease the concentration V_{As}. The reduction of As vacancies on the surface of the growing semiconductor as well as in the bulk is achieved by higher As partial pressures, i.e. higher V/III flux ratios. As the V/III ratio becomes larger the Si donor concentration increases along with the free carrier concentration. The increase of the free carrier concentration appears as steps in Fig. 5.18. The incorporation of amphoteric impurities is further discussed in Chap. 7.

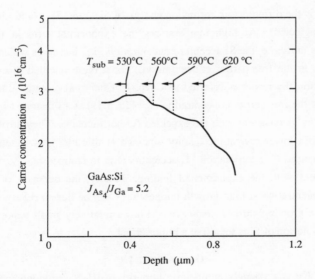

Fig. 5.19. Effect of substrate temperatute T_{sub} on the free electron concentration of Si-doped GaAs grown by MBE at a constant V/III flux ratio (after Chai et al., 1981).

The free carrier concentration of Si-doped GaAs as a function of depth for different growth temperatures is shown in Fig. 5.19. (Chai et al., 1981). The substrate temperature was lowered during growth from 620 °C to 530 °C in several temperature steps. Figure 5.19 shows that the free carrier concentration increases as the growth temperature is lowered. The authors argue that the increase is expected for amphoteric impurities. A high substrate temperature causes an increase in the As_4 desorption rate from the surface resulting in an increased V_{As} concentration. The increase in V_{As} concentration shifts the reaction equilibrium of the chemical reaction Eq. (5.13) to the right-hand side of the equation, which increases the concentration of Si acceptors, Si_{As}. As a result, the carrier concentration decreases at high substrate temperatures in agreement with the experimental result shown in Fig. 5.19.

Another explanation of the decrease in free carrier concentration at high growth temperatures is based on the different formation energies of Si_{Ga} and Si_{As}. Obviously Si_{Ga} has a larger formation energy than Si_{As} since Si is predominantly a donor. Denoting the difference in formation energy as ΔE ($\Delta E > 0$) the ratio of Si acceptor to Si donor concentration is given by

$$[Si_{As}]/[Si_{Ga}] \propto \exp(-\Delta E/kT) \qquad (5.14)$$

where ΔE is the difference in formation energy of Si_{Ga} and Si_{As} (and the entropy term is neglected). At high temperatures the exponential term in Eq. (5.14) approaches unity, i.e. the Si acceptor concentration Si_{As} increases which leads to a decrease in the free carrier concentration. For a more detailed discussion of compensation the reader is referred to Chap. 7. Chai et al. pointed out that the decrease of the free carrier concentration (see Fig. 5.19) at a substrate temperature of 530 °C is inconsistent with the expected further increase. They attributed the decrease to a lower crystalline quality obtained at this low growth temperature and to enhanced C incorporation. C acceptors tend to compensate the Si donors which agrees with the experimental finding. Finally, the decrease of the free carrier concentration at high growth temperatures can be due to reevaporation of Si from the growing surface. However, Si has a relatively small vapor pressure and reevaporation of Si is therefore a minor effect.

Germanium, Ge

Summary: Ge is a strongly amphoteric impurity in III–V semiconductors. Both p- and n-type conductivity can be obtained depending on substrate temperature and V/III flux ratio during MBE growth. AuGe alloys are used for ohmic contact metallization to n-type semiconductors.

Germanium incorporates on either the cation site or anion site of a III–V semiconductor. It is an acceptor and a donor on the anion and cation site, respectively. The Ge acceptor ionization energy in GaAs was determined to be 40 meV by low-temperature photoluminescence measurements (Ashen et al., 1975). Ge is a shallow hydrogenic donor on the cation site and has an ionization energy of 4–6 meV in GaAs and InP.

Germanium is a strongly amphoteric impurity in GaAs. It was shown that MBE-grown GaAs doped with Ga can be either p-type or n-type (Cho and Hayashi, 1971). The authors further showed that the incorporation of Ge strongly depends on the surface chemistry, i.e. on the (i) V/III flux ratio and (ii) the growth temperatures. Under As-rich growth conditions, Ge incorporates as a donor; correspondingly, under Ga-stabilized growth conditions, i.e. a V/III flux ratio close to unity, Ge incorporates as an acceptor.

Similarly, the variation of the growth temperature allows one to change the conductivity type of Ge-doped GaAs. At growth temperatures below the congruent sublimation temperature (which is 630 °C for GaAs), the epitaxial surface is As stabilized as long as V/III > 1. As a result, Ge incorporates as a donor. However, for growth temperatures that exceed the congruent evaporation temperature, Ga-stabilized surfaces are obtained. As a result, Ge incorporates as an acceptor on anion sites. Cho and Hayashi (1971) demonstrated that pn junctions can be grown using Ge as the single impurity species. The conversion from p- to n-type was achieved by an increase of the V/III flux ratio.

Künzel et al. (1982) investigated the effect of the two arsenic vapor species As_2 and As_4 as well as of the substrate temperature on the incorporation of Ge in GaAs. The experimental result of this study is shown in Fig. 5.20 which depicts the compensation $[Ge_{As}]/[Ge_{Ga}]$ as a function of reciprocal substrate temperatures for the two species As_2 and As_4. Compensation increases strongly at higher growth temperatures. The high growth temperatures result in increased As reevaporation from the epitaxial surface and thus a reduced As coverage of the surface. As a result, Ge increasingly occupies As sites. The authors further showed different compensation for As_4 and As_2 evaporation. As_2 species were generated by thermally cracking As_4. Künzel et al. (1982) argued that crystals grown with the more reactive As_2 have higher As surface coverage and therefore less compensation. It is interesting to note that the two solid lines intersect unity compensation, i.e. $[Ge_{As}]/[Ge_{As}] = 1$, at a temperature of 600–620 °C which is slightly below the congruent sublimation temperature for GaAs of 630 °C. The compensation of Ge and its dependence on the growth conditions preclude the use of Ge as a donor or acceptor for MBE-grown III–V semiconductor devices.

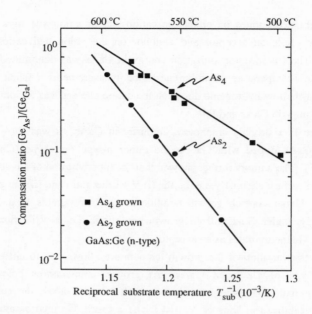

Fig. 5.20. Compensation of MBE-grown n-type Ge-doped GaAs as a function of substrate temperature. The use of As_2 dimers results in a reduction of auto-compensation compared to growth with As_4 tetramers (after Künzel et al., 1982).

Eutectic AuGe (88% Au, 12% Ge) is frequently used in ohmic contact metallization for III–V semiconductors. Alloying cycles of 420 °C for 20 s are sufficient to diffuse Ge into GaAs where it acts as donor. The chemistry and technology of alloyed AuGe ohmic contacts was reviewed by Braslau (1981). The author showed that in addition to Ge diffusion, the formation of domains with different chemical composition occurs in AuGe contacts.

Tin, Sn

Summary: Tin is a weakly compensating shallow donor in III–V semiconductors. Sn strongly segregates and diffuses at growth temperatures exceeding 500 °C.

Tin occupies cation sites in III–V semiconductors and is a hydrogenic shallow donor with 4–6 meV ionization energy in GaAs and InP. However, Sn was reported to incorporate as an acceptor in GaSb (Chang et al., 1977). Even though there are no reports on Sn acceptor incorporation in GaAs it is quite likely that a small fraction of Sn does indeed incorporate as acceptor in all III–V semiconductors.

The major drawback of Sn impurities in GaAs is the tendency to strongly

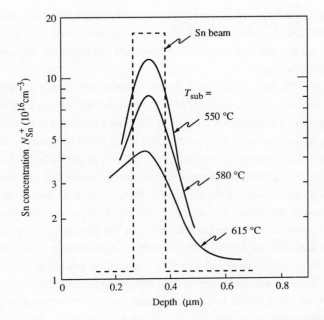

Fig. 5.21. Measured doping profiles of Sn-doped GaAs grown by MBE at different substrate temperatures. The dashed line is the ideal doping profile (without any dopant redistribution) calculated from the Sn beam intensity (after Cho, 1975).

redistribute during growth. Diffusion as well as surface segregation contribute to the redistribution, which was first discovered by Cho (1975). Cho analyzed Sn doping profiles in GaAs grown by MBE at different substrate temperatures. The measured Sn profiles are shown in Fig. 5.21 along with the calculated Sn flux profile indicated by the dashed line. The calculated Sn profile was obtained from the Sn beam intensity. It was assumed that all Sn impurity atoms impinging on the growing surface would incorporate into the crystal, i.e. a unity sticking coefficient was assumed.

The experimental result of Fig. 5.21 shows that Sn redistribution occurs for MBE growth temperatures of 550–615 °C. The impurity profile obtained from the 550 °C growth exhibits symmetric broadening towards the substrate side as well as towards the epitaxial surface. The symmetric broadening indicates that diffusion is the dominant redistribution mechanism. The impurity profile obtained from the 615 °C growth is clearly skewed towards the surface, i.e. the slope of the trailing edge of the profile is much steeper than the slope of the leading edge indicating surface segregation of Sn impurities. Further experiments of Cho (1975) with multiple Sn doping pulses demonstrated that Sn surface segregation (rather than diffusion) is the dominant redistribution mechanism at high growth temperatures.

Ploog and Fischer (1978) used Auger electron spectroscopy (AES) and secondary ion mass spectrometry (SIMS) to study Sn surface segregation. The study revealed that a significant amount of Sn, about three orders of magnitude larger than the portion being incorporated into the growing film, was detected on the surface and the outermost atomic layers (< 20 Å) of the epitaxial film. The segregation of Sn occurred at substrate temperatures as low as 490 °C. Ploog and Fischer (1978) also found that the segregation increases with Sn concentration.

Even though the strong redistribution rules out Sn as a useful donor for epitaxially grown III–V semiconductors devices, Sn doping of GaAs provides insight into and understanding of impurity surface segregation in semiconductors. Wood and Joyce (1978) proposed a surface rate limiting process in which the Sn is incorporated from a surface layer, not directly from the molecular beam. Harris et al. (1984) modeled the Sn surface segregation quantitatively assuming a lower energy of Sn atoms on the surface as compared to the bulk. At high growth temperatures, the Sn atoms segregate to the low-energy surface. Thermal equilibrium distribution of Sn is achieved for temperatures exceeding 600 °C. Growth at lower temperatures imposes a kinetic limitation on the Sn movement, and as a consequence, Sn atoms are trapped in the bulk. A further discussion of surface segregation of impurities in semiconductors is found in Chap. 8.

Lead, Pb

Lead has a large covalent atomic radius and is not expected to be very soluble in III–V semiconductors. There is only one report of Pb doping in GaAs (Wood, 1985a and 1985b). Lead doping of GaAs did not result in an increase of the free carrier concentration.

5.3.4 Group-VI dopants

The group-VI impurities include O, S, Se, Te, and Po and are called the chalcogens. Polonium has a comparatively large atomic radius and is therefore not expected to incorporate easily in typical III–V semiconductors. Oxygen on the other hand forms deep donor levels in GaAs. Incorporation of oxygen into any III–V semiconductors should therefore be avoided. The remaining useful group-VI impurities are S, Se, and Te. They occupy anion sites in III–V semiconductors and are shallow donors. Group-VI donors are not amphoteric. However, group-VI impurities can be electrically inactive in III–V semiconductors. All group-VI impurities have comparatively high vapor pressures. The vapor pressure decreases with increasing atomic number. That is, S has the highest and Te the lowest vapor pressure of the three elements S, Se, and Te. As a consequence of the high vapor pressure, group-VI impurities tend to reevaporate from the epitaxial surface. Furthermore, the incorporation of chalcogens depends strongly on the growth temperature.

Oxygen, O

Oxygen forms deep donor levels in III–V semiconductors. In GaAs two donor levels 400 meV and 670 meV below the conduction band edge were reported (see Chap. 9). Oxygen has a strong affinity to Al-containing compounds such as $Al_x Ga_{1-x} As$ and $Al_x In_{1-x} As$. It is desirable to avoid any oxygen incorporation.

Sulfur, S

Summary: Sulfur is a shallow donor in III–V semiconductors. The impurity has a high vapor pressure and the incorporation depends strongly on the epitaxial growth temperature.

Sulfur occupies anion sites in III–V semiconductors and is a shallow donor with 6 meV ionization energy in GaAs. It is used frequently for III–V semiconductors in which Si is a strongly amphoteric donor, e.g. in GaSb. Poole et al. (1988) studied S doping of GaSb and demonstrated that incorporated S in epitaxial GaSb has near 100% electrical activity. Free carrier concentrations exceeding 10^{18} cm^{-3} were achieved.

Sulfur has a high vapor pressure even at relatively low temperatures. This characteristic imposes serious restrictions on the maximum bake-out temperatures of MBE systems. Typical bake-out temperatures are in the range of 150 °C to 250 °C; at such temperatures S evaporates and would deplete an elemental S effusion cell during bake-out. In order to avoid this depletion, Davies et al. (1981) developed an electrochemical $Ag_2 S$ cell located within a conventional Knudsen cell. The solid-state electrochemical cell contained $Pt/Ag/AgI/Ag_2 S/Pt$ where the two Pt metallization layers serve as electrodes. Upon application of a bias to the electrochemical cell, elemental sulfur is generated which is then evaporated in the heated effusion cell. This method of S generation is compatible with the ultra-high vacuum of an MBE system; the cell is bakeable to approximately 350 °C without depletion of S from the cell. Davies et al. (1981) demonstrated low $(5 \times 10^{15}$ $cm^{-3})$ as well as high $(2 \times 10^{18}$ $cm^{-3})$ fully active S donor concentrations in GaAs.

Other compounds used for S doping of GaAs during MBE include lead sulfide, PbS. Wood (1978) showed that, upon evaporation of PbS from an effusion cell, S incorporates into the GaAs epitaxial layer while Pb does not, probably due to its large atomic radius. Doping levels ranging from 10^{16} cm^{-3} to 5×10^{18} cm^{-3} were demonstrated with the PbS cell. The major drawback of the PbS cell is its high vapor pressure which can result in a depletion of the effusion cell even during bake-out at 200 °C.

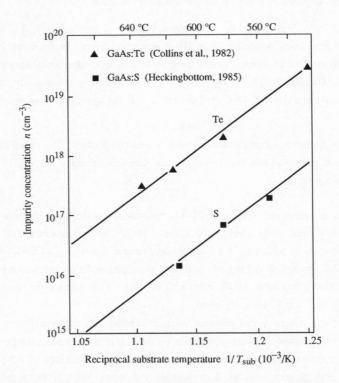

Fig. 5.22. Free electron concentration of S- and Te-doped GaAs grown by MBE as a function of reciprocal substrate temperature. The sulfur and tellurium doping impurities are obtained from S$_2$ and SnTe beams, respectively (after Collins et al., 1982 and Heckingbottom, 1985).

The high vapor pressure of S makes the incorporation of S strongly dependent upon the epitaxial growth temperature. Heckingbottom (1985) showed that the concentration of incorporated S in GaAs varies by one order of magnitude in a growth temperature interval of approximately 50 °C. The experimental carrier concentration of S-doped GaAs is shown in Fig. 5.22 as a function of substrate temperature. Heckingbottom attributed the reduction of S doping concentration at elevated substrate temperatures to the reevaporation of S from the growing surface. He also suggested that gallium sulfide is formed at higher growth temperatures according to $2Ga + \frac{1}{2}S_2 \leftrightarrow Ga_2S$ where Ga_2S is a gaseous volatile compound.

The high S vapor pressure and the strong dependence of the S doping concentration on the substrate temperature make the accurate control of S doping difficult. These characteristics make S clearly inferior to Si in GaAs. For other materials, such as GaSb in which Si is strongly amphoteric, S seems to be the better choice.

Selenium, Se

Summary: Selenium is a high vapor pressure shallow donor in III–V semiconductors. Se has high electrical activity with no tendency to compensate.

Selenium occupies anion lattice sites in III–V semiconductors and forms shallow, hydrogenic donors. It has an ionization energy of 6 meV below the conduction band edge in GaAs. Selenium and other chalcogens have a high vapor pressure which makes them difficult to use in an ultra-high vacuum environment. Effusion cells with elemental Se baked at temperatures of 250 °C would release considerable amounts of Se. Therefore elemental Se effusion cells are not used in MBE growth systems. Selenides such as lead selenide (PbSe) have much lower vapor pressures as compared to elemental Se (Hirama, 1964). Wood (1978) used PbSe for n-type doping in GaAs using a conventional effusion cell. The author showed that only Se but not Pb is incorporated during MBE growth and that the Pb does not interfere with the electrical and structural properties of GaAs grown by MBE at substrate temperatures of 480 °C. Even the coverage of GaAs with multiple atomic layers of lead did not result in incorporation of Pb. The chemical incorporation reaction proposed is given by $PbSe + V_{As} \leftrightarrow Pb\uparrow + Se_{As}$ where V_{As} and $Pb\uparrow$ are an As vacancy and a volatile lead atom, respectively.

The electrical characteristics of Se-doped GaAs revealed no evidence of electrical neutrality or autocompensation of the Se impurities (Wood, 1978). Instead, the Se impurities were assumed to be highly efficient. Electrically active Se doping levels as high as 3×10^{18} cm^{-3} were demonstrated. The growth of multiple doping pulses did not reveal evidence of excessive impurity redistribution during growth at 490 °C. However, an increase of the growth temperature to 560 °C resulted in drastically decreased free carrier concentrations. The decrease in free carrier concentration was attributed to reevaporation of PbSe from the GaAs surface.

It appears that the major drawback of Se as a doping impurity is its high vapor pressure which makes its handling in a ultra-high vacuum environment difficult. Furthermore, the strong dependence of the free carrier concentration on the substrate temperature makes control of the growth parameter complicated. Se doping is therefore used only if Si doping is not feasible. In compounds such as GaAsSb or GaSb, Si is strongly amphoteric. In these compounds, Se doping (along with other chalcogens) is a viable alternative to Si doping.

Tellurium, Te

Summary: Tellurium is a non–hydrogenic, shallow donor in III-V semiconductors. Te has both high electrical activity and a high vapor pressure and its incorporation depends strongly on growth conditions especially the growth temperature.

Tellurium occupies anion sites in III–V semiconductors with an ionization energy of 30 meV in GaAs. The ionization energy is much larger than expected for hydrogen-like donors in GaAs which have typical ionization energies of 5 meV. The larger ionization energy is probably due to a central cell correction of the big Te atom in the GaAs lattice.

Tellurium has not been evaporated from effusion cells containing elemental Te. Elemental Te is avoided in the ultra-high vacuum environment of an MBE system due to its high vapor pressure. Instead, Te is evaporated from compounds that have a lower vapor pressure such as SnTe (Collins et al., 1982), PbTe (Subbanna et al., 1988) and $Sb_2 Te_3$ (Chiu et al., 1990). Tellurium doping was done predominantly in Sb containing III–V compounds such as GaSb and $GaAs_{1-x}Sb_x$ due to the strong amphoteric nature of Si in these compounds (Chang et al., 1977; Yano et al., 1978).

At doping concentrations $N_{Te} \ll 10^{18}$ cm^{-3} Te is highly active. Collins et al. (1982) and Chiu et al. (1990) showed that the free carrier concentration follows the Te concentration for $N_{Te} \ll 10^{18}$ cm^{-3}. However at a high doping concentration of $N_{Te} \geq 10^{18}$ cm^{-3} the electrical activity of Te is reduced. At a Te concentration of 5×10^{19} cm^{-3} in GaAs the free carrier concentration was only 4×10^{18} cm^{-3} indicating that less than 10% of the Te atoms were electrically active.

The high vapor pressure of Te makes the incorporation of Te strongly dependent on the growth conditions. The concentration of Te in GaAs is shown in Fig. 5.22 as a function of the growth temperature (Collins et al., 1982). High Te concentrations exceeding 10^{19} cm^{-3} are achieved at the low growth temperature of 530 °C. As the growth temperature is increased to 630 °C the Te concentration drops by two orders of magnitude. A strong temperature dependence of the Te incorporation was also observed in GaSb by Chiu et al. (1990). However, the authors found that Te incorporation is insensitive to the growth temperature for $T < 540$ °C. Efficient incorporation of Te occurred for temperatures < 540 °C.

Polonium, Po

Polonium is an impurity with a large atomic radius. It is not expected to be very soluble in III–V semiconductors.

6

Gaseous doping sources

Elemental impurities cannot be used as doping sources in vapor-phase epitaxial growth systems because such impurities would assume the ambient temperature in the reactor and condense on the reactor walls. Vapor-phase epitaxial (VPE) growth techniques therefore require gaseous doping compounds, i.e. chemical gases transporting the doping impurity. Such gaseous doping compounds are called *doping precursors* and are mandatory for growth techniques with a viscous gas flow such as VPE. Growth techniques with molecular flow can employ either elemental doping sources or doping precursors. Chemical-beam epitaxy (CBE) frequently employs gaseous doping compounds. The use of precursors in CBE is motivated by parasitic chemical reactions between the organometallic sources used for group-III elements and elemental doping sources which result in a degradation in elemental source purity (see Sect. 6.1). Thus the two techniques, VPE and CBE, use doping precursors for different reasons.

The use of non-elemental chemical precursors allows a wide selection of chemicals. The most obvious requirement for these chemicals are (i) a sufficient vapor pressure at room temperature (ii) thermal decomposition of the chemical at the growth temperature, and (iii) no parasitic chemical reactions before and after thermal decomposition. These requirements can be satisfied by either *inorganic* precursors, e.g. silane or disilane, or *organometallic* precursors, e.g. dimethylzinc (DMZn).

Dopant incorporation from precursors can be extremely complicated and is controlled by a multitude of parameters. Chemistry, transport by convection and diffusion, thermodynamics, and fluid dynamics govern the incorporation of doping impurities. Even though considerable progress has been made in the understanding of doping incorporation, Razeghi (1989) concluded that the exact chemical decomposition pathways in OMVPE are not yet clearly understood.

It is useful to keep in mind the general chemical characteristics of the doping precursors containing alkyl groups. Alkyls are hydrocarbons with varying length of the carbon chain. The most frequently used alkyls are methane, ethane, and propane, i.e. CH_4, C_2H_6, and C_3H_8, respectively. The first general characteristic of alkyls is that their stability decreases with the length of the carbon chain. For example, ethane is more stable against thermally activated dissociation than propane. The second general characteristic is that the vapour pressure decreases with increasing length of the carbon chain. Methane is thus the alkyl with the highest vapor pressure. This characteristic also holds for organometallic compounds. For example, trimethylgallium (TMGa) has a higher vapor pressure than triethylgallium (TEGa). The same is valid for other organometallic compounds such as TMAl/TEAl or TMIn/TEIn. Furthermore, the two characteristics mentioned above are valid not only for carbon hydrides (alkyls) but also for silicon hydrides. That is, disilane (Si_2H_6) has a lower vapor pressure than silane (SiH_4).

The vapor pressure of the group-II, -IV, and -VI doping elements also follows some general tendencies. Among the group-II elements, Be, Mg, Zn, and Cd are most important. Within this group the vapor pressure increases with atomic weight, i.e. the vapor pressure is lowest for Be and highest for Cd (see, for example, Shapira and Lichtman, 1979). Reevaporation of doping impurities from a growing semiconductor surface is strongest for Cd. The group-IV impurities C, Si, Ge, and Sn share the same trend for the vapor pressure. At a given temperature, Sn has the largest vapor pressure and C the smallest vapor pressure of the group-IV elements. Finally, the group-VI elements used for doping are S, Se, and Te. Among these elements, the vapor pressure at a given temperature decreases with atomic weight. Thus, S has the highest vapor pressure and has the highest rate of reevaporation from a growing crystal surface.

In this chapter the use of gaseous doping compounds for epitaxial III–V semiconductor growth will be discussed. After the analysis of the general requirements for gaseous precursors, the doping homogeneity of OMVPE will be assessed. Finally, the properties of different doping compounds for p- and n-type doping of III–V semiconductors will be summarized.

6.1 General considerations

Doping impurities in vapor-phase epitaxially (VPE) grown materials are introduced by means of gaseous chemical compounds, the doping *precursors*. The use of chemical compounds rather than doping elements makes possible a wide selection of doping precursors. Among the most frequently used doping precursors are the *hydrides* (e.g. SiH_4, H_2S) and *organometallic* compounds (e.g. diethylzinc, DEZn; tetramethyltin, TMSn).

There is a multitude of requirements for chemical doping precursors. Ideal doping precursors (i) do not react in the gas phase of the VPE system before reaching the substrate, (ii) fully decompose at or near the growing crystal surface, (iii) incorporate into the crystal with unity probability, (iv) are electrically active in the semiconductor, and (v) do not exhibit the memory effect. A sufficiently *high vapor pressure* is required at or near room temperature for all doping precursors. The typical mole fraction of the doping compound in the gas phase is $10^{-10} - 10^{-1}$. In order to achieve such mole fractions, a high vapor pressure is advantageous. However, the transport of the doping precursor can be enhanced, if necessary. The first possibility is to heat the precursor source above room temperature. Excessive heating ($\gg 50 \degree C$) should be avoided since condensation of the precursor may occur in the gas lines or at the reactor walls (in cold wall reactors) before reaching the epitaxial crystal surface. A second possibility to increase the flow of the doping precursor is the use of a carrier gas bubbler (see Sect. 4.3).

It is desirable that dopants not exhibit any *memory effect* (Sect. 4.4) which complicates the growth of abrupt doping profiles. A strong memory effect has been observed for the doping precursors hydrogen selenide (H_2Se) and diethyltellurium (DETe) by Lewis et al. (1984) and Houng and Low (1986), respectively. The authors found an unexpected high background concentration after doping GaAs with DETe. Houng and Low (1986) found that adsorption of DETe onto the cold stainless steel tube walls gives rise to the doping memory effect. The authors showed that a reduction of the doping line surface area and heating of the doping lines to $60 \degree C$ reduces the memory effect. Previously, Lewis et al. (1984) demonstrated that H_2Se exhibits memory effects in OMVPE-grown GaAs. As a consequence, graded doping profiles were obtained. The memory effect was attributed to adsorption of H_2Se onto the internal walls of the system and subsequent desorption after the doping source was shut off.

Complete *thermal decomposition* is a desirable property of doping precursors which should occur at all growth temperatures, i.e. in the range $600-800 \degree C$ for the growth of GaAs. To illustrate the requirement of pyrolysis, consider the following examples. First, methane (CH_4) could be considered as a carbon doping source. However, methane is a very stable molecule. It is not thermally decomposed in the entire temperature range of interest. Therefore, methane cannot be used as a doping precursor in OMVPE. Second, silane (SiH_4) is a fairly stable molecule. Only a fraction of SiH_4 molecules are decomposed at typical growth temperatures of $600-800 \degree C$ (Kuech et al., 1984). Since the thermal decomposition efficiency depends exponentially on temperature, the incorporation of Si from a SiH_4 source is strongly growth temperature dependent. Such a strong temperature dependence is undesirable, since it does not allow the two parameters, i.e. doping and growth temperature, to be adjusted independently. Third, disilane (Si_2H_6) is a less stable molecule which decomposes into Si, H_2, and SiH_4 at the temperature range of interest ≤ 600 °C. The incorporation of Si from a disilane source is therefore approximately independent of growth temperature (Kuech et al., 1984). The above examples illustrate that complete pyrolysis is a desirable property for dopants. Other decomposition paths such as chemical reactions with other precursors or catalytic reaction are undesirable since they again lead to the interdependence of growth parameters. Such an interdependence should be avoided as much as possible to keep the growth and doping process well controlled.

Non-toxicity and non-flammability are desirable but not necessary conditions for doping precursors in OMVPE. In fact, many precursors are used despite their toxicity and flammability. Appropriate precautions, which are recommended by the manufacturers, are required when handling such doping precursors.

Doping precursors are required to not react in the gas phase of the growth system but only at the semiconductor surface. Premature, parasitic gas-phase reactions can alter the reaction pathways of doping precursors and complicate or even make impossible the detailed quantitative understanding of the doping process. It is finally desirable that doping incorporation is independent of reactor pressure and V/III flux ratio.

Doping precursors have also been used in chemical-beam epitaxy (CBE) which uses, as OMVPE, organometallics and hydrides for group-III and group-V elements, respectively. Elemental dopants were widely used in the 1980s for CBE growth (see e.g. Tsang, 1989). However, Skevington et al. (1990) found anomalous elemented source Si and Sn doping behaviour in InP grown by CBE. The authors used elemental source effusion cells for doping and found that the doping level decreases from growth run to growth run. The experimental result is depicted in Fig. 6.1 which shows the doping concentration of four consecutive growth runs of Si-doped InP. The effusion cell temperature, i.e. the nominal doping concentration during the four growth runs remained constant. Nevertheless, a clear decrease of the measured doping concentration from 5×10^{16} cm^{-3} to 2.8×10^{16} cm^{-3} is observed. The authors pointed out that the long-term aging process of the effusion cell caused major reproducibility problems, with the cell temperature requiring frequent adjustment to maintain a given doping level. Eventually, the required cell temperature became unacceptably high. When the effusion cell was removed, the Si charge was found

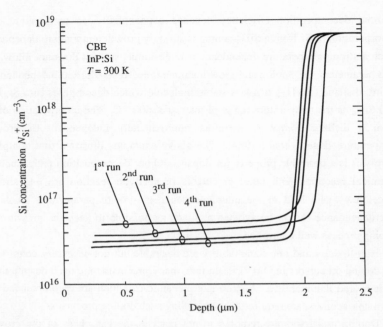

Fig. 6.1. Si doping profile in CBE-grown InP doped with an elemental Si effusion cell. The carrier concentration decreases monotonically in four consecutive growth runs (after Skevington et al., 1990).

to be covered by black and grey deposits. Skevington et al. (1990) also found reproducibility problems with Sn-doped InP. The Sn background concentrations began to vary, apparently unpredictably, up to an electron concentration in the 10^{18} cm^{-3} range. Even at low Sn effusion cell temperatures of 400 °C, the authors found a Sn background doping level in the 10^{17} cm^{-3} range. The Sn background doping level increased with the duration of the Sn effusion cell being heated to high temperatures.

The difficulties of elemental Si and Sn doping during CBE were attributed to chemical reactions between the organometallic group-III compounds with the charges of the heated doping effusion cells. Trimethylindium (TMIn) was used as the source material for cations. The authors proposed that silicon carbide (SiC) is formed as a result of the interaction of TMIn or alkyl radicals with the heated Si charge. Black carbon deposits were also found in the effusion cell. Furthermore, the authors proposed that an organometallic Sn compound was formed. Most likely is the formation of a methylated Sn compound (e.g. monomethyltin, dimethyltin), which is more volatile than Sn and thus can be incorporated into the epitaxial layer. The formation of tetramethyltin is ruled out on the basis of its small incorporation probability. Thus, the study of Skevington et al. demonstrates the necessity of a non-elemental doping source for CBE.

It is likely that the organometallic compounds also react with p-type impurities. Berylliumcarbide (Be_2C) is probably formed as a result of a chemical reaction between group-III organometallics and a Be cell heated to operating temperatures. Organometallic compounds tend to coat the ceramic and metallic parts of elemental effusion cells. Chemical reactions and short-circuits can result.

6.2 Doping homogeneity and uniformity

The homogeneity of epitaxial layers grown by hydride VPE or organometallic VPE can be categorized into three groups which are (i) doping uniformity, (ii) thickness uniformity, and (iii) composition uniformity. These three types of uniformities are interdependent. For example, thickness non-uniformity across a wafer automatically translates into doping non-uniformity. The three-dimensional doping concentration is inversely proportional to the epitaxial layer thickness if the doping incorporation rate is perfectly uniform across the epitaxial layer surface. Thus, a non-uniform doping distribution can occur due to layer thickness variations despite perfectly uniform dopant incorporation.

Composition non-uniformity is known as a changing composition of a ternary or quaternary alloy semiconductor across an epitaxial layer. Composition uniformity should not be confused with the stoichiometry of an epitaxial layer. Composition non-uniformities in hydride VPE and organometallic VPE system are usually small. The constituent precursor compounds of a ternary or quaternary semiconductor alloy are premixed, i.e. mixed before entering the reaction chamber. The premixing process usually provides near perfect mixing and negligible inhomogeneity of precursors in the gas phase.

Composition inhomogeneity can result in doping inhomogeneities as well. Consider a doping impurity whose incorporation strongly depends on the alloy composition. Such a strong dependence is observed for carbon which is amphoteric in $Ga_x In_{1-x} As$. Carbon occupies predominantly acceptor and donor sites in GaAs and InAs, respectively. Thus, a change in the alloy composition can result in doping inhomogeneity.

Finally, the doping non-uniformities can be a result of non-uniform doping incorporation. The physical origin for non-uniform doping incorporation can be threefold. First, the gas flow of the doping precursor can be non-uniform or even turbulent. Second, the doping gas can deplete along the gas flow direction. Such a depletion results in a smaller doping concentration at the exit side on the wafer as compared to the gas entry side. Third, doping non-uniformities can be the result of a variation of the temperature at the growing surface. The incorporation of some doping precursors is strongly temperature dependent. For example, the Si doping concentration is known to be strongly temperature dependent if derived from silane. Thus, a temperature variation across the wafer can result in non-uniform doping distribution. Furthermore, if the temperature of the growing crystal changes with time, a change in doping concentration results. Temperature and, in general, process stability are therefore essential to achieve homogeneous doping.

Low-pressure VPE reactors improve doping, thickness, and compositional homogeneity of epitaxial crystals. Low-pressure VPE reactors operate in a pressure range < 1000 mbar (< 750 Torr), with typical values of approximately 10-100 mbar (7.5-75 Torr). The gas velocity is increased in low pressure reactors which leads to a more homogeneous and laminar gas flow. Nevertheless, depletion effects are not eliminated in low-pressure VPE reactors i.e. depletion effects persist as a cause for non-uniform doping.

Doping homogeneity and uniformity were studied on an Aixtron low-pressure OMVPE system with a horizontal reaction chamber by Zilko and Schubert (1991). The reaction chamber can accommodate two 5 cm or one 7.5 cm wafer. Two 5 cm wafers were used for the homogeneity study, one wafer being upstream (near the gas entry port of the reactor) and one downstream (near the exhaust of the reactor). The results obtained from the upstream wafer were found to have better uniformity. The following results are from the upstream wafer. The substrates were not rotated during growth. The growth temperature was 640 °C, the growth rate was 2.3 μm/hr, and the V/III ratio was 94. Silane (SiH_4) diluted to 1% in hydrogen was used as the Si-doping precursor. The growth experiment was conducted at a pressure of 20 mbar (15 Torr) which has proven favorable for uniformity purposes. The epitaxial GaAs layers have a nominal doping concentration of 1.5×10^{17} cm^{-3}. Capacitance–voltage measurements were carried out at 300 K on 500 μm diameter Ti/Au Schottky contacts to determine the doping concentration.

The result of the lateral uniformity on a 5 cm Si-doped GaAs epitaxial upstream wafer is shown in Fig. 6.2. One series of measurements was carried out along the direction of the gas flow. A second series of measurements was carried out normal to the gas flow direction. The average free electron concentration is 1.15×10^{17} cm^{-3}. The highest and lowest concentrations measured on the wafer are 1.37×10^{17} cm^{-3} and 1.02×10^{17} cm^{-3}, respectively. The standard deviation of the doping concentration is 10% and is calculated from 100

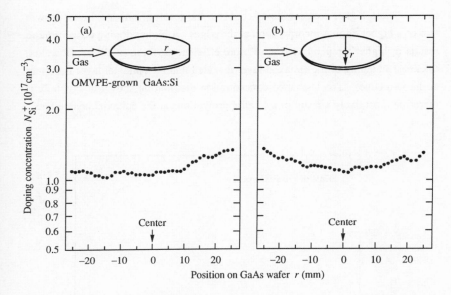

Fig. 6.2. Doping concentration of low-pressure OMVPE-grown n-type GaAs:Si along the diameter of a 5 cm wafer. The measurement points are (a) along the gas flow direction and (b) perpendicular to the gas flow direction.

measurement points.

The doping concentration measured along the gas flow direction has a clear upward trend near the exit side of the gas stream (see Fig. 6.2a). The increase is of a systematic nature which prevails from the center to the edge of the wafer. The increase is most likely due to depletion of the group-III precursor. The depletion of the group-III precursor results in a smaller growth rate near the downstream side of the wafer, since the group-III precursor determines the growth rate. Assuming a constant doping incorporation along the gas flow direction, the doping concentration increases due to the reduced thickness of the epitaxial layer. This interpretation is consistent with a thickness mapping of the epitaxial layer which shows a decrease of epitaxial layer thickness near the downstream side of the wafer. Note that a doping precursor depletion effect cannot explain the increase of the doping concentration. Doping precursor depletion is unlikely due to the small distribution coefficient of SiH_4, which is typically $\ll 10^{-2}$.

The doping concentration perpendicular to the gas stream is U-shaped. Higher doping concentrations are achieved at the edges than at the center of the wafer. These higher concentrations could be the result of either a lower growth rate or a higher doping incorporation at the edges. A higher doping concentration results at higher temperature due to more efficient cracking of SiH_4. However, thickness mapping of the epitaxial layer revealed that the epitaxial layer is thinner at the two edges. The U-shaped concentration distribution shown in Fig. 6.2b is therefore most likely a result of a smaller growth rate at the epitaxial layer edges.

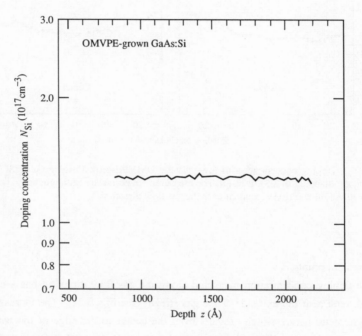

Fig. 6.3. Doping concentration of low-pressure OMVPE-grown n-type GaAs:Si versus depth.

The doping concentration of the OMVPE-grown GaAs is shown in Fig. 6.3 as a function of depth. Capacitance-voltage measurements were employed to measure the net active Si donor concentration ($N_{Si}^+ - N_{Si}^-$) as a function of depth. Figure 6.3 reveals no systematic dependence of the doping concentration on depth. The average doping concentration is 1.29×10^{17} cm^{-3}. The variation of the doping concentration with depth is $\leq 1\%$ for all locations. The constant doping concentration with depth indicates excellent long-term stability and process control and no transient effects with long time constants.

Several reports in the literature indicate a doping uniformity similar to the results discussed above. Gersten et al. (1986) reported on (i) thickness, (ii) doping, and (iii) compositional uniformity of a large-scale OMVPE multiwafer "chimney" reactor with 90 wafers (4.6 cm × 7.6 cm). The authors found a doping uniformity of ±10% which was equal to the thickness uniformity. An Al-compositional uniformity in $Al_x Ga_{1-x} As$ of ±3% was achieved. The authors emphasize the importance of accurate growth temperature control which improved the doping and thickness uniformity.

Hayafuji et al. (1986) reported on highly uniform growth of GaAs and $Al_x Ga_{1-x} As$ in a multiwafer OMVPE reactor (Cambridge Instruments MR-200). They demonstrated a variation of growth rate $\leq 5\%$ along the gas flow direction for two 5 cm diameter wafers under optimized carrier gas (H_2) flow conditions for GaAs and $Al_x Ga_{1-x} As$ growth. The variation among twenty 5 cm diameter wafers was $\leq 8\%$. The variation of Al mole fraction in $Al_x Ga_{1-x} As$ was $\leq 10\%$ along the gas flow direction and among ten 5 cm diameter wafers.

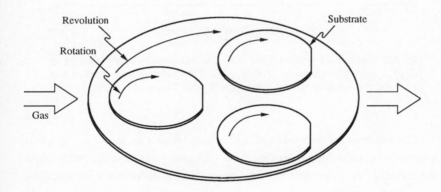

Fig. 6.4. Planetary motion including rotation and revolution of three substrates in a multiwafer OMVPE reactor. Planetary motion provides improved doping, thickness, and composition uniformity.

Excellent uniformity results were reported by Tanaka et al. (1987) using a multiwafer OMVPE system with 'planetary motion' of the wafers. The growth system was of the horizontal reactor type and can accommodate three 5 cm substrates. The planetary motion of the three substrates is illustrated in Fig. 6.4 as used by Tanaka et al. (1987). Each of the three substrates rotates around its own symmetry axis. Furthermore, a disk-shaped substrate holder rotates as well, subjecting the three wafers to a revolutionary motion. The disk-shaped substrate holder in turn is mounted on the graphite susceptor. The growth of Si-doped $Al_x Ga_{1-x} As$ was achieved by means of the precursors TMGa and TMAl for the cations, AsH_3 for the anions, and $Si_2 H_6$ for Si doping. The $Al_{0.28} Ga_{0.72} As$ growth rate was 3.5 Å/s and the V/III ratio was 70.

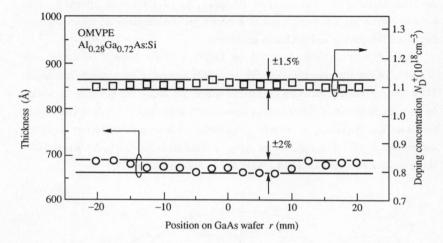

Fig. 6.5. Variation of epitaxial layer thickness and doping concentration of an $Al_{0.28}Ga_{0.72}As$ film grown by OMVPE. The 5 cm substrate was subjected to rotary and revolutionary motion during growth (after Tanaka et al., 1987).

The uniformity of thickness and doping concentration are shown in Fig. 6.5 as a function of position on one of the 5 cm substrates (Tanaka et al., 1987). Both the epitaxial layer thickness and the doping concentration were measured by capacitance–voltage measurements. The uniformity of the thickness and the carrier concentration of the n-type $Al_{0.28} Ga_{0.72} As$ film is better than ±2.0% and ±1.5%, respectively. The uniformity data is thus comparable to wafers grown by molecular-beam epitaxy. The variation of the thickness and the doping concentration among the three epitaxial layers grown simultaneously in the reactor is ±1.0%. The results demonstrate that impressive uniformities can be achieved in low-pressure OMVPE systems with planetary rotation and revolution of the substrates.

A theoretical model of growth rates as a function of the position on a wafer grown by OMVPE was developed by van de Ven et al. (1986) and Giling and van de Ven (1987). The authors considered a horizontal OMVPE reactor with a rectangular cross section and a laminar gas flow. They found that the growth rate decreases along the gas flow axis. The rate of decrease was well described by an $x^{1/3}$ power law in the laminar gas flow domain where x is the position on the wafer along the gas flow direction. Depletion of the gas phase was identified as the reason for the reduced growth rate. Note that if the depletion of the gas phase were indeed the only reason for non-uniformities, planetary motion of the substrate could greatly reduce non-uniformities of doping, thickness, and composition.

6.3 Gaseous doping precursors

This section covers gaseous doping precursors used during epitaxial growth of III–V semiconductors. The doping precursors are classified according to their column in the periodic table, i.e. Sect. 6.3.1 Group-IIa dopants, Sect. 6.3.2 Group-IIb dopants, Sect. 6.3.3 Group-IV dopants, and Sect. 6.3.4 Group-VI dopants. These sub-sections emphasize the chemical characteristics of the doping precursors rather than the property of the resulting doping impurities in the semiconductor which are covered in Chap. 5.

Gaseous doping precursors are used during epitaxial growth with organometallic and hydride vapor-phase epitaxy (OMVPE and VPE) and chemical-beam epitaxy (CBE). In principle, both growth techniques can use the same doping precursors. However, the CBE growth temperature is typically 100–200 °C lower than the VPE growth temperatures. The pyrolysis of gaseous doping precursors depends exponentially on the substrate temperature. The dopant incorporation is therefore less efficient in CBE owing to the lower growth temperatures. In other words, thermally less stable doping precursors are desirable for CBE growth as compared to VPE growth. Finally, for a general discussion of impurity characteristics (saturation, autocompensation, etc.) the reader is referred to Chaps. 7 and 8.

6.3.1 Group-IIa precursors

Among the group-IIa elements, Be and Mg precursors are most popular.

Be precursors

Diethylberyllium, (DEBe, C_2H_5Be) has been used as a Be precursor for OMVPE (Bottka et al., 1984) as well as for CBE (Weyers et al., 1990). DEBe is a well-behaved precursor with desirable incorporation characteristics. The Be incorporation into GaAs using DEBe is insensitive to growth temperature and V/III flux ratio (Parsons et al., 1986). The characteristics indicate that DEBe decomposes at growth temperatures, does not reevaporate from the growing crystal surface, and fully incorporates into the growing layer. Doping concentrations ranging between 10^{15} cm^{-3} and 10^{20} cm^{-3} were achieved with good surface morphology. The hole concentration in OMVPE-grown $Al_{0.12}Ga_{0.88}As$ is shown in Fig. 6.6 as a function of the DEBe mole fraction in the gas phase (Bottka et al., 1984). The V/III flux ratio during growth was six. DEBe was found to decompose readily and Be was concluded to occupy acceptor sites. The authors confirmed the efficient decomposition and incorporation of DEBe by a comparison with silane which required a ten times larger mole fraction in the gas phase to yield the same carrier density. For a DEBe mole concentration of 5×10^{-8}, the authors measured a net hole concentration in the mid 10^{18} cm^{-3} range, regardless of the V/III ratio used during the growth. The authors concluded that no competing chemical parasitic reactions occur simultaneously, which would hinder the incorporation of Be into the lattice.

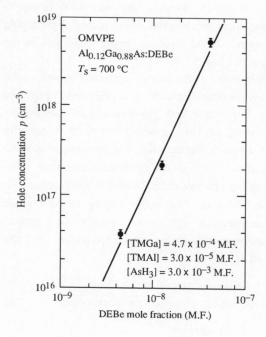

Fig. 6.6. Hole concentration in OMVPE-grown $Al_{0.12}Ga_{0.88}As$ as a function of DEBe mole fraction in the gas phase. A V/III flux ratio of six was used during growth (after Bottka et al., 1984).

The functional dependence of the hole concentration versus the DEBe mole fraction is shown in Fig. 6.6. The hole concentration depends superlinearly on the DEBe mole fraction. An increase of the DEBe mole fraction by a factor of 10 results in an increase of the free hole concentration by a factor of 100. Such a superlinear increase of the Be incorporation from DEBe was also found by Parsons et al. (1986). An increase of the hole concentration by a factor of 30 was found for a tenfold increase of the DEBe flow rate controlled by a H_2 bubbler.

The authors suggest that impurity compensation at low concentrations could explain the superlinear dependence. Dopant source gas manifolding was proposed as an alternative explanation.

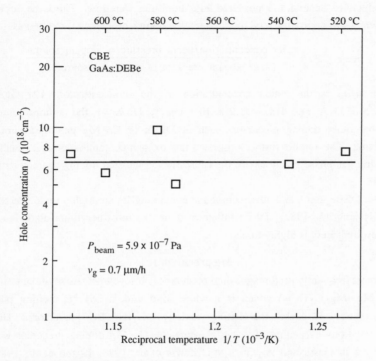

Fig. 6.7. Hole concentration (T = 300 K) of p-type GaAs grown by chemical-beam epitaxy using DEBe as a doping precursor as a function of reciprocal growth temperature (after Weyers et al., 1990).

The free hole concentration of Be-doped GaAs is shown in Fig. 6.7 as a function of the reciprocal growth temperature. The epitaxial GaAs was grown by chemical-beam epitaxy (Weyers et al., 1990) at a growth rate of 0.7 μm/hr using DEBe as the p-type dopant precursor. The hole concentration exhibits the desirable property of being independent of the growth temperature which makes DEBe a suitable precursor for CBE. The authors concluded that the doping efficiency of DEBe is close to unity, that DEBe is efficiently pyrolyzed on the surface, and that Be is subsequently incorporated into the lattice. The experiments were all performed at relatively low temperatures (520–620 °C) at which desorption of Be has not been observed. The authors concluded that these characteristics make DEBe a near ideal doping precursor for GaAs and $Al_x Ga_{1-x} As$.

Problems were encountered when DEBe was used as p-type precursor in InP during growth by CBE (Weyers et al., 1990). First, the doping level was not reproducible. Second, the measured hole mobilities were low. Third, the doping efficiency was approximately 10^{-3}. Here, the doping efficiency is defined as

$$\eta_{Doping} = \frac{\text{carrier concentration/beam pressure of dopant source}}{(4/a_0^3)/\text{beam pressure of group-III source}} \quad (6.1)$$

where $4/a_0^3$ is the cation concentration of the semiconductor. For GaAs, $a_0 = 5.6533$ Å, i.e. $4/a_0^3 \cong 2.2 \times 10^{22}$ cm^{-3}. However, the authors pointed out that other doping precursors such as DEZn or $Cp_2 Mg$ exhibited similar problems. The authors further suggested that oxygen contamination and possible parasitic gas-phase reactions are the cause for the unusual doping characteristics of InP.

Be, DEBe, BeO, and other compound containing Be are highly toxic (Parsons and Krajenbrink, 1983). DEBe inflames in air like most beryllium dialkyls and the resulting BeO is highly toxic.

Mg precursors

The most frequently used magnesium precursor is biscyclopentadienylmagnesium $(Cp_2 Mg, Mg(C_2 H_5)_2)$ which is a white solid with a 281 °C melting point temperature. It lacks the strong toxicity of the Be precursors. Hole concentrations exceeding 10^{19} cm^{-3} using the $Cp_2 Mg$ doping precursor were achieved in GaAs and $Al_x Ga_{1-x} As$ (Lewis et al., 1983; Kozen et al., 1986). Furthermore, Mg has not been reported to diffuse as strongly as e.g. Zn in the high concentration regime.

Another Mg precursor, bismethylcyclopentadienylmagnesium $(MCp_2 Mg, Mg(C_2 H_4 - CH_3)_2)$, was reported by Timmons et al. (1986). This compound is structurally similar to $Cp_2 Mg$ but has methyl groups connected to each carbon ring. The melting point of $MCp_2 Mg$ is 27 °C permitting a liquid source at moderate heating powers. Hole concentrations exceeding 10^{19} cm^{-3} were achieved in GaAs.

Doping experiments with $Cp_2 Mg$ revealed that the Mg concentration in GaAs increases monotonically up to substrate temperatures of approximately 580 °C (Weyers et al., 1990). At temperatures exceeding 600 °C an exponential decrease of the free hole concentration was observed. The decrease of the free carrier concentration was quantitatively modeled in terms of an Arrhenius characteristic with an activation energy of 5 eV. The reduction of the free hole concentration was attributed to reevaporation of Mg from the crystal surface. The same activation energy of 5 eV was found for elemental Mg doping in GaAs grown by molecular-beam epitaxy (Wood et al., 1982) confirming the hypothesis of Mg reevaporation. The sensitivity of Mg incorporation on the growth temperature is an undesired property which makes accurate process control mandatory.

The incorporation of Mg from MCp_2Mg is superlinear. Timmons et al. (1986) measured the free carrier concentration as a function of the H_2 carrier gas flow through the liquid MCp_2Mg doping source. For free carrier concentrations in the $10^{17} - 10^{19}$ cm^{-3} range, the concentration increased superlinearly with the carrier gas flow. The increase of the carrier gas flow by a factor of 3 resulted in a hole concentration increase by a factor of 10.

MCp_2Mg and Cp_2Mg were found to adsorb on the reactor walls of OMVPE systems (Timmons et al., 1986; Roberts et al., 1984). Such an interaction of the precursor with the reactor walls results in memory effects and makes abrupt doping profiles difficult to achieve. Timmons et al. (1986) alleviated this problem by injecting MCp_2Mg near the quartz reactor and by using a run/vent approach. Using this approach, the feasibility of top-hat doping profiles with constant maximum concentration and abrupt turn on and off were demonstrated.

6.3.2 Group-IIb precursors

Among the group-IIb elements, Zn precursors are the most frequently used. These are well studied and the studies illustrate the interdependence of growth and doping parameters in OMVPE. Several organometallic Cd precursors are used in OMVPE. In general, the Cd doping properties were found to be inferior to those of Zn.

Zn precursors

Zinc is the most popular p-type doping impurity in OMVPE-grown III–V semiconductors. Zinc precursors are well studied and much is known about the interdependence of Zn doping precursors and crystal growth parameters. Dimethylzinc (DMZn, $(CH_3)_2Zn$) and Diethylzinc (DEZn, $(C_2H_5)_2Zn$) are used as precursors (see for example, Bass and Oliver, 1977; Manasevit and Thorsen, 1972).

Thermal decomposition of DMZn and DEZn is believed to be close to unity at substrate temperatures as low as 500 °C. Weyers et al. (1990) estimated the doping efficiency (see Eq. (6.1)) of DEZn to be approximately 0.3 at a temperature of 520 °C. Glew (1984) assumed near-complete pyrolysis of DMZn and DEZn at temperatures $\geq$ 500 °C. Thus, effective thermal decomposition and effective incorporation of Zn can be obtained at low growth temperatures.

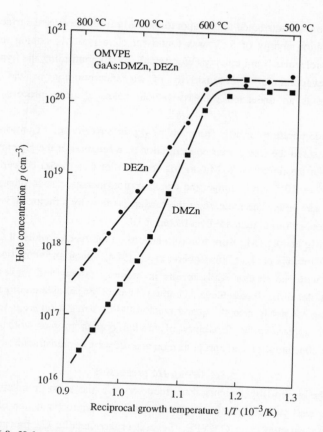

Fig. 6.8. Hole concentration of OMVPE-grown GaAs doped with DMZn and DEZn as a function of reciprocal growth temperature (after Glew, 1984).

For higher growth temperatures, the incorporation of Zn becomes strongly dependent on growth temperature. Figure 6.8 shows the hole concentration as a function of the growth temperature of OMVPE-grown GaAs (Glew, 1984). The hole concentration decreases rapidly with increasing substrate temperatures. Glew (1984) evaluated the activation energies associated with the slopes of the curves shown in Fig. 6.8. In the temperature range 575 °C to 675 °C, activation energies of 4.0 eV and 3.3 eV were obtained for DMZn and DEZn, respectively. In the temperature range 675 °C to 800 °C, activation energies of 2.7 eV and 2.1 eV were obtained for DMZn and DEZn, respectively. Similar values were reported by Bass and Oliver (1977).

The rapid decrease of the free hole concentration in Zn-doped GaAs grown at elevated temperatures can be attributed to Zn reevaporation from the crystal surface. Stringfellow (1989) pointed out that the hole concentration shown in Fig. 6.8 is inversely proportional to the vapor pressure of Zn, which increases by several orders of magnitude between 600 °C and 800 °C. Zinc doping is thus limited by desorption of Zn from the growing crystal surface.

Zinc incorporation also depends sensitively on the V/III flux ratio. Glew (1984) and Bass and Oliver (1977) reported the dependence of the free carrier concentration on the V/III flux ratio for DMZn and DEZn doping precursors. An increase of the free hole concentration was found for high V/III flux ratios. The enhanced hole concentration was attributed to an enhanced Zn incorporation on cation sites as concluded from the law of mass action. However, intermediate reaction steps involving As_4 and As_2 species cannot be excluded.

Conflicting reports were given on the dependence of Zn incorporation on the DMZn and DEZn mole fraction in the gas phase of the reactor. Glew (1984) found a sublinear dependence of DMZn incorporation in GaAs and a superlinear dependence of DEZn incorporation in GaAs. A previous report by Bass and Oliver (1977) found a linear dependence for the DMZn precursor in GaAs. For DEZn, a sublinear (Aebi et al., 1981), a linear (Keil et al. 1982), and a superlinear (André et al., 1977; Hallais, 1978) dependence of Zn incorporation on DEZn mole fraction were reported.

The complex interdependence of growth parameters and doping parameters in OMVPE growth was beautifully demonstrated in a series of experiments by Glew (1984). The author investigated the dependence of the growth rate on doping parameters such as doping species and doping precursor mole fraction in the gas phase. It is a desirable property of doping precursors that they not influence the growth rate, alloy composition (in ternary and quaternary semiconductors), etc. However, due to parasitic reactions, catalytic reactions, and complicated reaction pathways, interdependence of growth parameters and doping parameter results.

Figure 6.9 shows the dependence of the OMVPE GaAs growth rate on the reciprocal substrate temperature for the two doping precursors DMZn and DEZn (Glew, 1984). The dependence is characterized by three different regions. First, a low temperature regime in which the growth rate increases with temperature. Second, an intermediate growth temperature regime where the growth rate is constant. Third, a high temperature regime in which the growth rate decreases. The increase of the growth rate in the low temperature regime (500–575 °C) is attributed to the partial thermal cracking to TMGa (Nishizawa and Kurabayashi, 1983). With increasing temperature, the pyrolysis of TMGa is more complete and results in a higher growth rate. In the intermediate temperature region, the growth

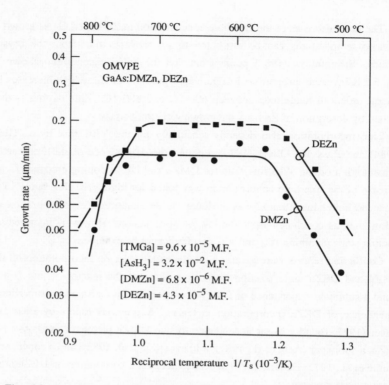

Fig. 6.9. Growth rate of OMVPE-grown GaAs as a function of reciprocal substrate temperature. The growth rate depends on the doping compound (DMZn or DEZn) used. The mole fraction (M.F.) of the gases are: [TMGa] = 9.6 x 10^{-5} M.F., [AsH$_3$] = 3.2 x 10^{-2} M.F., [DMZn] = 6.8 x 10^{-6} M.F., [DEZn] = 4.3 x 10^{-5} M.F., with a total flow of 3800 cm^3/min (after Glew, 1984).

rate is constant but clearly dependent on the doping species. A lower growth rate is found for the TMGa precursor as compared to the TEGa precursor. The transition to the high temperature regime depends critically on the doping species employed. Figure 6.9 reveals a transition temperature of 775 °C for DMZn and a lower temperature of 725 °C for DEZn. The origin of the qualitatively and quantitatively different dependence of DMZn and DEZn is not understood in detail. It is likely that parasitic cross reactions between products of the pyrolysis are the origin of this different growth rate behavior.

The dependence of the growth rate in OMVPE-grown GaAs on the doping precursor mole fraction in the gas phase is shown in Fig. 6.10 (Glew, 1984). The four growth temperatures 625, 650, 700, and 750 °C, and the two doping

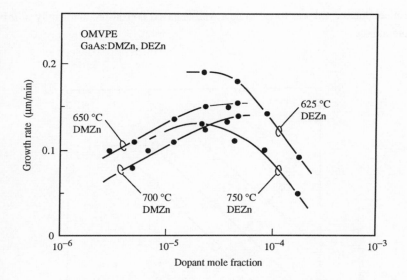

Fig. 6.10. Growth rate of GaAs grown by OMVPE as a function of the dopant concentration for the precursors DMZn and DEZn. The mole fractions in the gas phase are $[TMGa] = 9.6 \times 10^{-5}$ M.F., $[AsH_3] = 3.2 \times 10^{-2}$ M.F., with a total flow of 3800 cm^3/min (after Glew, 1984).

precursors DMZn and DEZn were used. Inspection of Fig. 6.10 reveals that the growth rate first increases with the Zn mole fraction and then, for higher Zn contents, decreases with Zn mole fraction. No conclusive explanation for the dependence of the growth rate on the dopant mole fraction was offered. A maximum of the growth rate is obtained at a dopant mole fraction of $2 \times 10^{-5} - 5 \times 10^{-5}$ which is similar to the maximum in growth rate for Sn-doped GaAs (Chang et al., 1983). Glew (1984) suggested that the decrease in growth rate at high Zn mole fractions is due to the interference of the high Zn concentration on the growing GaAs surface. In addition, parasitic, and catalytic gas phase reactions and reactions at the susceptor and substrate should be considered for the complicated dependence of the growth rate.

Cd precursors

Dimethylcadmium (DMCd, $(CH_3)_2$ Cd) was employed by a number of research groups as a doping precursor for Cd (see, for example Blaauw et al., 1987; Glade et al., 1989; Matsumoto et al., 1989; Matsumoto et al., 1990; Nelson and Westbrook, 1984). The use of the DMCd doping precursor leads to incorporation of Cd on cation sites of the III–V lattice and to p-type conductivity of the semiconductor. Cd was found to be a well-behaved dopant predominantly in InP compounds.

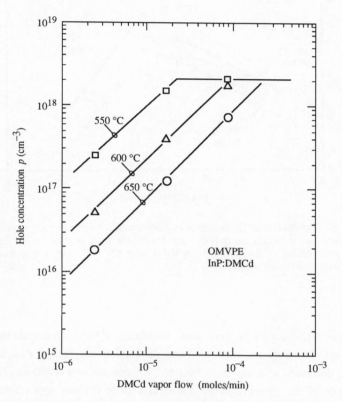

Fig. 6.11. Free hole concentration of DMCd-doped InP grown by OMVPE at 550 °C, 600 °C and 650 °C (after Nelson and Westbrook, 1984).

The incorporation of Cd from the DMCd precursor is limited by desorption of Cd from the growing surface for temperatures ≥ 550 °C. The free carrier concentration of InP grown by OMVPE is shown in Fig. 6.11 as a function of dopant mole fraction for three different growth temperatures (Nelson and Westbrook, 1984). The carrier concentration decreases by more than one order of magnitude as the growth temperature is increased from 550 °C to 650 °C. The decrease in hole concentration can be fully explained by Cd desorbing from the growth surface. The data shown in Fig. 6.11 does not reveal if all DMCd is decomposed during growth. Nevertheless, hole concentrations in the 10^{18} cm^{-3}

range can be achieved with Cd doping. Nelson and Westbrook (1984) ruled out that the change in doping concentration with growth temperature is due to a changing growth rate. Although the growth rate changes slightly with growth temperature, the decrease in doping concentration shown in Fig. 6.11 is too dramatic to be explained by a temperature-dependent growth rate.

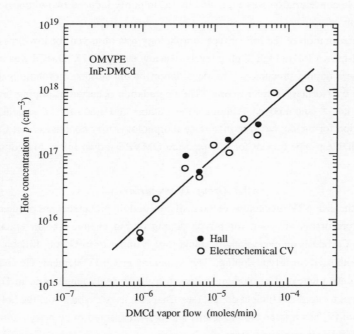

Fig. 6.12. Hall carrier concentration measured by the Hall effect and electrochemical capacitance–voltage profiling of OMVPE-grown InP as a function of DMCd mole fraction (after Nelson and Westbrook, 1984).

The hole concentration in OMVPE-grown InP depends linearly on the DMCd mole fraction in the gas phase. Figure 6.12 shows the free hole concentration measured by the Hall effect and electrochemical capacitance–voltage profiling as a function of the DMCd mole fraction in the gas phase for a growth temperature of 650 °C. A strictly linear dependence of the free hole concentration on the DMCd flow is obtained over more than two orders of magnitude in flow rate. Nelson and Westbrook (1984) confirmed the linear dependence by additional secondary ion mass spectrometry measurements. The authors concluded that the properties of DMCd discussed above make Cd a well-behaved and very useful p-type dopant for (adduct) OMVPE.

A linear dependence of the Cd acceptor concentration on the DMCd mole fraction was also found by Matsumoto et al. (1989) at temperatures of 680 °C during OMVPE growth. However, the linear dependence was not observed at higher temperatures. At a growth temperature of 720 °C the authors found an exponential dependence (rather than a linear one) of the acceptor concentration on the DMCd mole fraction in the gas phase. The unusual exponential dependence of the hole concentration was explained by Cd impurity induced two-dimensional nucleation.

A deterioration of the InP surface morphology was observed for low-pressure OMVPE-grown InP at high doping levels (Blaauw et al., 1987). DMCd was used as a p-type doping precursor. The degradation of the surface morphology was assessed by Normarsky microscopy. The degradation occurred at doping levels of 10^{17} cm^{-3} and above. Although the authors did not offer a conclusive explanation for the degradation in surface morphology, they concluded that Cd is not a suitable p-type dopant for low-pressure OMVPE-grown InP at high doping levels.

6.3.3 Group-IV precursors

Among the group-IV precursors, carbon, silicon, and tin precursors are of interest. Carbon precursors are used for p-type doping during chemical-beam epitaxial growth. Carbon is an exceptionally stable p-type impurity in GaAs. Silicon and tin are both used for n-type doping. The remaining group-IV elements Ge and Pb are of minor interest. Germanium is known to be strongly amphoteric in GaAs and cannot be used as efficient dopant (see Chap. 5). Lead, which is at the bottom of column IV, has a large atomic radius and is not expected to be a very soluble impurity in III–V semiconductors.

C precursors

The use of trimethylgallium ($(CH_3)_3$ Ga, TMGa) in III–V semiconductors containing Ga during VPE and CBE results in a considerable incorporation of C. The efficient incorporation of C from TMGa can be exploited by using TMGa as a Ga- and simultaneously as a C-doping precursor. Very high C acceptor concentrations are possible with the TMGa precursor. Carbon is known to be a stable p-type impurity in GaAs which makes the TMGa precursor an attractive choice for highly p-type doped GaAs.

It was demonstrated in the 1970s that the choice of the organometallic Ga precursor affects the unintentional C background concentration of GaAs grown by OMVPE (Seki et al., 1975; see also Fraas et al., 1984). The first systematic comparative study of the TMGa and TEGa precursors with respect to C incorporation was made by Pütz et al. (1986) for GaAs grown by CBE. The authors considered the Ga–C bond strength in TMGa and TEGa for a qualitative understanding of the C incorporation in the two cases.

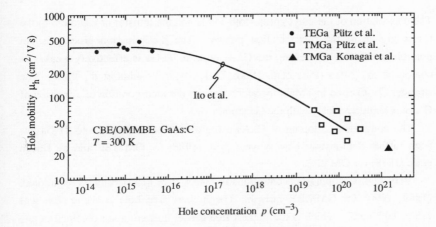

Fig. 6.13. Mobility as a function of hole concentration in C-doped GaAs grown by OMMBE using trimethylgallium (TMGa) and triethylgallium (TEGa) as Ga precursor (after Pütz et al., 1986; Konagai et al., 1990; Ito et al., 1984).

The sensitivity of C incorporation into GaAs on the choice of the Ga precursor is illustrated in Fig. 6.13. Pütz et al. (1986) used TMGa and TEGa as Ga precursors during CBE. Drastic differences were detected for the TMGa and the TEGa precursor with respect to the C doping concentration. The TEGa precursor results in moderately pure GaAs with residual background concentrations on the order of 10^{15} cm^{-3}. The background concentration was assumed to be C. The TMGa precursor, in contrast, results in very high C concentrations in the 10^{19} – 10^{20} cm^{-3} doping range. The high background doping concentration originates from the C contained in the alkyl radicals. The hole mobilities shown in Fig. 6.13 and their comparison with the best experimental mobility values for p-type GaAs (solid line, Wiley, 1975) indicate that the C-doped GaAs is not heavily compensated and is of high quality. These experimental results suggest that controlled C doping can be achieved by simultaneously using the two precursors TMGa and TEGa and by adjusting the TMGa/TEGa ratio to obtain the desired C-doping concentration. Doping concentrations ranging from 10^{14} cm^{-3} to 10^{21} cm^{-3} can be achieved using the two precursors.

Pütz et al. (1986) analyzed and discussed the chemistry of the TMGa and TEGa precursors which result in the dramatically different C incorporation. The effective strength of the Ga–C bond is weaker in TEGa as compared to TMGa. The dissociation of an ethyl group in TEGa occurs according to the reaction

$$(C_2H_5)_3 Ga \leftrightarrow (C_2H_5)_2 GaH + CHCH_3 . \tag{6.2}$$

The dissociation of the ethyl group from TEGa and the transfer of a H atom to the Ga is known as the β-elimination process. The β-elimination process is not possible in TMGa. Therefore, the Ga–C bond in TEGa is effectively weaker as compared to TMGa (Pütz et al. 1986; Tuck, 1982; Yoshida et al., 1985). The stronger Ga–C bond in TMGa therefore leads to the incorporation of a fraction of C atoms simultaneously with the Ga atom.

The controlled C doping of GaAs using a balance of TMGa and TEGa has been further demonstrated by Weyers et al. (1986) for OMMBE and by Kuech et al. (1988) for OMVPE.

Very high C doping concentrations in GaAs were reported by Konagai et al. (1989, 1990) for OMMBE growth. The highest free hole concentration was 1.5×10^{21} cm^{-3} which is the highest free carrier concentration obtained in any III–V semiconductor. The corresponding Hall mobility and resistivity are 22 cm^2/V s and 1.9×10^{-4} Ω cm, respectively. This data point is included in Fig. 6.13. The high C concentration puts the GaAs lattice under tensile stress and results in a reduction of the lattice constant by -0.5% at a C concentration of 10^{21} cm^{-3}. Konagai et al. (1990) pointed out that the measured reduction in lattice constant is quantitatively in agreement with Vegard's law (see Sect. 12.3.2).

Low growth temperatures are required to achieve high C concentrations in GaAs. Figure 6.14 shows the free hole concentration of (001) and (111) oriented C-doped GaAs as a function of reciprocal temperature. The highest doping concentrations are achieved at low temperatures of 450 °C. In addition to the doping concentration, the growth rate changes from 2.5 μm/h to 0.5 μm/h for a temperature of 600 °C and 450 °C, respectively. Konagai et al. (1989) proposed that the TMGa molecule is less likely to thermally dissociate at low growth temperature which leads to an enhanced incorporation of C, i.e. to high C doping concentrations.

Other doping precursors proposed for C doping of III–V semiconductors during OMMBE are ethene (C_2H_2) and trimethylarsenic $((CH_3)_3As$, TMAs). Weyers et al. (1990) showed that both ethene and TMAs are stable molecules that are not pyrolyzed effectively at typical growth temperatures and do not result in significant C incorporation. Thermal pyrolysis in a cracker cell upon entering the CBE or OMMBE growth system would result in a more efficient C incorporation from the ethene and TMAs precursors.

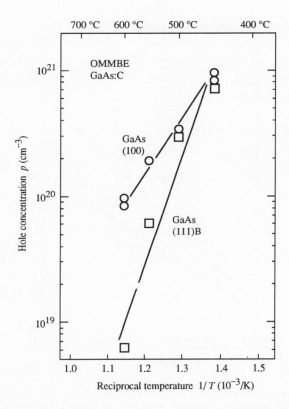

Fig. 6.14. Hole concentration of C-doped GaAs as a function of reciprocal growth temperature. TMGa is employed as both doping percursor and as group-III precursor (after Konagai et al., 1990).

Si precursors

A number of Si doping precursors are used for Si doping including silane (SiH_4), disilane (Si_2H_6), and triethylsilane ((C_2H_5)$_3$ SiH, TESiH). The precursors are used for n-type doping of III–V semiconductors in which Si is a donor.

The most obvious Si doping precursor is silane (see, for example, Bass, 1979), which initially proved to be a well-behaved precursor. Ohno et al. (1984) demonstrated, using silane as a Si doping precursor, that abrupt Si doping profiles can be achieved in GaAs. However, subsequent reports showed that the incorporation of Si from silane depends strongly on the growth temperature. A higher growth temperature yields increased Si incorporation in the semiconductor. The temperature dependence of Si incorporation is shown in Fig. 6.15 for

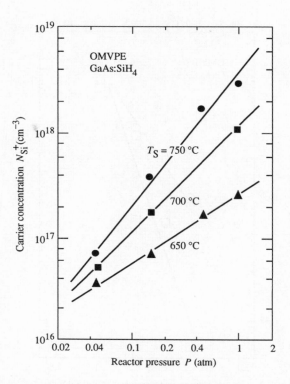

Fig. 6.15. Electron concentration of OMVPE-grown GaAs as a function of reactor pressure for different growth temperatures. Silane (SiH_4) is used as doping precursor (after Field and Ghandhi, 1986a).

OMVPE-grown GaAs doped using SiH_4 as a doping precursor (Field and Ghandhi, 1986a). The experimental data shows that the free carrier concentration (Si concentration) increases with growth temperature. The increase suggests that SiH_4 is not pyrolyzed in the entire temperature range. As the substrate temperature is increased, more Si is incorporated due to a more efficient thermal decomposition of silane. Field and Ghandhi (1986a) also developed a thermodynamical model which predicts the increasing Si concentration with reactor pressure depicted in Fig. 6.15.

The incomplete pyrolysis and the resulting dependence of the Si concentration on the growth temperature is an undesirable property of a doping precursor. Since the incomplete pyrolysis stems from the thermal stability of the SiH_4 molecule, a less stable molecule could alleviate this problem. Veuhoff et al. (1985) proposed that disilane (Si_2H_6), which is less stable than silane, should

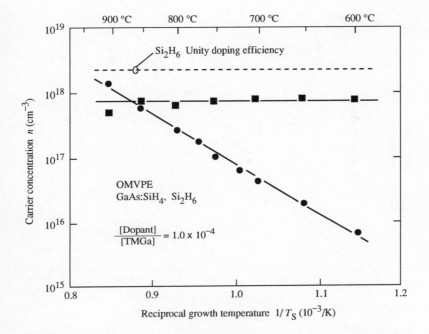

Fig. 6.16. Carrier concentration of OMVPE-grown GaAs as a function of reciprocal growth temperature for the two doping compounds silane (SiH_4) and disilane (Si_2H_6). The dashed line indicates unity doping efficiency for disilane. The ratio between the dopant and TMGa mole fraction is 1×10^{-4} (after Kuech et al., 1984).

therefore be used as a Si doping precursor. Previously, Kuech et al. (1984) demonstrated that the Si incorporation from the disilane precursor is independent of growth temperature in the temperature regime studied. The experimental results are shown in Fig. 6.16 which reveals that the incorporation of Si from the disilane precursor is independent of the growth temperature for the temperature interval 600 °C to 900 °C. The dopant mole fraction is kept constant for the entire temperature range studied. The experimental results further show that the incorporation efficiency (see Eq. (6.1)) of Si from the precursor disilane is approximately 0.5, i.e. much higher than the incorporation efficiency of Si from silane which can be $< 10^{-2}$.

A further advantageous property of the Si incorporation from disilane is the linearity of the incorporation. Figure 6.17 shows the free electron concentration as a function of the disilane mole fraction (Kuech et al., 1984). The free electron

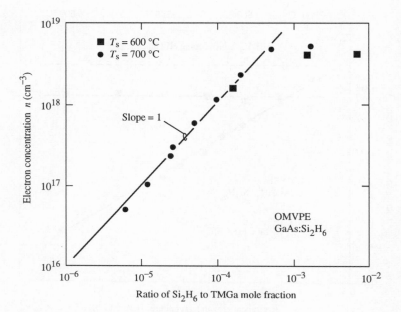

Fig. 6.17. Carrier concentration of OMVPE-grown GaAs as a function of Si_2H_6 concentration in the gas phase. A similar dependence was found for disilane (Si_2H_6) doping in $Al_xGa_{1-x}As$ (after Kuech et al., 1984).

concentration follows the dopant mole fraction in a linear fashion. The evaluation of the experimental data revealed a slope of unity. The incorporation of Si from the disilane precursor thus has several clear advantages namely (i) high pyrolytic efficiency, (ii) high incorporation efficiency, and (iii) an incorporation efficiency which is linearly dependent on the precursor mole fraction.

While advantageous properties were found for Si-doped GaAs and $Al_xGa_{1-x}As$, Si doping of InP using silane was reported to be problematic (Hsu et al., 1986). As for GaAs, the authors found a strong temperature dependence of the Si incorporation from silane. In the temperature range of 575 °C – 650 °C, an incorporation efficiency (see Eq. (6.1)) of 4.5×10^{-3} – 9.5×10^{-3} was found. Furthermore, the luminescence intensity decreased drastically with Si doping of InP. A reduction of more than one order of magnitude was found for Si doping levels in the 10^{17} cm^{-3} range. Concomitant with the increased Si concentration was the appearance of a broad 1.06 μm line which is indicative of midgap defect states.

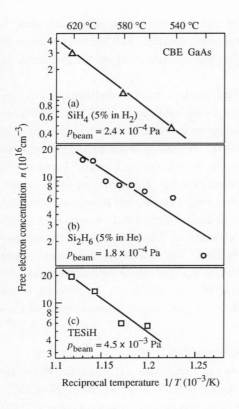

Fig. 6.18. Free electron concentration in n-type GaAs doped with (a) SiH_4 (5% in H_2), (b) Si_2H_6 (5% in He), and (c) TESiH grown by chemical-beam epitaxy as a function of reciprocal substrate temperature (after Weyers et al., 1990).

Gaseous Si precursors are also candidates for doping in CBE due to the doping anomalies observed with elemental doping effusion cells (Skevington et al., 1990). The authors reported the formation of SiC in the hot effusion cells which made controlled doping impossible. This problem can be overcome by gaseous doping precursors. Weyers et al. (1990) investigated the precursors silane, disilane, and triethylsilane (TESiH) for GaAs doping during CBE. The growth temperature dependence of the free carrier concentration is shown in Fig. 6.18 for the three doping precursors silane, disilane, and TESiH. The experimental results are summarized as follows. First, the free carrier concentration increases exponentially with reciprocal temperature for all of the three doping precursors. Second, silane results in a smaller free carrier concentration as compared to disilane. The experimental results indicate that the incorporation of Si from silane, disilane, and TESiH is limited by incomplete pyrolysis in the temperature range 540 – 620 °C. The Arrhenius behavior displayed in Fig. 6.18 is indicative of such a pyrolysis limited incorporation. The experimental results further indicate that the incorporation of Si from silane is less effective than the incorporation from disilane. The weaker efficiency of silane can be attributed to the greater thermal stability of the SiH_4 molecule as compared to the Si_2H_6 molecule.

The three doping precursors silane, disilane, and trimethylsilane do not have all the desirable properties required for CBE applications. The lack of complete pyrolysis of the three precursors is due to the lower growth temperatures used during CBE of typically 400 − 600 °C. A solution to the incomplete pyrolysis could be thermal decomposition of the Si precursor in a cracker furnace upon entering the ultra-high vacuum reaction chamber. Upon cracking the precursor, elemental Si would reach the growing crystal surface and incorporate with high efficiency as known from elemental Si doping during MBE. Weyers et al. (1990) also mentioned the possibility of a significant difference between Si precursor decomposition during OMVPE and CBE. The authors suggested that disilane (during OMVPE) is decomposed in the gas phase before reaching the semiconductor surface. They suggested that disilane is decomposed at the semiconductor surface during CBE. However, experimental evidence was not presented to support this hypothesis.

Sn precursors

Tetramethyltin $((CH_3)_4 Sn$, TMSn) and tetraethyltin $((C_2H_5)_4 Sn$, TESn) are used as doping precursors for Sn doping in III–V semiconductors. Both TMSn and TESn have been used for OMVPE growth (see, for example, Roth et al., 1984 and Parsons and Krajenbrink, 1984). TESn has also been used for Sn doping during chemical beam epitaxy. Tin doping in GaAs is known to be problematic for elemental effusion cell doping during MBE (see Chap. 5). Strong diffusion and segregation of Sn are some of the commonly observed phenomena. However, in atmospheric pressure OMVPE the adverse properties are not as pronounced.

Parsons and Krajenbrink found that Sn-doped GaAs using TESn during atmospheric pressure OMVPE has well-behaved characteristics. These characteristics include (i) good control of the Sn concentration over four orders of magnitude (ii) no surface accumulation of Sn, and (iii) weak Sn diffusion which made possible the growth of abrupt doping profiles; (iv) Sn-doped GaAs did not exhibit any measurable memory effect; the study further revealed that (v) the Sn incorporation does not depend on the growth temperature in the range 650 − 700 °C and that (vi) the Sn incorporation depends linearly on the TESn mole fraction in the gas phase over four orders of magnitude of Sn concentration.

A degradation of the GaAs surface morphology was found for high Sn doping levels (Yakimova et al., 1984). Tin doping using a TMSn precursor at concentrations $N_{Sn} \geq 10^{19}$ cm^{-3} resulted in a drastic increase of the etch pit density and a degradation of the surface morphology. The authors attributed the degradation to increased interstitial Sn incorporation at high doping concentrations. This interpretation was supported by a simultaneous saturation of the electrical activity in highly Sn-doped GaAs which occurred at $N_{Sn} \geq 10^{19}$ cm^{-3}.

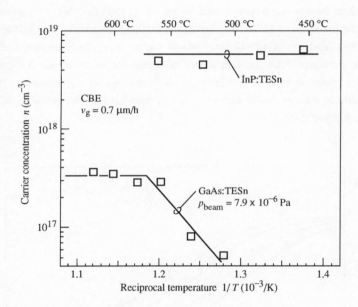

Fig. 6.19. Electron concentration in InP and GaAs grown by chemical-beam epitaxy as a function of reciprocal growth temperature. Tetraethyltin (TESn) is used as n-type doping percursor (after Weyers et al., 1990).

The incorporation of Sn from TESn in GaAs and InP during growth by CBE is more complicated. Figure 6.19 shows the free carrier concentration of Sn-doped GaAs and InP as a function of reciprocal growth temperature (Weyers et al., 1990). The free carrier concentration increases with the growth temperature for temperatures ≤ 570 °C. The increase indicates that the incorporation is limited by pyrolysis. At higher temperatures ($T > 570$ °C) the carrier concentration remains constant. The constant carrier concentration would suggest that TESn is completely decomposed at $T > 570$ °C. However, Weyers et al. (1990) concluded that the low incorporation efficiency (see Eq. (6.1)) of only 2×10^{-3} argues against such a complete decomposition. Instead the authors suggest that reevaporation of Sn from the crystal surface balances out the more efficient pyrolysis at higher temperatures. The two processes have similar activation energies, i.e. 1.5 eV and 1.35 eV for pyrolysis and reevaporation, respectively. The similarity of the activation energies indicates that pyrolysis and reevaporation can indeed balance.

Figure 6.19 shows a constant Sn incorporation in InP grown by CBE as a function of temperature. Tetraethyltin is used as a Sn doping precursor. It is unlikely that the low temperatures used for growth of InP ($T < 560$ °C) result in a complete pyrolysis of TESn. The low incorporation efficiency of 10^{-4} (Weyers et al., 1990) and the relatively low growth temperature argue against such a complete pyrolysis. The constant carrier concentration of Sn-doped InP could stem also from a balance of reevaporation and pyrolysis. However, this explanation is not conclusive due to a lack of data for the reevaporation of Sn from the InP surface.

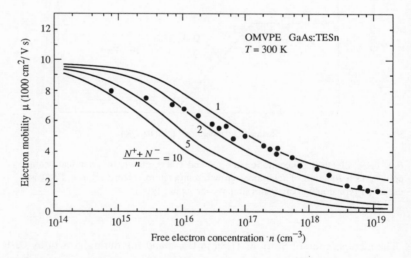

Fig. 6.20. Room-temperature Hall electron mobility as a function of free carrier concentration in Sn-doped GaAs grown by OMVPE (after Parsons and Krajenbrink, 1984).

Sn is known to autocompensate in GaAs only to a small extent. Electron mobilities are therefore expected to be high in Sn-doped GaAs. The electron mobilities of Sn-doped GaAs grown by OMVPE are shown in Fig. 6.20 (Parsons and Krajenbrink, 1984). The doping precursor used in this study is tetraethyltin. Also included in Fig. 6.20 are theoretical mobilities for four different compensation ratios (solid lines). The uppermost of the four curves is for uncompensated material, i.e. $n = N^+$. The experimental curves closely follow the uncompensated line for doping concentrations in the 10^{17} cm^{-3} range. At low doping concentrations ($n \cong 10^{15}$ cm^{-3}), the mobilities are lower than the uncompensated curve, probably due to unintentional background impurities. At the high concentration end ($n \cong 10^{19}$ cm^{-3}) the mobilities are lower than the uncompensated curve probably due to autocompensation of Sn. The autocompensation of group-IV impurities at high doping concentration is a general phenomenon observed for all group-IV impurities and will be further discussed in Chap. 7.

6.3.4 Group-VI precursors

Among the group-VI elements, sulfur, selenium, and tellurium are of interest. The remaining elements of group VI, oxygen and polonium, are of no interest. Oxygen forms donor levels located around midgap in GaAs and other III–V semiconductors. Polonium has a large atomic radius and is not expected to be soluble in III–V semiconductors.

Group-VI impurities have high vapor pressures. Among the three elements S, Se, and Te, sulfur has the highest vapor pressure. Descending to elements further down the group, the vapor pressure decreases. Nevertheless, the vapor pressures of Se and Te are still higher than, for example, Si. The high vapor pressure of group-IV elements leads to reevaporation of impurities from the growing semiconductor surface. The donor concentration of GaAs indeed decreases with increasing growth temperature if doped with any of the group-VI impurities. The decrease can be attributed to reevaporation of impurities from the crystal surface.

S precursors

Hydrogen sulfide (H_2S) is the most common doping precursor used for S doping of III–V semiconductors. It is a toxic gas, which can be easily detected by its characteristic smell resembling that of rotten eggs. Sulfur occupies anion sites in III–V semiconductors and is a shallow donor.

Sulfur is characterized by the highest vapor pressure of S, Se, and Te. The possibility of thermally activated desorption of S from epitaxial surfaces is therefore of prime interest. The free carrier concentration of S-doped GaAs grown by OMVPE is shown in Fig. 6.21 as a function of reciprocal growth temperature (Field and Ghandhi, 1986b). Four different pressures of 0.046, 0.15, 0.46, and 1.0 atmospheres were used during the growth experiments. The experimental finding shows a definite decrease of the free electron concentration at high growth temperatures. At low pressures, the decrease in electron concentration exceeds one order of magnitude for a temperature interval of only

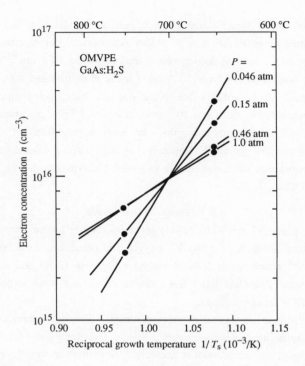

Fig. 6.21. Free electron concentration in OMVPE-grown GaAs as a function of reciprocal substrate temperature. Hydrogen sulfide (H_2S) is used as n-type doping precursor (after Field and Ghandhi, 1986b).

100 °C. The mole fractions of TMGa, AsH_3, and H_2S were kept constant during these experiments. Considering the high vapor pressure of S, it is very likely that the decrease of the free carrier concentration is due to a decrease in S incorporation which in turn is caused by increased reevaporation of S from the crystal surface.

Field and Ghandhi (1986b) also pointed out the relevance of the As vacancy concentration on the crystal surface. They assumed that the chemisorption and adsorption energies depend on the total coverages of H, S, AsH, and As_2. Since these coverages depend on temperature, a change in S incorporation is expected. The authors further commented on the pressure dependence of the S incorporation. They argued that the rate of desorption of S is limited by diffusion. The diffusion of S is faster and more efficient at low reactor pressures. Thus, a stronger decrease of the S concentration at low pressures can be understood on the basis of diffusion.

The doping behavior of S in GaAs grown by chloride vapor-phase epitaxy was investigated by Heyen et al. (1977). H_2S was used as the S doping precursor at a growth temperature of 750 °C. The study revealed that at a constant set of parameters the incorporation of S followed linearly the H_2S mole fraction in the gas phase over a wide range of concentrations. Further experiments showed, however, that the S incorporation also depends on the pressures of other reactants in the growth system including AsH_3, GaCl, and HCl.

In a subsequent study Veuhoff et al. (1981) confirmed the linear dependence of S incorporation and the H_2S mole fraction in the gas phase for chloride VPE grown, $\langle 001 \rangle$ oriented GaAs. The incorporation behavior was found to be different for GaAs grown on the (111) GaAs surface. For this orientation the authors found a sublinear dependence of the S incorporation versus H_2S mole fraction. The functional dependence of the S incorporation was found to follow a square-root dependence of the H_2S mole fraction.

The efficiency of S incorporation was found to depend on the V/III ratio during GaAs OMVPE growth (Bass and Oliver, 1977). The efficiency (see Eq. (6.1)) of S incorporation decreases with increasing V/III ratio. Since S occupies anion sites it competes with As. A high As/Ga flux ratio is therefore expected to reduce S incorporation. Hydrogen sulphide is gaseous at room temperature. Adsorption of H_2S at the walls of gas tubes, gas manifold, and reactor walls is not likely during doping with H_2S. Any memory effects are therefore not expected for H_2S doping and have not been reported.

Se precursors

Hydrogen selenide is the most commonly used doping precursor for Se doping in III–V semiconductors. Selenium occupies anion sites and is a shallow donor. The memory effect observed with H_2Se doping during OMVPE poses a constraint to the versatility of the H_2Se precursor.

The analysis of the doping characteristics of H_2Se in GaAs revealed that the precursor has a number of well-behaved characteristics. Glew (1982) showed that the incorporation of Se into GaAs follow a superlinear relationship with respect to the H_2Se mole fraction in the gas phase. The superlinear relationship was well described by a power law with an exponent of 3/2, i.e. $N_{Se} \propto [H_2Se]^{3/2}$ where the brackets represent the concentration (mole fraction) of H_2Se in the gas phase of the reactor. The 3/2 power law relation was observed at growth temperatures ranging from 650 °C to 800 °C and also for different V/III flux ratios during growth. The highest doping concentration achieved was 10^{19} cm^{-3} in GaAs.

The incorporation of Se strongly decreases with increasing V/III ratio. Glew (1982) found an approximately linear decrease of the Se concentration in the

GaAs crystal with increasing V/III flux ratio. A decrease of N_{Se} by a factor of 10 was found for an increase of the V/III flux ratio from 10 to 100. The sensitivity of Se incorporation to the V/III ratio can be understood on the basis of the law of mass action. Since Se occupies anion sites, it competes with As for lattice sites. Thus an increase of the V/III ratio leads to a reduction of N_{Se}.

One drawback of the H_2Se precursor for OMVPE is the occurrence of Ga_2Se_3 complexes during GaAs growth (Glew, 1982). It was assumed that a parasitic gas phase reaction above the susceptor forms Ga_2Se_3. The molecule Ga_2Se_3 was assumed to be incorporated as one entity into the crystal. Certainly the Ga_2Se_3 complexes do not act as shallow donors in GaAs.

Another major drawback of the H_2Se precursor for OMVPE is the pronounced occurrence of the ***memory effect***. The memory effect makes abrupt changes in the doping profile difficult. Furthermore, it results in an enhanced background Se impurity concentration in nominally undoped material. The memory effect of Se doping with the H_2Se precursor in OMVPE was investigated in detail by Lewis et al. (1984). They attributed the occurrence of the memory effect to the adsorption of H_2Se onto stainless steel surfaces of the gas manifold and onto the glass surfaces of the reactor. After termination of the doping process, H_2Se desorbs slowly from these surfaces and results in Se doping of nominally undoped material. The authors proposed several ways to reduce the memory effect. First, a low growth rate shortens the Se doping tails in the semiconductor after the intentional Se doping is terminated. Second, the reduction of the tubing length from the H_2Se gas cylinder to the reactor reduces the adsorption area and thus the memory effect. Third, the moderate heating of the H_2Se gas manifold and of the reactor walls prevents the adsorption of the precursor on the stainless steel and glass walls.

Te precursors

Diethyltellurium ($(C_2H_5)_2Te$, DETe) has been used as Te doping precursor for OMVPE growth of GaAs and InP (see for example: Houng and Low, 1986; Hsu et al., 1986). Tellurium incorporates on the anion site of III–V semiconductors and is a shallow donor. The DETe is a well-behaved precursor except for the memory effect which was found in DETe doping of GaAs (Houng and Low, 1986). DETe is also a candidate for n-type doping in CBE (Weyers et al., 1990).

Well-behaved doping characteristics of the DETe precursor in OMVPE-grown GaAs were found by Houng and Low (1986). The free electron concentration is shown in Fig. 6.22 as a function of the DETe mole fraction in the vapor. Free electron concentrations in the $10^{16} - 10^{18}$ cm^{-3} range were achieved. The relationship between the free carrier concentration (i.e. Te concentration) and the

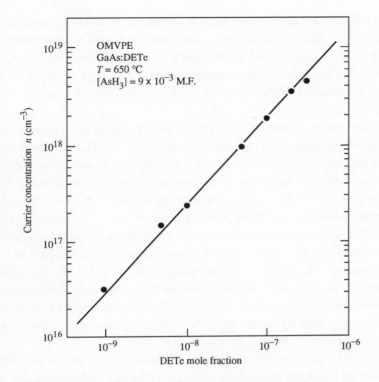

Fig. 6.22. Carrier concentration at 300 K versus DETe mole fraction in the gas phase of OMVPE-grown GaAs (after Houng and Low, 1986).

DETe mole fraction is linear which usually allows well-controlled doping of GaAs. The authors also evaluated the distribution coefficient for GaAs doping with DETe. The *distribution coefficient* is defined as $k = x_s/x_v$ where x_s is the fraction of lattice sites in the solid filled with Te impurity atoms and x_v is the ratio of DETe concentration and AsH_3 in the vapor phase of the reactor. Qualitatively the distribution coefficient is a measure of how efficiently Te is incorporated as compared to As. Note that the distribution coefficient is not identical but related to the doping efficiency defined in Eq. (6.1). However, if all incorporated impurities are electrically active, then the doping efficiency equals the distribution coefficient. If $k = 1$, for example, then Te has the same probability as As of being incorporated into the growing GaAs lattice. The authors inferred a distribution coefficient of $k = 8$ from the experimental data of Fig. 6.22.

The large distribution coefficient of $k = 8$ found for DETe doping of GaAs indicates that DETe is effectively decomposed at a growth temperature of $T = 650$ °C. Pyrolysis is not a limiting mechanism for the incorporation of Te. However, at lower growth temperatures of $520 - 580$ °C a limitation of Te doping due to a lack of pyrolysis was found for GaAs and InP grown by CBE (Weyers et al., 1990).

Houng and Low (1986) also investigated the dependence of Te incorporation on growth parameters. A decreasing incorporation of Te was found with increasing arsine pressure in GaAs growth by OMVPE. Assuming that the free carrier concentration equals the Te concentration, a power law dependence of $N_{Te} \propto [AsH_3]^{-0.71}$ was found. The decrease of N_{Te} with increasing arsine concentration can be qualitatively explained on the basis of the law of mass action. Tellurium competes with As for the anion sites in the zincblende lattice. An increase of the AsH_3 mole fraction reduces the probability of Te incorporation. The authors also investigated the influence of the growth temperature on the doping concentration of the DETe precursor. The increase of the growth temperature from 600 °C to 750 °C resulted in a decrease of the free carrier concentration by approximately 20%. This weak decrease is probably due to reevaporation of Te from the growing surface.

The DETe precursor was used for n-type doping of InP (Hsu et al., 1986). The authors found that the carrier concentration increased superlinearly with the DETe mole fraction. The superlinear increase contrasts the (linear) DETe doping characteristics of GaAs shown in Fig. 6.22. Hsu et al. (1986) state that the physical origin of the superlinear dependence of Te incorporation with DETe mole fraction is unknown. A distribution coefficient of $k = 36$ was inferred from the doping characteristics of Te in InP. A comparison of the distribution coefficients of S, Se, and Te revealed that $k_{Te} > k_{Se} > k_S$. This qualitative dependence is consistent with the vapor pressure of Te, Se, and S, which increases for elements with smaller atomic weight within column VI of the periodic system.

High free electron concentrations were achieved for Te-doped InP. The electron mobilities of highly Te-doped InP grown by OMVPE are shown in Fig. 6.23 (Hsu et al., 1986). Concentrations of 10^{19} cm^{-3} with a mobility exceeding 1000 cm^2/Vs were achieved. The solid curve in Fig. 6.23 is a fit to the data points with the highest mobilities. The figure also shows that the electron mobilities obtained with Si doping are consistently lower than the mobilities obtained from Te-doped GaAs. The lower mobilities of Si-doped GaAs are probably due to autocompensation of Si which is known to be amphoteric.

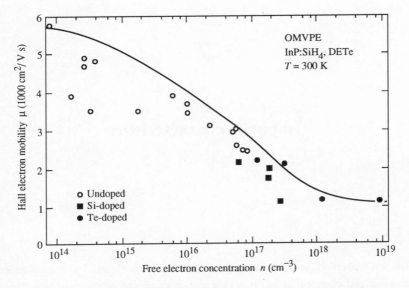

Fig. 6.23. Room-temperature electron mobility of OMVPE-grown unintentionally doped, Te-doped (diethyltellurium), and Si-doped (silane) InP as a function
of carrier concentration (after Hsu et al., 1986).

The DETe precursor was also used for doping of GaAs and InP during CBE
(Weyers et al., 1990). The authors found an increasing doping efficiency (see
Eq. (6.1)) with growth temperature for GaAs and InP growth. The authors
attributed this increasing doping efficiency to a more effective pyrolysis at higher
growth temperature. Note that the growth temperatures during CBE are generally
lower than during OMVPE. Weyers et al. (1990) employed GaAs growth
temperatures of $500 - 620 \, ^\circ C$ and InP growth temperatures of $470 - 560 \, ^\circ C$.
At low temperatures, the thermal dissociation of DETe molecules is less likely.
Furthermore, the very low pressures and the resulting molecular flow regime
make gas-phase reactions less likely in CBE as compared to VPE.

7

Impurity characteristics

The incorporation and activation of impurities is governed by a number of physical and chemical laws. It is essential to know and apply them in order to predict experimental functional dependencies of impurity incorporation. For example, the incorporation of many impurities depends on the epitaxial growth temperature. How does the electrical activity, compensation, etc. depend on this growth temperature? How are the impurity characteristics influenced by the V/III flux ratio? These are some of the questions that are of interest. Furthermore, some incorporation characteristics are found only for certain impurity elements. Autocompensation is prominent among group-IV impurities. In order to minimize autocompensation, it is desirable to know the functional dependences of amphoteric impurity incorporation.

Another example of impurity characteristics is the *doping efficiency*. This is defined as the ratio of the free carrier concentration and the dopant concentration. The doping efficiency is unity for an ideal dopant element and an ideal doping procedure. However, it can be quite low. For example, the doping efficiency of impurities incorporated by implantation can be below 10% before activation. The understanding of the principles governing the doping efficiency is therefore desirable.

Finally, very high doping concentrations become increasingly important as the spatial dimensions of semiconductor structures shrink. What are the highest doping concentrations achievable in III–V semiconductors? What limits the maximum impurity concentration? These are some of the topics and questions that will be addressed in this chapter.

Dopants can be incorporated into semiconductors by using several doping techniques including (i) doping during epitaxial growth, (ii) doping by implantation, and (iii) doping by diffusion. The nature of the incorporation can strongly affect the impurity characteristics. For example, the electrical activity of dopants is very different for doping during epitaxial growth or doping by implantation. The different processes of impurity incorporation are therefore considered first.

7.1 Doping incorporation during growth

The incorporation of doping impurities during epitaxial crystal growth depends on the nature of the epitaxial growth method. Dopant incorporation is quite different for the two prominent epitaxial growth techniques molecular-beam epitaxy (MBE) and organometallic vapor-phase epitaxy (OMVPE). Doping incorporation in OMVPE is dominated by the chemistry and thermodynamics of the doping and growth process. In MBE, doping incorporation is mainly controlled by kinetics such as impurity arrival rate, reevaporation rate etc. Therefore, the incorporation of impurities will be discussed separately for the two growth techniques. The law of mass action is very useful for the evaluation and prediction of trends in impurity incorporation and will be discussed first.

The *law of mass action* is frequently invoked to derive characteristics of impurity incorporation. The dependence of impurity incorporation on growth temperature and on V/III flux ratio are frequently analyzed in terms of the law of mass action. Due to the importance of the law of mass action, the basic properties of this law will be reviewed. A more exhaustive description of the law can be found in textbooks (see, for example, Kittel and Kroemer, 1980; Kroger, 1973; see also Sect. 3.4). Consider the chemical reaction of the starting reagents A and B and of the reaction products C and D.

$$A + B \leftrightarrow C + D \tag{7.1}$$

where A, B, C, and D can be elements or compounds, i.e. general products of a chemical reaction. The chemical reaction at constant temperature is associated with a change in energy of the system which is given by

$$\Delta G = \Delta H - T\Delta S \tag{7.2}$$

where ΔH represents the changes of different contributions to the energy (i.e. chemical, mechanical, electrostatic, etc.), T is the temperature, and ΔS is the change in entropy occurring in the reaction of Eq. (7.1). The term ΔG is *Gibbs free energy* of the system. The term ΔH in Eq. (7.2) is called *enthalpy*, formation energy, heat of formation, or heat content. It describes the total energy set free by the chemical reaction.

The quantity ΔH is the total change in enthalpy incurred during the chemical reaction of Eq. (7.1). ΔH can be written as the sum of different contributions to the enthalpy, that is

$$\Delta H = \sum_j \Delta E_j + P\Delta V \tag{7.3}$$

where P and V are the pressure and the volume of the system and $P\Delta V$ is the mechanical work done by the system. Examples for contributions to the change in internal energy ΔE_j are as follows.

1. Chemical binding energies. The formation energy of chemical bonds of the components A and B versus C and D (see Eq. (7.1)) can be different.

2. Electrostatic and magnetostatic energies. Electrostatic and magnetostatic energy densities are given by $\varepsilon \mathcal{E}^2$ and μH^2 where $\mathcal{E}$ and H are the electric and magnetic field, respectively.

3. Electronic energies. If electronic transitions are involved in Eq. (7.1), they also enter the change in energy.

4. Elastic energies. For impurity incorporation in a semiconductor, elastic energies can be important, especially if the impurity has a substantially different size as compared to the lattice atom it substitutes for. Furthermore, steric hindrance of impurities having a different valence structure than atoms of the host crystal contributes to the elastic energy.

The second term of Eq. (7.2) includes the change in entropy ΔS. The entropy is given by

$$S = k \ln w \tag{7.4}$$

where k is Boltzmann's constant. The parameter w is called the *disorder parameter*; it is the probability that a system will exist in the state it is in, relative to all the states it could be in. Equation (7.4) connects the macroscopic (or thermodynamic) quantity of entropy with a statistical or microscopic quantity, the probability. Any chemical reaction of a system that starts in one equilibrium state and ends in another equilibrium state *increases* the entropy of the system (Kittel and Kroemer, 1980).

Considering the chemical reaction of Eq. (7.1) and the change in total free energy given in Eq. (7.2), the **law of mass action** is given by

$$\frac{[C][D]}{[A][B]} = K = ce^{-\Delta G/kT} \tag{7.5}$$

where the brackets [...] represent the concentrations of the compounds and c is a constant. Thus the ratio of the products of the concentrations of initial and final reaction products is a constant at any given temperature. The law of mass action applies to *diluted* chemical solutions and to electronic 'reactions' in *non-degenerate* semiconductors. In the following sections, the law of mass action will be applied to study impurity incorporation during MBE and OMVPE.

7.1.1 Impurity incorporation in MBE

It is frequently argued that the growth of III–V semiconductors by MBE is mainly controlled by kinetics and that thermodynamics is not applicable to MBE growth. At low growth temperatures, incorporation of atoms would be limited by kinetic barriers that make it impossible for atoms to reach their thermal equilibrium positions. The lack of thermal equilibrium does not, however, apply to semiconductor *surfaces*. Heckingbottom (1985) showed that it is reasonable to assume that thermal equilibrium is indeed reached at growing semiconductor surfaces. At a growth temperature of 900 K, the vibrational frequency of an atom can be obtained from $\hbar\omega \cong kT$, which yields an attempt rate of $f = kT/2\pi\hbar \cong 10^{13} \ s^{-1}$. It is further assumed that a hop-attempt for surface diffusion is successful with the probability $p \cong \exp(-E_a/kT)$ where E_a is the activation barrier for the diffusion hop. Using an approximate activation energy of 1 eV for a surface hop (Heckingbottom, 1985; Neave and Joyce, 1983), one obtains the number of successful diffusion hops per second as $fp = 5 \times 10^7 s^{-1}$. At a growth rate of one monolayer per second atoms on the surface can complete 5×10^7 diffusion hops before they are buried by atoms impinging on the growing surface. The large number of diffusion hops makes it likely that lattice atoms as well as impurity atoms reach low-energy sites, i.e. their equilibrium positions. It is thus very reasonable to assume that the semiconductor surface is in equilibrium at growth temperatures of 900 K (627 °C).

Once the atoms are incorporated into the growing crystal the atoms become practically immobile. The activation energy for Ga self diffusion in GaAs was estimated to be 6 eV (Tan and Gösele, 1988; Goldstein, 1961). Calculating the frequency of successful diffusion hops according to $f_s = kT/2\pi\hbar \exp(-6 \ eV/kT)$ one obtains at $T = 900$ K a frequency of $f_s = 5 \times 10^{-21} s^{-1}$ which means that the atoms are practically immobile upon being incorporated into the growing semiconductor.

The clearly different regimes of surface and bulk have led to the concept of **local equilibrium** (Heckingbottom, 1985). The concept of local equilibrium

suggests that *equilibrium is reached at the surface but not in the bulk.*
Thermodynamics can thus be applied to the surface during crystal growth. On the
other hand, structures already incorporated in the semiconductor have high kinetic
barriers which results in (meta) stability of the structure. Such meta-stability is
evidenced by atomically abrupt doping profiles and interfaces which can be
grown by MBE.

The concept of local equilibrium cannot be applied to low growth
temperatures. Using again $f_s = kT/2\pi\hbar \exp(-E_a/kT)$ for the rate of successful
diffusion hops, the rate is obtained as $f_s \cong 2/s$ for a growth temperature of 400 K
(127 °C) and an activation energy of $E_a = 1$ eV. Obviously thermal equilibrium
positions cannot be reached at this growth temperature unless the growth rate is
exceedingly small. The above discussion of surface equilibrium, surface
diffusion, etc. has been kept brief. A detailed discussion of surface processes in
epitaxy was given by Kreuzer (1991).

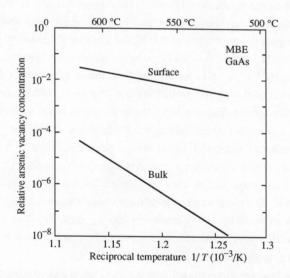

Fig. 7.1. Relative arsenic vacancy concentration on a GaAs surface and in GaAs
bulk as a function of substrate temperature for a constant arsenic surface
concentration (after Heckingbottom, 1985).

Surface chemistry of epitaxially grown III–V semiconductors plays a crucial
role in impurity incorporation. The surface chemistry is determined by a number
of factors, most importantly the abundance of adsorption (physisorption and
chemisorption) products, the species of adsorption products, and the

concentration of group-III and group-V vacancies. In addition, the concentration of vacancies depends (i) on the V/III flux ratio and (ii) on the growth temperature. For GaAs, for example, the As coverage and the concentration of As vacancies depend strongly on the growth temperature. The vapor pressures of Ga, As_2, and As_4 on a GaAs surface were measured by Arthur (1967) and reviewed by Cho and Arthur (1975). The As_2 and As_4 vapor pressures are strongly temperature dependent. The vapor pressure of Ga is smaller than that of As at growth temperatures of 500-600°C. Both elements have the same pressure, when evaporating from a GaAs surface at about 630 °C, which is called the *congruent evaporation temperature*. As a result of the strong temperature dependence of the As vapor pressure, the concentration of As vacancies also depends strongly on temperature. The concentration of As vacancies is shown in Fig. 7.1 as a function of temperature (Heckingbottom, 1985). Figure 7.1 reveals that the As vacancy concentration increases by more than one order of magnitude for a temperature interval of 100 °C. Analogously, at a constant V/III flux ratio the As coverage decreases rapidly with increasing temperature. The temperature dependence of the As vacancy concentration will be used in the further analysis of impurity incorporation.

Impurities ideally incorporate only on substitutional sites. Any incorporation as defect, complex, or on an interstitial site is undesirable. We now consider the conditions that result in incorporation on substitutional sites. As an example, we consider an impurity I that incorporates in GaAs predominantly on a substitutional As site (I_{As}). We also consider the second possibility that the impurity incorporates on an interstitial site (I_{Int}) and neglect all other possibilities. The atomic reaction is then written as

$$Int + I_{As} \leftrightarrow I_{Int} + V_{As} \tag{7.6}$$

where Int and V_{As} represents an interstitial site and an As vacancy, respectively. Application of the law of mass action to Eq. (7.6) yields

$$\frac{[I_{Int}][V_{As}]}{[I_{As}][Int]} = ce^{-\Delta E/kT}. \tag{7.7}$$

The concentration of interstitial sites in a crystal, [Int], is very large and is not significantly altered if some of the impurities occupy interstitial sites. We can therefore rewrite Eq. (7.7) as

$$[I_{Int}]/[I_{As}] = [V_{As}]^{-1}c^*e^{-\Delta E/kT} \tag{7.8}$$

where $c^* = c[Int]$ is a constant.

Two important consequences can be inferred from Eq. (7.8). First, in order to incorporate impurities predominantly on As sites, a large As vacancy concentration is advantageous. This can be achieved in terms of a low V/III flux ratio during growth. This result can be generalized as follows: *The incorporation of impurities on group-V (group-III) sites can be enhanced by a low V/III (high V/III) flux ratio.* Low V/III flux ratios can be advantageous for the incorporation of e.g. S, Se, Te, and C which occupy anion sites. On the other hand, a high V/III flux ratio can be advantageous for e.g. Be-acceptors, Zn and Si which occupy cation sites. It is worthwhile noting that interstitials were considered in Eq. (7.6) as electrically inactive impurities. Similar equations can be written for other electrically inactive configurations such as complexes, impurity pairs, or other defects. Considering other non-active impurity configurations, one arrives at the same conclusion with respect to the V/III flux ratio.

The second consequence of Eq. (7.8) concerns the temperature dependence of the incorporation. Eq. (7.8) predicts that $[I_{Int}]/[I_{As}] \propto \exp(-\Delta E/kT)$, where ΔE is the change in free energy in the reaction of Eq. (7.6). In order to maximize the concentration of impurities on the desired As site, $[I_{As}]$, growth must occur at a low temperature. At low temperature, the exponential term $\exp(-\Delta E/kT)$ is small, i.e. the concentration of impurities on the desired As sites, $[I_{As}]$, is large (see Eq. (7.8)). Consider now the other extreme of very high growth temperatures. For such temperatures (which may be impractical for other reasons) the exponential term $\exp(-\Delta E/kT)$ approaches unity, i.e. the concentration of interstitial impurity atoms will be high. That is, the thermal energy kT is so large that a small energy gain ΔE does not represent a strong driving force for impurities to occupy the 'right' substitutional As site.

The law of mass action thus indicates that low growth temperatures are advantageous to drive impurities on the lowest energy site which is, in the above example, a substitutional site. The lower limit of growth temperatures is not arbitrary but given by kinetic limitations. That is, the growth temperature must be sufficiently high to enable impurities to diffuse on the crystal surface in order to find low energy sites. Growth temperatures that are too low would impose a kinetic limitation on the impurity motion and prevent the impurities from equilibrating at the lowest energy sites.

It has been a general experimental observation that the epitaxial growth of III–V semiconductors at very high temperatures favors a multitude of defects (see, e.g. Razeghi, 1989) including impurity-related defects. This tendency can be explained on the basis of the law of mass action. Most impurity-related defects have lower formation energies than the formation energy of a substitutionally incorporated impurity (less energy is gained by forming an impurity-related defect compared to forming a substitutional impurity). Since the probability of defect creation is proportional to the exponential term of the law of mass action, $\exp(-\Delta E/kT)$, high temperatures will generally increase the formation of defects. Let us consider an example. Assume the difference in formation energy of a specific defect and a substitutional impurity, ΔE, equals 0.5 eV. Consider further the two growth temperatures $T_1 = 500$ K and $T_2 = 1000$ K. The probability of forming the defect at T_2 compared to T_1 increases by a factor of $\exp(\Delta E/kT_1 - \Delta E/kT_2) \cong 330$. That is, for the specific example illustrated above, the concentration of defects increases by about a factor of 330 at the higher of the two growth temperatures.

A further effect of high growth temperatures is the preferred reevaporation of the group-V constituent. The elements P, As, and Sb have a higher vapor pressure than their group-III counterparts. That is, the surface coverage of compound semiconductors with group-V constituents decreases rapidly at high growth temperatures. It can be shown by rate equations that the surface coverage of e.g. GaAs with As is inversely proportional to the As vapor pressure. Consequently, impurity surface reactions that are affected by the group-V element concentration on the surface are strongly influenced by the growth temperature via the group-V surface coverage.

As an example we consider the amphoteric impurity Si in GaAs and consider the exchange reaction on the surface

$$Si_{Ga} + V_{As} \leftrightarrow Si_{As} + V_{Ga} \tag{7.9}$$

where Si_{Ga} and Si_{As} represent Si on a Ga site (donor) and As site (acceptor), respectively; V_{As} and V_{Ga} represent an As and Ga vacancy, respectively. The Si acceptors (Si_{As}) compensate the Si donors (Si_{Ga}). This effect is called autocompensation. The *autocompensation ratio* is defined as the ratio of Si-acceptor concentration to Si-donor concentration, i.e. $[Si_{As}]/[Si_{Ga}]$. Application of the law of mass action yields

$$\frac{[Si_{As}][V_{Ga}]}{[Si_{Ga}][V_{As}]} = c \, \exp(-\Delta E/kT) \, . \tag{7.10}$$

This equation has two terms which strongly depend on temperature, which are $[V_{As}]$ and the exponential term. Both terms increase exponentially at higher temperatures. Since the Ga reevaporation rate is smaller, we assume, within a limited temperature interval, that $[V_{Ga}] \approx$ const. Thus the temperature dependence of $[V_{As}]$ and of the exponential term result in increased autocompensation at high growth temperatures. Such an increased autocompensation has indeed been observed experimentally by Chai et al. (1981)

who found increased autocompensation of Si in GaAs at low V/III flux ratio and high growth temperatures. Qualitatively similar findings were made for Ge- and C-doped GaAs (Künzel et al., 1982; Hoke et al., 1991a and 1991b). A further discussion of autocompensating impurities is given in Sect. 7.3.

The requirement of low growth temperatures is especially important at high doping concentrations. To illustrate the importance of low growth temperatures we consider again the law of mass action and attribute the energy difference, ΔE, to the difference in formation energy between the desired *substitutional* configuration and an undesired, electrically *non-active* configuration, e.g. an impurity complex. (Entropy is neglected since it is usually a small part of the free energy). Next, it will be shown that the difference in formation energy between the two configurations decreases at very high doping concentrations. As mentioned previously, the formation energy is composed of several parts including chemical bond energies, mechanical energies, and electronic energies. Due to band-filling, the electronic energy contribution to the formation energy increases with the free carrier concentration. Electrons entering the conduction band must be excited to the (degenerate) Fermi level rather than to the bottom of the conduction band. That is, electrons (in an n-type semiconductor) require an energy of (see Chap. 3)

$$E_{\mathrm{F}} - E_{\mathrm{D}} \cong E_{\mathrm{F}} - E_{\mathrm{C}} \cong \frac{\hbar^2}{2m^*}(3\pi^2 n)^{2/3} \tag{7.11}$$

where E_{F}, E_{D}, and E_{C} are the Fermi energy, donor level energy, and conduction band edge energy, respectively. The expression on the right-hand side of Eq. 7.11 is valid for extreme degeneracy in an isotropic parabolic conduction band. Equation 7.11 shows that the electronic energy increases with the free carrier concentration with $n^{2/3}$. At high electron concentrations, the Fermi energy can be several hundred meV above the bottom of the conduction band. Thus, the electronic energy required for incorporating a donor increases at high doping concentrations due to the increase in Fermi energy. See Sect. 7.4 for a more detailed discussion of the electronic energies. A second factor which may increase the energy required to incorporate a substitutional impurity at high doping concentrations is elastic strain energy. High doping concentrations tend to increase the elastic strain which can result in an increase in energy to incorporate substitutional impurities (see Chap. 7.5).

The decrease in the difference of formation energies, ΔE, makes electrically non-active impurity defects more likely. In order to reduce the probability of defect formation, low growth temperatures are required, especially at high doping concentrations. This is because the impurity defect is formed with a probability

$\propto \exp(-\Delta E/kT)$ which is smallest at low temperatures. It is, however, important to keep in mind that the growth temperature cannot be arbitrarily low. The growth temperature must be sufficiently high to enable the growing semiconductor surface to reach thermal equilibrium.

Another aspect of low growth temperatures is the higher group-V coverage occurring at low growth temperature due to the strong temperature dependence of the vapor pressure of group-V elements. Group-VI donor impurities (S, Se, Te) which occupy group-V sites, compete with anions. Thus, low growth temperatures may affect the efficient group-VI impurity incorporation due to the high group-V atom coverage at low growth temperatures.

It has been found experimentally for many impurities that the highest electrically active doping concentrations are indeed achieved at low growth temperatures. Chai et al. (1981) found that the highest free carrier concentrations in Si-doped GaAs are obtained at low growth temperatures. Under optimized low temperature conditions, free carrier concentrations as high as 1.6×10^{19} cm^{-3} were reached in Si-doped GaAs (Malik et al., 1992). Hamm et al. (1989) reported very high Be doping levels in Ga$_{0.47}$In$_{0.53}$As at growth temperatures as low as 366 °C. Similar results were reported for Sn-doped Ga$_{0.47}$In$_{0.53}$As (Kawaguchi and Nakashima, 1989; Panish et al., 1990a).

The incorporation of impurities has been modeled in terms of simple kinetic rate equations. Wood et al. (1982), and Poole et al. (1988) took into account an impurity arrival rate on the epilayer (per cm^2 per s), J_I, an impurity surface density, C_I, an impurity desorption rate from the surface, DC_I, and an incorporation rate into the epitaxial layer, KC_I. The change in surface concentration is then given by

$$\frac{dC_I}{dt} = J_I - DC_I - KC_I . \tag{7.12}$$

The authors further assumed that the impurity desorption rate is thermally activated, i.e. $D = D_0 \exp(-E_a/kT)$ where E_a is the activation energy for desorption. The incorporation constant was assumed to be proportional to the growth rate, v_g, according to $K = K' v_g$, where K' is a constant.

Under steady-state equilibrium conditions, the concentration of impurities on the surface does not change, i.e. $dC_I/dt = 0$. Under such steady state conditions one obtains (Poole et al., 1988)

$$J_I = C_I D_0 \, e^{-E_a/kT} + K' C_I v_g \tag{7.13}$$

which allows one to calculate the incorporated impurity concentration in the growing semiconductor according to

$$N_I = (J_I/v_g)/[1 + (D_0 e^{-E_a/kT}/K' v_g)] \tag{7.14}$$

where J_I/v_g is the maximum impurity concentration achievable at the growth rate v_g and the impurity flux J_I. In order to achieve the highest doping concentration J_I/v_g, the desorption rate must be zero.

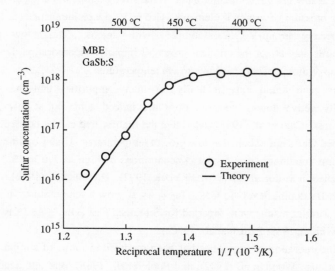

Fig. 7.2. Experimental and theoretical S concentration in GaAs grown by MBE as a function of reciprocal growth temperature. The theoretical curve is obtained from a set of rate equations (after Poole et al., 1988).

Poole et al. (1988) calculated the incorporation of sulfur into GaSb using the rate equation derived above. The result of their experimental and calculated S concentration is shown in Fig. 7.2 as a function of the reciprocal growth temperature. Experimental and theoretical points are in agreement. The parameters used for the calculation were as follows. Maximum impurity concentration (without desorption): $J_I v_g = 1.3 \times 10^{18}$ cm^{-3}, $(D_0/K' v_g) = 1.24 \times 10^{24}$, and $E_a = 3.5$ eV. The authors pointed out that the choice of the activation energy for S reevaporation of $E_a = 3.5$ eV is similar to the value of $E_a = 3.1$ eV found for S reevaporation from the GaAs surface reported by Andrews et al. (1983).

7.1.2 Impurity incorporation in OMVPE

The incorporation of doping impurities in OMVPE is more complicated as compared to MBE. It was shown in the previous section that the rate equation for dopant incorporation consists of different terms, which include the arrival rate, the reevaporation rate, and the incorporation rate. Although the basic rate equation is also valid for OMVPE, the process of doping incorporation is more complicated. In OMVPE, doping precursors are chemical *compounds* rather than doping *elements* as in MBE. The chemical doping compounds in OMVPE may, for example, not react efficiently and may not decompose if they are very stable molecules. Furthermore, the incorporation of doping impurities may depend on the nature of the majority constituents, i.e. the group-III and group-V compounds. Parasitic reactions between the majority constituents and the minority constituents (i.e. doping precursors) can influence the doping incorporation as well.

The **thermodynamic distribution coefficient** K_t is used to quantitatively describe the efficiency of dopant incorporation. The distribution coefficient is defined as the ratio of the mole fraction of the doping impurity in the semiconductor and the partial pressure of the doping precursor in the gas phase of the reactor at the gas–semiconductor interface, i.e. (Stringfellow, 1985, 1986, 1989)

$$K_t = x_D^s / P_D^i \qquad (7.15)$$

where the superscripts s and i refer to 'solid' and 'interface', respectively. The doping composition x_D^s is dimensionless and is the fraction of sublattice sites that are occupied by a doping impurity atom. The thermodynamic distribution coefficient has thus the dimension of reciprocal pressure. In addition to the thermodynamic distribution coefficient given by Eq. 7.15, the compositional distribution coefficient is frequently used (see, for example, Kisker and Zawadzki, 1988). The **compositional distribution coefficient** K_c is simply the ratio of the doping mole fraction in the solid semiconductor and of the doping precursor in the gas phase of the reactor, i.e.

$$K_c = x_D^s / x_D^g = x_D^s / (P_D^g / P_{III}^g) \ . \qquad (7.16)$$

Let us be more specific about the quantities x_D^s and x_D^g. For dopants occupying group-III sites, x_D^s is the fraction of group-III sites in the solid occupied with doping impurities. An analogous statement is valid for impurities occupying group-V sites. For dopants occupying group-III or group-V sites, x_D^g is the ratio of doping precursor and group-III precursor molecules. The x_D^g is thus the ratio of partial pressures of the doping and the group-III precursor. Note that K_c is dimensionless. Doping precursors with $K_c \ll 1$ do not readily incorporate. For example, C does not readily incorporate from CH_4 due to the thermal stability of the methane molecule. Doping impurities with $K_c \gg 1$ readily incorporate into the semiconductor. A distribution coefficient of $K_c \cong 1$ is desirable for the sake of process controllability. The use of the thermodynamic versus the compositional distribution coefficient has advantages and disadvantages. The *compositional* distribution coefficient can be easily determined from the initial source gas doping composition and the doping concentration in the semiconductor. However, the compositional distribution coefficient depends on many other parameters such as pressure, fluid dynamics, properties of the stagnant layer, etc. The thermodynamic distribution coefficient is more appropriate to rigorously model the doping incorporation during OMVPE. The *thermodynamic* distribution coefficient depends on the chemical processes at the interface only. However, it is more complicated to determine the true thermodynamic distribution coefficient experimentally.

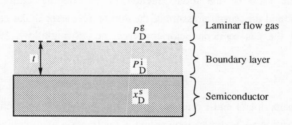

Fig. 7.3. Schematic illustration of the laminar flow region, the stagnant boundary layer of thickness t, and the epitaxial layer grown by OMVPE. The partial pressures of the doping precursor are P_D^g and P_D^i in the gas phase and at the interface, respectively. The doping mole fraction in the solid is x_D^s.

A further complication of the OMVPE doping is given by the *fluid dynamical processes* occurring in OMVPE reactors. These fluid dynamical properties result in a stagnant gas layer in the vicinity of the growing semiconductor surface. The stagnant boundary layer was discussed and illustrated in Chap. 6. The transport of doping precursor molecules and of majority precursor constituents is of clear interest for the modeling of doping incorporation. Stringfellow (1985) considered purely diffusive transport of precursor molecules through the stagnant boundary layer. It was further assumed that no turbulences occur in the source gases, the boundary layer or at their interface. The diffusion transport flux through the boundary layer of thickness t, schematically illustrated in Fig. 7.3, can be expressed as

$$j = \frac{D_D (P_D^g - P_D^i)}{RTt} \tag{7.17}$$

where D_D is the diffusion coefficient of the doping precursor, R is the gas constant, and T is the temperature. The partial pressures of the doping precursor are P_D^g in the gas phase above the stagnant boundary layer and P_D^i at the interface between the boundary layer and the semiconductor.

Pyrolysis of doping precursors or of majority constituents can either occur in or near the boundary layer or on the growing semiconductor substrate during OMVPE. Tsang (1989) pointed out that this is in contrast to CBE where the pyrolysis occurs only at the epitaxial crystal surface. The lack of knowledge of the exact location of pyrolysis further complicates the modeling of the doping incorporation process in OMVPE.

Stringfellow (1989) developed a model for the incorporation of a single impurity species in a binary III–V semiconductor and assumed that (i) the group-III element determines the growth rate and that the group-III element is depleted at the semiconductor–gas interface, (ii) that the V/III flux ratio is larger than unity, and (iii) that the diffusion coefficients of all gaseous precursor molecules through the boundary layer are equal. These assumptions allow one to obtain on the basis of Eq. (7.17) a simple expression for the doping concentration in the semiconductor, i.e.

$$x_D^s = \frac{j_D}{J_{III}} = \frac{P_D^g - P_D^i}{P_{III}^g} \tag{7.18}$$

where j_D and j_{III} are the fluxes of the doping precursor and the group-III precursor through the stagnant boundary layer, respectively. Note that the validity of Eq. (7.18) requires complete pyrolysis of all constituents. Inserting the thermodynamic distribution coefficient (see Eq. (7.15)) into Eq. (7.18) yields

$$x_D^s = \frac{P_D^g - (x_D^s/K_t)}{P_{III}^g} \tag{7.19}$$

where thermodynamic equilibrium conditions are assumed. Stringfellow (1989) differentiated between two cases.

In the first case, the vapor pressure of the impurity is low and all dopants reaching the crystal surface are incorporated, i.e. the dopant is depleted at the interface $P_D^i \ll P_D^g$. Equation (7.18) then yields the simple expression

$$x_D^s = P_D^g/P_{III}^g . \tag{7.20}$$

In this case the dopant distribution coefficient is inversely proportional to P_{III}^g; i.e.

inversely proportional to the growth rate. This case is called the *mass-transport-limited* case, since the incorporation of dopants is limited by the transport of dopants through the boundary layer.

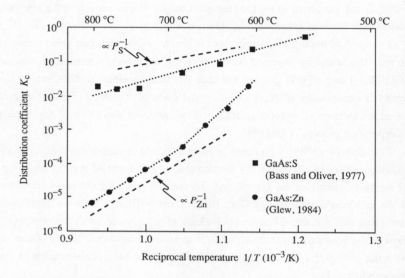

Fig. 7.4. Compositional distribution coefficient versus reciprocal growth temperature for S- and Zn-doped GaAs grown by OMPVE. The dotted lines represent a fit to the experimental data. The dashed lines are the reciprocal vapor pressures of S and Zn in arbitrary units (after Stringfellow, 1986).

In the second case, the dopant vapor pressure is high and most of the dopant is reevaporated from the epitaxial crystal surface. For this condition, $P_D^i \cong P_D^g$ and one obtains from Eq. (7.15)

$$x_D^s = K_t P_D^g . \tag{7.21}$$

It is of practical value to compare Eq. (7.21) with the compositional distribution coefficient of Eq. (7.16). Elimination of x_D^s from the two equations yields

$$K_c = K_t P_{III}^g \tag{7.22}$$

which is frequently used to describe the incorporation of doping impurities. In the second case, the thermodynamics at the epitaxial crystal-to-gas interface determines the incorporation of doping impurities. Furthermore, the distribution coefficient is independent of P_{III}^i and the growth rate. The dependence of the dopant incorporation on the group-V partial pressure was also considered

(Stringfellow, 1986, 1989).

Doping impurities with a high vapor pressure tend to reevaporate before they are incorporated into the semiconductor. Since the vapor pressure depends on temperature, the distribution coefficient of such dopants decreases with increasing temperature. It is obvious that the distribution coefficient of such impurities is inversely proportional to the vapor pressure of the impurity element. Experimental data of Glew (1984) and Bass and Oliver (1977) on the Zn and S dopant incorporation in GaAs were used by Stringfellow (1986) for a comparison between the distribution coefficients of these dopants and their vapor pressure. The comparison in shown in Fig. 7.4. The dotted lines are a fit to the experimental points. The dashed lines reflect the temperature dependence of the inverse vapor pressure. Comparison of the dotted (experimental) and dashed (theoretical) lines in Fig. 7.4 reveal that the two functional dependences are in good qualitative agreement. The agreement suggests that the incorporation of high vapor pressure doping elements such as Zn and S are indeed determined by reevaporation of the impurities from the epitaxial crystal surface. However, the kink in the experimental data of Zn indicates that the vapor pressure is not the only quantity controlling the doping incorporation.

7.2 Doping incorporation by implantation

Ion implantation is a powerful technique for incorporating doping impurities into III–V semiconductors. In this technique, charged impurity atoms are accelerated in an electric field and penetrate into the semiconductor target. Typical acceleration energies range from 10 keV to 500 keV. Typical doses of impurity atoms implanted in III–V semiconductors range from 10^{11} cm^{-2} to 10^{15} cm^{-2}. The scattering processes of impurity ions in the solid cause considerable structural damage which results in a small electrical activity of implanted impurities. Subsequent high-temperature anneals of the semiconductors reduce the damage and can increase the electrical efficiency of the impurity ions to near 100%. Typical annealing cycles are 700 to 1100 °C for 5 to 100 s. Ion implantation is also used to form highly resistive regions in semiconductors using hydrogen ions (protons) and oxygen ions. Hydrogen implantation leaves defect traces which, if not annealed, act as deep traps. Oxygen forms deep levels in III–V semiconductors and renders the material highly resistive. Several reviews have become available on ion implantation in III–V semiconductors and post-implantation annealing (see, for example, Anholt et al., 1988; Gill and Sealy, 1986; Sharma, 1989).

Ion implantation makes possible controlled doping with lateral selectivity. The lateral selectivity can be achieved by either scanning and blanking the ion beam or by implantation masks. The use of implantation masks allows one to obtain lateral doping distributions that are self-aligned with respect to the edge of the mask. Furthermore, the implantation dose and implantation depth can be precisely controlled. Excellent uniformities can be obtained over large areas of the semiconductor. A drawback of doping by ion implantation is the inherent width of the doping profiles in longitudinal and lateral direction. The finite width of the doping distribution is caused by the statistical nature of scattering processes of doping ions with lattice atoms (see, for example, Feldman and Mayer, 1986). Abrupt doping profiles are not feasible at high acceleration energies, i.e. large implantation depths.

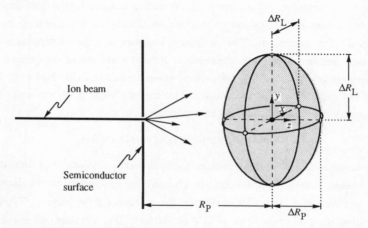

Fig. 7.5. Schematic illustration of the impurity distribution resulting from a focused ion beam propagating along the z-axis. The projected range of the implantation is R_p and the projected straggle is ΔR_p. The lateral straggle (along the x- and y-directions) is ΔR_L.

The spatial distributions of impurities implanted into the semiconductor can be described, in the simplest approximation, by gaussian distributions (see, for example Pickar, 1975). The basic implantation geometry of an infinitesimal small beam diameter is shown in Fig. 7.5. Ions propagate along the z-direction and enter the semiconductor at the origin of the cartesian coordinate system. The mean depth of the impurities is called the ***projected range***, R_P. The position standard deviation of the doping distribution in the projected and the lateral direction are ΔR_P and ΔR_L, respectively; ΔR_P and ΔR_L are called ***projected straggle*** and ***longitudinal straggle***, respectively. The quantities are illustrated in Fig. 7.5. The distribution of ions implanted into a semiconductor is thus given by

(Pickar, 1975)

$$N(x,y,z) = \frac{N_t}{(2\pi)^{3/2}\Delta R_P \Delta R_L^2} \exp\left[-\left(\frac{z-R_P}{\sqrt{2}\,\Delta R_P}\right)^2\right] \exp\left[-\frac{x^2+y^2}{(\sqrt{2}\,\Delta R_L)^2}\right]$$

(7.23)

where $N(x,y,z)$ is the ion concentration (per cm^3) and N_t is the total number of implanted ions. Thus $\int\int\int N(x,y,z)\,dxdydz = N_t$. Note that the distribution of Eq. (7.23) can be expressed in the cylindrical coordinates r and z with $r^2 = x^2 + y^2$.

If the ion beam is scanned uniformly over a wafer, the resulting impurity distribution depends only on the z-direction not on the x- and y-directions. The doping concentration is then given by

$$N(z) = \frac{N^{2D}}{\sqrt{2\pi}\,\Delta R_P} \exp\left[-\left(\frac{z-R_P}{\sqrt{2}\,\Delta R_P}\right)^2\right]$$

(7.24)

where N^{2D} is the implantation dose, i.e. the density of implanted ions (per cm^2).

The gaussian function of Eq. (7.24) is a simple approximation for true doping distribution. The two parameters R_P and ΔR_P, which are the median and the standard deviation of the gaussian distribution, can be determined from measured impurity depth profiles of ion implanted semiconductors. Secondary ion mass spectrometry is frequently used for such an impurity depth profiling. The depth profiling yields a measured value for the impurity concentration, $N(z)$, which in turn allows one to determine the projected range and projected straggle. The median and the standard deviation of a given sample are calculated according to (see, for example, Kreyszig, 1972; Ryssel and Ruge, 1986; Ryssel and Biersack, 1986; Selberherr, 1984)

$$R_P = \langle z \rangle = (N^{2D})^{-1} \int_0^\infty z\, N(z)\,dz ,$$

(7.25)

and

$$\Delta R_p = \sqrt{\langle z^2 \rangle - \langle z \rangle^2}$$

(7.26a)

where

$$\langle z^2 \rangle = (N^{2D})^{-1} \int_0^\infty z^2 \, N(z) \, dz \,, \tag{7.26b}$$

and $\langle z \rangle^2$ is obtained by taking Eq. (7.25) to the second power.

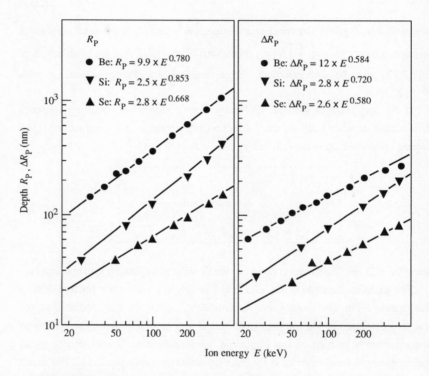

Fig. 7.6. Measured projected range (R_p) and projected straggle (ΔR_p) for Be, Si and Se implanted into GaAs at ion energies ranging from 20 keV to 400 keV. The six equations for R_p and ΔR_p are power-law fits in which R_p, ΔR_p and E are measured in nm and keV, respectively. The impurities with a dose of 1 x 10^{13} cm^{-2} were implanted into bare wafers at a tilt and rotation angle of 7° and 9°, respectively (after Anholt et al., 1988).

Experimental values for the projected range and the projected straggle of Be, Si, and Se ions implanted into GaAs are shown in Fig. 7.6 as a function of the ion energy (Anholt et al., 1988). The projected range and the projected straggle are found to follow a power-law dependence. The equations shown in Fig. 7.6 are valid if the units of the energy E are in keV and the units of R_P and ΔR_P are in nm. The projected straggle is typically 0.2 to 0.5 times the projected range.

Impurity profiles of ion-implanted III–V semiconductors are frequently asymmetric (i.e. non-gaussian) and skewed towards or away from the semiconductor surface. The skewedness of the implanted profiles is characterized by the parameter γ (Anholt et al., 1988) given by

$$\gamma = \langle (z - R_P)^3 \rangle / \Delta R_P^3 . \tag{7.27}$$

Positive values of γ of the order of unity and larger describe distributions decreasing approximately exponentially at large depths. Negative values of γ describe distributions skewed towards shallower depths.

Monte Carlo computer simulations were employed to calculate the distribution of ions implanted in semiconductors (see, for example, Biersack and Haggmark, 1980; Biersack and Ziegler, 1982; Cristel et al., 1980; Gibbons et al., 1975; Ziegler et al., 1985). The numerical Monte Carlo calculations are performed due to the lack of reliable analytic formulas for the ion distribution in implanted semiconductors. The traces and final positions of 25 H^+ ions implanted into GaAs at an ion energy of 50 keV are shown in Fig. 7.7. The plot was obtained using the TRIM Monte Carlo Code (Ziegler et al., 1991).

If the Monte Carlo calculation shown in Fig. 7.7 is performed with several thousand ions, a reliable ion distribution is obtained. The distributions of Si and Be in GaAs and InP are shown in Fig. 7.8 for implantation energies ranging from 10 keV to 500 keV. 50 000 ions were used for each distribution calculated by TRIM (Ziegler et al., 1991). The implantation depth increases with decreasing atomic number for the same implantation energy. Si and Be are popular doping impurities used for n-type and p-type doping, respectively. Hydrogen implantation (see Chap. 9) is used to render the semiconductor highly resistive. Such proton bombardment is frequently used, for example, in semiconductor lasers to confine the current flow to a well-defined area.

Donor and acceptors implanted into III–V semiconductors are not fully activated after the implantation. The lack of activation is due to (i) damage introduced during the implantation process and (ii) non-substitutional sites occupied by the donors and acceptors. The degree of activation is characterized by the *activation efficiency* η_a. The activation efficiency is defined as the ratio of free carrier density (n, p) and implanted ion density (N_D, N_A), that is

$$\eta_a = \frac{n^{2D}}{N_D^{2D}} ; \; \eta_a = \frac{p^{2D}}{N_A^{2D}} \tag{7.28}$$

for n-type and p-type implantation, respectively. The superscript 2D (two-dimensional) refers to carrier and ion densities per unit area. The activation efficiency after implantation but before annealing can be as low as 10%.

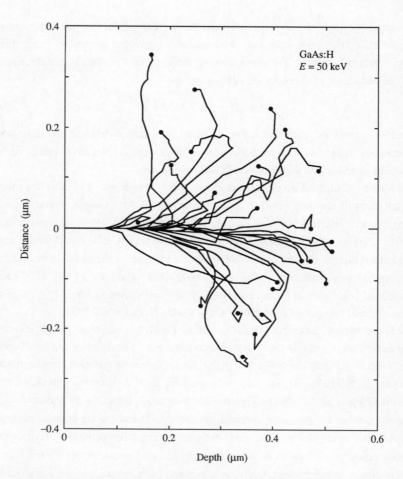

Fig. 7.7. Collision cascades of 25 hydrogen ions in GaAs involving many elastic and inelastic scattering events with the GaAs lattice. All hydrogen ions enter the lattice at the same point.

Ion implantation is usually performed with the implant target at room temperature. However, implantation at higher temperatures may be advantageous for a high donor and acceptor activation. Woodhouse et al. (1984) reported improved activation of Se in InP for an implantation target heated to 180 °C as compared to room temperature implantation.

Post-implantation annealing is used to reduce the concentration of defects and to increase the dopant activation. Rapid thermal annealing, i.e. a short, high-

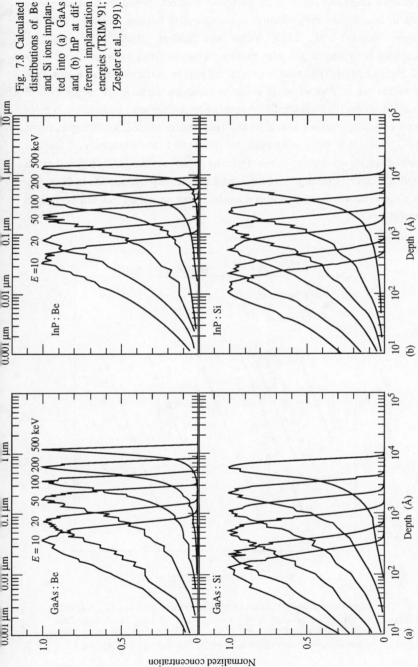

Fig. 7.8 Calculated distributions of Be and Si ions implanted into (a) GaAs and (b) InP at different implantation energies (TRIM '91; Ziegler et al., 1991).

temperature annealing cycle is the preferred method. Rapid thermal annealing results in less dopant redistribution as compared to furnace annealing (see, for example, Vescan et al., 1982; Maier and Selders, 1986). The annealing atmosphere is usually a gas with reducing characteristics such as forming gas (85% N_2/15% H_2). The annealing cycle can also be performed in an atmosphere with excess As or P pressure, in order to avoid the preferential evaporation of anions (As, P, or Sb) from the III–V semiconductor surface. In addition to rapid thermal annealing, furnace annealing and laser annealing (see, for example, Ferris et al., 1979) have been employed for implanted semiconductors. Furnace annealing yields lower activation efficiencies as compared to rapid thermal annealing. Laser annealing is mostly used for annealing of implants in Si at very high temperatures. In III-V semiconductors, laser annealing has not been used extensively.

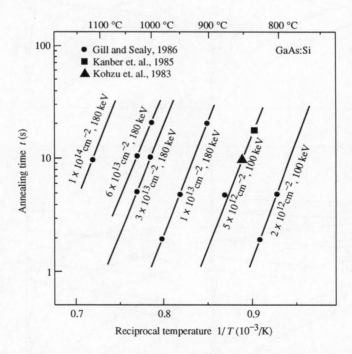

Fig. 7.9. Optimum annealing times and temperatures (for high activation and low diffusion) for Si-implanted GaAs with doses of $2 \times 10^{12} cm^{-2}$ to $1 \times 10^{14} cm^{-2}$ and implantation energies of 100 keV to 180 keV (after Gill and Sealy, 1986, reprinted by permission of the publisher, The Electrochemical Society, Inc.).

The annealing parameters, i.e. temperature and time, must be optimized in order to achieve maximum activation efficiency of implanted impurities. It is desirable to select a long annealing time and a high temperature, in order to anneal out the damage created by implantation and to maximize the activity of the dopant. On the other hand, short annealing times and low temperatures avoid excessive diffusion of the doping impurity. The optimum annealing times for GaAs implanted with Si donors at different implantation energies and doses are shown in Fig. 7.9 (Gill and Sealy, 1986). The plot also contains data points of Kanber et al. (1985) and Kohzu et al. (1983). Inspection of Fig. 7.9 shows that the optimum annealing temperature increases with both the implantation energy and the implantation dose. The requirement of higher temperature can be qualitatively explained by the larger damage caused by higher implantation energies and doses. Figure 7.9 further illustrates that optimum annealing times range from 2 to 20 s for GaAs:Si.

Dopants used for intentional p- and n-type doping have different characteristics which depend on the chemical nature of the implanted impurities. Sharma (1989) summarized the characteristics of Si, Se, Be, Mg, and Hg in InP. In the following, the characteristics of these dopants are summarized. The reader is also referred to Chap. 5 for further discussion.

Silicon is used as an n-type dopant in many III–V semiconductors. Si, as are all group-IV impurities, is amphoteric. The Si activation efficiency prior to annealing is low. A significant increase of the activity occurs during annealing (Bahir et al., 1989). Rapid thermal annealing is the preferred activation process in order to reduce anion out-diffusion. Furthermore, rapid thermal processing was reported to reduce the Si diffusion tail (Krautle, 1988). Relatively little redistribution is generally observed for Si-implanted GaAs and InP during annealing (Bahir and Merz, 1988).

Selenium (Se) is not amphoteric (in contrast to Si), and has a high activation efficiency. Peak electron concentrations $> 10^{19}$ cm^{-3} were obtained in InP (Donnelly and Hurwitz, 1980; Gill and Sealy, 1984). On the other hand, Se is a heavier atom than Si and causes more damage during the implantation process. Hot implants were used to minimize the generation of damage during implantation (Woodhouse et al., 1984). Furthermore, Se is a faster diffuser than Si which results in broader doping profiles after annealing.

Beryllium is a popular p-type dopant used for implantation. Be is a stable impurity at low and intermediate doses ($\leq 10^{13}$ cm^{-2}). At high doses (e.g. 10^{14} cm^{-2}), strong redistribution occurs during annealing in GaAs and InP (see, for example, Wilson et al., 1988). Furthermore, Be is a small atom and causes less damage as compared to other p-type dopants (Sharma, 1989). It is interesting to note that As and P co-implantation results in a higher electrical activation of Be (Wang, 1987). A higher anion concentration reduces the probability of Be occupying anion sites, i.e. enhances the probability of substitutional, active Be on cation sites.

Magnesium has been used as a p-type dopant in III–V semiconductors. It is heavier than Be and therefore causes more damage. High temperatures during implantation help to reduce the damage incurred during Mg implantation (Inada et al., 1981).

Ion implantation is used for a wide range of purposes in electronic III–V semiconductor structures and devices including (i) implantation for ohmic contact formation (ii) self-aligned implantation in field-effect transistors to reduce source and drain parasitic resistances, (iii) implantation of the electron-channel in field-effect transistors, (iv) formation of pn-junctions in junction field-effect transistors, (v) base regions in hetero and homo bipolar transistors, and (vi) highly conductive sub-collector regions in bipolar transistors. In photonic III–V semiconductor devices, ion implantation is frequently used to create highly resistive regions for current confinement. Oxygen and hydrogen ions (protons) are used for such isolation implantations.

7.3 Doping incorporation by diffusion

Doping impurities can be incorporated into III–V semiconductors by diffusion from a surface reservoir into the bulk. This method had great importance in the infancy of III–V device processing (see, for example, Weimer, 1962; Heime, 1967). However, more powerful doping techniques such as doping during epitaxial growth or by ion implantation rendered doping incorporation by solid-state diffusion less important. The major drawback of doping incorporation by diffusion are the limited possibilities for determining the shape of the doping profile; the concentrations of doping profiles created by diffusion are highest at the surface and continuously decrease into the bulk. Doping profiles with near arbitrary shape oppose the very nature of the diffusion process. Thus, doping incorporation by diffusion will be discussed in brevity due to the limited freedom offered by this doping technique and due to its limited use in present III–V processing technology.

The basic experimental configuration for impurity diffusion is shown in the inset of Fig. 7.10. The dopant reservoir can be considered as infinite, since it is assumed that only a minor fraction of doping impurities diffuses into the semiconductor. The fickian diffusion equation (see Chap. 8) must be solved for

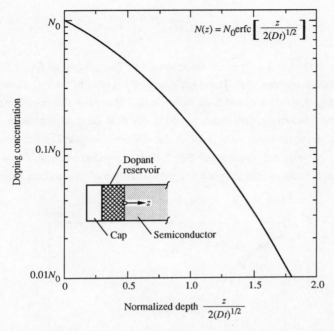

Fig 7.10. Complementary error function (erfc) describing the diffusion of impurities with surface concentration N_0 from an 'infinite' reservoir into a semiconductor. The inset shows an experimental layer sequence consisting of protective cap, dopant reservoir, and semiconductor.

$$N(z) = N_0 \quad \text{for} \quad z < 0 \quad \text{and} \quad t \gtrless 0, \qquad (7.29a)$$

$$N(z) = 0 \quad \text{for} \quad z > 0 \quad \text{and} \quad t < 0. \qquad (7.29b)$$

Note that the concentration of impurities on the surface equals N_0 independent of the time. The solution of the fickian diffusion equation for these boundary conditions is

$$N(z) = N_0 \ \text{erfc} \left(\frac{z}{2\sqrt{Dt}} \right) \qquad (7.30)$$

where erfc is the complementary error function and D is the diffusion coefficient of the doping impurity in the semiconductor. The complementary error function can be expressed in terms of an integral

$$\operatorname{erfc}(\alpha) = 1 - \operatorname{erf}(\alpha) = \frac{2}{\sqrt{\pi}} \int_{\alpha}^{\infty} e^{-t^2} \, dt \tag{7.31}$$

where $\alpha = \frac{1}{2} z/\sqrt{Dt}$ and erf is the error function. The integral of Eq. (7.31) can be easily solved numerically. The doping impurity profile of the erfc function is shown in Fig. 7.10 on a normalized depth scale. Note that the complementary error function decreases with depth more rapidly than the exponential function. Frequently, the square root $\sqrt{Dt}$ is called the **diffusion length**, L_D. At the depth of $z = L_D = \sqrt{Dt}$, the abscissa of Fig. 7.10 has a value of $\frac{1}{2} z/\sqrt{Dt} = \frac{1}{2}$ at which the concentration has dropped to approximately half its interface value N_0.

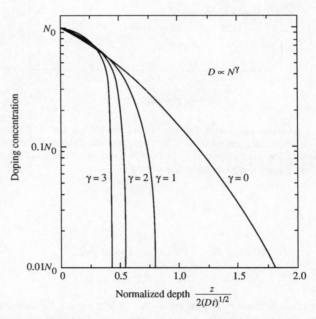

Fig. 7.11. Profile of impurities diffused into a semiconductor from an "infinite" reservoir with surface concentration N_0. The diffusion coefficient is assumed to depend on the concentration according to the power law $D \propto N^\gamma$ with $\gamma = 0, 1, 2$ and 3.

Equation (7.30) is valid for impurities with a diffusion coefficient that is independent of concentration. For many experimental cases, however, the diffusion coefficient depends on the concentration (see Chap. 8). The diffusion from an infinite reservoir with a concentration dependent diffusion coefficient was numerically solved by Weisberg and Blanc (1963). The result of the diffusion calculation is displayed in Fig. 7.11 for a diffusion coefficient that depends on concentration according to a power law, i.e. $D \propto N^{\gamma}$ with $\gamma = 0, 1, 2$, and 3. The profiles become increasingly abrupt for higher values of γ.

The simple theoretical analysis provided above illustrates that doping in terms of impurity diffusion from an infinite reservoir has only two degrees of freedom which are N_0 and $L_D = \sqrt{Dt}$. If N_0 is a fixed value such as the solubility limit, only one degree of freedom, namely the diffusion length remains. Thus, diffusion is much less versatile as compared to the more flexible techniques of doping during epitaxial growth and by implantation.

The high temperatures required for doping incorporation by diffusion can deteriorate the surface and near-surface crystalline quality of the semiconductor. If III–V semiconductors are heated to high temperatures, preferential evaporation of group-V elements occurs. The loss of P, As, or Sb has several consequences such as non-stoichiometry, increase in point defect (vacancy) concentration, formation of group-III precipitates, and degradation of the surface morphology. The loss of group-V elements can be avoided by (i) appropriate counter pressures in the diffusion furnace or (i) coating of the sample with a capping layer (see Fig. 7.10) that prevents the out-diffusion of group-V elements.

Various capping layers were employed to reduce the out-diffusion of group-V elements from the semiconductor. Singh et al. (1988) employed SiO_2, borosilicate glass, phosphosilicate glass, and Al_2O_3 to prevent preferential As loss from a heated GaAs surface. The same layers were also employed by Camlibel et al. (1984) for capping InP and $Ga_{0.47}In_{0.53}As$ during Zn diffusion. Al_2O_3 has also been used as a capping layer by Schmitt et al. (1984) for Zn diffusion in InP. Silicon nitride (Si_3N_4) films are frequently used as a capping layer on GaAs and InP based compounds, particularly for diffusion and activation processes following ion implantation (see, for example, De Souza et al., 1990; Valco et al., 1989). An alternative to separate capping layers are single layers that act as both dopant reservoir and capping layers. Emulsions have been employed for this purpose (Arnold et al., 1980, 1984). Such emulsions can be spun on by a conventional resist spinner (Heime, 1989). Upon heating, the emulsions solidify forming a solid mixture of SiO, SiO_2 and the diffusing element. The SiO and SiO_2 fulfill two purposes, namely passivation of the surface and preventing group-V out-diffusion.

Doping incorporation by diffusion can be done in an *open* or *closed system*. The use of capping layers discussed above makes possible the use of 'open' systems, i.e. conventional furnaces with an inert or reducing atmosphere surrounding the wafer. In contrast, a 'closed' system can contain the wafer in a sealed ampoule which is either evacuated or contains an inert gas. In addition, the closed system can contain a solid group-V element (P, As, or Sb). Upon heating the ampoule, the group-V element sublimes and generates a group-V-rich atmosphere which reduces out-diffusion of group-V elements from the wafer.

Diffusing impurity elements can be provided from either the gas phase or the solid phase. In the case of gas phase provision, the doping precursor flows over the sample, pyrolyzes, and incorporates into the semiconductor. In the case of solid phase provision, the diffusing element is contained in a solid film on the wafer. Inorganic materials have been employed for the diffusion source. Examples for the solid sources are fluorides (ZnF_2, see Camlibel et al., 1984), nitrides ($Zn_2 N_3$, see Arnold et al., 1984), and phosphides ($Zn_3 P_2$, see Schmitt et al., 1984). In addition to unmasked diffusion of an entire wafer, masked diffusion was also reported (Arnold et al., 1984). The authors used $Al_2 O_3$ as a mask. However, problems arose due to crystallization of the diffusion source at the edges of the $Al_2 O_3$ mask which made subsequent photolithographic processes difficult.

The technology of *ohmic contacts* in III–V semiconductors is partially based on the diffusion of doping impurities from the surface into the bulk. It is quite clear from several reviews of ohmic contact technology in III–V semiconductors that processes other than diffusion play a major role (see, for example, Braslau, 1981; Kuphal, 1981). Detailed studies show that new phases with varying domain size appear after the alloying process. The metallization used for ohmic contact formation in III–V semiconductors is summarized in Table 7.1.

Table 7.1. *Metallization for ohmic contacts*

Metallization	Typical composition	Conductivity
AuGe	88%/12%	n-type
AuSn	95%/5%	n-type
AuBe	99.1%/0.9%	p-type
AuZn	95%/5%	p-type

Frequently multilayer metallizations are used to provide a better surface morphology and lower contact resistances after annealing. As an example, we consider the AuGe/Ni/Au metallization for n-type semiconductors. Typical thicknesses are 1000 Å, 500 Å, and 2000 Å for the eutectic AuGe, Ni, and Au, respectively. Ni has a sufficiently low surface tension preventing the underlying

AuGe film from 'balling up'. Without the Ni, AuGe contacts have the tendency to form islands during the alloying process (balling up) which degrades the electrical characteristics of the contact. The final Au top layer provides a low resistance contact pad suitable for bonding and probing.

7.4 Amphotericity and autocompensation

Group-IV impurities such as C, Si, Ge, and Sn can occupy substitutional cation as well as anion sites in III–V semiconductors. These impurities are shallow donors and shallow acceptors on group-III and group-V sites, respectively. This characteristic of group-IV impurities in III–V semiconductors is called *amphotericity*. The origin of this word lies in the Greek word *amphora* which represents an ancient jar with two identical handles opposing each other; the word *amphoteric* means 'having two characters'. The two-sided characteristic of amphoteric impurities is undesirable and results in a partial or full compensation of donors by acceptors or vice versa. *Autocompensation* occurs if the compensation is due to the same chemical species of impurities, i.e. for amphoteric group-IV impurities in III–V semiconductors. It is obvious that the characteristics of amphotericity and autocompensation are uniquely related to C, Si, Ge, and Sn in III–V semiconductors and cannot occur in elemental semiconductors such as Si. In the following, the incorporation of an amphoteric impurity into a compound semiconductor is considered. The discussion is based on the law of mass action. A discussion of the thermodynamics of amphoteric impurities has also been given by Teramoto (1972) and by Munoz-Yague and Baceinedo (1982).

The degree of compensation of amphoteric impurities in III–V semiconductors depends on the doping concentration. It is generally found that the degree of compensation of such amphoteric impurities increases at high doping concentrations. Next, the dependence of the compensation on the doping concentration is analyzed quantitatively. As an illustrative example, Si-doped GaAs is considered. However, the results of the analysis have general validity and thus apply to all amphoteric impurities (C, Si, Ge, Sn) in III–V semiconductors. The incorporation of Si into the host GaAs is assumed to occur during epitaxial growth. Three different types of reactions can be distinguished for such an incorporation, namely the (i) host chemical reactions, the (ii) impurity incorporation reactions, and the (iii) electronic reactions. The host chemical reactions for incorporation of Ga and As from the gas phase are given by

$$Ga_{gas} + V_{Ga} \leftrightarrow Ga_{Ga} \tag{7.32a}$$

$$\tfrac{1}{2}As_{2,gas} + V_{As} \leftrightarrow As_{As} \tag{7.32b}$$

$$Ga_{gas} + \tfrac{1}{2}As_{2,gas} \leftrightarrow GaAs \tag{7.32c}$$

where V_{Ga} and V_{As} are a Ga vacancy and an As vacancy, respectively. The formation of point defects (e.g. Ga_{As}) is neglected. Amphoteric impurities can be incorporated either on anion or on cation sites. The impurity incorporation reactions are then given by

$$Si_{gas} + V_{Ga} \leftrightarrow Si_{Ga} \leftrightarrow Si_{Ga}^{+} + e \tag{7.33a}$$

$$Si_{gas} + V_{As} \leftrightarrow Si_{As} \leftrightarrow Si_{As}^{-} + h \tag{7.33b}$$

where e and h represent a free electron and hole, respectively. Note that any incorporation of Si into interstitial sites is neglected. We finally have the electronic reaction

$$e + h \leftrightarrow 0 . \tag{7.34}$$

In an n-type semiconductor this generation/recombination reaction is associated with an energy of magnitude $E_F - E_V$.

We assume that the Si donor and acceptor impurities are ionized, i.e. we restrict our considerations to room temperature or typical growth temperatures. The sum of the impurity incorporation reactions (Eqs. (7.33a and b)), taking into account the electronic reaction (Eq. (7.34)), yields

$$Si_{Ga}^{+} + 2e + V_{As} \leftrightarrow Si_{As}^{-} + V_{Ga} . \tag{7.35}$$

Finally, elimination of V_{Ga} and V_{As} by means of the host chemical reactions (Eqs. (7.32)) yields

$$Si_{Ga}^{+} + 2e + As_{As} + Ga_{gas} \leftrightarrow Si_{As}^{-} + Ga_{Ga} + \tfrac{1}{2}As_{2,gas} . \tag{7.36}$$

We apply the law of mass action to this reaction and obtain

$$\frac{N_A^{-} \, [Ga_{Ga}] \, [\tfrac{1}{2}As_{2,gas}]}{N_D^{+} \, n^2 [As_{As}][Ga_{gas}]} = K \tag{7.37}$$

where the brackets [...] represent the *concentrations* of the reagents and $N_A^{-} = [Si_{As}^{-}]$ and $N_D^{+} = [Si_{Ga}^{+}]$. The equilibrium constant K depends only on temperature. Equation (7.37) can be simplified by using $[Ga_{Ga}]/[As_{As}] = 1$ which is valid for stoichiometric GaAs. Further simplification results by assuming that the V/III flux ratio is constant and independent of the doping level. Then Eq. (7.37) can be written as

$$\frac{N_A^-}{N_D^+} = K^* n^2 \tag{7.38}$$

where K^* is a constant. Thus, the compensation ratio N_A^- / N_D^+ increases with the square of the free carrier concentration. The equation illustrates that the compensation ratio becomes *vanishingly small* for sufficiently small free carrier concentrations, i.e. $N_A^- \ll N_D^+$ for small N_D^+. Next we consider the high concentration limit and substitute $n = N_D^+ - N_A^-$ into Eq. (7.38) which yields

$$\frac{N_A^-}{N_D^+} = K^* (N_D^+ - N_A^-)^2 \tag{7.39}$$

Analysis of this equation reveals that $N_D^+ > N_A^-$ for all concentrations. Simultaneously the equation can only be satisfied at very high concentrations if $N_D^+ \cong N_A^-$. As a result, the compensation ratio approaches $N_A^- / N_D^+ \leq 1$ at high doping concentrations (see also Sect. 7.5.4). Thus, strong compensation is expected to occur at high doping concentrations for all amphoteric impurities in III–V semiconductors.

Note that the above considerations are valid only if equilibrium conditions are indeed established on the surface as well as in the bulk of the semiconductor. Furthermore, the reaction paths of the impurities must be complete, i.e. all relevant reaction paths must be included. If, for example, a major fraction of impurities were to incorporate on interstitial sites, then this reaction path would have to be included as well.

The above considerations are valid if the entire system is in equilibrium. During actual epitaxial growth, the crystal *surface* is in equilibrium, while the *bulk* may be far from equilibrium (see Sect. 7.1). Before an impurity is incorporated into the semiconductor, the impurity atom is chemisorbed to a (reconstructed) lattice site on the semiconductor surface. If the surface site is either an anion or cation site, the following exchange reaction can be written for Si on a GaAs *surface*.

$$\overset{\Delta G_s}{Si_{As} + V_{Ga} \leftrightarrow Si_{Ga} + V_{As}} \tag{7.40}$$

where ΔG_s is the Gibbs free energy, i.e. the sum of formation energy and the change in entropy $(T\Delta S)$ that the system experiences during the surface reaction. The equilibrium concentration of the reaction products is determined by the law of mass action

$$\frac{[Si_{As}][V_{Ga}]}{[Si_{Ga}][V_{As}]} = K_s = c \exp\left[-\frac{\Delta G_s}{kT}\right] \tag{7.41}$$

where the brackets [...] refer to the concentration of the respective quantity and c is a constant. Since Si in the above example is a doping impurity, its concentration is much smaller than the concentration of both As and Ga vacancies on the surface. Because of the small impurity concentration, the vacancy concentration, $[V_{As}]$ and $[V_{Ga}]$ is not changed significantly during the reaction of Eq. (7.40). Thus the vacancy concentration can be considered as constant and the law of mass action can be rewritten as

$$\frac{[Si_{As}]}{[Si_{Ga}]} = \frac{[V_{As}]}{[V_{Ga}]} c \exp\left[-\frac{\Delta G_s}{kT}\right] \cong \text{const.} \tag{7.42}$$

The equation illustrates the relevant factors for impurity incorporation. First, the **energy factor** $\exp(-\Delta G_s/kT)$ determines the degree of autocompensation of amphoteric impurities. That is, if $\Delta G_s/kT$ is a large quantity, autocompensation will be negligible for a wide range of growth conditions. An example for an impurity dominated by energies is C in GaAs; the autocompensation of this impurity depends little on the growth conditions (see Chap. 5 and Hoke et al., 1991a and 199b). Second, the **kinetic factor** $[V_{As}]/[V_{Ga}]$ which depends on the growth technique and the growth conditions, determines the incorporation and compensation of amphoteric impurities, provided the energy factor does not dominate. An example for an impurity dominated by kinetics is Ge in GaAs; the autocompensation of Ge depends strongly on the growth conditions, such as the V/III ratio and growth temperature (see Chap. 5 and Künzel et al., 1982). It is useful to keep in mind that the incorporation of impurities is a very complex process. Molecular dynamics calculations (Gilmer, 1992) show that the concentrations of Si_{As} and Si_{Ga} on the surface are not necessarily transferred into the bulk since Si in the near-surface region (1-3 monolayers) has a higher diffusivity than in the bulk. In the near-surface region, Si can change sites more easily than in the bulk.

The incorporation of amphoteric impurities on cation sites occurs preferably *at high V/III-flux ratios*. High V/III-flux ratios reduce $[V_{As}]$ in Eq. (7.42) which reduces the ratio $[Si_{As}]/[Si_{Ga}]$, i.e. increases $[Si_{Ga}]$. This tendency has been confirmed experimentally for the impurities C, Si, and Ge, and probably applies also to Sn (see Chap. 5). A second method to decrease $[V_{As}]$ is the use of an As-dimer source rather than an As-tetramer source. As_2 molecules are known to be more reactive than As_4 molecules which results in the reduction of As vacancies on the surface. The increased incorporation of impurities on group-III sites for the use of As_2 sources has been demonstrated for Si and Ge (see Chap. 5). The influence of the growth temperature is twofold. First, *low growth temperatures decrease the exponential factor* $\exp(-\Delta G_s/kT)$. As a result, the law of mass action can be dominated by the energy factor which increases the Si_{Ga} concentration (less autocompensation). Experimental observations agree with this tendency for GaAs:Si, since the highest free electron concentrations (i.e. minimal autocompensation) are obtained at low growth temperatures (see Chap. 5). Second, *low growth temperatures reduce the concentration of group-V vacancies*. The group-V concentration at the surface increases at low growth temperatures due to a reduction of reevaporation of group-V atoms from the surface. In Si-doped GaAs, decreasing autocompensation is expected at low growth temperatures, i.e. decreased $[V_{As}]$. This trend is indeed observed experimentally (see Chap. 5). Note that the two effects of the growth temperature result in reduced autocompensation for GaAs:Si. In contrast, for GaAs:C the two effects counteract. C predominantly occupies group-V sites in GaAs. A reduction of the growth temperature increases C_{As} due to the energy term $\exp(-\Delta G_s/kT)$. However, the concomitant decrease of V_{As} occurring at low growth temperatures makes C_{Ga}, i.e. autocompensation, more favorable. At present, sufficient experimental data are not available to assess the dominance of either of the two effects.

The incorporation of amphoteric impurities can be very different, if equilibrium conditions are achieved in the *bulk* rather than only on the *surface*. Several exchange reactions can occur in the bulk including

$$Si_{As} + V_{Ga} \overset{G_{b1}}{\leftrightarrow} Si_{Ga} + V_{As} \tag{7.43a}$$

$$Si_{As} \overset{G_{b2}}{\leftrightarrow} V_{As} + Si_{Ga} + I_{Ga} \tag{7.43b}$$

$$Si_{Ga} \overset{G_{b3}}{\leftrightarrow} V_{Ga} + Si_{As} + I_{As} \tag{7.43c}$$

where I_{Ga} and I_{As} represent a Ga and As interstitial, respectively. Note that Eq. (7.43a) is formally identical to Eq. (7.40). However, the change in free energy in the bulk, ΔG_b, can be different from the change in free energy for the surface reaction (Eq. (7.40)). The two other reactions (Eqs. (7.43b and c)) are so-called '*kick-out*' reactions. In these reactions a lattice atom is 'kicked out' of the lattice site into a interstitial position by the impurity atom. Other reactions, e.g. in which the impurity is kicked out by an interstitial lattice atom, can occur as well but will not be further considered. Note that the change in energy, is different for the three Eqs. (7.43a, b and c). Application of the law of mass action to Eqs. (7.43) yields

$$\frac{[Si_{As}][V_{Ga}]}{[Si_{Ga}][V_{As}]} = K_{b1} = c_1 \exp\left[-\frac{\Delta G_{b1}}{kT}\right],\qquad(7.44a)$$

$$\frac{[Si_{As}]}{[V_{As}][Si_{Ga}][I_{Ga}]} = K_{b2} = c_2 \exp\left[-\frac{\Delta G_{b2}}{kT}\right],\qquad(7.44b)$$

$$\frac{[Si_{Ga}]}{[V_{Ga}][Si_{As}][I_{As}]} = K_{b3} = c_3 \exp\left[-\frac{\Delta G_{b3}}{kT}\right].\qquad(7.44c)$$

The equations allow one to estimate trends for dopant incorporation under different growth conditions. For example, autocompensation is minimized for the reaction of Eq. (7.44a) at low concentrations of V_{As}. Low concentrations of V_{As} decrease $[Si_{As}]$ in Eq. (7.44b) but increase $[Si_{As}]$ in Eq. (7.44c). Note that $[Si] = [Si_{Ga}] + [Si_{As}]$ is assumed to be valid in Eqs. (7.43) and (7.44). Note also that the reactions of Eqs. (7.43a) and (7.44a) were estimated to be the most prominent exchange reactions (Munoz-Yague and Baceinedo, 1982). The above considerations illustrate that the incorporation of impurities depends on the location at which equilibrium is achieved, i.e. if equilibrium is achieved on the surface, in the bulk, or both. This can explain why different growth methods result in drastically different incorporation of amphoteric impurities. For example, Si is predominantly a donor in MBE and OMVPE grown GaAs, while Si is an acceptor in liquid-phase epitaxy (LPE) grown GaAs. Since the surface chemistry (especially the V/III flux ratio) is very different for these growth techniques, differences in impurity incorporation are expected as well.

Note that the charge states of the reaction products in Eqs. (7.40) and (7.43) have been neglected. The charge states of impurities on the semiconductor surface are unknown. In addition, the charge states of vacancies in the bulk have been calculated (Baraff and Schlüter, 1985) but have not been determined experimentally. Assuming the charge states V_{As}^0, V_{Ga}^0, Si_{Ga}^+, and Si_{As}^- in Eq. (7.43a) and taking into account the electronic reaction $e + h \leftrightarrow 0$, results in the same dependence of the compensation ratio on the free carrier concentration as given in Eq. (7.38), i.e. $N_A^- / N_D^+ \propto n^2$.

The incorporation of doping impurities is not expected to depend on the epitaxial growth *orientation*. Under equilibrium conditions, the compensation can be inferred from Eq. (7.44a) for impurities already incorporated in the bulk. Obviously, Eq. (7.44a) does not depend on the orientation of the growing crystal face. Nevertheless, experimental studies on the incorporation of Si impurities into GaAs grown along different crystallographic orientations revealed a clear dependence of the Si doping characteristics on the direction of the growth axis. Ballingall and Wood (1982) reported very different Si doping incorporation characteristics for epitaxial growth on $\langle 001 \rangle$, $\langle 011 \rangle$, and $\langle 111 \rangle$B oriented GaAs. The authors found that Si incorporates as a donor in $\langle 001 \rangle$ GaAs and as acceptor and donor for crystals with a (011) growth face. Low growth temperatures resulted in n-type conductivity of $\langle 011 \rangle$ oriented GaAs, while higher growth temperatures (> 550 °C) resulted in p-type conductivity. Growth on the (111) B face resulted in highly resistive material. Subsequent studies on Si doping incorporation were published by Bose et al. (1988) and Lee et al. (1989a, 1989b, 1990). Kadoya et al. (1990) reported Si doping on the (111)A face of GaAs during epitaxial growth. The conductivity was found to be n-type for a growth temperature of 480 °C and/or high V/III flux ratios. P-type conductivity was found for high growth temperatures and low V/III flux ratios. P-type conductivity for Si doping of the GaAs (111)A surface was also found by Miller and Asbeck (1987). All groups reporting growth of $\langle 111 \rangle$ oriented GaAs by MBE obtained textured or rough surface morphologies indicating that epitaxial growth on these surfaces is non-trivial.

It is useful to visualize the microscopic nature of the (001), (011), (111)A, and (111)B surfaces of III–V semiconductors. These different surfaces are shown in Fig. 7.12 where we have neglected any surface reconstruction. The (001) GaAs growth surface consists of either a layer of As atoms or a layer of Ga atoms; the (001) surface is a polar surface. In contrast, the (011) GaAs surface is non-polar. The (111)A and (111)B surfaces are again polar surfaces. While the (001) surface strongly favors Si donor incorporation, the two other surfaces favor an amphoteric incorporation of Si. Inspection of the surfaces shown in Fig. 7.12 indicates the reason for this peculiar Si incorporation characteristic. A Si impurity atom chemisorbed at the different surfaces has two bonds to host surface atoms for the (001) surface but only one bond to a host surface atom for the (011), (111)A, and (111B) surface. The weaker bonds of the impurity atom to the semiconductor

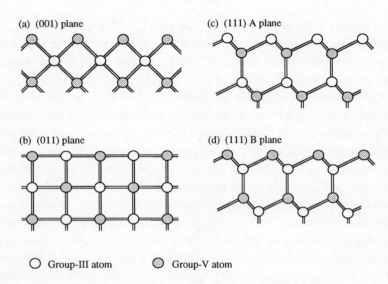

Fig. 7.12. Schematic illustration of the atomic structure of different surfaces of
III–V semiconductors (sideview; surface reconstruction neglected).

surface make site switching more likely. The weak surface binding with only one
bond makes the difference of Si_{Ga} and Si_{As} formation energy less relevant on the
(011) and (111) surface as compared to the (001) surface. It has been indeed
observed experimentally that the incorporation and amphotericity of Si in GaAs
depends strongly on the growth condition for the (011) and (111) face (Ballingall
and Wood, 1982; Miller and Asbeck, 1987; Kadoya et al., 1991). All reports
indicate that high V/III flux ratios and low growth temperature favor Si donor
incorporation while low V/III flux ratios and high growth temperatures favor Si
acceptor incorporation. This finding can be fully explained on the basis of the
law of mass action. High V/III flux ratios and low growth temperatures decrease
the concentration of As vacancies which results in a decreased $[Si_{As}]$, i.e. less
autocompensation of the Si donors as inferred from Eq. (7.42).

Assume that thermal equilibrium is reached in the bulk and that the
concentration of vacancies and defects in the bulk does not depend on the
epitaxial growth face. Under these assumptions, the incorporation of an
amphoteric impurity (e.g. Si) is not expected to depend on the growth orientation.
To explain the experimental facts which are in contrast to this expectation, two
possibilities should be considered. First, the concentration of vacancies may
depend on the orientation of the growth surface. Such a difference in vacancy and

defect concentration may occur because of the different surface chemistry and the different morphology of the epitaxial layer grown on different faces. Differences in the vacancy concentration result in different compensation ratios as evident from Eq. (7.42a). Second, while thermal equilibrium is achieved on the surface, it may *not* be achieved in the bulk at sufficiently low growth temperatures. While the surface equilibrium is determined by Eq. (7.42), the bulk equilibrium is given by Eq. (7.44a). Note that the free energies, ΔG_s and ΔG_b, are different for the surface and the bulk. If equilibrium is achieved only on the surface but not in the bulk (in agreement with Sect. 7.1), then impurities remain on their original site when incorporated in the bulk, i.e. remain metastable over long periods of time.

If amphoteric impurities are implanted into III–V semiconductors, it is desirable to keep autocompensation minimal. After implantation and subsequent activation (annealing) some fraction of the implanted impurity atoms may not be electrically active. Consider again the example of Si in GaAs; the Si impurities are either donors, acceptors, or electrically inactive (e.g. Si–Si pairs, Si clusters, etc.). The free electron concentration is then given by

$$n = [\text{Si}] - 2[\text{Si}_{\text{acceptor}}] - [\text{Si}_{\text{inactive}}] \tag{7.45}$$

where the brackets refer to the concentrations, and [Si] is the total Si concentration in the semiconductor. Assuming that the Si donor concentration is given by

$$[\text{Si}_{\text{donor}}] = [\text{Si}] - [\text{Si}_{\text{acceptor}}] - [\text{Si}_{\text{inactive}}] \tag{7.46}$$

the free electron concentration from Eqs. (7.45) and (7.46) is

$$n = N_D - N_A \tag{7.47}$$

where $N_D = [\text{Si}_{\text{donor}}]$ and $N_A = [\text{Si}_{\text{acceptor}}]$. The law of mass action (Eq. (7.44a)) indicates that a reduction in As vacancies decreases the autocompensation of Si in GaAs. *Co-implantation* of As in GaAs can reduce As vacancies in order to decrease the Si concentration on Ga sites. The amphotericity of other impurities can also be reduced by co-implantation. For example, the co-implantation of Ga in GaAs decreases the concentration of C donors in C-implanted GaAs. There are several reports on co-implantation of III–V semiconductors with As or P to improve the activation of impurities which preferentially occupy group-III sites (Kim et al., 1987; Wang, 1987, 1989). The authors report an improvement in the activation of Be-implanted InP. Note that Be is not an amphoteric impurity. Nevertheless, the co-implantation increases [Be$_{\text{In}}$], which results in improved activation. The activation of Si implanted into GaAs is improved upon co-implantation with As (Tell, 1991). The improvement in activation is usually relatively small and is therefore not used in the production of III–V semiconductor devices. Another method to improve Si activation in implanted GaAs is the employment of As overpressures during the activation anneal (see, for example, Pearton, 1988). The overpressure prevents As out-diffusion from the near surface region of GaAs and thus reduces the formation of As vacancies. As a result, a higher Si activation is achieved.

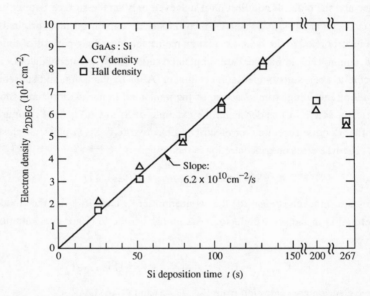

Fig. 7.13. Ionized impurity density (measured by capacitance–voltage technique) and free electron density (measured by the Hall effect) of GaAs doped with Si during growth interruption versus Si deposition time. The free carrier and the ionized Si density decrease simultaneously at high doping densities.

Autocompensation in III–V semiconductors doped with amphoteric impurities *increases at high doping concentration.* The increase can be quite drastic as in the case of GaAs doped with Si in the 10^{19} cm^{-3} range (see, for example, Maguire et al., 1987). For other impurities, e.g. C in GaAs the increase with doping concentration is less pronounced. An illustrative example of compensating and inactive Si in δ-doped GaAs is shown in Fig. 7.13 (Schubert et al., 1990a), which shows the free electron density as a function of Si deposition time. The Si impurities were evaporated onto the GaAs surface during growth interruption with a deposition rate of 6.2×10^{10} cm^{-2}/s. The free carrier density was measured by the Hall effect and by capacitance–voltage (CV) measurements. The experimental results of Fig. 7.13 show that the free electron density does not follow the Si doping density at densities $N_{Si} > 8 \times 10^{12}$ cm^{-2}. If compensation effects were absent, the experimental densities would follow the solid line in Fig. 7.13 which represents unity activation. However, both, the Hall density and the CV density decrease at high doping density. At a Si density of $N_{Si} = 1.66 \times 10^{13}$ cm^{-2} a free electron density of only 5.5×10^{12} cm^{-2} is measured indicating a compensation ratio of $N_A^- / N_D^+ = N_A^- / (N_{Si} - N_A^-) \cong 50\%$.

The maximum free carrier concentration shown in Fig. 7.13 is $n_{max}^{2D} = 8 \times 10^{12}$ cm^{-2}. This two-dimensional density corresponds to a three-dimensional concentration of $n = (n^{2D})^{3/2} \cong 2.2 \times 10^{19}$ cm^{-3}. Note that this three-dimensional concentration is higher than the highest free carrier concentration reported for homogeneously Si-doped GaAs of 1.8×10^{19} cm^{-3} (Ogawa and Baba, 1985).

Beall et al. (1988) arrived at similar conclusions about the saturation and decrease of the free carrier density in highly Si-doped GaAs, doped during growth interruption. The authors concluded that Si incorporates at high densities as electrically inactive Si clusters, Si–Si pairs and Si acceptors. Local-mode spectroscopy was used to identify the nature of Si impurities in δ-doped GaAs at high concentrations. The local-mode spectrum is shown in Fig. 7.14 for as-grown and annealed (600 °C, 30 min) GaAs. The local-mode spectrum allows one to identify the presence of several Si configurations including substitutional Si_{Ga}, compensating Si_{As}, and Si–Si pairs. Figure 7.14 reveals that the concentration of substitutional Si on Ga sites, Si_{Ga}, increases drastically upon annealing. Beall et al. (1989) explained the experimental results as follows. On the as-grown sample the doping profile is narrow with very high concentrations ($\cong 2.5 \times 10^{20}$ cm^{-3}). Most of the Si impurities are electrically inactive or compensating at high concentrations. Upon annealing, the concentration decreases due to diffusion of Si impurities out of the original impurity plane. As a result of the lower concentration, the largest fraction of Si occupies substitutional lattice sites ($[Si_{Ga}] = 2 \times 10^{13}$ cm^{-2}).

The amphoteric and compensating behaviour of group-IV impurities is a general observation made by many research groups (see also Chap. 5). The compensation is most pronounced at the highest Si concentrations. The most notable compensation ratios were reported by Greiner and Gibbons (1984). At a Si concentration of 2×10^{20} cm^{-3} the authors found an electrically active Si donor concentration of 2×10^{18} cm^{-3}, which corresponds to a compensation ratio of approximately 99%.

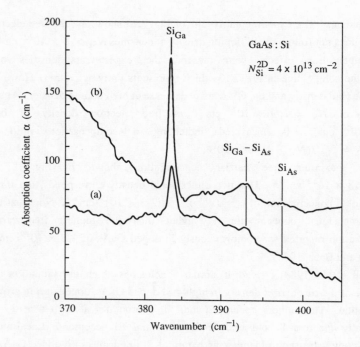

Fig. 7.14. Local vibrational mode (LVM) spectra for a [001] oriented GaAs epitaxial layer doped with Si of density 4×10^{13} cm^{-2} deposited during growth interruption. Spectrum (a) and (b) are obtained on the as-grown sample and after annealing (600 °C, 30 min), respectively (after Beall et al., 1989).

The increase in autocompensation of amphoteric impurities at high doping concentrations, which has been observed experimentally, agrees with the trend predicted by Eqs. (7.38) and (7.39). A quantitative analysis of experimental data on autocompensation has, however, not been performed.

It is useful to consider the total formation energy of an amphoteric impurity to be composed of a *chemical* contribution and an *electronic* contribution. The chemical contribution to the formation energy is due to the formation of bonds of the impurity in the host as well as strain and steric effects. The electronic contribution is due to the electronic processes associated with the donor and acceptor formation. For the creation of an ionized donor in an n-type semiconductor, the electron must occupy an empty conduction band state, i.e. the electron must be lifted from the donor level to the Fermi level. Thus, the electronic formation energy for a donor is given by

$$E_{\text{electronic}}^{\text{Donor}} = -(E_F - E_D) . \tag{7.48a}$$

If an acceptor level is formed in an n-type semiconductor, an electron from the conduction band can recombine onto the acceptor level. The electronic energy *gained* by an acceptor formation is thus

$$E_{\text{electronic}}^{\text{Acceptor}} = E_F - E_A . \tag{7.48b}$$

The total formation energy is the sum of the chemical and electronic part. The *difference* in formation energies between a donor and an acceptor is then given by

$$\Delta E = E_{\text{chem}}^{\text{Donor}} - E_{\text{chem}}^{\text{Acceptor}} + E_{\text{electronic}}^{\text{Donor}} - E_{\text{electronic}}^{\text{Acceptor}}$$

$$= \Delta E_{\text{chem}} - 2E_F + E_D + E_A$$

$$\cong \Delta E_{\text{chem}} - 2(E_F - E_C) - E_g \tag{7.48c}$$

where we assumed that $E_C - E_D \cong 0$ and $E_C - E_A \cong E_g$. The relative occupancy of cation and anion sites by amphoteric impurities depends on the difference in formation energies between the two sites. Assuming a Boltzmann distribution for the occupancy yields

$$\frac{N_A^-}{N_D^+} = K_B \ e^{-\Delta E/kT} \tag{7.49}$$

where ΔE is given by Eq. (7.48c) and K_B is a constant. Insertion of Eq. (7.48c) into Eq. (7.49) and using non-degenerate statistics for the conduction band, i.e. $E_F - E_C = kT \ln (n/N_c)$, yields

$$\frac{N_A^-}{N_D^+} = K_B^* \ n^2 \ e^{-\Delta E_{\text{chem}}/kT} . \tag{7.50}$$

Note that the equation has the identical dependence on the free carrier concentration as Eq. (7.38) which is, of course, not accidental.

The chemical formation energies of an amphoteric impurity are different for cation and anion site occupation. One can estimate the difference in formation energy from the different bond strengths of the impurity with the constituent anions and cations of the semiconductor. The bond strengths are known for some impurities, e.g. the C–Ga and the C–As bonds (Hartley and Patai, 1982) which are 61 kcal/mole $\cong$ 2.65 eV/atom and 57 kcal/mole $\cong$ 2.47 eV/atom, respectively. For tetrahedrally bonded C in GaAs, the difference in chemical binding energy is thus $\Delta E \cong 4 \times (2.65 \text{ eV} - 2.47 \text{ eV}) = 690 \text{ meV}$. However, this energy must be used with some caution because isolated binding energies are not necessarily representative for the tetrahedrally bonded C in GaAs since steric and other effects are not taken into account.

Another estimate of the difference in formation energy of amphoteric impurities between cation and anion site occupancy can be obtained from concentration measurements of the two species. Photoluminescence measurements can be used to obtain an estimate of the impurity concentrations on the different sites; that is, acceptors and donors of the same amphoteric impurities have distinctly different luminescence signals. The quantitative evaluation of luminescence spectra from Si-doped GaAs at a low doping level of $N_{Si} = 10^{16}$ cm^{-3} reveals the presence of $10^{12} - 10^{14}$ cm^{-3} Si acceptors (Harris and Schubert, 1990). Since $N_A/N_D \propto \exp(-\Delta E/kT)$, the difference in formation energy is estimated to be on the order of 300–500 meV.

7.5 Maximum doping concentration

The scaling of the spatial dimensions of semiconductor devices necessitates that the doping concentrations are increased as well. Certainly, the maximum attainable doping concentration has fundamental limits which depend on the doping method, the host semiconductor and the doping element. Equally important is the electrical activity of impurities at high doping concentrations. That is, high impurity atom concentrations are useful only if complete or near complete electrical activity is achieved. We therefore focus on the maximum *electrically active* impurity concentration rather than on the maximum impurity concentration. Table 7.2 summarizes the highest free carrier concentrations measured for various impurity elements and III–V semiconductors.

Table 7.2 shows that the maximum doping concentration depends strongly on the semiconductor hosting the impurities and on the impurity element. The highest doping concentrations are in the 10^{18} to 10^{20} cm^{-3} range. The highest value has been reported for GaAs:C with a hole concentration of 1.5×10^{21} cm^{-3}. Since GaAs has an atom concentration of $8/a_0^3 = 4.4 \times 10^{22}$ cm^{-3}, the electrically active C concentration corresponds to approximately 3% of the atom concentration of the host.

In the following sections, we consider several mechanisms that limit the maximum electrically active impurity concentration in semiconductors. The physical origins of the limitations discussed are quite different. The models are based on a concentration dependent diffusion coefficient (Sect. 7.5.1), coulombic repulsion (Sect. 7.5.2), defect models (Sects. 7.5.3 and 7.5.4), autocompensation (Sect. 7.5.5), and the solubility limit (Sect. 7.5.6).

7.5.1 Diffusion limited maximum concentration

The concentration of impurities in semiconductors can be limited to some maximum value, if the diffusion coefficient of impurities increases *drastically* at high concentrations. To illustrate this mechanism, assume that the diffusion coefficient of an impurity is very high for concentrations exceeding some critical concentration, N_{crit}, while it is much lower for concentrations below the critical concentration. Regions in the semiconductor with $N > N_{crit}$ are then characterized by strong diffusive motion of impurities. Redistribution occurs until impurities reach regions where $N < N_{crit}$, i.e. where the diffusion is slow.

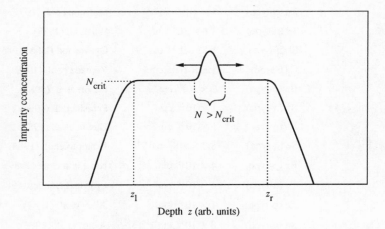

Fig. 7.15. Schematic illustration of a one-dimensional doping profile which locally exceeds a critical concentration N_{crit}, above which rapid diffusion occurs. Impurity atoms strongly diffuse until $N \leq N_{crit}$ for all values of z.

The concept of diffusion limited maximum concentration is illustrated in Fig. 7.15 which shows a one-dimensional impurity profile. The impurity profile has three distinct regions, that is, regions where $N < N_{crit}$, $N = N_{crit}$, and $N > N_{crit}$. The regions where $N \leq N_{crit}$ are assumed to be stable against rapid diffusion. In contrast, the region where $N > N_{crit}$ (i.e. $z_l \leq z \leq z_r$) is assumed to be highly diffusive. Impurities in this region strongly redistribute until they reach the regions $z < z_l$ and $z > z_r$. That is, strong diffusion occurs until $N \leq N_{crit}$ in the entire semiconductor.

A diffusion coefficient that increases with concentration is obtained from several models. The substitutional-interstitial diffusion model (see Chap. 8) predicts a power-law dependence of the diffusion coefficient on concentration,

Table 7.2. *Maximum free carrier concentrations in III–V semiconductors*

Semiconductor	Impurity	Concentration	Reference
$Al_{0.25}Ga_{0.75}As$	Si (n-type)	4×10^{18} cm^{-3}	Drummond et al. (1982)
$Al_{0.30}Ga_{0.70}As$	Be (p-type)	2×10^{19} cm^{-3}	Fujita et al. (1980)
$Al_{0.80}Ga_{0.20}As$	Be (p-type)	9×10^{18} cm^{-3}	Fujita et al. (1980)
$Al_{0.48}Ga_{0.52}As$	Sn (n-type)	2×10^{19} cm^{-3}	Cheng et al. (1981)
$Al_{0.50}In_{0.50}P$	Be (p-type)	3×10^{18} cm^{-3}	Nakajima et al. (1989)
GaP	Si (n-type)	8×10^{18} cm^{-3}	Baillargeon et al. (1989)
	Zn (p-type)	5×10^{19} cm^{-3}	Mottram et al. (1971)
GaAs	Te (n-type)	2.6×10^{19} cm^{-3}	Wu et al. (1988)
	Si (n-type)	1.8×10^{19} cm^{-3}	Ogawa and Baba (1985)
	C (p-type)	1.5×10^{21} cm^{-3}	Yamada et al. (1989)
	Be (p-type)	2×10^{20} cm^{-3}	Lievin et al. (1988)
$Ga_{0.47}In_{0.53}As$	Sn (n-type)	1×10^{20} cm^{-3}	Panish et al. (1990a)
	Be (p-type)	5×10^{20} cm^{-3}	Hamm et al. (1989)
$Ga_{0.5}In_{0.5}P$	Se (n-type)	8.3×10^{18} cm^{-3}	Iwamoto et al. (1984)
	Be (p-type)	4×10^{19} cm^{-3}	Nakajima et al. (1989)
GaSb	Te (n-type)	6×10^{18} cm^{-3}	Chin and Bonner (1982)
	Zn (p-type)	1×10^{20} cm^{-3}	Aliev et al. (1967)
InP	Sn (n-type)	8×10^{19} cm^{-3}	Astles et al. (1973)
	Sn (n-type)	1×10^{20} cm^{-3}	Panish et al. (1990b)
	Be (p-type)	7×10^{18} cm^{-3}	Benchimol et al. (1989)
	Zn (p-type)	2×10^{18} cm^{-3}	Tsang et al. (1991)
InAs	Te (n-type)	3×10^{19} cm^{-3}	Harrison and Houston

e.g. $D \propto N^2$ or $D \propto N^3$. However, such a dependence is too weak to explain the rigid upper limit for the impurity concentration observed in many III–V semiconductors.

A model that predicts a much stronger dependence of the diffusion coefficient on the impurity concentration was proposed by Mathiot and Pfister (1982, 1984) for impurities in Si. The model is based on the combination of the conventional vacancy/interstitial diffusion and diffusion in a percolation cluster. Recently, Larsen and Larsen (1990) tested the percolation/vacancy/interstitial model for several impurities in Si and found very good agreement with experimental results. The percolation/vacancy/interstitial model assumes that an infinite percolation cluster is formed when the impurity concentration exceeds a critical concentration N_{crit}. The impurity/vacancy potential is reduced in the semiconductor by the quantity ΔE_A. The enhancement of the impurity diffusion for $N > N_{crit}$ is due to this reduced binding energy; the enhancement factor F is equal to $\exp(\Delta E_A/kT)$. Mathiot and Pfister (1982, 1984, 1989) expressed the impurity diffusion coefficient as

$$D_{eff} = p_\infty D_{percolation} = (1 - p_\infty)D \qquad (7.51)$$

where D is the regular diffusion coefficient of the impurity (which may be concentration dependent) and $D_{percolation}$ is the diffusion coefficient of the percolation cluster. The probability of an impurity belonging to a percolation cluster is given by p_∞ which approaches unity at high concentrations. The percolation diffusion coefficient is proportional to the enhancement factor, i.e. $D_{percolation} \propto F = \exp(\Delta E_A/kT)$. The authors extracted the enhancement factor F and the critical concentration N_{crit} from experimental impurity profiles. Critical concentrations on the order of $N_{crit} \cong 10^{20}$ cm^{-3} and enhancement factors F of several thousands were found.

Although Mathiot and Pfister (1982) developed the percolation cluster model for Si, it is likely that it can also be applied to semiconductors other than Si. The original percolation cluster model was not intended to explain the maximum concentration in semiconductors. It is, however, quite evident that for sufficiently large enhancement factors F the critical concentration N_{crit} limits the maximum concentration of impurities in semiconductors.

7.5.2 Coulombic repulsion

The repulsive coulombic interaction of impurities within the depletion region of a growing semiconductor can limit the incorporation of impurities during growth. As a result, a certain maximum impurity concentration cannot be exceeded during epitaxial growth (Schubert et al., 1992). The model is of general nature and applies to semiconductors whose Fermi level is pinned at the surface.

The region near to the surface of a semiconductor is usually depleted of free carriers. The free carrier depletion is a result of the pinning of the Fermi level at a semiconductor surface. The pinning of the Fermi level is the result of electronic states (so-called Bardeen states) which are energetically located close to the middle of the fundamental gap for most of the III–V semiconductors. For GaAs, the Fermi-level pinning is found for different surface orientations, e.g. the $\langle 001 \rangle$, $\langle 011 \rangle$, and $\langle 111 \rangle$ surface orientation. Chiang et al. (1983) showed that the Fermi level at a GaAs surface is pinned at 0.55 ± 0.1 eV above the valence band maximum for all reconstructed surfaces studied by the group. Fermi-level pinning was found to be insensitive to temperature in the temperature range $20-500$ °C, consistent with the fact that the thermal energy kT is much smaller than other relevant energies (Nanichi and Pearson, 1969).

The coulombic energy in an n-type semiconductor is the sum of donor–donor, donor–electron, and electron–electron interaction energies. Since the concentration of free carriers can be neglected in the depletion region, the coulombic interaction energy is given by all donor–donor interactions, i.e.

$$E = \frac{1}{2} \sum_{i \neq j} \frac{e^2}{4\pi\varepsilon r_{ij}} \tag{7.52}$$

where ε is the permittivity of the semiconductor and r_{ij} is the separation of two donor ions. The sum is carried out for all impurities in the depletion region. Note that the coulombic interaction energy is reminiscent of the Madelung energy of ionic crystals. Unscreened Coulomb potentials can be used in Eq. (7.52) due to the low free carrier concentration in the depletion region. Screened Coulomb potentials must be used to calculate the interaction in the neutral part of the semiconductor. The screened interaction is, however, weaker. In any case, the coulombic interaction favors large distances between impurities.

The driving force can be calculated using the depletion approximation. The band diagram of a depleted n-type semiconductor surface is shown in Fig. 7.16. The Fermi level is pinned at an energy $e\phi_B$ below the conduction band edge. Impurities within the depletion region experience an electrostatic driving force towards the surface which results in a drift motion of the impurities with a velocity

$$v_d = \mu \mathcal{E} \tag{7.53}$$

where $\mathcal{E}$ is the field inside the depletion region and μ is the mobility of ionized impurities. The mobility can be expressed in terms of the diffusion coefficient using the Einstein relation $\mu = eD/kT$. The effect of the surface field was shown to redistribute impurities towards the surface (Schubert et al., 1990b). The maximum electric field occurs at the surface and is given by

$$\mathcal{E}_{max} = (2eN_D\phi_B/\varepsilon)^{\frac{1}{2}} . \tag{7.54}$$

The equation shows that the driving force is proportional to the square root of the doping concentration.

At some doping concentration the surface driving force becomes so strong that impurities are no longer incorporated into the crystal. Impurities then drift

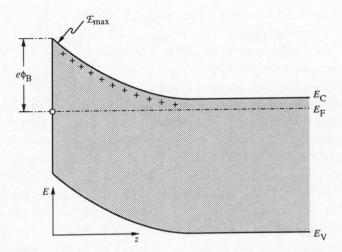

Fig. 7.16. Illustration of the band diagram of an n-type semiconductor whose Fermi level is pinned at the semiconductor surface. Ionized impurities in the space charge region interact repulsively due to their Coulomb potentials.

along the growth direction with the growing semiconductor surface. The condition can be quantified by assuming that the growth rate (i.e. growth velocity) equals the drift velocity

$$v_d = v_g \ . \tag{7.55}$$

The doping concentration at which this condition is fulfilled is obtained from Eqs. (7.53), (7.54), and the Einstein relation ($\mu = eD/kT$) and is obtained as

$$N_{max} = \left[v_g \ \frac{kT}{eD} \right]^2 (\varepsilon/2e\phi_B) \tag{7.56}$$

This concentration represents the maximum achievable doping concentration in semiconductors. Incorporation of impurities at a concentration higher than N_{max} is not possible, since impurities would drift to the surface at a faster rate than the growth rate.

We now consider the example of Be-doped GaAs. The diffusion coefficient of Be in GaAs has been determined to be $\cong 3 \times 10^{-16}$ cm^2/s at typical growth temperatures of $500 - 600$ °C (Schubert et al. 1990c). With a growth rate of

0.8 μm/hr = 2.2Å/s and a barrier height of $e\phi_B$ = 0.7 eV, the maximum doping determined from Eq. (7.56) is $N_{max} \cong 1.5 \times 10^{20}$ cm^{-3}. This value is in good quantitative agreement with experimentally achieved maximum concentrations of Be in GaAs (see Table 7.2).

The equation for the maximum doping concentration (Eq. (7.56)) also predicts how the doping concentration depends on (i) growth rate and (ii) the diffusion coefficient. Impurities with lower diffusion coefficients are expected to yield higher doping concentrations. This trend is indeed found for acceptors in GaAs. Carbon doping in GaAs is possible up to $\cong 10^{21}$ cm^{-3}; C is also the impurity with the lowest diffusion coefficient. Beryllium doping is possible up to $\cong 10^{20}$ cm^{-3} and Be is a moderately fast diffuser. Zinc doping is possible up to $\cong 10^{19}$ cm^{-3}; at the same time, Zn is known to be a fast diffuser. The dependence of the maximum doping concentration on the growth rate has not yet been thoroughly studied in III–V semiconductors. Initial experiments in Si grown by molecular-beam epitaxy at different growth rates indicate that N_{max} increases with enhanced growth rate in agreement with the theoretical expectation (Gilmer and Gossmann, 1993).

The coulombic interaction of impurities in the depletion region of a semiconductor has been calculated using Monte Carlo techniques (Schubert et al., 1992). In the calculation, the interaction of impurity ions and charged carriers trapped in surface states is considered. A calculated impurity profile is shown in Fig. 7.17. The calculation simulates a growth process in which impurities are deposited on the non-growing surface of a semiconductor. Subsequently, epitaxial growth is resumed. The calculation demonstrates that the rate of impurity incorporation is limited to some maximum value. If this maximum concentration is locally exceeded, repulsive interaction drives more impurities back to the semiconductor surface. In addition, the repulsive interaction prevents impurities from being incorporated from the surface into the bulk. Both mechanisms stabilize the impurity concentration in the bulk.

A further result of the repulsive Coulomb interaction is that impurities are not distributed randomly in the semiconductor crystal. It is usually assumed that impurities are distributed according to the **Poisson distribution** (see, for example, Shockley, 1961). However, closely spaced impurity configurations are unfavorable at high impurity concentrations due to repulsive Coulomb interaction. Inspection of the impurity distribution calculated by the Monte Carlo technique revealed that the distribution is non-poissonian. Specifically, configurations with two or more closely spaced impurities occur less frequently than expected on the basis of the Poisson distribution.

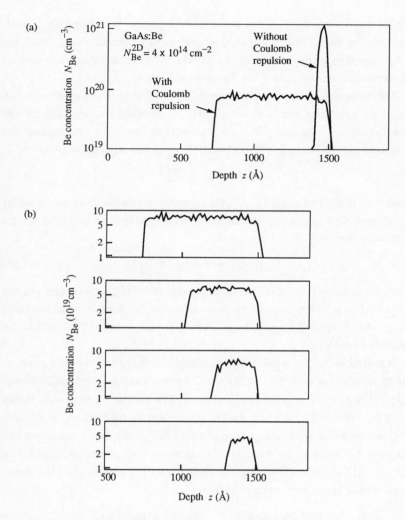

Fig.7.17. Calculated impurity distribution in a semiconductor doped during growth interruption (a) with and without taking into account Coulomb correlation effects. A growth rate of 1Å/s and a diffusion coefficient of $2 \times 10^{16} \text{cm}^2/\text{s}$ was used in the calculation. (b) Calculated profiles for different Be doses.

7.5.3 Compensating native defects

Native defects in III–V semiconductors such as vacancies, interstitials, and antisite defects can have compensating characteristics. That is, many of the native defects are positively charged in p-type semiconductors and negatively charged in n-type semiconductors. The change in the charge state of these defects reduces the concentration of free carriers. Furthermore, the abundance of such compensating native defects can increase with doping concentrations. Thus, for sufficiently high concentrations of the defects, the free carrier concentration in semiconductors can be limited to a maximum value.

The formation energy and the charge states of several native defects was calculated by Baraff and Schlüter (1985). As an example, we consider the formation of a Ga vacancy and a Ga interstitial by moving a substitutional Ga atom to an interstitial site in a GaAs lattice. The reaction equation is given by

$$0 \leftrightarrow V_{Ga} + I_{Ga}^{Td} \tag{7.57}$$

where V_{Ga} is a Ga vacancy; I_{Ga}^{Td} is a Ga interstitial which occupies an interstitial site of tetrahedral symmetry. The equilibrium concentration of the native defects is obtained from the law of mass action

$$[V_{Ga}][I_{Ga}^{Td}] = c \exp(-\Delta G/kT) \tag{7.58}$$

where the brackets [...] refer to the concentration of the respective defect and ΔG is the change in Gibbs free energy associated with the reaction. The formation energy ΔE is typically much larger than the entropy term $T\Delta S$ which can therefore be neglected.

It proved useful to represent the total formation energy as a sum of a *chemical* energy term and an *electronic* energy term. Assume that the defect has the charge state Z with $Z = -2, -1, 0, +1, +2$ and so on. The electronic energy term is then given by $-(E_F - E_D)Z$ where E_F and E_D are the Fermi-level and the defect-level energy and Z is the charge state of the defect. If the defect can accept two electrons $(Z = -2)$, the reduction in electronic defect energy amounts to $2(E_F - E_D)$ since two electrons can drop from the Fermi level to the defect level. The defect reaction energy can be written as

$$\Delta E(Z, E_F) = \Delta E(Z) - E_F \sum_n \sigma(E_F - E_{D,n}) \tag{7.59}$$

where $\Delta E(Z)$ is the *chemical* part of the formation energy and the second term of the right-hand side describes the *electronic* part of the energy. The sum of the second part is carried out over all energy levels, $E_{D,n}$, of the defect below the Fermi level E_F; σ is the step function which is zero for $E_F < E_{D,n}$ and unity for $E_F \geq E_{D,n}$. If the defect contains Z occupied levels and the occupancy remains constant in the range of Fermi levels considered, then Eq. (7.59) simplifies to (Baraff and Schlüter, 1985)

$$\Delta E(Z, E_F) = \Delta E(Z) - E_F Z \tag{7.60}$$

where Z is the charge state of the defect. The energy $\Delta E(Z)$ is the total energy of the defect in the charge state Z and is independent of E_F.

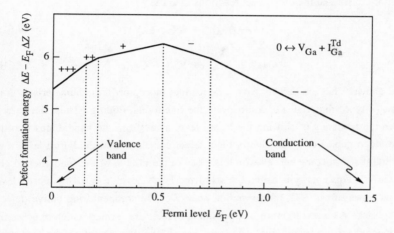

Fig. 7.18. Total reaction energies, $(\Delta E - E_F \Delta Z)$, for the defect reaction $0 \leftrightarrow V_{Ga} + I_{Ga}^{Td}$, where I_{Ga}^{Td} represents an interstitial Ga atom with tetrahedral symmetry surrounded by four As atoms (after Baraff and Schlüter, 1985).

The total defect formation energy for the Ga vacancy/interstitial defect in GaAs is shown in Fig. 7.18 as a function of the Fermi level. The reference level $E_F = 0$ is at the top of the valence band. Figure 7.18 illustrates two important features of the native defect. First, the charge state becomes more negative as the Fermi level increases. For $E_F \cong E_V$ (p-type GaAs), the defect is positively charged, i.e. traps holes; for $E_F \cong E_C$ (n-type), the defect traps electrons. The defect thus has a *compensating* nature regardless of the conductivity type of the semiconductor. Second, the total defect formation energy continuously decreases for either highly n-type or p-type conductivity. According to the law of mass action (Eq. (7.58)), the abundance of defects increases for high doping concentrations where $E_F \cong E_C$ or $E_F \cong E_V$, i.e. a lower formation energy prevails. Both of these characteristics clearly contribute to the compensation of intentional shallow impurities in semiconductors. The compensating nature was also found for other native defects in GaAs such as antisite defects and As vacancies (Baraff and Schlüter, 1985). Furthermore, the formation energy of compensating defects was found to decrease for Fermi levels close to the band edges (see Fig. 7.18). This characteristic was found for defects that are likely to occur in As-rich GaAs (e.g.

As$_{Ga}$) as well as in Ga-rich (e.g. Ga$_{As}$) GaAs.

Walukiewicz (1989) proposed that the nature of compensation native defects explains the free carrier saturation occurring at high doping concentrations. The author considered the two defect reactions in GaAs

$$V_{Ga} + As_{As} \leftrightarrow As_{Ga} + V_{As} \tag{7.61a}$$

$$Ga_{As} + V_{Ga} \leftrightarrow V_{As} + Ga_{Ga} \tag{7.61b}$$

and showed that the defects have a donor-like (acceptor-like) characteristic in p-type (n-type) material, i.e. compensate the intentional doping. Thus, the defects have the tendency of shifting the Fermi level away from the band edges into the forbidden gap. For a sufficiently high defect concentration, the Fermi level may be stabilized at some energy which is close to the middle of the fundamental gap in GaAs. This energy is called the stabilized Fermi level, E_{FS}. If the Fermi level moves away from E_{FS}, the formation energy of the compensating native defects decreases. As a result, more compensating defects are formed which opposes the movement of the Fermi level. At some Fermi level the rate of defect generation equals the increase in free carrier concentration which results in a saturation of the free carrier concentration.

The position of the stabilized Fermi level is of critical importance to the maximum achievable concentration. If E_{FS} is already close to the conduction band, the Fermi level can move into the conduction band without an abundance of compensating defects being formed. As a result, high values of n/N_c can be achieved, where N_c is the effective density of states at the bottom of the conduction band. If, however, E_{FS} is far below the conduction band edge, an abundance of compensating defects are formed, if the Fermi level moves into the conduction band. As a result, only low values of n/N_c can be achieved. Similar conclusions can be drawn for the position of E_{FS} relative to the valence band edge in p-type semiconductors.

A systematic study of the maximum carrier concentrations in semiconductors was published by Tokumitsu (1990). Figure 7.19 shows the stabilization position of the Fermi level in several elemental and compound semiconductors. The stabilization position is within the forbidden gap for most of the semiconductors; E_{FS} is within the conduction band for InAs. The author found that high values of n/N_c were achieved in semiconductors in which $E_C - E_{FS}$ is small; these semiconductors include InAs, Ga$_{0.47}$In$_{0.53}$As, and InP. On the other hand, high values of p/N_v were achieved in semiconductors in which $E_{FS} - E_V$ is small such as GaSb, Ge, Si, and GaAs. The experimental data was fitted to an exponential function. The maximum carrier concentration was expressed by

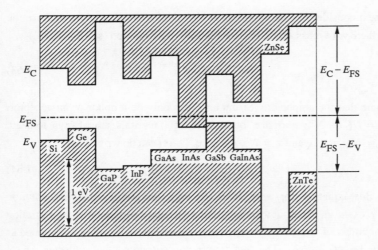

Fig. 7.19. Schematic illustration of the Fermi-level stabilization energies within the band gap of different semiconductors (after Tokumitsu, 1990).

$$n/N_c = 2.7 \times 10^3 \ e^{-5.5(E_C - E_{FS})} \ (\text{n-type}) , \qquad (7.62a)$$

$$p/N_v = 4.0 \times 10^2 \ e^{-6.1(E_{FS} - E_V)} \ (\text{p-type}) . \qquad (7.62b)$$

These empirical relations yielded good agreement with many experimental values for the maximum free carrier concentrations.

The model of compensating native defects is a general model in which the species of the impurity does not enter the model. Differences found between different impurity elements in the same semiconductor cannot be explained by the model. For example, the maximum hole concentration achievable in p-type GaAs varies between 10^{19} cm^{-3} (GaAs:Zn) and 10^{21} cm^{-3} (GaAs:C). Such differences are not expected on the basis of the model.

7.5.4 Limitations for amphoteric dopants

Amphoteric impurities are known to autocompensate at high impurity concentrations. The amphoteric impurities C, Si, Ge, and Sn can occupy either cation or anion sites in the zincblende lattice on which they act as donors or acceptors, respectively. The conditions of impurity incorporation can usually be selected in such a way that the incorporation of impurities occurs predominantly on one preferred site. For example, C occupies predominantly acceptor sites in GaAs when incorporated during epitaxial growth.

We next consider the incorporation at high doping concentrations and use the result of the law of mass action for the impurity incorporation (see Eq. (7.38))

$$\frac{N_A^-}{N_D^+} = K^* n^2 . \tag{7.63}$$

We assume that the amphoteric Si impurity can only be a donor or an acceptor, i.e. $N_{Si} = N_D^+ + N_A^-$, and that both types of impurities are always ionized. Using $n = N_D^+ - N_A^-$ and $x = N_A^- / N_{Si}$, Eq. (7.63) can be written as

$$x = K^* N_{Si}^2 (1 - 2x)^2 (1-x) . \tag{7.64}$$

Solving this equation for very large Si concentrations yields the solution $x = N_A^- / N_{Si} = \frac{1}{2}$. Thus, autocompensation increases with doping concentration until the number of impurities incorporated as donors equals those incorporated as acceptors, i.e. $N_A^- \cong N_D^+$. The free carrier concentration then *saturates*, i.e. a maximum of the free carrier concentration is reached.

For a given saturation concentration the constant K^* in Eq. (7.63) can be determined. Using a saturation concentration of 10^{18} cm^{-3} for Si in GaAs and $N_A^- \cong N_D^+$, one obtains $K^* \cong 10^{-36}$ cm^6. Using this value for the constant in Eq. (7.63) allows us to quantitatively estimate the compensation ratio at lower concentrations. As an example, we choose $n = 10^{16}$ cm^{-3} and obtain $N_A^- / N_D^+ = 10^{-4}$. This ratio is lower than the compensation ratio determined by optical techniques (Lu et al., 1990; Kamiya and Wagner, 1977) but in reasonable agreement with a recent study by Harris and Schubert (1990).

Another characteristic of amphoteric impurities is the tendency of pair formation. The pair formation can be driven by several effects including chemical, strain or coulombic effects. Coulombic interaction effects can be quantified in a straightforward way. Assume a semiconductor with randomly distributed positive donor ions and randomly distributed electrons. The coulombic interaction energy is zero for such a neutral semiconductor.* If, however, two impurities, one being a donor and the other one being an acceptor, form a closely spaced pair, a reduction in coulombic energy of $e^2 / 4\pi \varepsilon r$ results. For a nearest neighbor pair $r \cong 2.4$ Å, which yields an energy of $\cong 460$ meV. Considering the magnitude of this energy, pairing of amphoteric impurities is a quite attractive mechanism. (The chemical bond energies and strain energies may also change drastically for pairs but this is neglected here). The probability of impurity-pair formation increases *quadratically* with the impurity concentration

---

* Note that electrons are not distributed randomly due to correlation effects and screening. As a result the net coulombic interaction energy is *cohesive* for any doped semiconductor.

since *two* impurities are required to form a pair. Thus, the pair formation is expected to increase with impurity concentration. At some concentration all impurities incorporate as paired species. As a result, the free carrier concentration saturates or may even decrease for increasing doping concentration. Such a decrease has been observed in Si-doped GaAs (See Chap. 5).

It is necessary to compare the changes in electronic and coulombic energies with the difference in formation energies. As shown in Sect. 7.4, the difference in formation energies between the two sites of the GaAs lattice is typically several hundred meV. For C in GaAs the difference was estimated to be $\Delta E \cong 690$ meV. Other amphoteric impurities such as Si, Ge, and Sn have probably lower values of ΔE due to their higher atomic weight, i.e. lower reactivity. An energy difference of several 100 meV for ΔE is indeed comparable in magnitude to the increase in electronic energy and to the coulombic pair formation energy. The roughly comparable magnitude in energy suggests that the two mechanisms can impose a limit on the maximum free carrier concentration in III–V semiconductors doped with amphoteric impurities.

7.5.5 Concentration limited by DX-centers

A model for the maximum free electron concentration based on the DX-center was proposed by Theis et al. (1988). The model was proposed for GaAs in which DX-centers are known to form localized states which are in resonance with the conduction band. Substitutional donors can undergo a transition in which the initially substitutional donor moves to a threefold-coordinated interstitial site (Chadi and Chang, 1988). The displacement of the shallow donor from its substitutional site is accompanied by the breakage of one donor atom bond and the transfer of the electrons to the DX-center. The energy levels of shallow donors and DX-centers are schematically illustrated in Fig. 7.20. Shallow donor levels are indicated as well as the DX-levels which are within the conduction band. For increasing free carrier concentration the Fermi level rises until it coincides with the DX-level. At this point a substitutional donor undergoes lattice relaxation and captures two electrons according to (Chadi and Chang, 1988)

$$2D^0 \leftrightarrow 2D^+ + 2e \leftrightarrow D^+ + DX^- \qquad (7.65)$$

where D and DX represent the shallow donor and the DX-center configuration. The properties of the DX-center will be discussed in more detail in Chap. 9.

The Fermi level cannot exceed the DX-center level since electrons of additional donors are captured by donors undergoing the transition to DX-centers.

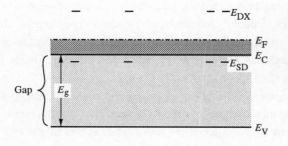

Fig. 7.20. Schematic illustration of shallow donor levels and DX levels of substitutional Si impurities in GaAs.

Additional electrons are captured according to Eq. (7.65). Using degenerate statistics, the maximum free electron concentration is given by (see Chap. 3)

$$n_{\max} = \frac{1}{3\pi^2}\left[\frac{2m^*(E_{\mathrm{DX}} - E_{\mathrm{C}})}{\hbar^2}\right]^{3/2} \tag{7.66}$$

where E_{DX} is the energy of the donor in the DX-configuration. As an example, we assume $E_{\mathrm{DX}} - E_{\mathrm{C}} = 200$ meV, $m^* = 0.067\ m_0$, and a single, isotropic, parabolic conduction band minimum; the maximum free carrier concentration is then obtained from Eq. (7.66) as $n_{\max} \cong 7 \times 10^{18}$ cm^{-3}. Theis et al. (1988) extrapolated a maximum free electron density in GaAs limited by the DX-configuration to be $1.5 \times 10^{19} - 2 \times 10^{19}$ cm^{-3}.

The energy level of the DX-configuration depends on the doping concentration. Etienne and Thierry-Mieg (1988) found that the DX-level moves away from the conduction band minimum at high doping concentrations. The authors further found in Al$_x$Ga$_{1-x}$As that the DX-center and the persistent photoconductivity effects (see Chap. 9) *decrease* rather than increase with doping density. These findings together with the fact that strong persistent photoconductivity has not been found in highly doped GaAs indicate that the DX-center is not the dominant mechanism limiting the free carrier concentration in n-type GaAs.

7.5.6 Solubility and strain

The *solid solubility limit* is not well known for impurities in III–V semiconductors. In contrast, the solid solubility limit of impurities in the group-IV semiconductor Si have been well documented (Beadle et al., 1985) for shallow donors and acceptors. The solid solubility limits the concentration of impurities in Si and applies to high-temperature growth from the melt. However, epitaxial III–V semiconductors are rarely grown from the melt (except in liquid-phase epitaxy) Most epitaxial III–V semiconductors are grown from the gas phase at lower temperatures such as $400-700$ °C. It is questionable if the solubility limit can be applied at such low temperatures. The solid solubility is a property of equilibrium which is not necessarily reached at low temperatures.

The solid solubility limit of one compound A in another compound B is defined as the maximum concentration of A in B at which A is homogeneously distributed in B without any precipitates. The presence of precipitates of a doping element in a semiconductor can thus be used as an indicator that the solid solubility limit has been exceeded. Precipitates can be detected in III–V semiconductors, for example, by transmission electron microscopy (TEM) or by secondary ion mass spectrometry (SIMS). There have been a number of reports on the presence of precipitates or impurity clusters in III–V semiconductors, for example Si precipitates in GaAs (Chai et al., 1981), or Be precipitates in $Ga_{0.47} In_{0.53} As$ (Hamm et al., 1989). However, the occurrence of precipitations depends strongly on the growth conditions such as substrate temperature and V/III flux ratio. It is not useful to establish a certain concentration as a solubility limit if this 'limit' depends strongly on growth conditions. Furthermore, impurity concentrations were found to be limited to some concentration without the observation of precipitates (Schubert et al., 1990c; Panish, 1990). The limitation of impurity concentrations by the 'solubility limit' has to be considered with caution.

Elastic *strain* is another physical mechanism that can limit the impurity concentration in semiconductors. An impurity atom with a different atomic size than the host lattice atom either exerts *compressive* or *tensile* strain on the semiconductor lattice, if the impurity is larger or smaller than the lattice atom, respectively. To accommodate the impurity on a lattice site, the strain energy must be provided during the incorporation process. If the strain energy (per impurity atom) strongly increases with impurity concentration, then impurities may not incorporate at all above some critical concentration. In order to evaluate the concentration dependence, the interaction energy must be calculated. Morgan (1972) investigated the elastic interaction between two defects separated by a distance r. The author found that the strain interaction energy is proportional to r^{-3}. The absolute magnitude of this energy is not known in covalent materials. Lannoo and Bourgoin (1981) assumed that it is similar in covalent materials and metals. An interaction energy of <10 meV was found for a defect separation of 10 Å. This relatively small interaction energy indicates that elastic strain can limit the impurity concentration only in the extreme concentration regime, i.e. for concentrations $\geq 10^{21}$ cm^{-3} ($r \leq 10$ Å) where the interaction energy assumes higher values.

8

Redistribution of impurities

Spatial redistribution of impurities in semiconductors occurs at sufficiently high temperatures. That is, impurities may not remain on the site of their initial incorporation. As discussed in the previous chapter, the diffusive redistribution can even be used to incorporate impurities from the surface into the bulk of the semiconductor. However, impurity redistribution is most frequently an undesired effect; it leads to the smearing of doping profiles or, for example, can result in the displacement of the location of a pn-junction.

Redistribution of impurities can be *isotropic* or *anisotropic*. Conventional diffusion is an example for isotropic redistribution. On the other hand, anisotropic redistribution of impurities occurs in the presence of energy gradients. Such gradients can be the result of electric potential gradients, strain gradients, surfaces, etc. Preferential redistribution of impurities occurs along the energy gradient, i.e. towards regions of lower energy.

The vicinity of a semiconductor surface can lead to preferential redistribution of impurities. Some impurities tend to remain on the surface during epitaxial growth and do not readily incorporate into the bulk. The impurities *segregate* to the *surface*. Surface segregation is an undesired effect since it complicates the growth of well-defined impurity distributions. The effect of preferential redistribution of impurities (in the vicinity of a surface) towards the surface is called *surface migration*. This can be the result of potential gradients or defect gradients in the vicinity of a surface.

The mechanisms of impurity redistribution can be quite different in nature. *Substitutional diffusion* occurs if an impurity hops from one lattice site to a second, vacant lattice site. During *substitutional–interstitial diffusion*, an impurity leaves its substitutional site and hops to an interstitial lattice site, and then again to a vacant lattice site. The impurity on the interstitial site may also **kick-out** a regular lattice atom and take its place. The exact mechanism of diffusion plays, as we shall see later, an important role in the dependence of the diffusion mechanism on impurity concentration.

8.1 Diffusion and brownian motion

Diffusion of impurities and defects in semiconductors is the result of random hops of the impurity atoms and defects. The random hops are thermally activated. The thermally stimulated random motion of particles is known as **brownian motion** and was first observed for small carbon particles floating on water. The similarity of brownian motion and solid-state diffusion was pointed out by Einstein (1905, 1906) who developed the *theory of brownian motion*. An exhaustive review on the theory of brownian motion was given by Risken (1984).

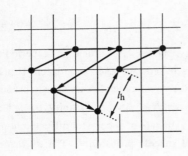

Fig. 8.1. Illustration of diffusion hops of an impurity atom between substitutional sites of a lattice. The average hop distance in the illustration is approximately two lattice constants.

The random movement of an impurity in the lattice matrix of a semiconductor is schematically illustrated in Fig. 8.1. The impurity hops from lattice site to lattice site. The **average hop length** is denoted as l_h and is approximately two lattice constants for the example shown in Fig. 8.1. We now consider a single, successful hop of an impurity from one lattice site to a second, *vacant* lattice site. The diffusion hop can be characterized by an **attempt rate** and a **success probability**. The attempt rate is the number of attempts (per unit time) of the impurity to leave its lattice site. The attempt rate is given by the vibrational frequency, i.e. each vibration of the impurity atom represents one attempt to leave the lattice site. The attempt frequency is given by

$$\hbar\omega = \hbar 2\pi f = (3/2) \, kT \, . \tag{8.1}$$

At a temperature of 600 °C, the attempt rate obtained is $f = 2.7 \times 10^{13} \text{ s}^{-1}$.

The probability that the attempt is successful is modeled by an energy barrier

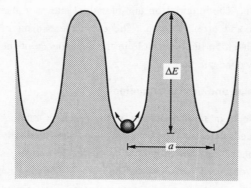

Fig. 8.2. Diffusion of impurity atoms in a periodic crystal potential with period a occurs via random jumps to adjacent energy minima. The energy barrier between the minima is ΔE.

that the impurity atom must overcome in order to leave the original lattice site. The success probability is given by

$$p = e^{-\Delta E/kT} \tag{8.2}$$

where ΔE is the magnitude of the energy barrier. A particle in a periodic potential is schematically shown in Fig. 8.2. In order to reach an adjacent lattice site, an energy barrier of magnitude ΔE must be overcome. For an energy barrier of vanishing magnitude, $\Delta E = 0$, the success probability equals unity, $p = 1$, i.e. each attempt would be a successful attempt. The rate of successful diffusion hops is given by the product of attempt rate and success probability. The **diffusion constant D** (also called diffusion coefficient) of an impurity is defined as (see, for example, Risken, 1984)

$$D = l_h^2 \, fp \tag{8.3}$$

where fp is the rate of successful diffusion hops and l_h is the average hop length. Using Eq. (8.2) the diffusion constant can be written as

$$\boxed{D = D_0 \, e^{-\Delta E/kT}} \tag{8.4}$$

where $D_0 = l_h^2 f$. The temperature dependence of D_0 is much weaker as compared to the dependence of the exponential term $\exp(-\Delta E/kT)$. It is therefore frequently assumed that D_0 is a constant, i.e. independent of temperature. Comparison of $D_0 = l_h^2 f$ with Eq. (8.3) reveals that D_0 is the diffusion constant if all hop attempts were successful, i.e. $p = 1$.

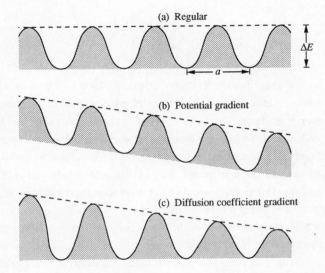

Fig. 8.3. Schematic energy diagram of a periodic crystal potential for (a) regular diffusion, (b) diffusion in a potential gradient (electric field), and (c) diffusion in a potential with a diffusion coefficient gradient, i.e. a spatially changing diffusion coefficient.

Pure diffusion is isotropic due to the random nature of the diffusion process. Anisotropic diffusion results in the presence of additional energy gradients. Consider a positively charged impurity in an electric field. A diffusion hop along the field distribution decreases the electrostatic energy by $e\mathcal{E}\,l_h$, while the diffusion against the field direction increases the energy by $e\mathcal{E}\,l_h$, where $\mathcal{E}$ is the electric field. Thus, a diffusion hop along the field direction is more likely than against the field direction. Diffusion events along the field direction are not impeded by the field. However, the probability for a successful attempt against the field direction is reduced by $\exp(-e\mathcal{E}\,l_h/kT)$ in analogy to Eq. (8.2). The preferential direction of the diffusion hops results in a net motion of the positively charged impurities along the field direction. The *drift velocity* of the impurities is characterized by a mobility μ and is given by

$$v = \mu\mathcal{E}. \tag{8.5}$$

The mobility μ is related to the diffusion constant according to

$$\mu = \frac{eD}{kT} \qquad (8.6)$$

which is called the **Einstein relation** (Einstein, 1905, 1906). It relates the random, isotropic diffusive motion with the directed motion in the presence of a potential energy gradient. The periodic crystal potentials associated with regular diffusion and with diffusion in a potential gradient are schematically shown in Fig. 8.3(a) and (b), respectively. It is evident that diffusion in a potential gradient occurs preferably to the low-energy side, i.e. right-hand side of Fig. 8.3(b). A potential resulting in a spatially varying diffusion constant is shown in Fig. 8.3(c). The height of the energy barriers changes, resulting in a spatially changing diffusion coefficient. Preferential diffusion towards the side with the lower barrier results. That is, diffusion occurs preferentially along the *diffusion constant gradient*. Diffusion in a material with a diffusion constant gradient will be analyzed quantitatively in the next section.

Up to now we have considered the diffusion process from an initial lattice site to a final site. It was implicitly assumed that the final site is vacant in order to be able to accommodate the diffusing impurity. The diffusion constant D has been used as a purely phenomenological constant, i.e. without considering the microscopic nature of the diffusion process. The microscopic mechanisms of diffusion processes are, however, of great importance. The dependence of the diffusion constant D on crystal parameters, for example the vacancy concentration, can be derived from the nature of the microscopic diffusion process. As an example, we consider the diffusion of an impurity on substitutional lattice sites. First, a vacant lattice site must be available for a successful diffusion hop. That is, the diffusion constant is proportional to the *concentration of vacancies*. Second, if there is no vacancy in the vicinity of the impurity, it cannot make a diffusion jump. However, a vacancy can diffuse into the vicinity of the impurity, thus enabling the impurity to make a successful hop. Therefore, the impurity diffusion constant is also proportional to the *vacancy diffusion constant* D_V (Manning, 1968; see also Sects. 8.3.1 and 8.6), i.e.

$$D \propto [V]D_V \qquad (8.7)$$

where [V] is the concentration of vacancies. Depending on the detailed diffusion mechanism, other factors may also enter. Different diffusion processes are discussed further in Sect. 8.3.

8.2 Fickian diffusion

In the simplest form of diffusion, impurities make random atomic jumps and the movement of impurities is *uncorrelated*. For such uncorrelated motion the movement of impurities does not depend on other impurities, that is the diffusion is *independent of the concentration* of the impurities in the semiconductor. If the atomic jumps are random and uncorrelated then the net motion of impurities is proportional to the concentration gradient. The impurity flux (per unit time and unit area) is then given by

$$j(x,y,z) = -D \ \nabla N(x,y,z) \tag{8.8a}$$

where $N(x,y,z)$ is the impurity concentration and D is the diffusion constant. Even though D depends on the temperature, D is a *constant* with respect to the spatial variables x, y, and z of the differential equation. For an impurity distribution which depends only on the z-direction of a cartesian coordinate system, the equation simplifies to

$$j(z) = -D \ \frac{\mathrm{d}}{\mathrm{d}z} \ N(z) \tag{8.8b}$$

which is **Fick's first law**. Note that $j(z)$ is the flux of impurities in the z-direction and the flux depends only on z. Fick's first law states that the impurity flux is proportional to the impurity gradient. If the gradient is zero (i.e. constant impurity concentration for all x,y,z), then there is no net movement of impurities ($j=0$) despite the many random atomic jumps which occur.

In the simplest case D is a constant and does not depend on the spatial position or on the impurity concentration. In this case, Eq. (8.8) is a linear differential equation of first order, and the diffusion process is referred to as **linear fickian diffusion**. However, the diffusion constant may depend on the position and on the impurity concentration, i.e. $D = D(N,z)$. This case is referred to as **non-linear fickian diffusion**. In both cases, the impurity flow is along the concentration gradient. The flow along the gradient is the result of random, uncorrelated hops. As an example, consider two regions 1 and 2 with impurity concentrations N_1 and N_2. If $N_2 > N_1$, there will be a net flow from 2 to 1 because there are more impurities in 2 than in 1 and consequently the hop rate $2 \rightarrow 1$ will be larger than $1 \rightarrow 2$.

In addition to Fick's first law, the **continuity equation** must be taken into account. For one-dimensional impurity distributions that depend only on the z-direction, the continuity equation is given by

$$\frac{\partial}{\partial z} j(z) = -\frac{\partial}{\partial t} N(z,t)$$

(8.9)

The equation states the fact that the increase (decrease) of impurities in any volume element equals the net flow of impurities into (out of) the volume element per unit time.

Insertion of Eq. (8.8) into Eq. (8.9) yields

$$\frac{\partial}{\partial t} N(z,t) = \frac{\partial}{\partial z} D \frac{\partial}{\partial z} N(z,t) .$$

(8.10a)

Under the assumptions that (i) D is a constant which does not depend on z, or that (ii) the dependence of D on z is much weaker than the dependence of N on z, Eq. (8.10a) simplifies to the linear, second-order partial differential equation

$$\frac{\partial N(z,t)}{\partial t} = D \frac{\partial^2 N(z,t)}{\partial z^2}$$

(8.10b)

which is known as *Fick's second law*. In the following we consider the solution of Fick's second law for some specific impurity distributions. Specifically, the cases of constant impurity density and constant impurity surface concentration will be analyzed.

We first consider the case of a *finite reservoir*, that is the total number of diffusing atoms does not change with time. Assume a semiconductor which has a sheet of impurities at $z = 0$ in the xy-phase. The two-dimensional density of the impurities is denoted as N^{2D} (in units of cm^{-2}). Such a doping distribution can be expressed in terms of the Dirac δ-function

$$N(z) = N^{2D} \delta(z) .$$

(8.11)

Upon diffusion, i.e. the random, uncorrelated motion of impurities, the impurity distribution broadens.

The solution of Fick's second equation with an initial δ-function distribution is given by the *gaussian distribution*. After impurities with a diffusion constant D diffuse for a time t in a semiconductor, the impurity distribution is given by

$$N(z) = \frac{N^{2D}}{\sqrt{2\pi} \sqrt{2Dt}} \exp\left[-\frac{z^2}{4Dt}\right]$$

(8.12)

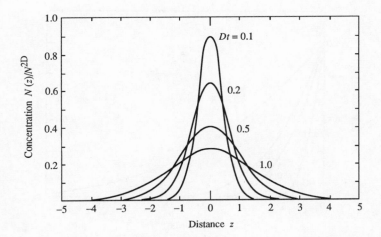

Fig. 8.4. Gaussian distribution of diffusing impurities for different values of Dt. For $t \to 0$ the impurity distribution becomes δ-function-like.

which is a solution of Fick's second law. The quantity $L_D = \sqrt{Dt}$ is called the *diffusion length*. The gaussian impurity distribution is shown in Fig. 8.4 for different values of $L_D = \sqrt{Dt}$. The standard deviation of the gaussian distribution, σ, is related to the diffusion length by $\sigma = \sqrt{2}\, L_D = \sqrt{2Dt}$. The full-width half-maximum (FWHM) of the gaussian distribution is given by FWHM $\cong 2.355\sigma \cong 2.355\sqrt{2Dt}$. Note that any doping distribution can be expressed as a superposition of δ-functions or gaussian functions. Since the diffusion processes are uncorrelated, the doping distribution after diffusion can be obtained by superposition of the diffused individual gaussian distributions.

Next we consider the diffusion of impurities in a semiconductor with *constant surface concentration*, i.e. the concentration of impurities at some location $z = 0$ (surface) is maintained at all times (infinite reservoir). Such conditions can be achieved experimentally with a thin film of impurities on a semiconductor (e.g. a Si film on GaAs). If the total amount of impurities diffusing into the semiconductor is small as compared to the amount of impurities in the surface film, then the surface concentration remains approximately constant. The initial conditions of the diffusion with a constant surface concentration are given by

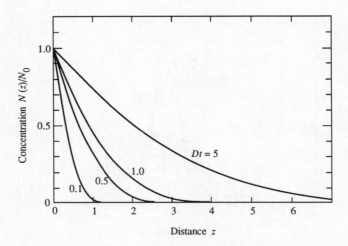

Fig. 8.5. Impurity concentration versus depth for different values of Dt. The impurity concentration at the surface $N(z = 0) = N_0$ is a constant. The impurity distribution is given by the error function complement (erfc).

$$N(z,t) = N_0 \quad \text{at } z = 0 \text{ for all } t \,, \qquad (8.13a)$$

$$N(z,t) = 0 \quad \text{at } z > 0 \text{ and } t = 0 \,. \qquad (8.13b)$$

The solution of Fick's second law with the constant surface concentration boundary condition is given by the error-function-complement (erfc). The concentration of diffusing impurities with diffusion coefficient D at the location z and the time t is given by

$$N(z,t) = N_0 \text{ erfc } \frac{z}{2\sqrt{Dt}} \qquad (8.14)$$

Examples of such diffusion profiles are shown in Fig. 8.5 for different values of Dt. The two analytical solutions of Fick's second law given above (Eqs. (8.12) and (8.14)) are the most relevant for practical diffusion problems. The distribution of impurities with more complex initial profiles can be synthesized by superposition of several simple (e.g. gaussian) distributions.

Fick's first and second laws were obtained for a diffusion constant that does not depend on the spatial position. This assumption may not be valid for all experimental conditions. In some cases, the diffusion constant may actually depend on the spatial location. Consider a semiconductor whose vacancy concentration increases in the near-surface region. Such an increased vacancy concentration in the vicinity of the surface can be due to out-diffusion of anions or cations, which can occur during annealing without a protective cap. Since the substitutional diffusion coefficient is proportional to the vacancy concentration, the diffusion coefficient is expected to increase in the near-surface region and thus depends on the spatial coordinate. For the one-dimensional, z-dependent case

$$D = D(z) . \tag{8.15}$$

Fick's laws need to be modified in the case of a diffusion coefficient which depends on the spatial coordinate. *Generalization* of *Fick's first law* yields

$$j(z) = -D(z) \frac{dN(z)}{dz} - N(z) \frac{dD(z)}{dz} . \tag{8.16}$$

The new second term of the right-hand side of the equation represents a net flux of impurities to regions with a smaller diffusion coefficient. Impurities in a highly diffusive region diffuse rapidly until they reach a slowly diffusive region. As a result, the equilibrium impurity concentration will be *higher* in the region with a *smaller* diffusion coefficient.

Application of the continuity equation to Eq. (8.16) yields the *second generalized fickian law* as

$$\frac{\partial N(z,t)}{\partial t} = \frac{\partial^2}{\partial z^2} [D(z) N(z,t)]$$

$$= D(z) \frac{\partial^2 N(z,t)}{\partial z^2} + 2 \frac{\partial D(z)}{\partial z} \frac{\partial N(z,t)}{\partial z} + N(z,t) \frac{\partial^2 D(z)}{\partial z^2} . \tag{8.17}$$

The third term of the sum is usually very small and can therefore be neglected. As an example, we consider the equilibrium impurity distribution in a semiconductor with a spatially varying diffusion constant. In equilibrium, $\partial N / \partial t = 0$, and consequently

$$\frac{\partial D(z)}{\partial z} N(z) = -D(z) \frac{\partial N(z)}{\partial z} . \tag{8.18}$$

The equation implies that a diffusion constant that decreases along some spatial direction results in an increased impurity concentration along that direction.

The linear fickian diffusion theory assumes uncorrelated, random diffusion hops of impurities. The assumption of uncorrelated motion is unlikely to be fulfilled at very high concentrations where pair formation may occur. Thus, in a general treatment, the diffusion coefficient must be considered to be concentration dependent. Some diffusion mechanisms indeed yield such concentration dependence (see substitutional–interstitial diffusion in the following section). In this case the first generalized fickian equation (Eq. (8.16)) applies. It can be written as

$$j(z) = -D(z) \frac{dN(z)}{dz} - N(z) \frac{dD[N(z)]}{dN(z)} \frac{dN(z)}{dz} . \qquad (8.19)$$

If D has a non-linear functional dependence on N or z, then Eq. (8.19) becomes non-linear. Diffusion processes following Eq. (8.19) are called *non-linear fickian* diffusion processes. There are no general analytic solutions to the non-linear diffusion equation. Usually, numerical and iterative techniques are employed to solve it. Weisberg and Blanc (1963) solved the equation for a linear, quadratic, and cubic dependence of D on the concentration N (i.e. $D \propto N$, $D \propto N^2$, and $D \propto N^3$). A frequently used numerical method to solve the non-linear diffusion equation is the *method of finite differences* (see, for example, Tuck, 1988). The analysis of impurity profiles originating from non-linear diffusion is not straight forward. A method frequently invoked to analyze such diffusion profiles is the *Boltzmann–Matano method*, which allows one to extract the concentration-dependence of D (see, for example, Tuck, 1988).

8.3 Diffusion mechanisms

The atomic mechanism for impurity diffusion in III–V semiconductors depends on the physical and chemical properties of the respective impurity and its host semiconductor. The magnitude of the diffusion process can depend on a large variety of parameters such as vacancy concentration, concentration of other point defects, atomic size of the impurity, free carrier concentration, position of the Fermi level, stoichiometry, etc. In the following, several diffusion mechanisms are discussed and their functional dependences are analyzed.

To illustrate the relevance of different characteristics, we consider two examples of diffusion processes, namely substitutional and interstitial diffusion of impurities, as schematically depicted in Fig. 8.6. In the substitutional diffusion (Fig. 8.6a) process, the impurity hops from one substitutional site to a second, vacant substitutional site. Since one vacancy is required for each diffusion hop, it is expected that the diffusion coefficient for this process is proportional to the vacancy concentration. In the interstitial process (Fig. 8.6b), an impurity hops from a first to a second interstitial site. It is reasonable to assume that atoms with a small atomic diameter can readily propagate via interstitial sites. The concentration of vacancies plays a minor role in interstitial diffusion. The two

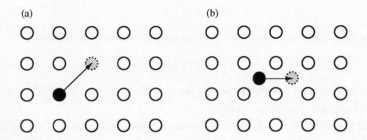

Fig. 8.6. Schematic illustration of (a) substitutional diffusion proceeding through vacancies and (b) interstitial diffusion proceeding through interstitial sites of the lattice.

examples illustrate that completely different dependences are expected for different types of diffusion processes. In the following sections, the characteristics of several diffusion processes are analyzed.

8.3.1 Substitutional diffusion

The diffusion process via vacant substitutional lattice sites is called substitutional impurity diffusion. It is assumed that impurities occupy only substitutional sites of the lattice and that an impurity hop occurs to a vacant, *adjacent* lattice site as illustrated in Fig. 8.6(a). Since *one* vacancy is required for *one* diffusion hop, the substitutional diffusion coefficient is proportional to the vacancy concentration, i.e.

$$D \propto [V] . \qquad (8.20a)$$

where the concentration of vacancies, $[V]$, is either the group-III or the group-V vacancy concentration depending on the site which the impurity preferentially occupies. The vacancy concentration depends on a number of parameters such as growth conditions, annealing conditions, stoichiometry of the semiconductor, Fermi level, etc. Consequently, the magnitude of the diffusion coefficient can vary for different diffusion conditions.

Next we consider an impurity with *all* adjacent lattice sites being occupied. Then the impurity cannot diffuse at all. However, not only the impurity, but also a lattice vacancy can diffuse. The probability that the vacancy diffuses to a site adjacent to the impurity is proportional to the vacancy diffusion constant D_V. Thus, the diffusion constant of the impurity is proportional to the product of the vacancy concentration and the vacancy diffusion constant, i.e.

$$D \propto [V]D_V \ . \tag{8.20b}$$

Finally the diffusion constant for substitutional diffusion can be written as

$$D = f[V]D_V \tag{8.20c}$$

where f is a constant which depends on the crystal structure of the semiconductor (see, for example, Manning, 1968).

Phillips (1973) concluded that the probability of substitutional diffusion does not depend on the atomic size of the diffusing impurity. The author pointed out that different impurities of the same group have similar activation energies in Si and Ge. It was concluded that the similarity in diffusion activation energy indicates that the atomic radii do not play an important role in the activation process. Even though this conclusion seems plausible, the irrelevance of the atomic radius has not been demonstrated in III–V semiconductors. Phillips (1973) further suggested that the long-range Coulomb interaction between charged impurities and charged vacancies influences the probability of substitutional diffusion of impurities. The author suggested that impurities and vacancies with opposite charge tend to associate (electrostatic attraction) resulting in a higher diffusion coefficient. In contrast, the long-range coulombic interaction between impurities and vacancies with the same charge tends to increase the diffusion activation energy, i.e. decrease the probability of diffusion. The role of the charge state of vacancies and other native defects will be further discussed in the following section.

8.3.2 Impurity-defect complex diffusion

Frequently impurities diffuse as impurity-defect complexes rather than as isolated impurity atoms. Two examples of defects which can participate in the diffusion of a complex are shown in Fig. 8.7. The *Schottky defect* (Fig. 8.7a) is a vacancy formed by the transfer of a lattice atom to the crystal surface. Schottky defects are more likely to occur in the vicinity of semiconductor surfaces. The *Frenkel defect* (Fig. 8.7b) is a vacancy–interstitial pair which is formed by moving a lattice atom to an interstitial site.

The diffusion of impurity-defect complexes has very different characteristics as compared to the diffusion of isolated atoms. Defects can assume different charge states. Shockley and Last (1957) showed that any flaw (defect) can be represented by a series of energy levels which can be occupied or empty depending on the position of the Fermi level. The defect can assume the charge states ..., 2+, 1+, 0, 1–, 2–, ... Depending on the charge state, the interaction of the impurity with the defect will be different. The diffusion of a complex formed by the impurity and the defect of charge state ..., 2+, 1+, 0, 1–, 2–, ... will be

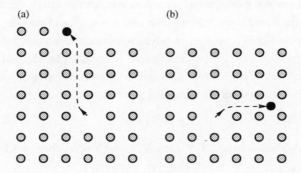

Fig. 8.7. Schematic illustration of point defects: (a) Schottky defect and (b) Frenkel defect.

different for each charge state of the impurity. Assuming that the diffusion coefficient of each complex is independent of other complexes, the effective diffusion coefficient is the linear combination of the diffusion coefficient of each complex (Shaw, 1975). As an example, we assume that the defect is a vacancy with charge states ..., V^{2+}, V^+, V^0, V^-, V^{2-}, ... For an *intrinsic* semiconductor, the effective diffusion is then given by

$$D_i = D_i^0 + D_i^+ + D_i^- + D_i^{2+} + D_i^{2-} + \cdots \tag{8.21}$$

where D_i^s is the diffusion coefficient of the complex formed by a vacancy of charge state s and the impurity under intrinsic conditions*. Under extrinsic conditions, the concentration of vacancies changes. If the concentration of vacancies is much smaller than the total concentration of impurities, then the diffusion coefficient for each complex is proportional to the concentration of the particular vacancy. The *effective* diffusion coefficient in then given by

$$D = D_i^0 \frac{[V^0]}{[V^0]_i} + D_i^+ \frac{[V^+]}{[V^+]_i} + D_i^- \frac{[V^-]}{[V^-]_i} + D_i^{2+} \frac{[V^{2+}]}{[V^{2+}]_i} + D_i^{2-} \frac{[V^{2-}]}{[V^{2-}]_i} + \cdots \tag{8.22}$$

where $[V^s]_i$ and $[V^s]$ are the concentrations of vacancies with charge state s under intrinsic and extrinsic conditions, respectively.

---

* The provision that a semiconductor be intrinsic *and* contain impurities seems to contradict. However, intrinsic conditions can be achieved in a doped semiconductor with $N_D = N_A$ and $n = p = n_i$, i.e. in a fully compensated semiconductor.

The relative concentrations of vacancies with charge state s depends on the position of the Fermi level. Fermi–Dirac statistics applies to the energy levels of defects. As the Fermi level rises in n-type semiconductors, more vacancy states become populated, i.e. the concentration of negatively charged vacancies increases with the Fermi level. Consider a neutral and a negatively charged vacancy. The ionization of the vacancy is given by

$$e + V^0 \leftrightarrow V^- \ . \tag{8.23}$$

The relative concentrations of V^0 and V^- are obtained from the law of mass action

$$\frac{[V^-]}{n[V^0]} = K(T) = ce^{-\Delta E/kT} \tag{8.24}$$

where $K(T)$ is the temperature-dependent equilibrium constant. Under intrinsic conditions one obtains

$$\frac{[V^-]_i}{n_i[V^0]_i} = K(T) \ . \tag{8.25}$$

Assuming that most of the vacancies are in the neutral state, i.e. $[V^0] \gg [V^-]$, then $[V^0]_i \cong [V^0]$; the combination of Eqs. (8.24) and (8.25) then yields

$$\frac{[V^-]}{[V^-]_i} = \frac{n}{n_i} \ . \tag{8.26}$$

Thus, the concentration of negatively charged vacancies increases with the free electron concentration. As an example, consider an impurity–vacancy complex that readily diffuses if the vacancy is negatively charged. The formation of the complex is then more likely in n-type semiconductors. As a result the diffusion characteristic depends on the free electron concentration, i.e. on the position of the Fermi level. In order to quantify the above result, Eq. (8.26) can be generalized for different charge states s of the vacancy.

$$\frac{[V^s]}{[V^s]_i} = \left(\frac{n}{n_i}\right)^s \ . \tag{8.27}$$

Insertion of the result into Eq. (8.22) yields the *extrinsic* diffusion coefficient

$$D = D_i^0 + D_i^+ \left[\frac{p}{n_i}\right] + D_i^- \left[\frac{n}{n_i}\right] + D_i^{2+} \left[\frac{p}{n_i}\right]^2 + D_i^{2-} \left[\frac{n}{n_i}\right]^2 + \cdots \ . \tag{8.28}$$

For n-type (p-type) semiconductors the terms containing the hole (electron) concentration can be neglected.

The energy levels of various defects in GaAs have been calculated (Baraff and Schlüter, 1985). The authors also estimated the concentration of defects for a given position of the Fermi level and predicted abundant native defects in either highly doped p-type or highly doped n-type material. Not only the defect states, but also the diffusion behaviour of the impurity-defect complex should be known in order to quantitatively estimate the relevance of impurity-defect complexes with respect to their diffusion characteristics. Unfortunately, the precise diffusion characteristics of impurity-defect complexes are unknown for III–V semiconductors. Nevertheless, the quantitative analysis given above allows one to appreciate and understand the complexity of the diffusion of impurity-defect complexes and their dependence on doping.

8.3.3 Interstitial diffusion

In the pure interstitial diffusion process, impurities hop from one interstitial site to another, and and never occupy substitutional lattice sites. Therefore, the interstitial diffusion process is of limited practical relevance for common group-IIa, -IIb, -IV, -VI impurities in III–V semiconductors, as the major fraction of these impurities occupies substitutional sites in the III-V zincblende lattice.

Interstitial sites in a zincblende lattice are of tetrahedral symmetry; either four anions or four cations are closest to the center of the interstitial void. Impurities with a small atomic radius are more easily accommodated in interstitial sites. A small atomic radius is also advantageous for diffusion hops from one interstitial site to another. In III–V semiconductors, hydrogen (H) occupies interstitial sites and diffuses interstitially (Pankove and Johnson, 1991). The interstitial diffusion mechanism and the small size allows H to diffuse rapidly in semiconductors, even at relatively low temperatures. For a further discussion of H in III–V semiconductors see Chap. 9. Lithium (Li) is a well-known interstitial diffuser in Si. Of atomic number 3, Li is a rather small atom.

8.3.4 Substitutional–interstitial diffusion

The substitutional–interstitial diffusion mechanism is one of the most important impurity diffusion mechanisms in semiconductors. It was proposed and analyzed by Frank and Turnbull (1956). It is assumed in the substitutional–interstitial model that the majority of impurities occupy substitutional sites. However, these substitutional impurities have a very small probability of diffusing substitutionally. It is further assumed that a minor fraction of impurities occupies interstitial sites. These interstitial impurities have a much higher diffusion constant since they are less strongly bound to the lattice. The diffusion process is then dominated by the highly mobile interstitials.

Any substitutional impurity atom can hop into an interstitial site according to

$$S \leftrightarrow I + V \tag{8.29a}$$

where S, I, and V represent the substitutional impurity, interstitial impurity and a vacancy, respectively. It is further assumed that the three species, S, I, and V, are neutral. The concentrations of S, I, and V can be inferred from the law of mass action provided the formation energy of the reaction is known. Usually [I] << [S], i.e. the concentration of interstitials is much smaller than the concentration of substitutional impurities.

The validity of Eq. (8.29a) is restricted to the case where S, I, and V are neutral. However, it is the very nature of substitutional impurities to be charged. As an example, we assume that the substitutional impurity is singly negatively charged (acceptor) and that the interstitial is positively charged. Then Eq. (8.29a) needs to be modified according to

$$S^- + 2h \leftrightarrow I^+ + V \tag{8.29b}$$

where h represents a positively charged hole. Under equilibrium conditions, the concentrations are given by the law of mass action

$$\frac{p^2[S]}{[I][V]} = K_1 = ce^{-\Delta E/kT} \tag{8.30}$$

where p is the hole concentration. With [I] << [S] and $p \cong$ [S] the equation simplifies to

$$K_1[S]^3 = [I][V] . \tag{8.31}$$

If the impurities diffuse through the crystal then [S] and [I] are a function of the spatial coordinate, z, and the time t. On the other hand the vacancy concentration [V] can be assumed to be constant for all spatial positions. Then there is no vacancy gradient and consequently no net diffusion of vacancies. It follows from Eq. (8.31) that the concentration of interstitials, [I], increases strongly with the impurity concentration. Recalling that interstitials diffuse much faster than substitutionals, it is evident that the effective diffusion coefficient increases with impurity concentration.

In order to quantify the increase of the diffusion coefficient, we apply Fick's second law to the substitutional–interstitial model. Note that the total impurity concentration is given by $N = $ [S] + [I]. The diffusion equation then reads

$$\frac{\partial[I]}{\partial t} + \frac{\partial[S]}{\partial t} = D_I \frac{\partial^2[I]}{\partial z^2} + D_S \frac{\partial^2[S]}{\partial z^2} \qquad (8.32a)$$

where D_I and D_S are the diffusion coefficients for interstitial and substitutional diffusion, respectively. Since $D_I \gg D_S$, the equation reduces to

$$\frac{\partial[I]}{\partial t} + \frac{\partial[S]}{\partial t} = D_I \frac{\partial^2[I]}{\partial z^2} . \qquad (8.32b)$$

The concentration of interstitials, $[I]$, can be replaced by $[S]$ by using Eq. (8.31), which gives

$$\left[\frac{3K_1[S]^2}{[V]} + 1 \right] \frac{\partial[S]}{\partial t} = D_I \frac{\partial}{\partial z} \left[\frac{3K_1[S]^2}{[V]} \frac{\partial[S]}{\partial z} \right] . \qquad (8.33)$$

If K_1 on the left-hand side of the equation is substituted with Eq. (8.31), the first term in the bracket becomes $3[I] [S]/p^2$ which is much smaller than one since $[I] \ll p$ and $p \cong [S]$. Equation (8.33) then simplifies to

$$\frac{\partial[S]}{\partial t} = \frac{\partial}{\partial z} \left[\frac{3K_1 D_I}{[V]} [S]^2 \frac{\partial[S]}{\partial z} \right] . \qquad (8.34)$$

Comparison of the coefficients of the equation with Fick's law (Eq. (8.10a)) yields the *effective* diffusion coefficient

$$D = \frac{3K_1 D_I}{[V]} [S]^2 . \qquad (8.35)$$

Note that the diffusion coefficient is proportional to the *square of the substitutional impurity concentration*. Since $N \cong [S]$, that is nearly all impurities are on substitutional sites, the diffusion coefficient is proportional to the square of the overall impurity concentration. As a result, diffusion occurs more rapidly at high concentrations as compared to lower doping concentration.

Inspection of Eq. (8.35) reveals that the 'power of two' ($D \propto N^2$) is a result of the difference of two in the charge states between the substitutional S^- and the interstitial I^+. If, for example, the substitutional were S^- and the interstitial were I^{2+}, then the diffusion coefficient were proportional to the *third* power of the doping concentration, i.e. $D \propto N^3$.

Note that the interstitial–substitutional model results in non-linear Fickian diffusion. The individual atomic jumps are correlated since the probability of a diffusion jump depends on the concentration of the impurities. By adopting the concentration-dependent diffusion constant of Eq. (8.35), the Fickian diffusion equation (Eq. (8.10a)) can be used.

Interstitial impurities can reoccupy a substitutional lattice site by (i) occupying a previously vacant lattice site or (ii) via the '*kick-out*' mechanism. In the 'kick-out' mechanism, an initially interstitial impurity displaces a substitutional lattice atom which is forced into an interstitial lattice site. The final configuration of the 'kick-out' reaction is a substitutional impurity atom and a interstitial host atom. Analogously, an interstitial host lattice atom can 'kick-out' an impurity atom and force it into an interstitial site. The formation energy of the 'kick-out' reaction is smaller than the formation energy of a Frenkel defect. Accordingly, the 'kick-out' reaction is a fairly common reaction in semiconductors.

8.3.5 Interchange diffusion

Interchange diffusion effects involve the interchange of positions of one impurity atom and one or several atoms of the host semiconductor. Two examples of interchange diffusion effects are illustrated in Fig. 8.8 which shows (a) a *direct interchange* diffusion hop and (b) a *cooperative interchange* diffusion hop. The direct interchange diffusion involves two neighboring atoms. Three or more atoms take part in cooperative interchange diffusion processes.

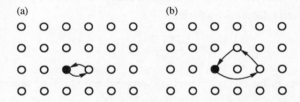

Fig. 8.8. Schematic illustration of (a) the direct interchange diffusion mechanism and (b) the cooperative interchange diffusion mechanism.

Interchange diffusion effects do not require vacant lattice sites or interstitials. Certainly, the diffusion of impurities via vacancies or interstitials is more effective as compared to the diffusion via interchange effects. However, when diffusion via interstitials or vacancies is unlikely, interchange diffusion may be the dominant diffusion mechanism. The probability of interchange diffusion hops is relatively low. The probability further decreases if the interchange process involves more than two atoms.

8.3.6 Combination of diffusion mechanisms

Each of the previous sections focused on a single, pure diffusion mechanism. Under realistic conditions a combination of several mechanism occurs. For example, pure substitutional and the substitutional–interstitial mechanism may occur simultaneously. It is reasonable to consider the different diffusion mechanisms as independent. A diffusion hop is a relatively seldom event, i.e. impurities predominantly occupy substitutional lattice sites. As a result, the different diffusion processes can be considered as *independent*. The overall diffusion process is then the *sum* of all individual diffusion processes. In this context it is useful to recall that the diffusion constants of different diffusion processes can differ by many orders of magnitude. It is likely that a single specific diffusion mechanism dominates the entire diffusion process. It is therefore frequently sufficient to consider only the dominant diffusion process despite the simultaneous occurrence of several diffusion processes.

8.3.7 Enhanced diffusion along imperfections

Spatially extended imperfections in semiconductors are preferred diffusion channels for impurities. The structural stability of the semiconductor is lower in the vicinity of imperfections. Such imperfections are, for example, grain-boundaries, interfaces between mismatched semiconductors, misfit dislocations, surfaces, or spatially extended clusters of defects. Due to the low structural perfection of extended defects, it is quite evident that the formation energy of point defects such as vacancies, interstitials, etc., is lower in the vicinity of the imperfection. As a result of the lower formation energy, the defects have a higher concentration at or in the vicinity of the structural imperfection. Diffusion processes including impurity diffusion processes are therefore more likely to occur in the vicinity of the imperfections, since the diffusion processes depend on the concentration of point defects as mentioned in the preceding sections.

A well-known example of enhanced diffusion along imperfections is *pipe diffusion*, which occurs along dislocation lines in semiconductors. The diffusion probability along the one-dimensional dislocation line is enhanced leading to a pipe-like distribution of impurities and to a preferential movement of impurities along the dislocation line.

8.4 Some specific impurities

The diffusion mechanisms of impurities in compound semiconductors are clearly more complicated and involved than those in the elemental semiconductors Si and Ge. The number of types of native defects is much larger in compound semiconductors as compared to elemental semiconductors. The native defects include group-III and group-V vacancies, group-III and group-V interstitials, and antisite defects which cannot occur in elemental semiconductors. In addition, compound semiconductors may suffer from non-stoichiometry. The occurrence of anion-rich or cation-rich semiconductors also depends on the epitaxial growth technique. While gas-phase epitaxial techniques work under anion-rich conditions, liquid-phase epitaxy grows under cation-rich conditions. Furthermore, if the impurity diffusion process occurs in the vicinity of the surface, out-diffusion of group-V elements may occur, which further complicates the nature of the impurity diffusion process. Due to the complexity of diffusion processes in III–V semiconductors, a detailed analysis of diffusion processes for specific impurities is presently not available. Therefore, this section primarily emphasizes the experimental findings for different impurities.

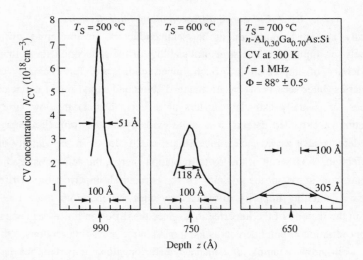

Fig. 8.9. Capacitance–voltage profiles of Si δ–doped $Al_{0.30}Ga_{0.70}As$ grown at 500, 600 and 700 °C by molecular-beam epitaxy.

Diffusion of impurities in semiconductors can occur during crystal growth or during post-growth annealing. The importance of diffusion during epitaxial growth is illustrated in Fig. 8.9 (Schubert et al., 1989). Figure 8.9 shows the capacitance–voltage profiles of Si δ-doped $Al_{0.30}Ga_{0.70}As$ grown by molecular-beam epitaxy at the three temperatures of 500 °C, 600 °C, and 700 °C. The broadening of the profiles from 51 Å to 305 Å illustrates that different growth temperatures can have a drastic effect on the impurity distribution by means of diffusion effects.

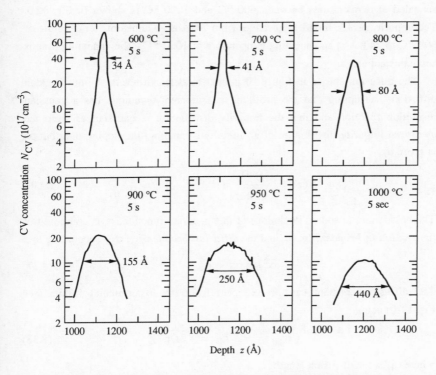

Fig. 8.10. Capacitance–voltage profiles of different pieces of an MBE-grown Be δ-doped GaAs sample annealed at temperatures of 600 to 1000 °C for 5 s.

Many diffusion experiments are performed with semiconductors coated with a film of impurities. Upon annealing, impurities diffuse into the semiconductor. However, this method is obviously very sensitive to the surface condition, i.e. surface contamination, out-diffusion of group-V elements, and other surface effects. A method less sensitive to surface effects employs δ-doped semiconductors (Schubert et al., 1988). In the case of δ-doped semiconductors the impurity reservoir is buried ≥ 1000 Å below the surface. Surface effects are therefore drastically reduced. Epitaxial growth under appropriate conditions (low temperatures) insures that the diffusion occurring during epitaxial growth is insignificant. Post-growth annealing of the δ-doped samples allows one to choose well-defined temperature profiles during different annealing cycles. The diffusion length and the diffusion coefficient can be inferred from the broadening of the doping profile. An example of impurity distribution measurements on GaAs:Be annealed at temperatures between 600 °C and 1000 °C is shown in Fig. 8.10. The impurity profile broadening ranges from 34 Å at an annealing temperature of 600 °C to 440 Å at an annealing temperature of 1000 °C. The annealing time is kept constant at 5 s.

The diffusion shown in Fig. 8.10 is next modeled in terms of linear fickian diffusion. Assuming that the profiling measurement technique has a gaussian resolution function and that the impurity distribution is gaussian as well, the *measured* impurity distribution of an initially δ-function-like doping distribution is given by

$$N(z) = \frac{N^{2D}}{\sqrt{2\pi}\,\sigma}\,e^{-(z-z_0)^2/(2\sigma^2)}\,. \tag{8.36}$$

The broadening (standard deviation) of this gaussian function, σ, is composed of the *resolution broadening*, σ_r, and the *diffusion broadening*, σ_{diff}, according to

$$\sigma^2 = \sigma_r^2 + \sigma_{diff}^2\,. \tag{8.37}$$

The diffusion constant can be extracted from the experimental results (see Fig. 8.10) using

$$\sigma_{diff} = \sqrt{2}\,L_D = \sqrt{2Dt} \tag{8.38}$$

where L_D is the diffusion length.

The experimental procedure described above (see also Fig. 8.10) avoids surface effects since the doped layer can be buried in the bulk. Note however, that the analysis using Eqs. (8.36) to (8.38) is valid for a concentration-independent diffusion constant. If the diffusion constant is concentration dependent, other methods, such as the Boltzmann–Matano technique, must be used. The following sections summarize the diffusion characteristics of technologically important impurities in III–V semiconductors.

8.4.1 Beryllium (Be)

Beryllium is one of the major doping impurities in III–V semiconductors. The first report on Be diffusion in GaP dates back to 1972 (Ilegems and O'Mara, 1972). The authors studied Be diffusion with concentrations $> 10^{20}$ cm^{-3}. At such high concentrations the diffusion coefficient strongly increases over its low-concentration value. A Boltzmann–Matano analysis of the 1000 °C anneal temperature indicated a coefficient of $\cong 10^{-7}$ cm^2/s for concentrations $N_{Be} > 10^{20}$ cm^{-3}. Subsequently Ilegems (1977) and McLevige et al. (1978) studied Be diffusion in GaAs grown by molecular-beam epitaxy. The epitaxial layers were annealed at 900 °C after growth. Diffusion constants at this temperature were $10^{-13} - 10^{-14}$ cm^2/s for concentrations of approximately 10^{19} cm^{-3}.

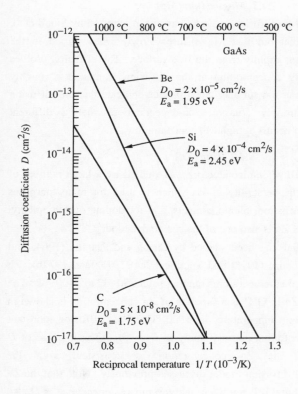

Fig. 8.11. Diffusion coefficients of C, Si, and Be in GaAs versus reciprocal temperature.

The diffusion constant of Be in GaAs was determined as a function of temperature in the range 600 °C $\leq T \leq$ 1000 °C (Schubert et al., 1990b). The concentration of Be was in the 10^{18} cm^{-3} range. The temperature-dependence of the diffusion constant is shown in Fig. 8.11 and can be described by an activation energy of $E_a = 1.95$ eV and $D_0 = 2 \times 10^{-5}$ cm^2/s using the equation $D = D_0 \exp(-E_a/kT)$. The values shown for Be in Fig. 8.11 agree with the data published by Ilegems (1977) and McLevige et al. (1978).

Beryllium redistributes strongly in GaAs at doping concentrations exceeding 10^{19} cm^{-3} (Schubert et al., 1990b). The redistribution was found to be directed towards the substrate as well as towards the surface of the epitaxial layers. Doping concentrations exceeding 10^{20} cm^{-3} could not be achieved at typical growth temperatures due to the strong Be redistribution (see also Sect. 5.3.1 and 7.5.2). Similar results were obtained for Be in InP and Ga$_{0.47}$In$_{0.53}$As (Panish et al. 1991).

8.4.2 Magnesium (Mg)

Experimental results on Mg diffusion in GaAs have been reported by Small et al. (1982). The authors point out that Mg diffuses slowly when buried in the semiconductor but diffuses rapidly from the free surface. The diffusion profiles had a rapidly decreasing concentration at the diffusion front and a roughly constant concentration near the surface. Such profiles indicate that the diffusion coefficient is either strongly concentration-dependent or that a different redistribution mechanism occurs at high Mg concentrations.

8.4.3 Zinc (Zn)

The diffusion of Zn in III–V semiconductors was studied by a large number of groups; this is due to the versatility of Zn which is used for many purposes including pn-junction formation, ohmic contacts, and for doping during epitaxy. The diffusion behavior of Zn is also one of the best understood.

Diffusion of Zn in GaP has been studied by Chang and Pearson (1964a and 1964b), Luther and Wolfstirn (1973) and Tuck and Jay (1977a and 1977b). Zn diffusion was investigated at temperatures ranging from 850 °C to 1100 °C. The diffusion coefficient of Zn in GaP as a function of Zn concentration is shown in Fig. 8.12. The diffusion temperature is 1000 °C. The diffusion constant increases with Zn concentration approximately as $D \propto N_{Zn}^2$. The increase of D with the square of the Zn concentration is consistent with the substitutional–interstitial diffusion model discussed previously. Note that the Zn concentrations are very high; it is not likely that the square-dependence of D also applies to the low-concentration regime.

Zn diffusion in GaAs has been reported by Cunnell and Gooch (1960) and by Goldstein (1961). These early studies already indicated that Zn diffusion does not follow simple Fickian diffusion with a constant diffusion coefficient. Subsequent work by Kadhim and Tuck (1972) showed that the diffusion constant of Zn in GaAs increases quadratically with Zn concentration. The experimental results are shown in Fig. 8.12 for an anneal temperature of 1000 °C. The diffusion experiments were carried out under excess arsenic pressure in order to avoid the

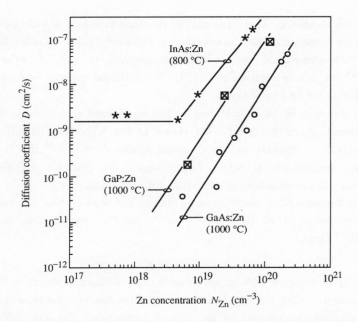

Fig. 8.12. Diffusion coefficient of Zn in GaP at 1000 °C (Chang and Pearson, 1964a and 1964b), Zn in GaAs at 1000 °C (Kadhim and Tuck, 1972), and Zn in InAs at 800 °C (Boltaks et al., 1969) as a function of Zn concentration.

out-diffusion of As from the specimen. Note again that the concentration range studied is very high ($10^{19} - 10^{20}$ cm^{-3}). The squared dependence of D on N_{Zn} has not been demonstrated at lower concentrations.

Zinc diffusion into InP was first reported by Chang and Casey (1964) at temperatures of $600 - 900$ °C. The authors analyzed the diffusion profiles using the Boltzmann–Matano method and concluded that the Zn diffusion coefficient is not constant but increases with Zn concentration. Subsequent work by Kundukhov et al. (1967) indicated that only a small fraction of the Zn atoms diffused into InP are electrically active; that is the major fraction (e.g. 90%) of the Zn occupies sites other than the substitutional In site. The low electrical activity of Zn in InP was confirmed by independent measurements (Tuck and Zahari, 1977a and 1977b) who found only 1% of the Zn atoms to be electrically active as acceptors. The physical origin of the low electrical Zn activity has been proposed to be due to Zn–V_P complexes, where V_P represents one or more phosphorus vacancies (Tuck and Hooper, 1975; Young and Pearson, 1970; Yamada et al.,

1983). The P vacancies were assumed to be generated during the annealing cycle. Despite the problematic electrical behaviour of Zn in InP, Tuck and Zahari (1977a and 1977b) determined the Zn diffusion constant in InP. A value of 2×10^{-8} cm^2/s was obtained at 800 °C. The diffusion constant was weakly dependent on the Zn concentration.

The diffusion of Zn in InAs was studied by Boltaks et al. (1969). Their experimental results for 800 °C are shown in Fig. 8.12. The Zn diffusion coefficient is a constant for 'low' concentrations ($< 5 \times 10^{18}$ cm^{-3}) and increases quadratically at higher concentrations; the quadratic increase is consistent with the substitutional–interstitial diffusion model.

Zn diffusion in AlSb was reported by Showan and Shaw (1969). The authors found values of D of $10^{-9} - 10^{-8}$ cm^2/s for Zn concentrations in the 10^{19} cm^{-3} range.

8.4.4 Carbon (C)

Carbon is an impurity with an exceptionally low diffusion constant in III–V semiconductors. This property makes C an attractive doping impurity in devices with stringent requirements of the doping distribution. The low diffusion constant of C provides long-term stability of the operating characteristics of such devices. The diffusion coefficient of C in GaAs is shown in Fig. 8.11 versus reciprocal temperature. The epitaxial samples used for the diffusion experiments have C concentrations in the 10^{18} cm^{-3} range. Comparison of the C diffusion constant with the Be diffusion constant reveals that the C coefficient is about 2 orders of magnitude lower in the temperature range of interest.

C diffusion in GaAs has been studied at 800 °C for epitaxial samples grown by flow-rate modulation epitaxy (Kobayashi, 1987). The concentration of C was in the mid 10^{17} cm^{-3} to the low 10^{18} cm^{-3} range. The authors inferred a diffusion constant of 2×10^{-16} cm^2/s in agreement with the data shown in Fig. 8.11.

8.4.5 Silicon (Si)

Silicon has a rather low diffusion constant in many III–V semiconductors. The diffusion coefficient of Si in epitaxial GaAs is shown in Fig. 8.11 for Si concentrations in the 10^{18} cm^{-3} range (Schubert et al., 1988). The Si diffusion constant is higher than the constant of C but lower than that of Be. The diffusion constant of Si in GaAs was also measured by Beall et al. (1989) at a temperature of 600 °C. The authors found $D \cong 5 \times 10^{-18}$ cm^2/s for Si doping in the 10^{18} cm^{-3} range. The authors noted that either substitutional Si diffusion or substitutional–interstitial diffusion could prevail. They further note that Si-pair

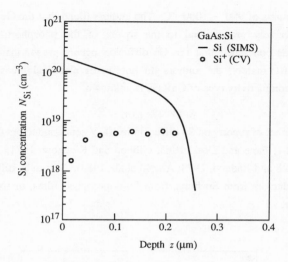

Fig. 8.13. Depth profile of total Si concentration (solid line) and electrically active Si concentration (circles) in GaAs. Silicon was diffused from a Si film on the surface into GaAs at 1050 °C for 3 s (Greiner and Gibbons, 1984).

diffusion is unlikely at sufficiently low concentrations.

Silicon diffusion experiments at very high Si concentrations were performed by Greiner and Gibbons (1984). The authors diffused impurities from a Si film into a GaAs layer. The samples were thermally annealed at 1050 °C for 3 s. The resulting Si profile measured by secondary ion mass spectrometry (SIMS) and electrochemical capacitance–voltage (CV) profiling are shown in Fig. 8.13. While SIMS measures the total Si atom concentration, CV-measurements yield the electrically active net electron concentration. Figure 8.13 reveals that 90 – 99 % of the total Si-atom concentration is electrically inactive or compensated. The authors concluded that the Si impurities diffuse predominantly as $Si_{Ga} - Si_{As}$ compensating pairs. These pairs are assumed to have a much larger diffusion coefficient than isolated Si impurities. The diffusion constant determined by the authors for Si-pair diffusion is indeed approximately two orders of magnitude higher than the Si diffusion data shown in Fig. 8.11.

8.4.6 Germanium (Ge)

Since Ge is strongly amphoteric in III–V semiconductors, Ge is not of high technological significance. There is only one report on Ge diffusion in III–V semiconductors. Schneider and Nebauer (1975) studied the diffusion of Ge in

GaP at temperatures of 900 – 1000 °C. The authors found that the Ge diffusion constant is inversely proportional to the square of the phosphorus ambient pressure. It was concluded that the Ge diffusion occurs predominantly via P vacancies. Unfortunately, the authors do not report electrical measurements. Therefore, the conductivity type of GaP:Ge is unknown.

8.4.7 Tin (Sn)

There are a number of reports on Sn diffusion in III–V semiconductors (Goldstein and Keller, 1961; Fane and Goss, 1963; Gibbon and Ketchow, 1971; Yamazaki et al., 1975; Tuck and Badawi, 1978; Arnold et al., 1984). Sn can be diffused into III–V semiconductors from Sn films, from Sn-doped glass films, or from a gas-phase precursor.

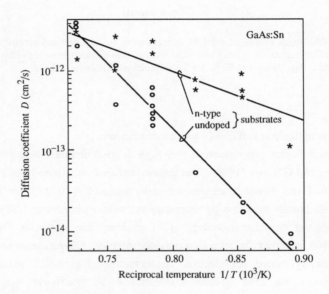

Fig. 8.14. Diffusion coefficient of Sn in GaAs as a function of reciprocal temperature using undoped and n-type GaAs substrates as starting materials (Tuck and Badawi, 1978).

The diffusion constant of Sn in GaAs is shown in Fig. 8.14 (Tuck and Badawi, 1978). Tin was diffused into GaAs from the gas phase. Two types of substrates were used, namely undoped and n-type substrates. The diffusion experiments were carried out under excess As pressure in order to avoid loss of As from the

GaAs substrates. The diffusion coefficients shown in Fig. 8.14 illustrate that the diffusion depends sensitively on the properties of the host. However, at high temperatures ($\cong 1100$ °C) there is no appreciable difference between the two substrates. The authors attribute the similarity of the diffusion constant at high temperatures to the similarity in electron concentration: At such high temperatures both types of semiconductor substrates are intrinsic, i.e. the free electron concentration is the same. It was concluded that most of the Sn occupies Ga sites and that Sn diffuses via Ga vacancies.

Tin diffusion was also studied in GaSb (Uskov, 1974) and in InSb (Sze and Wei, 1961). In InSb, the diffusion constant was determined in the temperature range 385–512 °C and was described by $D_0 = 5.5 \times 10^{-8}$ cm^2/s and an activation energy of $E_a = 0.75$ eV.

8.4.8 Sulfur (S)

Sulfur diffusion studies have been reported for several III–V semiconductors including GaAs (Vieland, 1961; Frieser, 1965; Kendall, 1968; Young and Pearson, 1970; Matino, 1974; Prince et al., 1986), GaP (Young and Pearson, 1970), InP (Dutt et al., 1984; Chin et al., 1983a, 1983b), and InAs (Schillmann, 1962). The diffusion of S in GaAs has been characterized by D_0 and an activation energy, E_a. The parameters are shown in Table 8.1. The diffusion constants for S in GaAs at 900 °C inferred from Table 8.1 are in the $10^{-12} - 10^{-13}$ cm^2/s range, i.e. 1–2 orders of magnitude larger than the diffusion constant for Si in GaAs.

Table 8.1. *Diffusion parameters for S in GaAs*

D_0 (cm^2/s)	E_a (eV)	Reference
1.2×10^{-4}	1.80	Vieland (1961)
2.6×10^{-5}	1.86	Frieser (1965)
1.6×10^{-5}	1.63	Kendall (1968)
1.85×10^{-2}	2.6	Young & Pearson (1970)
10.9	2.95	Matino (1974)
0.27	3.0	Prince et al. (1986)

8.4.9 Selenium (Se)

Diffusion of Se in GaAs has been studied by Khludkov and Lavrishchev (1976). Selenium was used to form a pn-junction in a p-type GaAs substrate. It was diffused from the semiconductor surface. The authors assumed that the Se surface concentration is approximately 10^{20} cm^{-3}. An erfc function was used to fit the impurity distribution in order to get an effective diffusion coefficient. It is questionable if fickian diffusion theory can be applied at such high concentrations, since correlation effects become increasingly important at high impurity concentrations.

Diffusion of Se in InSb has been reported by Rekalova et al. (1969 and 1971). Selenium was diffused from a surface film containing 1% Se and 99% In. Diffusion temperatures ranged from 350 °C to 500 °C in a reducing hydrogen atmosphere. The technique was used to form pn-junctions in p-type InSb. Diffusion of Se has also been studied in InAs (Schillmann, 1962) and in InP (Dutt et al., 1984).

8.4.10 Tellurium (Te)

Diffusion of Te in GaAs was studied by Karelina et al. (1974) in order to form a pn-junction in a p-type substrate. Annealing temperatures were 1000–1150 °C while simultaneously maintaining an ambient arsenic pressure. The authors concluded that Te diffuses mainly via As vacancies and divacancies.

The diffusion of Te in InSb was also reported by Rekalova et al. (1969 and 1971). The Te was diffused from a surface film containing 0.5% Te and 99.5% In. Diffusion temperatures ranged from 350 °C to 500 °C in a reducing hydrogen atmosphere. The technique was used to form pn-junctions in p-type InSb . Diffusion of Te has also been studied in InAs (Schillmann, 1962).

8.4.11 Chromium (Cr)

Chromium is a deep acceptor in GaAs and has an energy level close to the center of the gap. It is used for co-doping residual n-type GaAs in order to render the semiconductor semi-insulating. Chromium is an impurity of significant technological importance. There is an abundance of technical papers on the diffusion of Cr in GaAs; many of these publications report similar results.

Diffusion experiments in GaAs using an external gas-phase chromium source were carried out at 800 – 1100 °C by Tuck and Adegboyega (1979). The authors found that the Cr profiles after diffusion did not resemble erfc-function-like profiles as expected from linear fickian diffusion. It was found that the Cr diffuses deeply into the GaAs substrates, with diffusion lengths of several tens of microns for diffusion times of one hour. Diffusion lengths >>100 µm were achieved for diffusion times of several hours. (Note that the term 'diffusion length' is not defined for non-fickian diffusion).

Out-diffusion of Cr from Cr-doped GaAs substrates into epitaxial layers grown on top of the substrate has proven to be a significant problem. Linh et al. (1980) grew epitaxial layers by molecular-beam epitaxy on Cr-doped GaAs substrates and demonstrated a significant pile-up of Cr ($N_{Cr} \cong 10^{18}$ cm^{-3}) at the substrate–epilayer interface. Similar results were obtained with epilayers grown by organometallic vapor-phase epitaxy (Huber et al., 1982). Further out-diffusion experiments were reported by Deal and Stevenson (1986), Tuck and Adegboyega (1979), Vasudev et al. (1980), Watanabe et al. (1982), and Wilson (1982).

8.4.12 Iron (Fe)

Iron is a deep acceptor in InP and has an energy level close to the center of the gap. It is used for co-doping n-type InP in order to render the semiconductor semi-insulating. Iron has significant technological importance in InP.

The diffusion of Fe in InP has been studied by Shishiyanu et al. (1977) at temperatures of 720–940 °C and by Brozel et al. (1981, 1982) at a temperature of 900 °C. The experimental results revealed that Fe diffuses deeply into InP. Shishiyanu et al. (1977) fitted erfc functions to the experimental Fe profiles in order to obtain the diffusion coefficient. The authors estimated the diffusion constant $D \cong 10^{-9} - 10^{-10}$ cm^2/s at temperatures of 720–940 °C.

The out-diffusion of Fe from Fe-doped substrates into epitaxial InP layers was studied by Holmes et al. (1981), Oberstar et al. (1981) and Chevrier et al. (1980). The former group grew epitaxial InP on Fe-doped InP substrates using liquid-phase epitaxy. The Fe profiles were measured by secondary ion mass spectrometry (SIMS). The profiles were fitted with a erfc distribution in order to obtain the Fe diffusion constant in the temperature range 600 °C to 710 °C. The temperature dependence of the diffusion coefficient was expressed in terms of $D = D_0 \exp(-E_a/kT)$ with $D_0 = 6.8 \times 10^5$ cm^2/s and $E_a = 3.4$ eV. It should be noted that the activation energy of 3.4 eV is surprisingly high. Chevrier et al. (1980) also used SIMS to analyze Fe out-diffusion from InP into undoped epitaxial layers grown by vapor-phase epitaxy. The authors found that the Fe concentration in the epitaxial layer was almost as high as the Fe doping level in the substrate. In addition, a peak of the Fe concentration occurred at the epilayer–substrate interface. Such Fe peaks are not expected from an erfc distribution.

8.5 Surface segregation and surface migration

Many impurities have the tendency to redistribute preferentially towards the surface of the semiconductor. The redistribution is *anisotropic* which is in contrast to the isotropic redistribution occurring during diffusion. The preferential redistribution towards the surface can occur during bulk or epitaxial crystal growth. Surface segregation and surface migration are two mechanisms which both describe the redistribution of impurities towards the surface. **Segregation** describes the effect that impurities accumulate on the surface and are *segregated* from the crystal bulk. Segregating impurities mainly remain on the surface during epitaxial growth and only a minor fraction incorporates into the crystal. The term **migration** is used if impurities are incorporated in the crystal but redistribute within the crystal preferentially towards the crystal surface. In the following, three surface segregation/migration mechanisms will be discussed which are (i) zero-order surface segregation, (ii) first-order surface segregation, and (iii) Fermi-level-pinning induced surface migration.

8.5.1 Zero-order surface segregation

The incorporation of impurities into a growing semiconductor crystal can be described by a rate equation which takes into account the (i) arrival rate, (ii) desorption rate, (iii) incorporation rate of impurities, and (iv) the density of impurities on the growing semiconductor surface. The rate equation can be expressed as

$$\frac{dN_s}{dt} = J - k_d N_s - k_i^{(m)} N_s^m \tag{8.39}$$

where dN_s/dt denotes the change in impurity surface density, J is the incident impurity flux (per cm^2 and per s), $K_d N_s$ is the desorption flux, and $k_i^{(m)} N_s^m$ is the density of incorporated impurities per unit time, i.e. the incorporation rate of order m. For $m = 0$ the incorporation rate is $k_i^{(0)} N_s^0 = k_i^{(0)}$ which is a constant. Eq. (8.39) then simplifies to

$$\frac{dN_s}{dt} = J - k_d N_s - k_i^{(0)} \tag{8.40}$$

which is the rate equation describing *zero-order segregation*. The implications of the rate equation are as follows. First, since the incorporation rate is constant, the impurity concentration in the bulk (per cm^3) is constant as long as $N_s > 0$ and the growth rate remains constant. The impurity concentration in the bulk is then given by $N = k_i^{(0)}/v_g$ where v_g is the growth rate. The concentration in the bulk remains constant until $N_s = 0$, i.e. the surface impurity reservoir is depleted.

Zero-order surface segregation is schematically illustrated in Fig. 8.15. An impurity flux impinges on the surface during the time interval $t_1 \leq t \leq t_2$. The incorporation rate of impurities is lower than the incident flux and results in a bulk impurity concentration N. Impurities also accumulate on the surface in the time interval $t_1 \leq t \leq t_2$. (Desorption of impurities is neglected). After termination of the incident impurity flux, impurity incorporation continues from

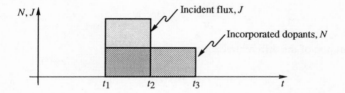

Fig. 8.15. Schematic illustration of incident dopant flux and concentration of incorporated dopants versus time for zero-order segregation. The time axis can be transformed into distance along the growth direction, z, by using $z = v_g t$, where v_g is the growth rate.

the surface reservoir until the surface reservoir is depleted ($N_s = 0$). The resulting impurity distribution is a top-hat distribution with a constant impurity concentration in the interval $t_1 \leq t \leq t_3$.

We now consider the case where $(J - k_d N_s) < k_t^{(0)}$. Then $dN_s/dt < 0$, i.e. the surface impurity reservoir is depleting. At some time the surface concentration becomes $N_s = 0$. Then Eq. (8.40) is no longer valid since N_s cannot become a negative quantity. Instead, the incorporation rate equals the arriving impurity flux diminished by the desorption rate.

Various physical mechanisms can be envisaged to account for the constant incorporation rate. For example, the constant incorporation rate can be given by the maximum achievable impurity concentration in the bulk as discussed in Chap. 7. Examples for zero-order segregation are found in the very high doping regime where the resulting impurity profiles are top-hat shaped (Schubert et al., 1990b; Hamm et al., 1989; Panish et al., 1990).

8.5.2 First-order surface segregation

First-order segregation of impurities on a semiconductor occurs if the rate of impurity incorporation is proportional to the surface density of impurity atoms. The appropriate rate equation (Eq. (8.39)) is obtained by selecting $m = 1$, i.e.

$$\frac{dN_s}{dt} = J - k_d N_s - k_t^{(1)} N_s . \tag{8.41}$$

We will now show that the impurity concentration *decreases exponentially* towards the surface after the incident impurity flux is terminated ($J = 0$). As a further simplification, assume that impurities do not desorb from the crystal, i.e. $k_d = 0$. Eq. (8.41) then reduces to

$$\frac{dN_s}{dt} = -k_t^{(1)} N_s .$$ (8.42)

The solution of the differential equation is given by

$$N_s(t) = N_s^0 e^{-k_t^{(1)} t}$$ (8.43)

where it is assumed that the surface concentration at $t = 0$ equals $N_s^0 = N_s(t = 0)$. The impurity concentration incorporated in the bulk (per unit time) is given by $N = -dN_s(v_g dt)^{-1}$. The impurity concentration in the bulk can hence be obtained by differentiation of Eq. (8.43)

$$N(t) = (k_t^{(1)} N_s^0 / v_g) e^{-k_t^{(1)} t} .$$ (8.44a)

The important result of Eq. (8.44) is that impurity concentrations decrease exponentially towards the surface. Note that the time-dependence of the equation can be easily transformed in a spatial dependence by the growth rate $v_g = z/t$. Using $k_t^{(1)} t = k_z z$, Eq. (8.44a) can be written as

$$N(z) = k_z N_s^0 e^{-k_z z} .$$ (8.44b)

The physical driving force for first-order surface segregation can be a lower surface energy. Hofman and Erlewein (1978) and Harris et al. (1984) modeled first-order surface segregation in terms of a lower surface energy. A schematic diagram of the total free energy of the impurity in the bulk and the surface is shown in Fig. 8.16 (Harris et al., 1984). The authors considered Sn in [001] GaAs; therefore the energy minima correspond to Ga planes in a GaAs lattice. The separation of the energy minima, a, is half a lattice constant, i.e. $a = \frac{1}{2}a_0$. It was assumed in the model that the activation energy E_2 (see Fig. 8.16) is so large that no hopping of impurity atoms to the adjacent minimum occurs. The activation energy E_1 was assumed to be sufficiently small such that an equilibrium distribution between the surface layer (s) and the layer $(s-1)$ was achieved. If the transition probability from the sub-surface layer to the surface is p_1 and the transition probability of the reverse process is p_2 than the relative impurity densities per layer are

$$\frac{N_{s-1}}{N_s} = \frac{p_2}{p_1} = A e^{-\Delta G/kT}$$ (8.45)

where ΔG is the free energy difference between the surface and the sub-surface layer and the constant A contains other factors such as the attempt frequency (vibrational frequency).

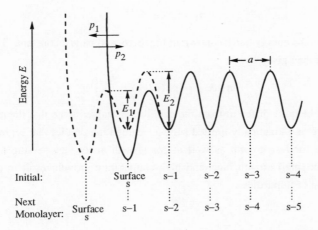

Fig. 8.16. Schematic representation of the energy of a Sn atom in GaAs as a function of depth in the vicinity of the GaAs surface. The energy minima represent Ga sublattice planes spaced by $a = a_0/2$ in the [001]-direction. The solid line is the initial potential energy whereas the dashed line is the potential energy after the growth of an additional GaAs monolayer (after Harris et al., 1984).

Harris et al. (1984) considered two cases for the segregation, i.e. (i) the equilibrium case and (ii) the kinetically limited case. If equilibrium is reached, Eq. (8.45) applies and gives the appropriate ratio of the impurity density in the surface and sub-surface layer. The constant $k_t^{(1)}$ is then obtained as

$$ak_t^{(1)} v_g = ak_z \cong A\, e^{-\Delta G/kT} \tag{8.46}$$

where $k_z v_g = k_t^{(1)}$ describes the spatial decrease of the doping profile (see Eqs. (8.44a) and (8.44b)). In this case the segregation is given only by the difference in free energy of the surface layer and the sub-surface layer. The equilibrium case applies to low growth rates and high temperature growth.

In the second case, impurities do not reach equilibrium. Impurities are buried under the top semiconductor layer and do not have sufficient time to equilibrate with the surface. The non-equilibrium case applies to fast growth rates or low growth temperatures. Assuming that the transfer of impurities between layers is kinetically limited, the probability for an impurity to hop to the surface is given by (Harris et al., 1984)

$$p_1 = p_0 \, e^{-E_1/kT} \tag{8.47}$$

where E_1 is the energy barrier shown in Fig. 8.16, and p_0 is a constant. The decay constant is then given by

$$\ln(ak_z) = p_0(a/v_g) \, e^{-E_1/kT} \tag{8.48}$$

which can be quite different from the equilibrium value. Note that the models of equilibrium and kinetically limited surface segregation predict that segregation is minimized for fast growth as well as for growth at low temperatures. On the other hand, small growth rates and high-temperature growth result in enhanced segregation of impurities.

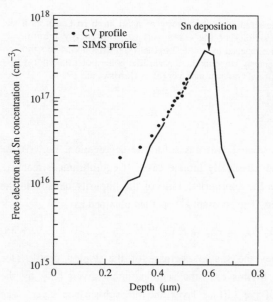

Fig. 8.17. Secondary ion mass spectrometry (SIMS) and capacitance–voltage (CV) profiles of GaAs δ-doped with Sn at a level of $3 \times 10^{12} \mathrm{cm}^{-2}$ during growth interruption (after Harris et al., 1984).

The model of kinetically limited segregation of impurities was used to explain the redistribution of Sn impurities in GaAs during growth by molecular-beam epitaxy (Harris et al., 1982 and 1984). The doping profile of epitaxially grown GaAs:Sn is shown in Fig. 8.17. Tin was deposited on a GaAs surface. Subsequently, a 0.6 μm thick GaAs top layer was deposited. The C-V and SIMS profiling techniques show a pronounced tail towards the surface. The decrease of the Sn concentration follows an exponential dependence which is in agreement with the model.

An alternative model for impurity surface segregation was proposed by Wood and Joyce (1978). They assumed that the *incorporation* of impurities from the surface is limited and that most of the impurities remain on the surface. In order to incorporate impurities from the surface, a large surface density must build up before a significant concentration of impurities is incorporated into the semiconductor. If indeed a surface reaction limits the incorporation of impurities, the incorporation rate is expected to strongly depend on the surface stoichiometry. Wood and Joyce (1978) investigated the system GaAs:Sn and proposed that the Sn incorporation rate depends on the Ga vacancy concentration which in turn is related to the V/III flux ratio. The authors found that the Sn incorporation increases with the group-V flux, which supports their assumption that the Sn incorporation depends on the Ga vacancy concentration. However, Harris et al. (1984) did not observe a clear dependence of the incorporated Sn concentration on the As_4 or As_2 flux. Therefore, they concluded that the model of surface rate-limitation to incorporation proposed by Wood and Joyce (1978) is incorrect.

8.5.3 Fermi-level-pinning induced surface drift

Surface effects can have a strong influence on the redistribution of impurities during epitaxial growth. In this section we consider the influence of Fermi-level pinning on the distribution of impurities near the semiconductor surface. The pinning of the Fermi level at a semiconductor surface results in the trapping of free carriers at the surface and the formation of a dipole field near the surface. The electric dipole field exerts a force on charged impurities which tend to drift to the surface, i.e. they redistribute towards the surface.

Fermi-level pinning at an interface of a semiconductor and a second medium, such as a metal, insulator, or vacuum, is a classic phenomenon in semiconductor physics. The Fermi level at the interface is frequently pinned at a specific energy $e\phi_B$ below the bottom of the conduction band. Bardeen (1947) attributed Fermi-level pinning to interface states which are of donor and acceptor type. Energetically, such Bardeen states can be in the conduction or valence band (e.g., for InAs); more frequently, the states are located close to the middle of the fundamental gap (e.g., GaAs). In an earlier model, Schottky (1938, 1940) and Mott (1938) proposed that the energy $e\phi_B$ of a metal–semiconductor contact equals the difference of the work function of the metal and the electron affinity of the semiconductor. The physical origin of Bardeen's interface or surface states was ascribed to metal-induced gap states (Heine, 1965) or to localized states at the semiconductor surface or interface. Such localized states could be due to atomic steps of the surface (Huijser et al., 1977) defects, (Spicer et al., 1975, 1980) or surface reconstruction (Ihm et al., 1983, Chiang et al., 1983). Different pinning energies were found for different crystal orientations, e.g., the $\langle 100 \rangle$ and $\langle 110 \rangle$ GaAs orientations (Kahng, 1964). Using angle-resolved photoemission of GaAs (100) surfaces, Chiang et al. showed that the Fermi level is pinned at 0.55 ± 0.1 eV above the valence-band maximum for all reconstructed surfaces studied by the group. They proposed that defect states associated with slight non-stoichiometry pin the Fermi level. Fermi-level pinning was found to be insensitive to temperature in the entire temperature range $20\ ^\circ\mathrm{C} \le T \le 500\ ^\circ\mathrm{C}$ (Nanichi and Pearson, 1969), consistent with the fact that the thermal energy kT is much smaller than other relevant energies.

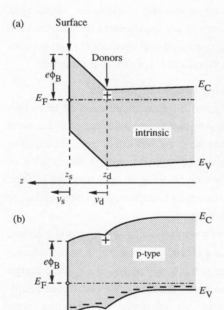

Fig. 8.18. (a) Band diagram of a semiconductor subject to Fermi-level pinning which contains a δ-function-like donor spike. (b) Band diagram of a semiconductor subject to Fermi-level pinning which contains p-type background doping and a δ-function-like donor spike. The coulombic dipole interaction between the surface and the donor layer can be reduced by p-type background doping.

The band diagram of a semiconductor with a Fermi level pinned at the surface is shown in Fig. 8.18(a). A sheet of impurities is located at $z = z_d$. Since the semiconductor is assumed to grow epitaxially, the surface moves with the velocity v_s (i.e. growth rate). Impurities in the sheet are subject to the electric field and drift along the growth direction with a velocity v_d. For a two-dimensional doping density of N_D^{2D}, the electric field of the dipole is given by

$$\mathcal{E} = eN_D^{2D}/\varepsilon, \tag{8.49a}$$

$$\mathcal{E} = \phi_B/(z_s - z_d) \tag{8.49b}$$

where e is the elementary charge, ε is the permittivity, and $z_s = v_s t$ and z_d are the position of the surface and the doped layer, respectively. Equation (8.49a) is valid if the doped layer is depleted of all free carriers, i.e., $z_s - z_d \leq \phi_B \varepsilon/eN_D^{2D}$, while Eq. (8.49b) is valid if the doped layer is partly depleted of free carriers, i.e., $z_s - z_d \geq \phi_B \varepsilon/eN_D^{2D}$. The drift velocity of dopants is given by

$$v_d = \frac{dz_d}{dt} = \mu\mathcal{E} \tag{8.50}$$

where μ is the ion mobility, which is obtained from the Einstein relation $\mu = De/kT$. Note that Eq. (8.50) and the Einstein relation apply to any diffusion mechanism, for example, the substitutional mechanism or the substitutional–interstitial mechanism, provided that the initial and final sites of the atomic-diffusion jump are substitutional sites on which the impurity is ionized.

For small distances between the doped layer and the surface, the dipole field is given by Eq. (8.49a) and the drift velocity of the dopants is given by

$$v_d = \frac{De}{kT}\frac{eN_D^{2D}}{\varepsilon}. \tag{8.51}$$

As an example, we consider the drift of Si impurities in $Al_x Ga_{1-x} As$ (Schubert et al., 1988, 1989). At a low growth temperature of 500 °C, the diffusion coefficient of Si in $Al_{0.30} Ga_{0.70} As$ is $D = 2.5 \times 10^{-16}$ cm^2/s and Eq. (8.51) yields $v_d = 0.1$ Å/s for the drift velocity, which is relatively small. However, at $T = 660$ °C, $D = 6 \times 10^{-15}$ cm^2/s, and the drift velocity is 2 Å/s, which is comparable in magnitude to the growth velocity (v_g) of 1.29 µm/h = 3.6 Å/s for $Al_{0.30} Ga_{0.70} As$ (0.9 µm/h for GaAs). Thus, significant surface migration is expected to occur at $T = 660$ °C.

To obtain a single differential equation for small $z_s - z_d$ and large $z_s - z_d$, the electric field given in Eqs. (8.49a) and (8.49b) is approximated by

$$\mathcal{E} = \left[\left(\frac{e}{\varepsilon}N_D^{2D}\right)^{-1} + \left(\frac{\phi_B}{z_s - z_d}\right)^{-1}\right]^{-1}. \tag{8.52}$$

which represents a lower limit of the true field. The differential equation then becomes

$$\frac{dz_d}{dt} = \frac{De}{kT} \left[\frac{\varepsilon}{eN_D^{2D}} + \frac{z_s - z_d}{\phi_B} \right]^{-1}, \qquad (8.53)$$

with $z_s = v_s t$.

The surface electric field can be reversed in direction by appropriate background doping. Figure 8.18(b) shows the band diagram of a semiconductor with a n-type doping spike embedded into p-type background doping. The band diagram shows that the surface electric field is reversed in direction and reduced in magnitude. It is thus possible to screen and reduce the surface electric field by appropriate background doping.

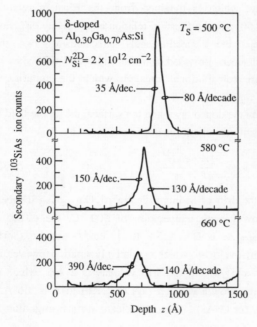

Fig. 8.19. Secondary ion mass spectrometry profile of Si in $Al_{0.30}Ga_{0.70}As$ grown at 500, 580 and 660 °C by molecular-beam epitaxy. Silicon was deposited during growth interruption to achieve a δ-function-like doping profile. SiAs with molecular weight 103 was used for detection.

The model of Fermi-level-pinning induced drift has been employed to explain experimental results on Si-doped $Al_{0.30}Ga_{0.70}As$ (Schubert et al., 1990a). Si-dopant profiles obtained from SIMS are shown in Fig. 8.19 for the three samples grown at 500, 580, and 660 °C. At a low growth temperature a sharp peak is observed. The slight asymmetric shape of the SIMS profile obtained at low growth temperature is due to the well-known 'knock-on' effect. As the growth temperature is increased, the profiles exhibit significant asymmetry. Surface migration of dopants is evident from the profiles, especially at $T_s = 660$ °C. The steepnesses of the leading and the trailing slopes of the SIMS spectrum are evaluated in terms of the length in which the Si signal decreases by 1 order of magnitude. The leading slope increases from 35 Å/decade to 390 Å/decade, indicating the drift of impurities towards the surface. The trailing slope increases from 80 Å/decade to 140 Å/decade indicating diffusion of dopants, which is expected to be symmetric with respect to both sides of the doping spike.

Qualitatively the same, but quantitatively weaker redistribution of impurities towards the surface is observed for Si in GaAs. The smaller redistribution of impurities in GaAs is consistent with the smaller diffusion constant found in GaAs as compared to $Al_x Ga_{1-x} As$.

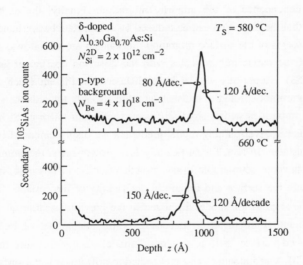

Fig. 8.20. Secondary ion mass spectrometry profile of Si in $Al_{0.30}Ga_{0.70}As$ grown at 580 and 660 °C. The epitaxial layer is also homogeneously doped with Be.

Fermi-level pinning at the semiconductor surface causes the doped layer to be depleted of free carriers and a localization of electrons in surface states. If the resulting dipole field is the driving force towards the surface, then this migration process can be reduced by screening the dipole field. Figure 8.20 shows two SIMS profiles in $Al_x Ga_{1-x} As$ epitaxial samples with p-type Be background doping. The concentration of Be is chosen to be $N_A = 4 \times 10^{18}$ cm^{-3} in order to compensate for the Si dopants within 50 Å. Figure 8.20 reveals that the segregation length is drastically reduced from 150 Å/decade to 80 Å/decade at a growth temperature of 580 °C. The decrease of the segregation length is attributed to the screening of the surface dipole caused by Fermi-level pinning. The trailing slope of the SIMS profile is changed insignificantly, indicating that diffusion and the 'knock-on' effect are not influenced by background doping. The SIMS profile for the sample grown at $T = 660$ °C shows the same qualitative trend as the sample grown at the intermediate temperature (see Figs. 8.19 and 8.20). The surface drift is reduced and the SIMS profile has a more symmetric shape, indicating that diffusion dominates the dopant redistribution. The experimental and theoretical results show that the surface migration process of Si in $Al_x Ga_{1-x} As$ can be consistently explained by the model of Fermi-level-pinning induced drift of impurities. The model is consistent with the substrate temperature dependence of the impurity migration. Finally, the experimental results show that the migration can be reduced by appropriate background doping.

Further aspects of the surface migration mechanism are as follows. First, the magnitude of the redistribution scales with the diffusion coefficient as inferred from Eq. (8.53). Impurities with a large diffusion coefficient (such as Sn in GaAs) are therefore expected to redistribute more strongly towards the surface as compared to impurities with a small diffusion coefficient (such as Si in GaAs). Second, the Fermi-level-pinning model predicts a *net* displacement of impurities toward the surface. In fact, for sufficiently low growth rates, the impurity drift velocity can become comparable to the growth velocity; in this case, impurities drift along with the surface and remain in the vicinity of the surface. This is in contrast to the segregation model discussed in the previous section, in which no net displacement of impurities is expected. Third, the model of Fermi-level-pinning induced surface drift is a *general* model. Based on this model, *all* impurities in III–V semiconductors are expected to drift towards the surface.

The influence of the Fermi level at the surface on the incorporation has been considered previously by a number of groups (Casey et al., 1971; Zschauer and Vogel, 1971; Panish, 1973; Wolfe and Stillman, 1973; Wolfe and Nichols, 1977). In liquid-phase epitaxy, the incorporation of impurities at the liquid/solid interface is controlled by the equilibrium between the liquid and the crystal surface. The incorporation is therefore influenced by the position of the Fermi level at the growing surface rather than in the bulk of the crystal (Panish, 1973). The effect of Fermi-level pinning on the amphoteric behavior of Sn in GaAs was studied by Wolfe and Stillman (1973). They found that the amphoteric behavior of Sn can be explained by a surface-state density of 3×10^{11} cm^{-2}. Wolfe and Nichols (1977) explained thin p-type regions often observed at the (n-type) epilayer/(n-type) substrate interface with the electric field associated with surface states and substrate doping. Surface states and substrate doping were believed to be the major cause of impurity gradients in epitaxial GaAs.

8.6 Impurity-induced layer disordering

Impurities have a profound effect on the self-diffusion of group-III elements and probably on group-V elements in III–V semiconductors. Laidig et al. (1981) found drastically increased Al–Ga diffusion in an AlAs/GaAs superlattice upon Zn diffusion. Zinc was diffused from a surface layer into the bulk. The inter-diffusion of the group-III elements of the superlattice was found even at moderate temperatures of 600 °C.

The diffusion of group-III and group-V elements in III–V semiconductors is known as *self-diffusion*. It is well known that the self-diffusion coefficient is very small in III–V semiconductors and that the corresponding activation energy for self-diffusion can be very high. For example, the Al–Ga interdiffusion activation energy was determined by Mei et al. (1987) in AlAs/GaAs structures to be $E_a = 4$ eV for doping concentrations $\geq 5 \times 10^{17}$ cm^{-3}. Tan and Gösele (1988) measured the activation energy for group-III self-diffusion in undoped (near-intrinsic) GaAs and obtained $E_a = 6$ eV. The large activation energies demonstrate that self-diffusion is a rather unlikely process. The temperature dependence of the Ga self-diffusion coefficient was measured by Palfrey et al. (1981). For Ga diffusion from a Ga film on the surface of a GaAs crystal, the authors determined a diffusion coefficient of

$$D = 3.9 \times 10^{-5} \; e^{-2.6 \; eV/kT} \; cm^2/s \; . \tag{8.54}$$

Note that the activation energy of 2.6 eV is much lower than the values obtained by Mei et al. as well as Tan and Gösele. However, a comparison is not straightforward since the diffusion experiments of Palfrey et al. were performed on non-stoichiometric material, i.e. GaAs with excess Ga. Furthermore, the residual impurity concentration in GaAs was probably different for the three reports. The self-diffusion of As in GaAs has been investigated by Palfrey et al. (1983) as well as by Goldstein (1961). The former group reports an activation energy of 3 eV for As self-diffusion in GaAs. In contrast, Goldstein estimates an activation energy of 10.2 eV for As self-diffusion in GaAs. Note that the two results are drastically different. Nevertheless, measurements by Palfrey et al. (1983) indicate that the As diffusion coefficient in GaAs is low, e.g. 5.2×10^{-16} cm^2/s at $T = 1000$ °C. It is thus well established that the self-diffusion coefficients in GaAs are very small under intrinsic conditions.

The self-diffusion in other III–V semiconductors is associated with a high activation energy as well. The activation energies for In and P diffusion in InP are 3.85 eV and 5.65 eV, respectively (Goldstein, 1961). In InAs, activation energies of 4.0 eV and 4.5 eV for In and As, respectively, were determined (Kato et al., 1969). Activation energies of 4.3 eV were determined for both In and Sb for self-diffusion in InSb (Kendall and Huggins, 1969). The above results confirm that self-diffusion in intentionally undoped III–V semiconductors is a relatively slow process.

The self-diffusion in III–V semiconductors is changed drastically at high impurity concentrations. The drastic change of self-diffusion is illustrated in Fig. 8.21 (Meehan et al., 1984). The figure shows an $Al_x Ga_{1-x} As/GaAs$ superlattice that has been disordered on the left-hand side by Si diffusion. The right-hand side was masked by $Si_3 N_4$ and, as a consequence, Si did not diffuse into the semiconductor. The superlattice is unaffected in this region by the annealing cycle of 10 h at 850 °C. Figure 8.21 illustrates that the inter-diffusion of group-III elements is strongly enhanced by impurity diffusion. Further experimental results have been summarized in an excellent review by Deppe and Holonyak (1988), which also summarizes experimental results on impurity-induced disordering by the donors Si, Ge, S, Sn, and Se as well as the acceptors Zn, Be, and Mg. Furthermore, disordering by (i) diffusion as well as (ii) implantation was described.

Next, we consider a physical model that accounts for increased self-diffusion in III–V semiconductors. As an example, we consider the group-III self-diffusion only and neglect group-V diffusion and follow the arguments of Deppe and Holonyak (1988). The self-diffusion of group-III lattice atoms must proceed through native defects of the crystal. Therefore, the self-diffusion rate will be dependent on the diffusion rate of the native defects, but since the concentration of native defects is typically much less than the concentration of group-III lattice atoms ($\cong 2 \times 10^{22}$ cm^{-3}), the group-III self-diffusion rate will also depend on the concentration of the requisite crystal defects. If we consider, for example, self-diffusion of the group-III atoms through group-III vacancies, then

$$D_{III} = f D_{V_{III}} [V_{III}] \tag{8.55}$$

where D_{III} is the group-III atom diffusion rate, $D_{V_{III}}$ is the group-III vacancy diffusion rate, V_{III} is a group-III vacancy, the brackets [ ] denote concentration, and the prefactor f contains information about the crystal structure (Manning, 1968) and the concentration of group-III lattice sites. Equation (8.55) shows that as $[V_{III}]$ increases so will D_{III}.

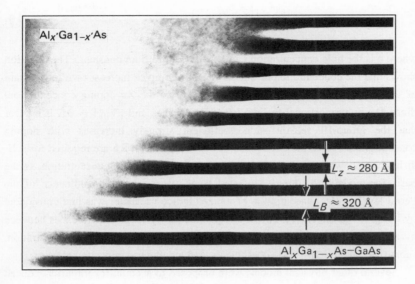

Fig. 8.21. Bright field transmission electron micrograph of a 40 - period $Al_{0.60}Ga_{0.40}As/GaAs$ superlattice ($L_B \approx 320$ Å, $L_z \approx 280$ Å) which has been disordered into bulk $Al_xGa_{1-x}As$ by Si diffusion (left side). The $Al_{0.60}Ga_{0.40}As/$ GaAs superlattice masked with Si_3N_4 has 'survived' the 850 °C (10 h) anneal (right side) (Meehan et al., 1984).

It has been known for a long time that if the point defects in a semiconductor possess various charge states, the equilibrium defect concentrations will depend on the crystal Fermi level (Longini and Greene, 1956). To illustrate further the influence of the Fermi level on the native defect concentration we consider the impurity–vacancy complex proposed by Laidig et al. (1982). They proposed that in the transfer of a Zn atom from a substitutional site to an interstitial site, an intermediate stage could occur, which consists of a complex formed by interstitial Zn and a group-III vacancy

$$Zn_i^+ + V_{III} \leftrightarrow (Zn_i V_{III})^+ \leftrightarrow Zn_{\overline{III}} + 2h \qquad (8.56)$$

where Zn_i^+ and $Zn_{\overline{III}}$ are a (positively charged) interstitial Zn atom and a (negatively charged) substitutional Zn atom, respectively. We now consider the middle and right-hand side of Eq. (8.56) and recall that at equilibrium, the ratio of the concentrations of the initial and final products of the reaction is a constant, i.e.

$$\frac{[(Zn_i\, V_{III})^+]}{[Zn_{\overline{III}}]\, p^2} = K \tag{8.57}$$

where p is the hole concentration and K is the equilibrium constant. The equation shows that the concentration of the Zn-vacancy complex increases with the square of the hole concentration, i.e. with the square of the Zn doping concentration. Since D_{III} is proportional to $[V_{III}]$ (see Eq. (8.55)), and $[V_{III}] \propto p^2$, it is clear that the group-III self-diffusion coefficient strongly increases with doping concentration. Thus, the occurrence of native defects which are required for self-diffusion and whose concentration depends on the doping concentration, is the basis for models on impurity-induced layer disordering. The influence of the Fermi level on the concentration of ionized lattice vacancies was first considered by Longini and Greene (1956) who explained the systematic differences between n- and p-type semiconductors with respect to lattice vacancy concentration, substitutional atom diffusion coefficients, and amphoteric impurity behavior.

Several other physical models were proposed to account for impurity-induced layer disordering which were summarized by Deppe and Holonyak (1988). Tan and Gösele (1987) assumed that both group-III and group-V vacancies can be charged either positively or negatively. They suggested that, in the case of p-type doping, group-III self-diffusion proceeds mainly because of positively charged group-III vacancies, while group-V diffusion proceeds because of positively charged group-V vacancies. Similarly, in the case of n-type doping group-III self-diffusion proceeds via negatively charged group-III vacancies and group-V diffusion via negatively charged group-V vacancies. Independently, at about the same time, Deppe et al. (1987) presented data showing that the diffusion of the Si impurity in GaAs is controlled in part by the crystal Fermi energy. Based on the dependence of the Si diffusion rate on the Fermi level and on the As vapor pressure over the crystal during annealing, Deppe et al. (1987) suggested that the Si diffusion, and the layer intermixing in $Al_x G_{1-x} As/GaAs$ quantum wells or superlattices that accompanies the Si diffusion, is due to donor vacancy complexes, $Si_{Ga} - V_{Ga}$. In related work, Kaliski et al. (1987) have presented experimental data on annealing of quantum wells containing either the donor Se or the acceptor Mg showing that for As-rich annealing conditions n-type quantum wells undergo considerable intermixing while p-type quantum wells remain stable. Under As-poor annealing conditions, the n-type quantum wells remain more stable while the p-type quantum wells undergo layer intermixing. Based on the fact that the layer intermixing depends on the type of doping (n or p) and As overpressure, the suggestion has been made that intermixing occurring in p-type quantum wells under As-poor annealing conditions is due to group-III interstitials, while the intermixing occurring in n-type quantum wells under As-rich conditions is due to group-III vacancies (Kaliski, 1987). Devine et al. (1987) have independently suggested that layer disordering via Be diffusion in $Al_x Ga_{1-x} As/GaAs$ superlattices depends on group-III interstitials. Previously, Willoughby (1983) has suggested that interstitials are involved in $Al_x Ga_{1-x} As/GaAs$ superlattice disordered by Zn diffusion.

9

Deep centers

Deep centers are chemical impurities, native defects, or a combination of both, which have at least one energy level in the forbidden gap. The energy level or levels of deep centers are far removed from the conduction and valence band edges, which is in contrast to shallow impurities. The phenomenology of deep centers was discussed and analyzed in Chap. 2. In the present chapter, the characteristics of some specific deep centers will be summarized.

The *DX-center* is of primary technological importance in n-type $Al_x Ga_{1-x} As$ devices such as transistors and lasers. The properties of this center have puzzled researchers for more than a decade. The *EL2 center* can cause undoped bulk-grown GaAs to be semi-insulating. The EL2 center is of importance for the growth of GaAs bulk crystals and for the production of substrates. *Hydrogen* can be used in two different ways to influence the electronic properties of III–V semiconductors. First, protons implanted into III–V semiconductors cause damage to the crystal; the damaged regions can trap carriers. Second, hydrogen can passivate acceptor as well as donor impurities. *Oxygen* is also used to render III–V semiconductors semi-insulating. *Chromium* and *iron* are deep impurities which are used to make GaAs and InP highly resistive. Both Cr and Fe are deep acceptors and compensate for residual shallow donors in the two semiconductors. They are used as intentional deep impurities in the bulk growth of GaAs and InP. Finally, rare-earth atoms such as Er, Pr, or Nd incorporated into semiconductors have energy levels which are not associated with either the conduction or valence band. Intra-impurity transitions of rare-earth impurities can be radiative. Such transitions are of interest for optoelectronic device applications.

9.1 DX-type centers

At the end of the 1970s it became clear that n-type $Al_x Ga_{1-x} As$ with Al mole fractions of $0.2 \leq x \leq 0.4$ has very unusual properties as compared to GaAs. These properties include a large donor activation energy, a strong sensitivity of the conductivity on optical radiation, as well as large barriers to emission and capture. These anomalous properties, which are not expected from an effective-mass-like donor, led Lang et al. (1979) to the conclusion that the deep center involves a donor atom which forms a complex with another constituent. The authors therefore designated the center as a *donor-complex* or DX-center. Subsequently, this name became associated with the doping properties of the deep donor in n-type $Al_x Ga_{1-x} As$.

The measurements of the free carrier concentration in n-type $Al_x Ga_{1-x} As$ revealed a *large thermal activation energy* of the DX-center in Te-doped $Al_x Ga_{1-x} As$ (Nelson, 1977; Lang and Logan, 1977), in Si-doped $Al_x Ga_{1-x} As$ (Schubert and Ploog, 1984; Chand et al., 1984) as well as in $Al_x Ga_{1-x} As$ doped with Se, Sn, etc. The large thermal ionization energy is evidenced by the dependence of the free carrier concentration on reciprocal temperature which is illustrated in Fig. 9.1 for Si-doped $Al_{0.25} Ga_{0.75} As$. The closed circles illustrate the electron concentration measured in the dark. The free electron concentration remains constant for temperatures below 150 K. The electron concentration increases at temperatures higher than 150 K indicating the thermal ionization of the deep donor center. Note that the ionization process continues even at room temperature, i.e. saturation of the free carrier concentration has not occurred even at $T = 300$ K. The increase of the carrier concentration at room temperature indicates that the DX-center has a large thermal activation energy. The saturation of the free carrier concentration at low temperatures ($T < 150$ K) was attributed to additional shallow donors which do not freeze out at low temperatures if the concentration of such shallow donors is sufficiently high (Schubert and Ploog, 1984). The authors evaluated the deep donor thermal activation energy and showed that the conventional formula frequently used to determine donor activation energies

$$ n \propto \exp \left[- \frac{E_{dd}}{2kT} \right] \tag{9.1} $$

can *not* be used if shallow and deep donors co-exist in a semiconductor. Instead, they showed that the deep donor activation energy, E_{dd}, must be determined according to

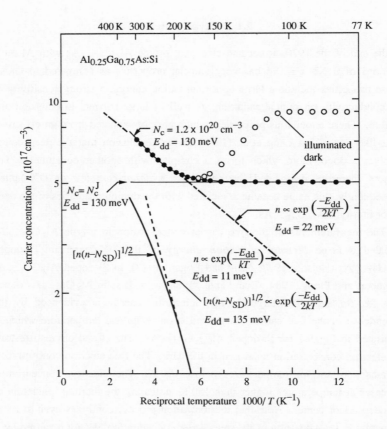

Fig. 9.1. Dependence of the Hall electron concentration in n-type Al$_{0.25}$Ga$_{0.75}$As:Si on inverse lattice temperature. Solid and open circles indicate experimental data measured in the dark and after illumination at low temperatures, respectively. The deep-donor thermal activation energy E_{dd} is evaluated by three methods. The simple relations $n \propto \exp(-E_{dd}/kT)$ and n $\propto$ $\exp(-E_{dd}/2kT)$ yield $E_{dd} = 11$ and 22 meV, respectively (dashed-dotted line). A more realistic value of $E_{dd} = 135$ meV is obtained using the relation $(n^2 - nN_{SD})^{1/2} \propto \exp(-E_{dd}/2kT)$ at low temperature indicated by the dotted line. A simulation of the carrier concentration versus temperature in terms of Fermi–Dirac statistics yielding $E_{dd} = 130$ meV for the thermal activation energy of the deep donor (solid curve) coincides with the experimental data only if a density of states much larger than that of the Γ valley is used. The experimental data cannot be fitted to the experimental data by means of the lower joint density of states N_c^J.

$$\sqrt{n(n - N_{SD})} \propto \exp\left[-\frac{E_{dd}}{2kT}\right] \qquad (9.2)$$

where N_{SD} is the shallow donor concentration ($N_{SD} = 5 \times 10^{17}$ cm^{-3} in Fig. 9.1). A deep donor thermal activation energy of $E_{dd} = 130$ meV was evaluated using this procedure.

The electronic properties of n-type Al$_x$Ga$_{1-x}$As change strongly as the Al mole fraction increases. Figure 9.2 shows the Hall carrier concentration of Si-doped n-type Al$_{0.32}$Ga$_{0.68}$As. Comparison with Fig. 9.1 yields that (i) the deep donor concentration has increased while (ii) the shallow donor concentration has decreased. Evaluation of the thermal activation energy yielded $E_{dd} = 135$ meV. Thus, the experimental results show that (i) the thermal activation energy of the deep DX-donor in n-type Al$_x$Ga$_{1-x}$As ($0.2 \le x \le 0.4$) is $E_{dd} \cong 130-140$ meV independent of the Al mole fraction, and (ii) that the relative concentration of deep DX-donors increases with Al mole fraction while simultaneously the shallow donor concentration decreases. Note that as a consequence of these properties, the free electron concentration at $T = 300$ K is *lower* than the Si impurity concentration. Several groups have interpreted the free electron concentration decrease as a lower ***doping-efficiency*** of n-type impurities in Al$_x$Ga$_{1-x}$As as compared to GaAs (see, for example, Ishibashi et al., 1982). The physical reason for the decreased carrier concentration at $T = 300$ K is the large thermal activation energy of the deep DX-donor which is much larger than kT at room temperature. Watanabe and Maeda (1984) independently came to the conclusion that shallow donors and DX-centers co-exist in Si-doped Al$_x$Ga$_{1-x}$As. Evaluation of the shallow and deep donor concentrations yielded that the shallow donor concentration decreases at $x \cong 0.25$ while simultaneously the deep donor concentration increases. They further concluded that the change in Hall activation energy is a result of the changing concentration of the two types of donors.

A different interpretation of the thermal activation energy of deep DX-centers has been proposed by a number of groups (see, for example, Ishibashi et al., 1982; Chand et al., 1984). The authors evaluated the temperature dependent Hall concentration measurements using

$$n \propto \exp\left[-\frac{E_{dd}}{kT}\right] \qquad (9.3)$$

and obtained deep donor activation energy that steeply increases from $E_{dd} \cong 10$ meV ($x = 0.20$) to $E_{dd} \cong 160$ meV ($x = 0.45$). Note, however, that even at small Al mole fractions ($x = 0.22-0.24$) where the donor energy is

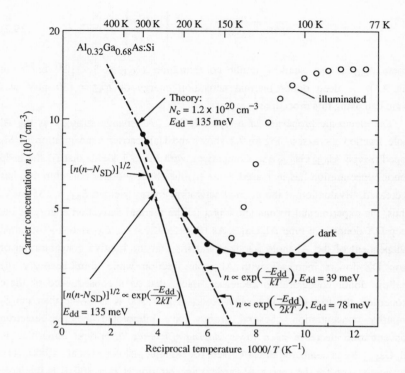

Fig. 9.2. Dependence of the Hall electron concentration in n-type $Al_{0.32}Ga_{0.68}As$:Si on inverse lattice temperature. Solid and open circles indicate experimental data measured in the dark and after illumination at low temperatures, respectively. The deep-donor thermal activation energy E_{dd} is evaluated by three methods. The simple relations $n \propto \exp(-E_{dd}/kT)$ and $n \propto \exp(-E_{dd}/2kT)$ yield $E_{dd} = 39$ and 78 meV, respectively (dashed-dotted line). A more realistic value of $E_{dd} = 135$ meV is obtained using the relation $(n^2 - nN_{SD})^{1/2} \propto \exp(-E_{dd}/2kT)$ at low temperature indicated by the dotted line. A simulation of the carrier concentration versus temperature in terms of Fermi–Dirac statistics yielding $E_{dd} = 135$ meV for the thermal activation energy of the deep donor (solid curve) coincides with the experimental data only if a density of states much larger than that of the Γ valley is used.

supposedly shallow (10–20 meV), the ionization of the deep donor is not complete even at room temperature. The discrepancy between low activation energy but incomplete ionization at room temperature was not addressed by the authors.

Another key characteristic of DX-centers in $Al_xGa_{1-x}As$ is the ***persistent photo-conductivity*** (PPC) effect. Upon illumination of n-type $Al_xGa_{1-x}As$ at low temperature, the carrier concentration increases. The increase in carrier concentration persists even after the illumination has been turned off. This phenomenon is known as persistent photoconductivity. The light used for illumination has a wavelength below the fundamental gap of $Al_x Ga_{1-x}As$; band-to-band excitation is therefore excluded. The Hall carrier concentrations after illumination are shown in Figs. 9.1 and 9.2 by open circles. The carrier concentration remains high and does not decay to its dark value even hours or days after termination of the photoexcitation. The high PPC concentration can be quenched by heating the sample to a temperature >150 K as shown in Figs. 9.1 and 9.2. The kinetics of the quenching were first measured by Nelson (1977) at temperatures of 60–80 K in Te-doped $Al_x Ga_{1-x}As$. The rate of the decrease of the carrier concentration increases at higher temperatures. The optical threshold energy for the PPC effect has been determined to be 0.85 eV ± 0.1 eV for Te-doped $Al_x Ga_{1-x}As$ (Lang et al., 1979) and 1.0 eV ± 0.1 eV for Si-doped $Al_x Ga_{1-x}As$ (Schubert et al., 1985). The large difference in thermal ionization energy and photoionization energy is well established for all n-type impurities in $Al_x Ga_{1-x}As$. The large difference between thermal ($\cong 0.1$ eV) and optical ($\cong 1$ eV) activation energy cannot be explained by conventional, effective-mass-type impurity models. It is worthwhile to note that the PPC effect has deleterious consequences for semiconductor devices containing n-type $Al_x Ga_{1-x}As$. Several efforts were undertaken to reduce or eliminate the PPC effect. Most notably, Baba et al. (1983) showed that the use of an AlAs/n-GaAs superlattice can eliminate PPC and improve the Si doping activation by a spatial separation of the Si impurities from the AlAs.

The model of *large lattice relaxation* has been successfully used to explain the peculiar properties of the DX-center in $Al_x Ga_{1-x}As$ (Lang and Logan, 1977; Lang et al., 1979; Jaros, 1982). The concepts of *small* and *large* lattice relaxation are briefly reviewed in the following. The local structure of an impurity or defect generally depends on the electronic configuration of the center. That is, if the electronic configuration of a defect is changed, the structural configuration changes as well, i.e. the defect configuration relaxes. To illustrate the interdependence of the electronic and the structural configuration, we consider the hydrogen molecule as an example. The energy of the bonding and anti-bonding state of a hydrogen molecule is shown in Fig. 9.3(a). The distance between the two protons depends on the electronic state of the molecule, i.e. the total energy is minimized at different core distances for electrons occupying either the bonding or the anti-bonding state. This property results in a difference between absorption and emission energies for electronic transitions between the bonding and anti-bonding state. The difference in absorption and emission energies is known as the Franck–Condon shift. This shift can be explained by assuming that the two

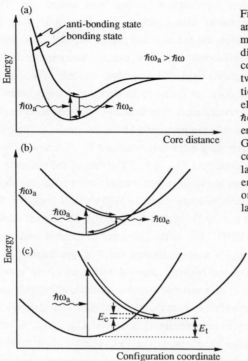

Fig. 9.3. (a) Bonding and antibonding state energy of a molecule as a function of distance between the two cores. The difference between the electronic absorption energy, $\hbar\omega_a$, and the electronic emission energy, $\hbar\omega_e$, is the Franck–Condon energy, $E_{FC} = \hbar\omega_a - \hbar\omega_e$. (b) Generalized configuration coordinate diagram for small lattice relaxation. (c) Generalized configuration coordinate diagram for large lattice relaxation.

nuclei do not move during the short time of the electronic transition due to the much heavier nuclear masses as compared to the electronic masses. Upon an electronic transition, the molecule 'slowly' moves to a new configuration, i.e. new core distance, vibrational frequency, etc. As a consequence, the emission energy, $\hbar\omega_e$, is smaller than the absorption energy, $\hbar\omega_a$ as illustrated in Fig. 9.3(a). The difference $\hbar\omega_a - \hbar\omega_e$ is called the **Franck–Condon energy** (see also Chap. 2).

The principle of small lattice relaxation is generalized in Fig. 9.3(b) which shows the defect energy for two electronic configurations as a function of the configuration coordinate. Without specifying the exact meaning of the configuration coordinate it is implied that the *structural* configuration of the impurity or defect is changed upon moving along the configuration coordinate. It is further frequently assumed that the total energy changes *quadratically* as the configuration is changed around the minimum energy position. This parabolic characteristic of the total defect energy is derived from Hooke's law. It is thus assumed that the elastic energy of the defect dominates. Note that the electronic transitions between the two electronic states are allowed for the case of small lattice relaxation.

This situation is changed drastically for *large lattice relaxations*. The configuration coordinate diagram corresponding to large lattice relaxation is shown in Fig. 9.3(c). The diagram shows two configurations of the defect. Excitation from the ground-state of the defect to the excited state requires a photon energy $\hbar\omega_a$. Once the defect has minimized its energy in the excited state, the defect *cannot return to the ground state* by means of an optical (i.e. 'vertical') transition. The excited state is thus a *metastable state*. Note that the excited state is metastable only in the model of large lattice relaxation (Fig. 9.3c) but not for small lattice relaxation (Fig. 9.3b).

While optical transitions are always 'vertical' in the configuration coordinate diagram, thermally excited transitions need not be. During thermal excitation, the defect is displaced from its equilibrium position by vibrational energy. In order for the defect to return from the metastable state to the ground state, it must overcome the thermal capture barrier. The energy of the thermal capture barrier is denoted as E_c in Fig. 9.3(c). Let us estimate the probability of overcoming a capture barrier of $E_c = 200$ meV at a temperature of $T = 4.2$ K. The probability of thermally overcoming the barrier (per attempt) is given by $\exp(-E_c/kT)$ which amounts to 10^{-240}, which is zero for all practical purposes. The large lattice relaxation model can consistently explain the large difference of the thermal and the optical excitation energy. The thermal excitation energy (shown as E_t in Fig. 9.3c) is the difference between the energy minima of the two defect configurations which can be much smaller than the energy required for a vertical optical transition.

Lang and Logan (1977) and Lang et al. (1979) have successfully applied the model of large lattice relaxation to the DX-center in Te-doped n-type $Al_x Ga_{1-x} As$. The schematic representation of the relevant energies of the DX-center and their representation in the configuration coordinate diagram are shown in Fig. 9.4(a) and (b) respectively. The authors determined the energies for Te-doped $Al_x Ga_{1-x} As$ as follows: Optical excitation, $E_o = 850 \pm 100$ meV; thermal excitation, $E_t = 100 \pm 50$ meV; thermal emission, $E_e = 280 \pm 30$ meV; thermal capture, $E_c = 180 \pm 20$ meV. For Si-doped $Al_x Ga_{1-x} As$ the energies can be summarized as follows (Schubert and Ploog, 1984; 1985; Chand et al., 1984; Watanabe et al., 1984; Mizuta et al., 1985; Künzel et al., 1983 and 1984): $E_o \cong 1$ eV, $E_t \cong 150$ meV, $E_e \cong 300$ meV, $E_c \cong 150$ meV. Other n-type impurities such as Se and Sn (see, for example, Tachikawa et al., 1984) have qualitatively similar but quantitatively slightly different characteristics than Te- and Si-doped $Al_x Ga_{1-x} As$. Lang (1986)

(a)

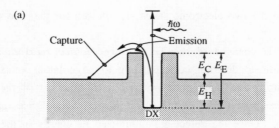

(b)

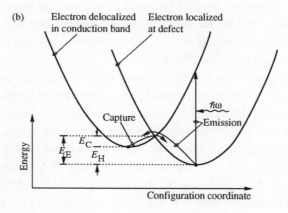

Fig. 9.4. (a) Energy barriers for transient capture (E_C), transient emission (E_E), Hall activation energy (E_H), and optical ionization energy for the DX-center in Te-doped $Al_xGa_{1-x}As$. (b) Schematic configuration coordinate diagram of the center (after Lang et al., 1979).

reviewed the energies of Se, Te, Si, and Sn donors in $Al_xGa_{1-x}As$ and S and Te donors in $GaAs_xP_{1-x}$.

Two atomic configurations were attributed to the unrelaxed and the relaxed state of the DX-center by Chadi and Chang (1988b, 1989), Zhang and Chadi (1990), and Chadi and Zhang (1991). The authors proposed that donors, e.g. a Si donor, can exist in two different configurations in $Al_xGa_{1-x}As$. One of the configurations is the simple substitutional donor which can be neutral or ionized. This configuration is shown for a Si donor on a cation site of $Al_xGa_{1-x}As$ in Fig. 9.5(a). In this normal substitutional site, the Si atom is 0.8 Å above the plane defined by any three adjacent As atoms. In the relaxed DX-configuration, the donor atom occupies an interstitial, threefold coordinated site. One of the four donor-arsenic bonds is broken as illustrated in Fig. 9.5(b). In this relaxed position, the Si atom is 0.2 Å below the plane defined by the three As atoms. The

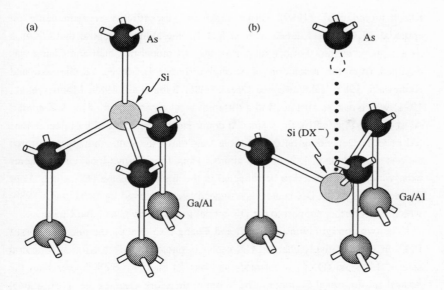

Fig. 9.5. (a) Substitutional Si donor on the cation site of an $Al_xGa_{1-x}As$ lattice. (b) DX-configuration of a Si donor in $Al_xGa_{1-x}As$. The Si donor occupies a threefold coordinated interstitial site and has three cations in its immediate vicinity.

total displacement of the Si is thus approximately 1.0 Å which is indeed a large displacement. Chadi and Chang (1988b) used ab initio self-consistent pseudopotential calculations to show that the DX-configuration is a stable configuration that is negatively charged by two electrons. The following reaction describes the transition from a substitutional donor to a DX-center

$$D^0 + e \leftrightarrow DX^-$$ (9.4a)

where D^0 and DX^- represent a neutral fourfold coordinated substitutional donor and a negatively charged DX-center, respectively. Thus, an ionized shallow donor can capture two electrons by assuming the DX-configuration, i.e.

$$D^+ + 2e \leftrightarrow DX^- .$$ (9.4b)

The capture of two electrons by the DX-center is supported by experiments of Fujisawa et al. (1990). The authors showed by co-doping of $Al_xGa_{1-x}As$ with Si and Ge that the number of electrons trapped by Ge impurities exceeds the number of Ge impurities, which favors a negative charge state of the DX-center. In contrast, a one-electron ground state of the DX-center has been concluded by

Khachaturyan et al. (1989). From magnetic susceptibility measurements on epitaxial $Al_x Ga_{1-x} As$ doped with Si and Te, the authors propose that DX is a paramagnetic center. Further support of the DX-model of Chadi and Chang was obtained from the resolution of several different DX-levels (Kaniewska and Kaniewski, 1985; Takikawa and Ozeki, 1985; Baba et al., 1989; Mooney et al., 1991) which can be attributed to a different local environment of the DX-center. As shown in Fig. 9.5(b), the displaced donor atom is surrounded by three cations (Al or Ga). Depending on the Al mole fraction of the semiconductor, the three cations can include 0, 1, 2, or 3 Al atoms. Thus, four distinct local environments exist which are expected to be reflected in the signature of the DX-center. Four distinct levels of the DX-center have indeed been observed by Baba et al. (1989) which lends further support to the DX-model proposed by Chadi and Chang.

The two configurations of Chadi and Chang also explain the phenomenon of PPC. If the negatively charged DX-center is photoionized, it assumes a neutral state. This state (DX^0) is *unstable* and has an energy 1.3 eV higher than the neutral substitutional Si donor. The Si donor therefore assumes the substitutional D^0 configuration. At sufficiently low temperatures, the Si cannot assume the negatively charged DX^--configuration due to the large capture barrier. The ability to explain the PPC effect lends support to the model of large lattice relaxation. Models proposing small lattice relaxation (see e.g. Mizuta et al., 1985; Mizuta and Kitano, 1987; Mizuta and Mori, 1988) cannot explain the PPC effect observed experimentally. Theoretical and experimental aspects of the DX-center in $Al_x Ga_{1-x} As$ and in $GaP_x As_{1-x}$ were reviewed by Wolford (1991). Further reviews on the DX-center were given by Jaros (1982), Mooney (1990), and Theis (1990).

The reaction describing the formation of the DX-center (see Eq. (9.4)) can also be written as

$$2D^0 \rightarrow D^+ + DX^- . \tag{9.5}$$

In n-type $Al_x Ga_{1-x} As$, this reaction is strongly exothermic due to the low energy of the DX^- configuration. Donor centers for which Eq. (9.5) is exothermic are called **negative-U centers**, where U is the effective electron–electron correlation energy of the two centers. The two major contributions to the electron–electron correlation energy are (i) the repulsive coulombic energy of the two electrons and (ii) the lattice distortion energy (Street and Mott, 1975). If the lattice distortion energy more than compensates for the electron repulsion, a *negative* effective correlation energy results. Since such a negative effective correlation energy applies to the DX-center, it is frequently referred to as a negative-U center.

The symmetry of the DX-center in Te-doped $Al_x Ga_{1-x} As$ was investigated by Peale et al. (1992a, 1992b) in terms of absorption spectroscopy of the DX-center. The authors used polarized light incident along various symmetry axes for excitation of the DX-center. They found no significant effect for different polarization directions (absence of dichroism) and concluded that either the dipole matrix elements of the DX-center are accidentally very isotropic, or that more likely there is *no* large symmetry-lowering distortion at this center. The absence of a symmetry-lowering distortion of the DX-center is in conflict with the broken-bond model of Chadi and Chang. Subsequently, Chadi (1992) proposed a new model in which substitutional donors in $Al_x Ga_{1-x} As$ have two distinct negatively charged DX-like deep-donor states. The first state has a broken-bond configuration while the second arises from a symmetric 'breathing mode' relaxation around the impurity. The latter breathing mode is a symmetric radially outward displacement of its four nearest-neighbors. Chadi (1992) proposed that the absence of dichroism, which has been observed experimentally, is consistent with the tetrahedrally symmetric breathing mode configuration of substitutional donors.

9.2 The EL2 defect

The EL2 defect has had a colourful history. It is the history of the dominant defect in GaAs which has been identified to be most probably the As-antisite defect, As_{Ga}, i.e. a native defect with no other constituents. As a complicating feature, the EL2 defect exhibits metastability due to lattice relaxation which is very similar to the metastability of the DX-center discussed in the previous section. The EL2 center is instrumental in rendering semi-insulating nominally undoped GaAs bulk crystals grown by the Czochralski technique. The name 'EL2' is due to Martin et al. (1977) who analyzed and classified many deep levels in GaAs.

The deep level transient spectroscopy (DLTS) signal of undoped n-type GaAs grown by organometallic vapor-phase epitaxy (OMVPE) is shown in Fig. 9.6 as a function of temperature (Samuelson et al., 1981). The DLTS spectrum reveals a major signal at $T = 390$ K which is the dominant defect level in GaAs. The negative amplitude of the DLTS signal indicates that the level is an electron trap. This dominant deep level is called 'EL2' or 'O-level' in the literature. For the spectrum shown in Fig. 9.6, the concentration of the EL2 level is approximately 1.25×10^{14} cm^{-3}. For VPE-grown GaAs, the concentration of the EL2 level varies between 10^{13} and 10^{16} cm^{-3} (Wagner et al., 1980; Ozeki et al., 1979). In bulk GaAs grown by the Czochralski technique, the EL2 defect can have similar concentrations. Holmes et al. (1982) showed for bulk GaAs that the EL2 defect

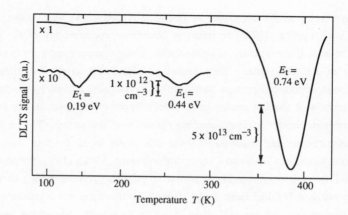

Fig. 9.6. Typical DLTS spectrum of unintentionally doped n-type GaAs grown by organometallic vapor-phase epitaxy (after Samuelson et al., 1981).

concentration varies between 5×10^{15} cm^{-3} and 1.7×10^{16} cm^{-3} depending on the As atom fraction in the melt. The EL2 level is particularly low in GaAs grown by molecular-beam epitaxy (MBE) under optimized growth conditions where the EL2 concentration is below 10^{13} cm^{-3}. Evaluation of the activation energy of the dominant deep donor yields $E_{dd} = 0.74$ eV below the conduction band edge (Samuelson et al., 1981). Evaluation of the ionization energy by other groups yielded $E_{dd} = 0.72...0.82$ eV. Figure 9.6 also exhibits DLTS signals at $T = 140$ K and $T = 270$ K. The concentration of these two levels is approximately 10^{12} cm^{-3}, i.e. much lower than the EL2 defect. The two low-concentration levels will not be further discussed.

The EL2 defect is, as will be shown later, a double donor which has significant technological importance for the growth of semi-insulating (SI) GaAs. GaAs bulk crystals grown by the liquid encapsulated Czochralski (LEC) technique contain a range of chemical impurities including Mn, Fe, B, C, Si, and S (Holmes et al., 1982). The dominant residual impurity is the shallow acceptor C with a concentration in the 10^{15} cm^{-3} range. If no deep traps were present in LEC-grown GaAs, the residual conductivity would be p-type with a resistivity of 0.1 to 10 Ω cm. However, many applications require resistivities of the substrate $> 10^7$ Ω cm, i.e. semi-insulating substrates. The occurrence of the EL2 defect allows one to obtain highly resistive GaAs for sufficiently high concentrations of EL2. The EL2 donor level located approximately 0.75 eV below the conduction band edge compensates for the shallow C acceptor impurities. For $N_{EL2} > N_C$, a fraction of the EL2 centers (i.e. N_C/N_{EL2}) will be ionized, while the remaining EL2 centers will be neutral (the ionization energy of 750 meV is too large for a thermal ionization process at room temperature). As a consequence, the Fermi level is pinned at the (upper) EL2 level which is approximately in the middle of the GaAs band gap. As a result, GaAs with sufficiently high concentration of EL2 has *near-intrinsic* characteristics.

Holmes et al. (1982) showed that the EL2 concentration can be controlled during LEC growth via the As/Ga composition in the melt. For their specific experimental conditions they obtained GaAs with p-type conductivity for an atomic As concentration below 0.475. For As concentrations above 0.475 the GaAs was semi-insulating. The dependence suggests that the EL2 defect is As related, e.g. an As_{Ga} antisite defect. For high As/Ga ratios, the concentration of EL2 defects increases which compensates the shallow acceptors. At low As/Ga ratios, the shallow C acceptor impurities are not fully compensated leading to residual p-type conductivity of the substrates. An increasing EL2 concentration with increasing V/III ratio was also found by Miller et al. (1977), Ozeki et al. (1979), Wagner et al., (1980) and Samuelson et al., (1981). The latter authors found a dependence of $N_{EL2} \propto p_{AsH_3}^{0.5}$.

The EL2 defect has been considered originally to be related to oxygen in GaAs. Increasing the partial pressure of oxygen in the ampoule (Ainslie et al., 1962) or adding Ga_2O_3 to the melt (Fertin et al., 1967) during GaAs growth by the horizontal Bridgman technique results in a drastic change of the electrical properties of the as-grown GaAs. The deep oxygen level was thought to compensate shallow impurities (Allen, 1960; Haisty et al., 1962). However, it was shown later on that the deep EL2 level is probably not related to oxygen (Wagner et al., 1980; Huber et al., 1979). Furthermore, oxygen contamination became a less severe problem due to the increased purity of source materials and the employment of chemically inert materials in growth reactors.

Among the many specific models proposed for the EL2 defect there are only two models that are directly supported by spectroscopic evidence. These are the isolated As_{Ga} antisite defect and the $As_{Ga} - As_i$ antisite–interstitial pair (Kaufman, 1989). At present, the As antisite defect, As_{Ga}, is considered to be the most probable physical origin for the EL2 center. Several groups have unequivocally established that the EL2 concentration increases with the partial pressure of the As precursor or the As mole fraction in the bulk (see, for example, Samuelson et al., 1981, Holmes et al., 1982, and references therein). Furthermore, Weber (1985, 1991) and Muszalski et al. (1991) showed that unusually high (up to 10^{20} cm^{-3}) concentrations of the EL2 defect are obtained in *GaAs grown at very low temperatures* by MBE. Such low growth temperatures are well known to yield non-stoichiometric, As-rich, GaAs. The non-stoichiometry is due to the lack of As reevaporation from the growing GaAs layer. (These growth conditions do not fulfill the three-temperature-criterion formulated by Guenter (1958) for stoichiometric growth of GaAs). As a result, an abundance of As_{Ga} defects are formed in GaAs grown at low temperatures.

It has also been argued that the EL2 defect could be related to an As antisite and As interstitial pair, $As_{Ga} - As_i$, or other As_{Ga}-related defects (Spaeth et al., 1990). However, several observations do not support the hypothesis of an $As_{Ga} - As_i$ pair. First, thermal treatment of GaAs containing the EL2 defect revealed that the EL2 defect is relatively stable at temperatures of $800 - 900$ °C. Makram-Ebeid et al. (1982) compared the stability of EL2 with Cr in GaAs. If one component of the EL2 were an interstitial As, As_i, the As_i would be expected to diffuse rapidly due to its interstitial position. However, such a rapid diffusion has not been observed. Furthermore, Baraff et al. (1988) investigated the $As_{Ga} - As_i$ pair defect in GaAs using electronic structure calculations. They found no plausible mechanism for binding the two components of the pair to each other. If there is indeed no plausible binding mechanism, the probability of the occurrence of $As_{Ga} - As_i$ pairs is much smaller than the experimentally observed EL2 concentrations. The authors also predicted an effective-mass-like shallow level for the $As_{Ga} - As_i$ pair which has not been found experimentally.

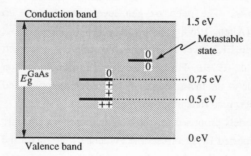

Fig. 9.7. Energy levels of the EL2 defect within the GaAs band gap. The EL2 defect has two donor levels at 0.5 eV and 0.75 eV above the valence band edge; the defect also has a metastable neutral level originating from a defect configuration with large lattice relaxation.

The schematic scheme of the electronic levels of the EL2 defect is shown in Fig. 9.7. The As has two excess electrons if occupying a Ga site, i.e. is a double donor. The lowest donor level is 0.5 eV above the top of the valence band. This donor level is usually not observed in DLTS measurements since it requires very high temperatures (>> 300 K) to be ionized. Evidence for the lowest EL2 donor level is obtained from luminescence and luminescence excitation spectroscopy which reveal EL2 related bands at 0.63 eV and 0.68 eV (Tajima, 1987; Tajima et al., 1987). The second donor level is approximately 0.75 eV above the valence band edge. In semi-insulating GaAs, the Fermi level is pinned at this level. The EL2 defect has also a third level which has metastable properties. The metastable level has a neutral charge state; its properties will be discussed below.

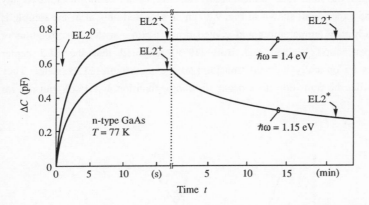

Fig. 9.8. Typical photocapacitance transients at 77 K for n-type GaAs containing the EL2 defect. Initially the EL2 defects are in the neutral (EL2^0) state. After illumination with $\hbar\omega = 1.15$ eV and 1.4 eV, the defects assume an ionized state (EL2$^+$). Long illumination with $\hbar\omega = 1.15$ eV quenches the capacitance indicating that the defect assumes again the initial neutral (metastable) state (after Vincent and Bois, 1978).

One of the key features of the EL2 defect is its metastability which has complicated the understanding of the defect. The metastability of the EL2 manifests itself at low temperatures in the following way. Upon photoexcitation the characteristic features associated with the EL2 disappear. These features include luminescence bands, absorption bands and electrical activation. The metastable characteristics of the EL2 were postulated based on the photocapacitance characteristics of GaAs shown in Fig. 9.8 (Vincent and Bois, 1978). The capacitance of an n-type GaAs sample with a Schottky contact is measured during illumination with different photon energies. Initially, all EL2 centers are in a neutral state (EL2^0). Upon illumination with $\hbar\omega = 1.15$ or 1.4 eV, the EL2 level becomes ionized (EL2$^+$) which results in an increase of the capacitance. Certainly, several transitions compete in the illumination process including EL2-to-CB and VB-to-EL2 excitation. That is, at sufficiently high illumination intensities the major fraction of EL2 is ionized while a small fraction is temporarily in a neutral state. Illumination with $\hbar\omega = 1.4$ eV yields a steady-state capacitance change while illumination with $\hbar\omega = 1.15$ eV yields a surprising result. The change of the capacitance decreases and approaches the initial capacitance value as shown in Fig. 9.8. The decrease of the capacitance can be explained *only if the final charge state of the EL2 is the same as the initial state*. Therefore, the final state, which is called EL2*, must be neutral. Furthermore, the final state is *metastable* since the center cannot be ionized any more. The experiment shown in Fig. 9.8 can be repeated only after the sample is heated to higher temperatures and cooled down again. In the interpretation of their experiments, Vincent and Bois (1978) assumed that the EL2 center undergoes an optically induced transition to the metastable EL2* state. They assumed that the transition does occur for photoexcitation at 1.15 eV but not for 1.4 eV.

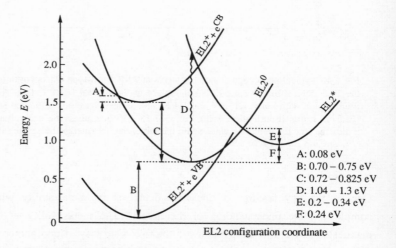

Fig. 9.9. Configuration coordinate diagram for the EL2 defect in the normal and the lattice-relaxed, metastable state. In the normal state the electron can be in the valence band (EL2$^+$ + e^{VB}), localized at the defect (EL2^0), or in the conduction band (EL2$^+$ + e^{CB}). The second donor level of the EL2 defect (0.5 eV above the VB, see Fig. 9.7.) is not shown (after Vincent and Bois, 1978).

The metastable state of the EL2 was further modeled using the configuration coordinate diagram (Vincent and Bois, 1978). A configuration coordinate diagram for the EL2 center is shown in Fig. 9.9, which shows the neutral state, EL2^0, the metastable state EL2*, and the singly ionized state, EL2$^+$ + e^{CB} or EL2$^+$ + e^{VB}. Note that the lower donor level, EL2^{++}, is not shown since it is irrelevant for the photocapacitance experiments discussed here. The lower donor level can be considered to be always occupied at sufficiently low photoexcitation intensities. Transition 'C' is the thermal ionization of the upper EL2 donor level; transition 'D' is the optical transition from the neutral state, $EL2^0$, to the metastable state, EL2*. In order to relax from EL2* to $EL2^0$, the energy barrier 'E' needs to be overcome. Since the energy barrier 'E' is on the order of 200–340 meV, this transition cannot occur at low temperatures, i.e. the center remains in the metastable EL2* configuration.

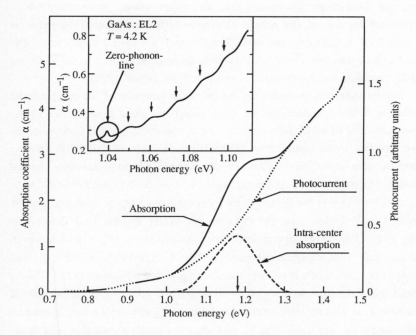

Fig. 9.10. Absorption coefficient and photocurrent versus photon energy for n-type GaAs containing the EL2 defect with a concentration of $2 \times 10^{16} cm^{-3}$. The inset shows the EL2 zero-phonon-line intra-center transition and five phonon replicas (after Kaminska, 1984).

Further evidence for the photon-induced $EL2^0$ to EL2* transition was obtained from absorption measurements which revealed the signature of the 1.15 eV intra-center transition (Kaminska et al. 1983) in n-type GaAs. A typical absorption measurement for GaAs containing the EL2 defect is shown in Fig. 9.10 (Kaminska, 1984; Dischler et al., 1986). After the sample is cooled in the dark, the absorption measurement reveals an absorption band starting at energies of 0.8 eV (solid line). The absorption band is completely bleached after sufficiently long illumination with $\hbar\omega = 1.18$ eV (dashed lines). The intra-center transition has its zero-phonon line at $\hbar\omega = 1.04$ eV as shown in the inset. Phonon replicas are observed at energies 11, 22, 33, ... meV higher than the zero-phonon line (Kaminska et al. 1983). The authors showed that intra-center absorption occurs at energies ranging from 1.04 eV to 1.3 eV with a peak absorption occurring at approximately 1.18 eV. Figure 9.10 also shows the photocurrent signal which is due to electronic transitions creating free electrons or holes. Comparison of the absorption signal with the photocurrent signal reveals that the intra-center absorption (i.e. absorption minus photocurrent) is a significant fraction of the overall absorption. The background absorption at $E > 0.8$ eV is probably due to EL2-to-CB and VB-to-EL2 excitations. The band-to-EL2 transitions have been shown to be much stronger than the EL2^0-to-EL2* transition (see, for example, Kaufmann, 1989; Baraff, 1992).

Previous theoretical studies on the electronic structure of the As$_{Ga}$ antisite defect in GaAs revealed that the As$_{Ga}$ defect has three bound states of A$_1$ symmetry, one of which falls deep into the fundamental gap and is occupied by two electrons in the neutral state of the defect (Bachelet et al., 1983). This A$_1$ state can also be occupied by a single electron resulting in a positively charged defect. The energy levels of the gap A$_1$ state of the defect were calculated to be $E_1 \cong 0.8$ eV (++/+) and $E_2 \cong 1.1$ eV (+/0) above the valence band edge and are about 0.3 eV higher than the experimental values for the EL2 defect (see Fig. 9.14). The energy separation between the two levels ($E_2 - E_1$), however, agrees well with experimental results. The A$_1$ symmetry of the gap state designates a *spherically symmetric* state (see, for example, Cotton, 1971). The A$_1$ states thus have the symmetry of the hydrogenic ls orbital. The calculations of Bachelet et al. also revealed an additional state of T$_2$ symmetry which is resonant with the conduction band. The T$_2$ symmetry designates a state that is invariant for $90°$ rotations in the planes of a cartesian coordinate system. Thus, the T$_2$ state has the same symmetry as atomic p orbitals.

The fine structure of the zero-phonon line of the EL2^0-to-EL2* transition was studied under uniaxial stress in order to assess the symmetry of this transition. Kaminska et al. (1985) and Bergman et al. (1988) showed that the transition is consistent with the symmetry of a A$_1 \rightarrow$ T$_2$ transition at a defect with overall tetrahedral (T$_d$) symmetry. On this basis Kaminska et al. concluded that the EL2 is the As$_{Ga}$ antisite defect. The tetrahedral symmetry of the EL2 defect was confirmed by Nissen et al. (1990, 1991) using photoluminescence studies under uniaxial stress. The authors used a transition from the conduction band to the A$_1$ midgap level of the As$_{Ga}$. No splitting was found under uniaxial stress, indicating that there was no orientational degeneracy to be lifted, and as a consequence, that there could be no defect closely attached to the As_{Ga}.

A remarkable microscopic physical model for the EL2 defect was proposed by Chadi and Chang (1988a) and independently by Dabrowski and Scheffler (1988, 1989) using self-consistent total energy calculations of the As_{Ga} defect. The calculations showed that, immediately after an $A_1 \rightarrow T_2$ transition at the neutral isolated As_{Ga}, the total energy of the system would be somewhat greater than the energy of a strongly distorted configuration. In this strongly distorted configuration, one of the four As–As bonds of the central As_{Ga} atom is broken and the As_{Ga} atom moves along a $\langle 111 \rangle$ direction into an interstitial position 0.29Å below the plane of its three remaining As neighbors. The other As atom whose bond was broken, remains in its original position. Thus, the distorted configuration gives rise to two threefold coordinated As atoms with different local environments. The As_{Ga} defect may spontaneously relax into this configuration, which was shown to exhibit metastability, after an optically induced $A_1 \rightarrow T_2$ transition. The normal configuration (fourfold coordinated or tetrahedral symmetry) and of the metastable configuration (threefold coordinated, interstitial As with broken bond) of the As_{Ga} antisite defect have a strong similarity to the D^0 and DX^- configuration of the DX center. To visualize the similarity it is useful to reinspect Fig. 9.5. If the Si atom in Fig. 9.5(a) is replaced with an As atom, the regular As_{Ga} antisite defect is obtained. The metastable configuration of the EL2 defect is obtained, if the Si (DX^-) atom in Fig. 9.5(b) is replaced by a As atom. In the metastable position, the As is approximately 0.3 Å below the plane of its three remaining As neighbors. The normal position of the As is approximately 0.9 Å above the plane. Thus, the As atom moves 1.2 Å along the $\langle 111 \rangle$ direction, away from its 'ideal' Ga substitutional position, and breaks one of its bonds by stretching it to $\cong 3.6 \text{ Å}$ (Chadi and Chang, 1988a). The breaking of the As–As bond may be understood as the placement of an electron in the antibonding state of the As_{Ga} donor defect and thereby weaken one of the existing bonds so much that the defect atom would move, breaking the bond holding it in place (Baraff, 1992). Note that the DX-configuration is negatively charged (DX^-), whereas the As in the same interstitial position is neutral. Thus, the normal-to-metastable transition is a neutral-to-neutral state transition. This feature of the center is in agreement with the experimental photocapacitance measurements shown in Fig. 9.8 which indicate that the $EL2^0 - EL2^*$ transition does not involve a change in charge state.

The physical and electronic properties of the EL2 center have been reviewed by Weber (1985), Martin and Makram-Ebeid (1986), Kaufmann (1989), and Baraff (1992).

9.3 Hydrogen

Hydrogen is not an electrically active substitutional impurity in III–V semiconductors. That is, if hydrogen were incorporated in an intrinsic, undoped semiconductor, the electronic properties of the semiconductor would not be changed by the hydrogen incorporation. However, hydrogen strongly interacts with shallow donor and acceptor impurities in III–V semiconductors. Hydrogen can neutralize such shallow impurities and consequently change the electronic properties of III–V semiconductors dramatically. In addition, hydrogen implantation creates defects which can significantly change the conductivity of the semiconductor. For example, semi-insulating characteristics of GaAs with resistivities exceeding $10^7 \, \Omega$ cm can be achieved by implantation of hydrogen. As a result, hydrogen is of significant technological importance for III–V device technology.

The two major methods of hydrogen incorporation are (i) by diffusion at relatively low temperatures and (ii) by implantation of H ions (protons). The resulting characteristics of the semiconductor strongly depend on the method by which hydrogen is incorporated. Therefore, the following sections are devoted to characteristics of semiconductors with hydrogen being incorporated by different methods, namely incorporated by *diffusion* (Sect. 9.3.1) and incorporated by *implantation* (Sect. 9.3.2). Other methods for hydrogen incorporation such as electrochemical techniques (Seager, 1991) will not be discussed. A final section (Sect. 9.3.3) is dedicated to the influence of residual hydrogen during *growth* by molecular-beam epitaxy.

9.3.1 Hydrogen incorporation by diffusion

Hydrogen can be incorporated from a H surface population into the bulk of the semiconductor by means of diffusion. Atomic (H) and molecular (H_2) hydrogen surface populations can be generated by a low pressure (0.1 to 100 Torr) glow-discharge, direct current (dc) or radio-frequency (rf) plasma reactor (Seager and Ginley, 1982). In the direct current type reactor, the semiconductor serves as the cathode resulting in a positive proton current towards the semiconductor. Proton current densities in the range of several mA/cm^2 impinge on the semiconductor surface at applied biases of few hundred volts. The voltage drop occurs in the cathode near region (Boenig, 1982) resulting in a substantial kinetic energy of protons which may lead to some H^+-ion damage in the surface near region of the semiconductor. Furthermore, electric field inhomogeneities may lead to H^+ current inhomogeneities, which in turn lead to a non-uniform H distribution.

Some of the above mentioned disadvantages are avoided in radio-frequency plasma reactors in which the sample is mounted 'downstream'. The 'downstream' configuration of the sample implies that the sample is spatially removed from the radio-frequency generator (see, for example, Pearton et al., 1987a). The 'downstream' formation effectively eliminates the direct bombardment damage discussed above. The lack of exposure to a high electric field also reduces the non-uniformity of the hydrogen surface population discussed above.

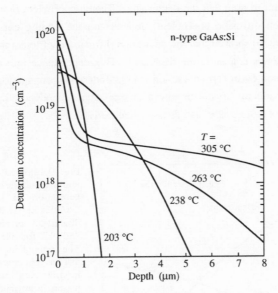

Fig. 9.11. Deuterium concentration profiles in bulk n-type GaAs:Si after exposure to a radio-frequency deuterium plasma (with power density of 0.2 W/cm^2) for 90 min at different temperatures (after Chevallier et al., 1991).

The deuterium (D) concentration as a function of depth for a 90 min exposure of GaAs to a radio-frequency D plasma is shown in Fig. 9.11 (Chevallier et al. 1991). The GaAs samples were heated to different temperatures resulting in markedly different D distributions. The authors showed that the D profiles can be fitted well by the erfc-function. The good fit indicates that the diffusion mechanism follows fickian laws. The fit further indicates that regular diffusion is the dominant mechanism by which D is incorporated into the semiconductor. At a temperature of 240 °C, the authors estimated the diffusion constant to be 3×10^{-12} cm^2/s. The temperature dependence of the hydrogen diffusion constant has been described by Chevallier et al. (1991) according to

$$D = D_0 \exp\left(-E_a/kT\right), \tag{9.6}$$

with $D_0 = 115 \text{ cm}^2/\text{s}$ and $E_a = 1.38$ eV. Note that the diffusion coefficient of H and D are many orders of magnitude higher as compared to substitutional impurities such as Si, Be, Sn, etc. The large diffusion coefficient of H indicates that the diffusion mechanism of H and D in GaAs is a very rapid one such as interstitial diffusion. Interstitial diffusion is indeed expected for H and D as the small size of these atoms allows them to easily propagate via interstitial sites. It is finally important to note that the deuterium diffusion coefficient (and probably also the hydrogen diffusion coefficient) depends on the doping concentration. Pearton et al. (1987b) showed that the deuterium diffusion coefficient increases as the intentional donor concentration decreases. This dependence indicates either some kind of donor–hydrogen interaction which hinders the propagation of H and D in the semiconductor crystal or else a charge-state dependence, i.e. that the ionized H diffuses more easily than does the neutral H.

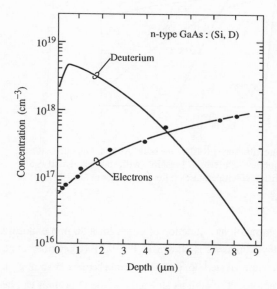

Fig. 9.12. Deuterium and free carrier concentration in n-type GaAs:Si exposed to a deuterium plasma at 250 °C for 90 min. The decreasing deuterium concentration in the near surface region is attributed to deuterium out-diffusion during sample cooling after plasma exposure (after Chevallier et al., 1986).

The donor–hydrogen or donor–deuterium interaction is further illustrated in Fig. 9.12 which shows the free electron concentration and the deuterium concentration versus depth (Chevallier et al., 1986). The deuterium was incorporated by exposing the sample to a radio-frequency deuterium plasma for 90 min at 250 °C. Before plasma exposure, the carrier concentration did not depend on depth. Figure 9.12 reveals that the free electron concentration clearly decreases with increasing deuterium concentration. This result indicates that hydrogen *passivates* or *neutralizes* shallow donors in semiconductors. The passivation process for Si donors can be written as (Chevallier et al., 1991)

$$H^0 + Si^+ + e \leftrightarrow Si-H \qquad (9.7)$$

where H^0 represents a neutral hydrogen atom.

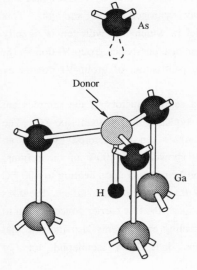

Fig. 9.13. Schematic representation of the group-IV donor–hydrogen complex with hydrogen in the AB site.

The commonly accepted atomic model for donor passivation by hydrogen is shown in Fig. 9.13. The substitutional Si atom is fourfold coordinated. The hydrogen is in an interstitial (AB) site and bound to the Si donor. One of the original Si–As bonds is broken leaving the As atom opposite the hydrogen atom with a lone pair of electrons. The electron count of the Si–H complex yields that the complex is neutral, i.e. H has passivated the donor atom.

Unambiguous information on the Si–H bond was obtained from local vibrational modes (LVM) of hydrogen bound to Si isotopes which clearly demonstrated the existence of the H–Si bond (Pajot et al., 1988b). The authors proposed that hydrogen is located in an antibonding position along a trigonal axis and that the Si is fivefold coordinated. However, Briddon and Jones (1989) subsequently showed that the fivefold coordination of Si is not stable. Large cluster calculations revealed that the Si–As bond opposite the hydrogen (see Fig. 9.13) is broken and that the Si is fourfold coordinated. The additional electron provided by the donor forms a lone pair with the dangling bond of the threefold coordinated As atom.

Although the atomic model shown in Fig. 9.13 has been verified for Si–H in GaAs, it is generally assumed that the model also applies to other donor impurities in III–V semiconductors. Experimentally, passivation of donors by hydrogen has been found for a large number of donors including Si, Ge, Sn, S, Se, and Te (Pearton et al., 1986). In addition, the activation energy for the recovery of the electrical activity, i.e. the dissociation of the donor–hydrogen bond, was found to be proportional to the donor–hydrogen bond strength. This interdependence further supports the model in which the hydrogen is directly bound to the donor atom. Note that hydrogen also passivates group-VI donors (S, Se, Te). A definite atomic model for the passivation of group-VI donors by hydrogen atoms has not yet been developed.

Hydrogen diffuses out of hydrogenated semiconductors, if the samples are heated to sufficiently high temperatures, typically larger than 400 °C. The original conductivity of hydrogenated samples can be restored by such an annealing cycle. McCluskey et al. (1989) showed that 100% of the original conductivity of a Si-doped GaAs sample can be restored upon heating to 400 °C for 5 min. The Si-doped layer was buried 3 μm deep below the surface. Previous hydrogen plasma exposure of the sample reduced the free carrier concentration to 1/3 of its original concentration. The annealing experiments also indicate that hydrogen does not favour incorporation into III–V semiconductors at temperatures exceeding approximately 400 °C.

Hydrogen also passivates shallow *acceptors* in III–V semiconductors. The atomic model for the substitutional group-II acceptor is shown in Fig. 9.14. The hydrogen atom is bound to an arsenic atom and occupies a BC-site between the (Be) acceptor impurity and the arsenic neighbors. The hydrogen atom thus satisfies the electron pair requirement of the arsenic neighbor. The entire hydrogen-impurity complex is neutral, i.e. hydrogen passivates the acceptor. To date, the atomic model shown in Fig. 9.14 is generally assumed not to be restricted to Be but to apply to all substitutional group-II acceptors.

Originally, the atomic model shown in Fig. 9.14 was shown to be valid for substitutional Be impurities. The Be–H complex has been studied by local vibrational mode (LVM) spectroscopy (Nandhra et al., 1988). The use of uniaxial stress during LVM spectroscopy allows one to lift orientational degeneracies and identify the symmetry of the complex. Low-temperature LVM measurements showed that the As–H bonds are aligned along the trigonal axis (Stavola et al., 1989). Although the measurements did not reveal if the hydrogen occupies the BC or the AB position, it is now commonly assumed that it occupies the BC site.

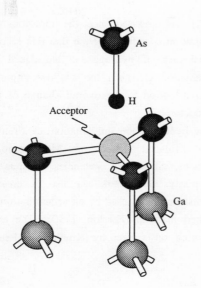

Fig. 9.14. Schematic representation of the group-II acceptor–hydrogen complex with hydrogen in the BC site.

The hydrogen BC site was originally proposed by Pankove et al. (1985) and has been confirmed by cluster calculations (Briddon and Jones, 1989).

Finally, hydrogen can also passivate group-IV acceptors such as C or Ge. Chevallier et al. (1988) proposed an atomic model in which the hydrogen is bound to the group-IV acceptor and is in a BC position between the group-IV acceptor and one of its group-III neighbors.

9.3.2 Hydrogen incorporation by implantation

Hydrogen implantation is a process that is widely used to render III–V semiconductors highly resistive. *Semi-insulating* properties with resistivities $\leq 10^7 \ \Omega$ cm can be achieved by hydrogen implantation as well. The implantation process allows one to choose the implantation depth which provides high flexibility to tailor the electrical properties of thin films.

The result of hydrogen implantation, which is frequently termed *proton* implantation, is twofold. First, high-energy protons create *defects* along their implantation path. These defects include point defects such as interstitials, vacancies, antisite defects, etc. Such defects commonly have levels within the fundamental gap which can trap free carriers. The semiconductor thus becomes highly resistive. Note that this mechanism is irrelevant if hydrogen is introduced by diffusion (see Sect. 9.3.1). Diffusion is an equilibrium process and usually does not introduce point defects. Second, hydrogen forms complexes with donor and acceptor impurities (see Sect. 9.3.1). As a consequence, the impurities are neutralized and the free carrier concentration is reduced. Note that this second mechanism is the only relevant mechanism if hydrogen is introduced by diffusion. However, at typical implantation energies, the shallow impurity passivation by hydrogen is a minor contribution to the overall change of the electrical characteristics of the semiconductor.

The simplest defects created during hydrogen ion implantation are Frenkel pairs, consisting of a vacancy and the displaced lattice atom. More complex defects such as divacancies, trivacancies, clusters of vacancies, clusters of interstitials, clusters involving an impurity, and others can also be created (Corbett, 1971). Line defects such as dislocations caused by an accumulation of point defects are common in implanted material (Pearton, 1990). The total number of atoms displaced by an incoming ion is given by (Kinchin and Pease, 1955)

$$N_d \cong E_n / 2E_d \tag{9.8}$$

where E_n is the total energy loss of implanted ions in nuclear collisions, and E_d is the displacement energy for a lattice atom. The displacement energy is approximately 15–20 eV in GaAs and has similar values in other III–V semiconductors. Note that E_n is usually much smaller than the primary ion energy. This is because most of the energy loss of primary ions is due to electronic collisions (i.e. collision of ions with the electron shell of lattice atoms) rather than nuclear collisions. In an example, Pearton (1990) estimated that a 200 keV H^+ ion creates about nine displaced atoms.

Experimental investigations on the defect creation in III–V semiconductors upon ion implantation confirmed the formation of a multitude of point defects (Wesch et al., 1989). The study revealed that amorphization of the semiconductor occurs via homogeneous defect nucleation. The authors further showed that the creation and annihilation of a defect depends strongly on the implantation temperature. 'Hot' implants, i.e. implantation at elevated target temperatures, tend to result in less defects due to the annihilation during the implantation process, e.g. the recombination of Frenkel defects. Tell et al. (1990) reported enhanced diffusion of Si impurities in GaAs subjected to H ion implantation. The increase in the Si diffusion coefficient was found to be as large as one order of magnitude. The enhanced diffusivity of Si in GaAs was attributed to enhanced defect concentration in the implanted material.

Hydrogen distributions obtained by implantation in GaAs and InP at different implantation energies are shown in Fig. 9.15. The distributions are calculated using the 'TRIM '91' computer code (Ziegler et al., 1991). The hydrogen atom

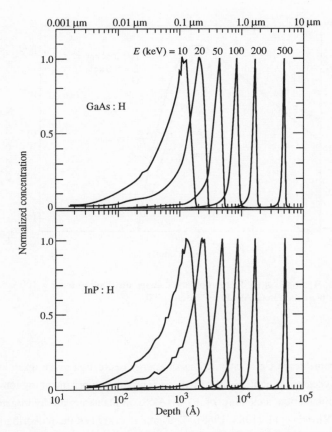

Fig.9.15. Concentration of implanted hydrogen versus depth in GaAs and InP for
different implantation energies (calculated with TRIM '91 code).

distribution shown in Fig. 9.15 does not represent the distribution of defects in the
crystal. It is evident that defects are created along the path of the atom, i.e. the
defect concentration is higher than the hydrogen atom concentration in the near-
surface region.

Pearton (1990) reported a quantitative comparison between the hydrogen ion
concentration and the defect concentration in GaAs implanted with 200 keV
protons. The comparison of hydrogen atom concentration and defect
concentration is shown in Fig. 9.16. The two concentrations exhibit a
qualitatively similar dependence. However, the damage distribution is closer to
the surface than the ion distribution. The damage profile shown in Fig. 9.16 was

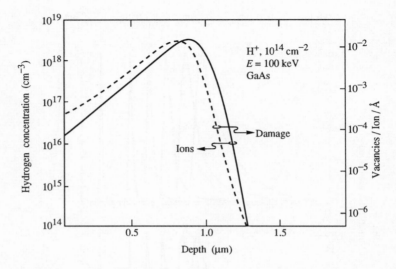

Fig. 9.16. Hydrogen atom distribution and damage distribution for $E = 100$ keV proton implantation in GaAs (after Pearton, 1990).

obtained from Monte Carlo calculations of energy deposition given up in atomic stopping processes, or equivalently, of vacancy production per incoming ion.

Post-implantation annealing of n^+-GaAs:Si implanted with hydrogen was studied by Zavada et al. (1985, 1988). The authors described the redistribution of hydrogen during the thermal annealing process by a diffusion constant whose temperature dependence was described by

$$D = D_0 \exp(-E_a/kT) \ , \tag{9.9}$$

with $D_0 = 1.5 \times 10^{-5}$ cm^2/s and an activation energy of $E_a = 0.62$ eV. The activation energy of 0.62 eV is much smaller than the activation energy for hydrogen diffusion without previous implantation (see Eq. (9.6)) which indicates that the defects created in the implantation process facilitate the diffusive motion of hydrogen.

Post implantation annealing of hydrogen-implanted III–V semiconductors has several consequences. First, hydrogen spatially redistributes as discussed above. Second, hydrogen–impurity complexes are formed at moderate annealing temperatures (200–500 °C). The formation of Si–H complexes in GaAs has been established by infra-red absorption spectroscopy (Pajot et al., 1988a). Third, at high temperatures ($T > 500$ °C) hydrogen diffuses strongly and eventually diffuses out of the semiconductor. The hydrogen–impurity bond is easily broken at such high temperatures. Fourth, defects created during the implantation process can be annealed out at sufficiently high temperatures. As a result, the conductivity of the semiconductor increases; for appropriate annealing conditions, the original conductivity of the semiconductor can be restored with a near 100% efficiency.

The possibility of rendering III–V semiconductors semi-insulating by hydrogen implantation made this technique very attractive for device applications. Steeples et al. (1980) showed that resistivities of $> 10^7 \, \Omega$ cm can be achieved in GaAs by proton implantation. Previously, proton bombardment was shown to be very efficient for pn-junction devices and Schottky contact isolation (Foyt et al., 1969). Proton implantation has also been used for optical waveguide formation (Garmire et al., 1972) and for laser fabrication (Van der Ziel et al., 1981). More recently, continuous wave operation was achieved in AlGaAs/GaAs quantum well lasers (Jackson et al., 1988), GaInAsP/InP buried ridge lasers (Kazmierski et al., 1989), and in AlGaAs/GaAs vertical cavity lasers (Tu et al., 1992) using proton implantation.

9.3.3 Role of hydrogen during growth

The quality of epitaxial crystals grown by molecular-beam epitaxy can be improved by introducing *atomic* hydrogen into the growth chamber. The improvement has been reported by several groups (see, for example, Calawa, 1978; Bachrach and Bringans, 1983; Sugiura et al., 1986; Pao et al., 1986; Kopev et al., 1989). However, the improvement is not due to the incorporation of hydrogen into the semiconductor but rather to chemical reactions of hydrogen with other residual gases in the growth chamber.

Several epitaxial growth systems have large quantities of H and H_2 in the reaction chamber. These growth techniques include GSMBE, OMVPE, and CBE. Hydrogen is introduced in these techniques via PH_3, AsH_3, or as a carrier gas. Hydrogen has also been intentionally introduced during molecular-beam epitaxial growth, i.e. a technique that usually neither requires nor generates hydrogen during growth. The hydrogen has been introduced either as molecular hydrogen (Calawa, 1978) or as dissociated hydrogen atoms (Bachrach and Bringans, 1983). However, H cannot be detected in the epitaxial layers by common analytical techniques such as secondary ion mass spectrometry if the growth temperature is $\geq 400 \, °C$. That is, hydrogen is not incorporated effectively at typical epitaxial growth temperatures.

The main advantage of hydrogen introduction during molecular-beam epitaxial growth is the chemical reaction of hydrogen with residual gases in the growth chamber. It has been pointed out (Calawa, 1978) that carbon (including CO and CO_2) and oxygen are contaminants in GaAs and other semiconductors. These residual contaminants are present in sufficient abundance to dope films in the $10^{13} - 10^{14}$ cm^{-3} range. For example, an impurity with a unity sticking coefficient and a partial pressure of about 10^{-12} Torr at the substrate surface would result in a concentration of unwanted impurities of 10^{17} cm^{-3} (Calawa, 1978). Both carbon and oxygen are electrically active impurities in GaAs which degrade the quality of the material. Because hydrogen is highly reactive with carbon and oxygen, the presence of H_2 and H during growth can be expected to affect the incorporation of these two impurities.

An improvement of the transport characteristics of GaAs has indeed been found experimentally upon introduction of hydrogen. Calawa (1978) found a reduction of the residual carrier concentration by a factor of 2 and a significant increase in carrier mobility. The author also found a type conversion from p-type to n-type upon hydrogen introduction. Pao et al. (1986) found similar improvements; in addition, a dramatic reduction in deep electron trap concentration was found. Kopev et al. (1989) reported a decrease in the density of oval defects in GaAs upon the introduction of hydrogen. Oval defects were attributed to GaO and Ga_2O_3 on the crystal surface; both types of oxides were reduced in density upon hydrogen introduction.

The incorporation of H into GaAs:C during epitaxial growth by CBE and OMVPE was analyzed by Stavola et al. (1991). Hydrogen was found to incorporate from the organometallic sources used as group-III precursors, from hydrides used as group-V precursors, and even from forming gas (15% H_2, 85% N_2) used as an annealing ambient. Hydrogen was found to readily incorporate into GaAs at growth temperatures ≤ 500 °C which resulted in a passivation of 5–50% of the C impurities. Post-growth annealing at 550 °C for 60 s in an inert ambient (He) was sufficient to dissociate the C–H complexes and electrically activate the C acceptors.

9.4 Chromium, iron, and oxygen

Compound semiconductors can be intentionally doped with deep impurities in order to render the semiconductor highly resistive. The ionization energy of such deep impurities is much larger than the thermal energy, kT, at room temperature. Deep impurities are therefore not thermally ionized at room temperature. Furthermore, shallow impurities can be compensated with deep centers yielding very low free carrier concentrations. Consider, for example a semiconductor with a shallow donor concentration N_D and a deep acceptor concentration N_A, with $N_A > N_D$. Electrons originating from the shallow donors will occupy the deep acceptor states rather than conduction band states. Since $N_A > N_D$, only a fraction of the deep acceptor states are occupied. As a result, the Fermi level is pinned at the deep acceptor level. Thus, intentional doping with deep impurities can drastically reduce the free carrier concentration.

If the energy level of a deep impurity is at or near the intrinsic Fermi level, i.e. in the vicinity of the middle of the forbidden gap, the free carrier concentration equals the intrinsic carrier concentration. The intrinsic carrier concentration is very small, e.g. $n_i \cong 10^7$ cm^{-3} in GaAs. Such low concentrations cannot be achieved by conventional methods, since the unintentional background impurity concentration usually exceeds the intrinsic concentration by many orders of magnitude. It is thus desirable that intentional deep impurities have an energy level close to the middle of the forbidden gap of the semiconductor.

The energy levels of chromium (Cr) and iron (Fe) in GaAs and InP are shown in Fig. 9.17. The energy levels closest to the midgap position are the Cr acceptor level in GaAs and the Fe acceptor level in InP. The lowest free carrier concentrations are indeed achieved with Cr-doped GaAs and Fe-doped InP.

Semi-insulating substrate materials are of technological importance for device isolation purposes. Materials with resistivities exceeding 10^7 Ω cm are called *semi-insulating*. Semi-insulating GaAs and InP can be obtained by means of doping with Cr and Fe, respectively, in order to compensate for shallow donor impurities. In addition, semi-insulating materials can also be grown by selecting growth conditions that favor the formation of native defects which in turn can trap free carriers. Highly resistive properties of GaAs (see Sect. 9.2) and InP (Hofmann et al., 1989) have been demonstrated *without* intentional doping. The EL2 defect has been identified as causing the highly resistive characteristic in GaAs. In InP, the relevant center causing high resistivity has not yet been identified.

9.4.1 Chromium (Cr)

Chromium is a deep acceptor in III–V semiconductors. In GaAs, the Cr acceptor energy is 0.79 eV above the valence band edge as shown in Fig. 9.17. The level is very close to the midgap position in GaAs which makes Cr an ideal impurity for the growth of semi-insulating GaAs. In InP, the Cr acceptor energy is 0.96 eV above the valence band edge which is not particularly close to the InP midgap position. Furthermore, Cr has a much lower solid solubility in InP than in GaAs, which necessitates very high purity starting material of InP to achieve high resistivity (Mullin et al., 1971). Chromium is therefore not used in InP to render the material semi-insulating.

Chromium has been widely used in bulk GaAs grown by the Czochralski

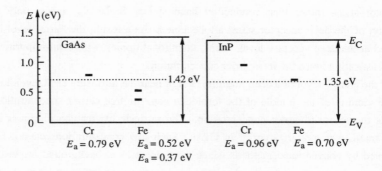

Fig. 9.17. Acceptor levels of Cr and Fe in GaAs and InP measured from the top of the valence band.

technique (see, for example, Lindquist, 1977). Residual n-type conductivity is obtained in GaAs grown in a quartz (SiO_2) crucible. The n-type conductivity results from Si impurities that are incorporated from the quartz walls into the bulk GaAs. Si impurities have typical concentrations in the low 10^{16} cm^{-3} range in such unintentionally doped GaAs. In order to compensate for the n-type conductivity, Cr is added to the melt at a concentration of typically 5×10^{16} cm^{-3}. The Cr concentration is thus higher than the shallow donor concentration which yields a complete compensation of the shallow donors. Cr-doped GaAs is also grown by the liquid encapsulated Czochralski (LEC) technique which has several advantages over the regular Czochralski technique without an encapsulant (Müller, 1987).

Cr-doped GaAs has also been grown by other bulk growth techniques such as the gradient freeze method. Orito et al. (1984) achieved resistivities exceeding 10^8 Ω cm in Cr-doped GaAs grown by this method. Silicon was found to be the dominant shallow impurity in nominally undoped GaAs. The residual Si concentration ranged between 2×10^{15} and 2×10^{16} cm^{-3}. The Cr concentration in the GaAs ranged between 1×10^{16} and 4×10^{16} cm^{-3} and increased with growth time due to Cr segregation effects.

Cr-doped GaAs grown by the horizontal Bridgman and the LEC technique have been compared by Miyazawa and Nanishi (1983). The authors found that the horizontal Bridgman grown material has an inhomogeneous distribution of electrical properties even at relatively low Cr concentrations in the melt. The LEC-grown material was found to be more uniform and suited for use in GaAs integrated circuit fabrication.

A major drawback of Cr-doped GaAs is the Cr diffusion and redistribution during high-temperature processing and during epitaxial growth. The redistribution of Cr has been discussed in Chap. 8. Redistribution effects can lead to non-uniform electrical characteristics of a wafer which in turn can affect device parameters such as the threshold voltage of an injection laser. Therefore, Cr-doped GaAs substrates are infrequently used at the present time. Instead, *undoped, semi-insulating GaAs* substrates grown by the LEC technique constitute the major part of the present GaAs substrate market. The semi-insulating electrical characteristic of the undoped GaAs substrate is due to the native EL2 defect (see Sect. 9.2) which compensates residual shallow impurities.

The electrical characteristics of undoped GaAs grown by the LEC technique have been investigated by Matsumura et al. (1983). The authors showed that the resistivity and the dislocation density are correlated. A low dislocation density was found in the highly resistive areas of the GaAs wafer. They found undoped semi-insulating GaAs superior over Cr-doped GaAs due to the lack of inhomogeneities and non-uniformities induced by Cr redistribution.

9.4.2 Iron (Fe)

Iron (Fe) is a deep acceptor in III–V semiconductors. In InP, the Fe acceptor energy is 0.70 eV above the valence band edge as shown in Fig. 9.17 (Iseler, 1979). The Fe level is very close to the midgap position in InP which makes it an ideal impurity for the growth of semi-insulating InP. In GaAs, the Fe acceptor energies are 0.52 and 0.37 eV above the valence band edge which is not particularly close to the GaAs midgap position. Iron is therefore not used in GaAs to render the semiconductor semi-insulating.

Fe-doped InP has semi-insulating properties with resistivities as high as 4.6×10^7 Ω cm (Kim et al., 1986). The authors also investigated the variations of the resistivities across 5 cm InP substrates. A V-shaped pattern of the resistivity was found along the $\langle 011 \rangle$ and the $\langle 01\bar{1} \rangle$ axes. The Hall mobilities on the wafers were 1700–3000 cm^2/V s. The low electron mobility was attributed to neutral impurity scattering.

Iron doping has been used for the growth of semi-insulating InP bulk crystals using the LEC technique (Shimakura et al., 1989), modified LEC techniques (Oda et al., 1991), the horizontal Bridgman technique (Kainosho et al., 1990), the vertical gradient freeze (VGF) technique (Monberg et al., 1987) and an improved Bridgman technique called electrodynamic gradient (EDG) technique (Monberg et al., 1988). All of these bulk crystal growth techniques allow one to obtain InP:Fe with semi-insulating electrical characteristics. The highest resistivities achieved exceed 10^8 Ω cm (Oda et al., 1991). The deep Fe acceptor impurities compensate for shallow donor impurities in InP. The dominant shallow donor impurity in bulk InP was found to be Si, which incorporates into the polycrystalline starting material from quartz boats and quartz ampoules (Oda et al., 1990). The Si-impurity concentration is typically in the low 10^{15} cm^{-3} range in nominally undoped single crystal InP. Consequently, a typical Fe concentration in the mid to high 10^{15} cm^{-3} is required in order to compensate for the shallow donors.

Iron has been used as a dopant during epitaxial growth to obtain highly resistive InP. Long et al. (1984) reported resistivities as high as 2×10^8 Ω cm for Fe-doped InP grown by organometallic vapour-phase epitaxy. A gaseous Fe doping precursor is required for the growth process. Long et al. (1984) used $Fe(CO)_5$ (pentacarbonyl iron) and $Fe(C_5H_5)_2$ (ferrocene) as precursors. Ferrocene is a solid at room temperature with a relatively low vapour pressure of 1.16 Pa (Speier et al., 1986).

Epitaxial Fe-doped InP has been used in semiconductor device structures such as transistors and lasers. Temkin et al. (1988) demonstrated $Ga_{0.47}In_{0.53}As/InP$ heterostructure field-effect transistors which employed Fe-doped InP as the gate insulator. The gate metallization was deposited directly on the epitaxially grown InP:Fe layer. Excellent extrinsic transconductances of 560 mS/mm were obtained on transistors with a 1.2 μm long gate (Chen et al., 1989). A device transit frequency of 24 GHz was extrapolated from the measurements. Fe-doped InP is also used as current blocking layer in semiconductor lasers emitting at $\lambda = 1.3$ μm using ferrocene as doping precursor (Dutta, 1989).

Iron impurities redistribute themselves during epitaxial growth and during post-growth processing (see, for example, Oberstar et al., 1981). The redistribution of Fe has been discussed in further detail in Chap. 8. Such Fe redistribution can be detrimental to semiconductor devices. In order to avoid Fe redistribution effects, it is desirable to grow *undoped, semi-insulating InP*. Klein et al. (1986) showed that highly resistive InP can be obtained from nominally undoped starting material by annealing. The authors obtained a resistivity of $3.6 \times 10^5 \Omega$ cm by using an annealing temperature of 900–940 °C and a phosphorus overpressure of 6 bar. Undoped InP with semi-insulating characteristics, i.e. $>10^7$ Ω cm, was reported by Hofmann et al. (1989). The high resistivity was obtained by annealing the InP in an ampoule at 900–950 °C. Phosphorus heated to 460–480 °C was located in the same ampoule. The duration of the annealing process was 80 hours. The authors attributed the semi-insulating characteristic of the InP to a deep acceptor with a concentration $<5 \times 10^{15}$ cm^{-3}. The microscopic physical nature of the deep acceptor remains unknown.

9.4.3 Oxygen

Oxygen (O) has been intentionally incorporated into III–V semiconductors during crystal growth (Ainslie et al., 1962) or by ion implantation (Favennec, 1976; Devenaud and Favennec, 1979). As a result, highly resistive materials have been obtained. The highest resistivities obtained in GaAs:O exceed 10^8 Ω cm. Originally, O impurities in GaAs were thought to create deep donor states at an energy of 0.75 eV below the conduction band edge (Gooch et al., 1961). This level has been labeled as the EL2 center and it was generally assumed that O impurities were the microscopic origin of the EL2 center (see Sect. 9.2). However, attempts to deliberately create the 0.75 eV trap in GaAs by intentional incorporation of $Ga_2 O_3$ during epitaxial growth proved unsuccessful (Otsubo et al. 1973; Lang and Logan, 1975). Clear evidence that O is not related to the EL2 defect was obtained from combined secondary ion mass spectrometry (SIMS) and deep level transient spectroscopy (DLTS) experiments on GaAs grown by the Bridgman technique (Huber et al., 1979). Even at the present time it is not clear which states are created by O in GaAs.

Implantation of O into GaAs yields semi-insulating material regardless of the original conductivity type. Resistivities of $10^7 - 10^8$ Ω cm were obtained from n-type and p-type GaAs implanted with O (Von Neida et al., 1989; Pearton, 1990). The high resistivity persists even after annealing at moderate temperatures (e.g. 500 °C). This result shows that O in GaAs does not create just a single level in GaAs such as a deep donor level or a deep acceptor level. In order to compensate shallow donors *and* shallow acceptors, O must create several states within the fundamental gap, or alternatively, form different complexes in GaAs which have one or several trap levels.

Oxygen is frequently implanted into GaAs for device isolation purposes. In addition to the highly resistive character of O-implanted GaAs, oxygen is more stable than hydrogen (see previous section). While the defects created by H implantation can be annealed out at moderate temperatures (e.g. 500 °C), oxygen and oxygen-related complexes are much more stable in GaAs. Annealing of GaAs:O at 500 °C does not restore the original conductivity of intentionally doped GaAs (Pearton, 1990). However, annealing of O-implanted GaAs at higher temperatures of 700–900 °C decreases the resistivity drastically for Zn-, Mg-, Cd-, Si-, and S-doped material. The decrease of the resistivity suggests that the O-induced damage is annealed out. For Be-doped GaAs implanted with O, the resistivity did not decrease substantially. Pearton (1990) proposed that Be–O complexes such as $Be_{Ga} - O_{As}$ may cause the lack of conductivity recovery upon annealing.

The role of O in GaP is well understood and has been reviewed by Dean (1986). The state introduced by O into the gap of GaP is a donor level located 0.4 eV below the conduction band edge. Low temperature (77 K) luminescence measurements revealed the existence of an 1.96 eV luminescence band whose intensity strongly increases under O-rich crystal growth conditions (Gershenzon and Mikulyak, 1962). The authors concluded that the 1.96 eV luminescence band originates from a donor-to-valence-band transition and that the microscopic origin of the donor probably is O occupying a substitutional P site. The O donor energy was estimated to be 0.4 eV, since the luminescence transition (1.96 eV) occurred 0.4 eV below the band gap energy (E_g = 2.36 eV) at T = 77 K. Oxygen in GaP also has technological importance. GaP is an indirect semiconductor with a very low electroluminescence efficiency. Incorporation of small amounts of O increases the luminescence efficiency drastically. The emission from GaP:O is in the visible part of the spectrum and has a green colour.

9.5 Rare-earth impurities

Lanthanides and actinides are rare-earth elements with incompletely filled 4f and 5f electron shells. The partially filled 4f electron shell of lanthanides is surrounded by closed $5s^2$ and $5p^6$ shells shells. If III–V semiconductors are doped with rare-earth atoms, the outer s and p shells effectively screen the inner 4f shell from the crystalline environment of the host. Radiative transitions can occur between 4f orbitals of lanthanides. The energy of intra-4f-shell transitions is nearly independent of the host semiconductor due to the screening of the 4f electron shells by the outer electrons. As a consequence, the intra-4f-shell transitions are spectrally pure and exhibit narrow linewidths that are only weakly (inhomogeneously) broadened by the host semiconductor. Furthermore, the energy of the intra-4f-shell transitions is independent of temperature due to the atomic nature of the transitions.

Doping of III–V semiconductors with lanthanides is motivated by the possibility of sharp emission lines whose energies are independent of temperature. Such doped structures are potentially useful for LEDs and lasers. Rare-earth impurities used for doping include Praseodymium (Pr), Neodymium (Nd), Erbium (Er), and Ytterbium (Yb) (Ennen and Schneider, 1985). Examples are Yb in InP (Zakharenkov et al., 1981), Yb in GaAs (Ennen et al., 1983a), Yb in GaP (Kasatkin et al., 1981a), Er in GaAs (Klein et al., 1991), Er in GaP and InP (Ennen et al., 1983b), Pr in GaP (Kasatkin et al., 1981b). Of special interest is the lanthanide Er which emits at a wavelength of approximately 1.55 μm. This corresponds to the wavelength of minimum loss in silica fibers. Er is also used for doping silica fibers (Miniscalco, 1991). Such Er-doped fibers serve as distributed optical amplifiers.

It has been assumed that Er is threefold positively charged (Er^{3+}) and occupies cation sites in III–V semiconductors (Ennen and Schneider, 1985). The electronic intra-$4f^{11}$-shell transition is then $^4I_{13/2} \rightarrow {}^4I_{15/2}$. More recently, a study on Er-ion-implanted GaAs showed that Er can exist in a number of different configurations including Er^{2+}, Er^{3+}, tetrahedrally symmetric interstitial sites, and as complexes of interstitial Er with defects or impurities (Klein et al., 1991). The study also found that the Er^{3+} photoluminescence line is comprised of a superposition of lines from different Er sites.

The luminescence intensity of rare-earth-doped III–V semiconductors is highest at low temperatures. The luminescence yield strongly decreases at higher temperatures, e.g. at room temperature. This phenomenon is known as *temperature quenching* of the luminescence. Taguchi et al. (1992) showed that the temperature quenching of the luminescence in Yb-doped InP is due to transfer to energy from the excited Yb atom back to an electronic excitation of the InP. The authors argued that electron–hole pairs created in InP by above-bandgap illumination excite the Yb atom via an intermediate bound excitonic state. At high temperatures, the intermediate excitonic state is formed by deexcitation of Yb atoms. Furthermore, the luminescence of rare-earth-doped III–V semiconductors does not increase linearly with the rare-earth concentration. At high concentrations, the luminescence intensity decreases, which is known as *concentration quenching* of the luminescence. The concentration quenching is probably due to the formation of complexes and pairs involving the rare-earth ions.

10

Doping in heterostructures, quantum wells, and superlattices

This chapter is devoted to the role of impurities in semiconductor structures which consist of different types of semiconductor materials. Semiconductor heterostructures, quantum wells and superlattices are structures in which the individual layers have spatial dimensions comparable to the carrier de Broglie wavelength. As a consequence, quantum effects cannot be neglected in such small semiconductor structures. Furthermore, the spatial dimensions of such structures can be comparable to the Bohr radius of impurities. As a consequence, the characteristics of impurities, e.g. ionization energy and wave function, are changed. The area of quantum semiconductor physics has gained much interest since the 1970s. The physical properties of several semiconductor quantum structures are closely related to the doping and to residual impurities in such structures. Among these structures are selectively doped heterostructures, doping superlattices, doped quantum wells, and doped quantum barriers, which will be discussed in this chapter.

10.1 Selectively doped heterostructures

Selectively doped heterostructures are structures which consist of a doped wide-gap semiconductor and an undoped narrow-gap semiconductor. Selectively doped heterostructures were first realized by Stormer et al. (1978) and Dingle et al. (1978) in an attempt to reduce scattering of carriers by ionized impurities. The electron mobilities obtained in $Al_x Ga_{1-x} As/GaAs$ heterostructures at low temperatures can exceed 10^7 cm^2/V s (Pfeiffer et al., 1989a).

The band diagram of a selectively doped n-type heterostructure is shown in Fig. 10.1. The structure consists of a doped wide-gap semiconductor (left side of

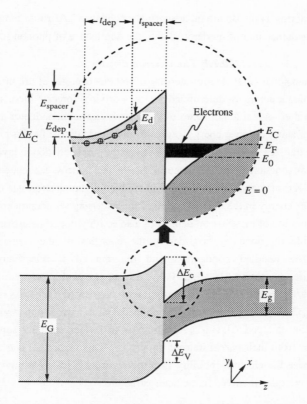

Fig. 10.1. Band diagram of a selectively doped heterostructure consisting of a doped wide-gap semiconductor and an undoped narrow-gap semiconductor. Due to the lower conduction band energy, electrons transfer from their parent donors to the narrow-gap semiconductor.

Fig. 10.1) and an undoped narrow-gap semiconductor (right side of Fig. 10.1). Since the conduction band edge of the narrow-gap semiconductor is assumed to be lower in energy, electrons originating from donors in the wide-gap semiconductor transfer to the narrow-gap semiconductor. The transferred electrons form a quantized, two-dimensional electron gas (2DEG) located at the interface. The two-dimensional character of the electron gas as well as the quantization of motion of the electrons in the z-direction (see Fig. 10.1) was demonstrated by Störmer et al. (1979). Electrons are spatially separated from their parent ionized impurities. As a consequence, ionized impurity scattering is reduced in such heterostructures as compared to a doped bulk semiconductor. The electron mobility is especially enhanced at low temperatures where ionized impurity scattering is the dominant scattering mechanism. At room temperature, the mobility enhancement is moderate due to the dominance of phonon scattering.

10.1.1 The electron density

The two-dimensional (2D) electron density of the electron gas at the interface of the two semiconductors is determined by two driving forces. First, electrons transfer from their parent donor states to the narrow-gap semiconductor due to the availability of states at lower energy. Second, an electric dipole field is created by the electron transfer whose charges consists of the depleted donor layer in the wide-gap semiconductor and the electron gas in the narrow-gap material. The transfer of electrons continues until the Fermi level of the electron gas coincides with the donor energy in the neutral region of the wide-gap semiconductor.

The calculation of the electron density of a selectivity doped heterostructure is most conveniently done by considering the energies of the electron states involved. The detailed conduction-band diagram of a selectively doped heterostructure is shown in Fig. 10.1. As an example, we consider an $Al_x Ga_{1-x} As/GaAs$ heterostructure consisting of a doped $Al_x Ga_{1-x} As$ region, an undoped $Al_x Ga_{1-x} As$ spacer, and an undoped GaAs narrow-gap region. The purpose of the undoped $Al_x Ga_{1-x} As$ spacer is to further spatially separate the free electrons from their parent ionized impurities. This separation was shown to further increase the electron mobility (Stormer et al., 1981c). We now consider the different energies involved in the heterostructure.

(i) **Donor ionization energy,** E_d: The thermal ionization energy of the shallow donor in GaAs and $Al_x Ga_{1-x} As$ is approximately 5 meV. At low temperatures, the Fermi level coincides with the donor level in the undepleted $Al_x Ga_{1-x} As$ side of the heterostructure.

(ii) **Depletion energy,** E_{dep}: Assuming a three-dimensional (3D) doping concentration N_D, the 2D density of transferred electrons is given by

$$n_{2DEG} = N_D \, t_{dep} \tag{10.1}$$

where t_{dep} is the (unknown) thickness of the depletion region. Any residual acceptor concentration is neglected. The magnitude of the electric field at the end of the depletion region can be obtained from Gauss's law

$$\mathcal{E} = \frac{e}{\varepsilon} N_D \, t_{dep} = \frac{e}{\varepsilon} n_{2DEG} . \tag{10.2}$$

The energy drop in the depletion region is then given by

$$E_{\text{dep}} = \frac{1}{2} e \, \mathcal{E} \, t_{\text{dep}} = \frac{e^2}{2\varepsilon} \frac{n_{\text{2DEG}}^2}{N_D} . \tag{10.3}$$

In the case of $N_D \to \infty$, i.e. for δ-doping, the depletion region thickness and energy drop approach zero, i.e. $t_{\text{dep}} \to 0$ and $E_{\text{dep}} \to 0$.

(iii) **Spacer energy, E_{spacer}:** The energy drop in the spacer region is given by

$$E_{\text{spacer}} = e\mathcal{E} \, t_{\text{spacer}} = \frac{e^2}{\varepsilon} \, n_{\text{2DEG}} \, t_{\text{spacer}} \tag{10.4}$$

where t_{spacer} is the thickness of the spacer region.

(iv) **Conduction band discontinuity, ΔE_c:** The conduction band discontinuity is a materials parameter. For the material system $Al_x Ga_{1-x} As/GaAs$ the conduction band discontinuity is given by

$$\Delta E_c \cong \frac{70}{100} \Delta E_g = \frac{70}{100} \, 1.247 \, \text{eV} \, x_{\text{Al}} \tag{10.5}$$

where it is assumed that 70% of the bandgap discontinuity occurs in the conduction band. The factor $1.247 \, \text{eV} \, x_{\text{Al}}$ expresses the change of the $Al_x Ga_{1-x} As$ band gap with Al mole fraction (Casey and Panish, 1978). For a 32% Al mole fraction, the conduction band discontinuity calculated from Eq. (10.5) equals 279 meV.

(v) **Subband energy, E_0:** The ground-state energy of the triangular well can be conveniently obtained by a variational calculation using the trial function

$$\psi(z) = A \, z \, e^{-\alpha z} \tag{10.6}$$

where α is the trial parameter and A is a normalization parameter. The wave function vanishes at $z = 0$, i.e. at the interface, as schematically shown in Fig. 10.2. The wavefunction decays exponentially in the triangular GaAs side of the barrier. The normalization condition $\langle \psi | \psi \rangle = 1$ yields the normalization constant $A = 2\alpha^{3/2}$ and thus the normalized wave function

$$\psi(z) = 2\alpha^{3/2} \, z \, e^{-\alpha z} \tag{10.7}$$

which is also called the Fang–Howard wave function (Fang and Howard, 1966).

Calculation of the energy expectation value for this wave function yields

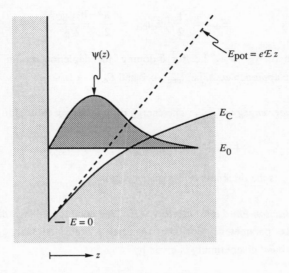

Fig. 10.2. Fang–Howard wave function, $\psi(z) \propto z \exp(-\alpha z)$, at the interface of an $Al_xGa_{1-x}As/GaAs$ heterostructure.

$$\langle E \rangle = \langle \psi | H | \psi \rangle$$

$$= \langle \psi \left| -\frac{\hbar^2}{2m^*} \frac{\partial^2}{\partial z^2} + E_{pot}(z) \right| \psi \rangle$$

$$= e\mathcal{E} \frac{3}{2\alpha} + \frac{\hbar^2}{2m^*} \alpha^2 \tag{10.8}$$

where the potential energy $E_{pot}(z) = e\mathcal{E}z$. Minimizing the expectation value of the energy yields the trial parameter α according to

$$\alpha = \left[\frac{3}{2} e\mathcal{E} \frac{m^*}{\hbar^2} \right]^{1/3}. \tag{10.9}$$

Insertion of the trial parameter α into Eq. (10.8) yields the ground-state energy of the triangular potential well

$$E_0 = \langle E \rangle_{\min} = \frac{3}{2} \left[\frac{3}{2} e\mathcal{E} \frac{\hbar}{\sqrt{m^*}} \right]^{2/3} \tag{10.10}$$

which can also be expressed in terms of the 2D electron density

$$E_0 = \frac{3}{2} \left[\frac{3}{2} \frac{e^2}{\varepsilon} n_{2\text{DEG}} \frac{\hbar}{\sqrt{m^*}} \right]^{2/3}. \tag{10.11}$$

(vi) Degeneracy energy of 2DEG, $E_\text{F} - E_0$. Due to the finite density of states, the Fermi level increases above the ground-state energy E_0 at high free carrier densities. The energy $E_\text{F} - E_0$ is obtained as

$$E_\text{F} - E_0 = n_{2\text{DEG}} / \rho_{\text{DOS}}^{2\text{D}} \tag{10.12}$$

where $\rho_{\text{DOS}}^{2\text{D}}$ is the two-dimensional density of states given by $\rho_{\text{DOS}}^{2\text{D}} = m^* / \pi \hbar^2$. The equation is valid for degenerate electron systems, where the Fermi–Dirac distribution can be approximated by a step function. The corresponding equation for the non-degenerate limit is given in Chap. 3.

The total energy balance of the heterostructure can be written as (see Fig. 10.1 for illustration)

$$E_\text{d} + E_\text{dep} + E_\text{spacer} + E_0 + (E_\text{F} - E_0) = \Delta E_\text{c} \tag{10.13}$$

where it is assumed that the Fermi energy is the same on both sides of the interface, i.e. the system is in electronic equilibrium. All the energies in Eq. (10.13) can be expressed as a function of $n_{2\text{DEG}}$ which is then the only variable in the equation. It is not possible to explicitly solve the equation for $n_{2\text{DEG}}$. However, the solution for $n_{2\text{DEG}}$ can be easily obtained by an iterative process.

Fully self-consistent calculations of the electron density in selectively doped heterostructures have been reported in the literature (Ando, 1982a and 1982b; Stern and Das Sarma, 1984; Vinter, 1984). In such self-consistent calculations, the Schrödinger and Poisson equations are solved in an iterative process and many-body effects may be taken into account. Nevertheless, the results obtained from the approximate calculation described above compare favorably with the much more elaborate self-consistent calculations. In self-consistent calculations, the solution of Poisson's equation is straightforward whereas the accurate solution of Schrödinger's equation is more elaborate. Therefore, several groups used approximate solutions of the Schrödinger equation, including the variational solution (Stern, 1983; Fang and Howard, 1966) and the WKB approximation (Ando, 1985).

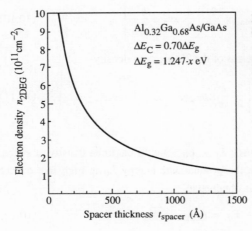

Fig. 10.3. Calculated electron density in selectively doped $Al_xGa_{1-x}As/GaAs$ hetero-structures as a function of spacer thickness, t_{spacer}.

As an example, the carrier density of the 2DEG calculated from Eq. (10.13) is shown in Fig. 10.3 as a function of the spacer thickness. The Al mole fraction used is $x = 32\%$ which leads to a conduction band discontinuity of $\Delta E_c = 279$ meV. Other parameters used are $\varepsilon/\varepsilon_0 = \varepsilon_r = 13.1$ and $m^* = 0.067 m_0$. The structure is assumed to be δ-doped which results in a zero depletion energy $E_{dep} = 0$. Figure 10.3 reveals that the electron density decreases for large spacer thicknesses. The electron density can also be calculated for different doping concentrations, N_D, which enter E_{dep} in Eq. (10.13). The values of n_{2DEG} are lower for homogeneous doping as compared to the δ-doped case shown in Fig. 10.3. Delta-doped heterostructures were shown to have higher electron densities (Schubert et al., 1987b) as well as higher electron mobilities (Schubert et al., 1989b) as compared to their homogeneously doped counterparts.

10.1.2 The electron mobility

In bulk semiconductors, the dominant scattering mechanism at low temperatures is ionized impurity scattering, which is strongly reduced in selectively doped heterostructures. The spatial separation of electrons from their parent ionized impurities reduces ionized impurity scattering making possible electron mobilities exceeding 10^6 cm^2/V s at low temperatures. In order to elucidate the mobility enhancement we will first discuss the role of impurity scattering in bulk semiconductors.

Ionized impurity scattering in semiconductors is governed by the same physical principles as Rutherford scattering. In both scattering processes the trajectory of a charged particle is diverted by the interaction with another charged particle. The coulombic interaction of two charged particles is strongest for a small distance between the interacting particles and a long interaction time. Such a long interaction time is given for slowly moving electrons, that is for a non-degenerate carrier gas at low temperatures. The low-temperature mobility is therefore a measure of the impurity and defect content in non-degenerately doped semiconductors.

The mobility due to ionized impurities was calculated by Conwell and Weisskopf (1950) and later by Brooks and Herring (Brooks, 1955). A review on ionized impurity scattering was given by Chattopadhyay and Queisser (1981). While Conwell and Weisskopf (CW) used unscreened Coulomb potentials, i.e. $V = e/4\pi\varepsilon r$, Brooks and Herring (BH) used screened Coulomb potentials, i.e. $V = (e/4\pi\varepsilon r)e^{-\lambda r}$, where λ is the inverse screening length.

The ionized impurity mobility in the Conwell–Weisskopf approximation is given by

$$\mu_{CW} = \frac{128\sqrt{2\pi}\,\varepsilon^2 (kT)^{3/2}}{N_{II}\, e^3 \sqrt{m}} \left[\ln\left(1 + \frac{(12\pi\varepsilon kT^2)}{N_{II}^{2/3}\, e^4} \right) \right]^{-1} \quad (10.14)$$

where N_{II} is the concentration of ionized impurities.

The Brooks–Herring approach, which includes the screening of impurities by free carriers, yields for the ionized impurity mobility

$$\mu_{BH} = \frac{128\sqrt{2\pi}\,\varepsilon^2 (kT)^{2/3}}{N_{II}\, e^3 \sqrt{m}} \left[\ln(1 + \beta_{BH}^2) - \frac{\beta_{BH}^2}{1 + \beta_{BH}^2} \right]^{-1}. \quad (10.15)$$

The parameter β_{BH} is given by

$$\beta_{BH} = 2\,\frac{m}{\hbar}\left[\frac{2}{m}3kT \right]^{1/2} r_D \quad (10.16)$$

where $r_D = (\varepsilon kT/e^2 n)^{1/2}$ is the Debye screening length, and n is the free carrier concentration. For degenerately doped semiconductors, the Debye screening length must be replaced by the Thomas–Fermi screening length and the thermal energy must be replaced by the Fermi energy.

Both the CW and the BH approach predict a temperature dependence of

$$\mu \propto T^{3/2}, \quad (10.17)$$

i.e. an increasing mobility with temperature. For a non-degenerate semiconductor the experimental mobility indeed approaches zero for $T \to 0$. In *degenerately* doped semiconductors, the Fermi velocity is larger than the thermal velocity at sufficiently low-temperatures and the mobility is expected to remain constant in the low-temperature regime. Note that a qualitative discrepancy exists between the CW and the BH approximation with regard to their density dependence. While the CW-mobility decreases continuously at high concentrations, the BH-mobility first decreases with impurity density but then increases again at very high impurity concentrations (Seeger, 1982). The increase in the BH-mobility is due to screening of the ionized impurity potentials.

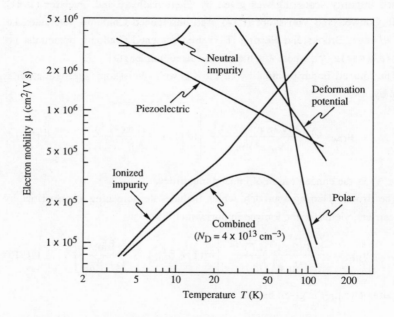

Fig. 10.4. Calculated electron mobilities due to different scattering mechanisms and combined mobility inferred from Matthiessen's rule in high purity GaAs ($N_D = 4 \times 10^{13}\,\text{cm}^{-3}$) as a function of temperature (after Wolfe et al., 1970).

The calculated electron mobility in moderately doped bulk n-type GaAs is shown in Fig. 10.4 (Wolfe et al. 1970). In addition to ionized impurity scattering, neutral impurity, piezoelectric, deformation potential, and polar phonon scattering are shown. The combined mobility is also included and is obtained from the sum of the individual scattering mechanisms according to $\mu^{-1} = \sum_i \mu_i^{-1}$ (Matthiessen's rule).

The major scattering mechanisms in selectively doped $Al_x Ga_{1-x} As$/GaAs heterostructures were reviewed by Stormer (1983) and are given by bulk and interface phonon-scattering, alloy scattering in the $Al_x Ga_{1-x} As$, neutral and ionized impurity scattering in the GaAs (unintentional impurities), remote impurity scattering (intentional impurities), interface roughness scattering, intersubband scattering, and real-space scattering. In the following, the scattering mechanisms will be discussed.

(1)   Scattering by bulk phonons and interface phonons occurs at all finite lattice and electron temperatures. However, since the two materials $Al_x Ga_{1-x} As$ and GaAs are very similar in density and dielectric constant, interface phonons are weak and electron scattering via these phonons is negligible. Calculations of the scattering rate due to bulk phonons show similar scattering rates for bulk material and heterostructures (Price 1981, 1984a, 1984b; Hess, 1979). Scattering by optical phonons dominates the mobility in selectivity doped heterostructures at temperatures ≥ 100 K. Above this temperature the mobility in selectivity doped heterostructures coincides with the mobility in high-purity GaAs.

(2)   Alloy scattering due to random distribution of Al and Ga in the alloy $Al_x Ga_{1-x} As$ is a minor scattering mechanism in selectively doped heterostructures. The random cation distribution leads to potential fluctuations which scatter charge carriers. The penetration of the electron wave function into the $Al_x Ga_{1-x} As$ is small (note that the penetration in the Fang–Howard wave function is assumed to be zero). Consequently, the scattering due to alloy fluctuations plays a minor role. Alloy scattering in bulk semiconductors increases with the Al mole fraction and reaches a maximum at $x = 0.5$. Experimental reports of the mobility in selectively doped heterostructures suggested an increase of the mobility with Al mole fraction (Stormer et al., 1981a) as well a decrease of the mobility (Drummond et al., 1982). The optimum Al content for selectively doped heterostructures was reported to be $x = 32\%$ (Pfeiffer, 1990).

(3)   Neutral and ionized unintentional impurities in the undoped GaAs and $Al_x Ga_{1-x} As$ layer contribute to the scattering in selectively doped heterostructures. However, the scattering by neutral impurities is much weaker than the scattering by ionized impurity atoms due to the lack of Coulomb charge. The background concentration in the undoped GaAs can be in the low 10^{13} cm^{-3} range. Higher background concentrations (e.g. 10^{16} cm^{-3}) were demonstrated to reduce the electron mobility drastically

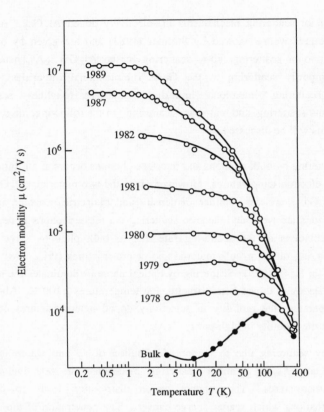

Fig. 10.5. Electron mobilities in selectively doped $Al_xGa_{1-x}As/GaAs$ heterostructures (open circles) as a function of temperature for different years. The mobility of bulk GaAs ($N_D = 2 \times 10^{17}$ cm^{-3}, solid circles) is shown for comparison (Pfeiffer et al., 1989a).

(Drummond et al., 1981a).

(4)   Remote impurity scattering occurs due to the Coulomb potentials of the intentional dopants in the $Al_xGa_{1-x}As$. Remote impurity scattering can be reduced by increasing spacer thicknesses, i.e. be moving the impurities further away from the electron gas. However, the increasing spacer thickness reduces the electron density (as shown in Fig. 10.3), which again results in a lower mobility due to a reduction in screening. Remote ionized impurity scattering dominates the mobility for small spacer thicknesses ($t_{spacer} < 300$ Å). On the other hand, remote ionized impurity scattering becomes irrelevant for very large spacer thicknesses. The highest

mobilities are obtained with spacer thicknesses on the order of 600 Å – 700 Å (Harris et al., 1986; Pfeiffer et al., 1989b). Remote impurity scattering in non-degenerate systems has the $\mu \propto T^{3/2}$ dependence. In a highly degenerate system, the Fermi velocity is larger than the thermal velocity. Furthermore, the Fermi velocity is independent of temperature. As a consequence, the mobility saturates at low temperatures. Walukiewicz et al. (1984) showed that for typical electron densities the mobility is temperature-independent for $T < 50$ K. However, for very high mobility samples the temperature range of constant mobility is shifted towards lower temperatures (see Fig. 10.5).

(5) Interface roughness scattering occurs due to the roughness of the interface between the $Al_x Ga_{1-x} As$ layer and the GaAs. The roughness of the interface depends on the growth conditions such as the growth temperature and the growth interruption time at the interface. Using 120 s growth interruptions at an $Al_x Ga_{1-x} As$/GaAs interface smooth islands of several 100 Å diameter can be obtained (Kopf et al., 1991). Nevertheless, even the existence of monolayer flat islands with a diameter of several hundred ångströms does not necessarily imply a reduction of interface roughness scattering: The Fermi wavelength is approximately 350 Å (at $n_{2DEG} = 5 \times 10^{11}$ cm^{-2}), i.e. comparable to the island sizes achievable by growth interruptions. The lateral correlation length of interface roughness in MBE grown GaAs/AlAs quantum wells was investigated by Noda et al. (1991) using mobility measurements. Lateral correlation lengths as large as 300 Å were deduced.

(6) Intersubband scattering becomes relevant at either sufficiently high temperatures or at high densities at which a second subband becomes populated. In selectively doped $Al_x Ga_{1-x} As$/GaAs heterostructures, the second subband becomes populated at an electron density of 6×10^{11} – 7×10^{11} cm^{-2} (Ando and Mori, 1979). Intersubband scattering does not open a new scattering channel (instead, all the above scattering mechanisms can contribute) but represents an increase in scattering probability due to an increase in phase space. At low temperatures when the second subband of the 2DEG becomes populated, quasi-elastic scattering can occur from one subband to the other giving rise to a sudden drop in mobility (Stormer, 1983). The drop in mobility due to intersubband scattering was estimated theoretically (Mori and Ando, 1980) and measured experimentally to be as large as 30% (Stormer et al., 1981a,b).

(7)   Real-space transfer scattering can occur at very high electric fields and is due to the transfer of electrons from the GaAs channel to the $Al_x Ga_{1-x} As$ layer. The mobility in the $Al_x Ga_{1-x} As$ is much lower than in the GaAs layer due to alloy scattering. As a result, the mobility is expected to drop dramatically at the onset of this transfer. Hess et al. (1979) who considered real-space transfer in semiconductor heterostructures concluded that negative differential conductivity can occur due to this transfer.

The electron mobility in $Al_x Ga_{1-x} As/GaAs$ heterostructure is shown in Fig. 10.5 as a function of temperature for the years 1978 through 1989 (Pfeiffer et al., 1989a). The improvement of the mobility over the years reflects (i) the introduction of a spacer layer, (ii) improved vacuum in the MBE process, (iii) use of δ-doping, (iv) higher purity sources, and (v) improved substrate quality and preparation.

10.2 Doping aspects of selectively doped heterostructures

The distribution and concentration of dopants in selectively doped heterostructures determines many of its transport properties. To achieve, for example, high mobilities, thick spacer layers are required in order to minimize remote ionized impurity scattering. To achieve, for example, high electron densities, high doping concentrations are required. In this section, the *optimum dopant distribution* for (i) maximum electron density in heterostructures and (ii) maximum electron mobility in heterostructures is analyzed. Furthermore, the differences in *normal* and *inverted heterostructures* will be discussed. The properties of normal heterostructures ($Al_x Ga_{1-x} As$ grown on top of GaAs) are different from inverted heterostructures (GaAs grown on top of $Al_x Ga_{1-x} As$). The differences are mainly due to the different doping distributions in the structures.

10.2.1 Heterostructures with maximum electron density

High electron densities in selectively doped heterostructures are most desirable for field-effect transistors. The transconductance (amplification) of field-effect transistors increases with the electron density. The electron density increases with decreasing spacer thickness and reaches a maximum for zero spacer thickness. The effect of different doping distributions on mobility must be studied in samples with the *same spacer thickness*.

The conduction band diagram of two selectively doped heterostructures with the same spacer thickness is shown in Fig. 10.6. The figure depicts (a) a δ-doped and (b) a homogeneously doped heterostructure, respectively. The δ-doped heterostructure can be considered to be the limit of homogeneous doping for

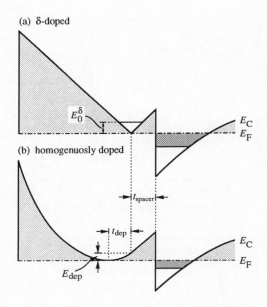

Fig. 10.6. Conduction band diagram of (a) a δ–doped and (b) a homogeneously doped heterostructure. The thickness of the undoped spacer region is the same in both heterostructures. The selectively δ-doped heterostructure yields the highest densities of the two-dimensional electron gas.

$N_D \to \infty$. This suggests that δ-doping results in the highest electron density. The thickness of the depletion region, t_{dep}, and the associated energy drop (see Eq. 10.3), E_{dep}, approach zero in the limiting case of δ-doping. In order to maximize the free carrier concentration, the terms on the left-hand side of Eq. (10.13) need to be minimized whereas the term on the right-hand side of Eq. (10.14) (i.e. ΔE_c) needs to be maximized. Delta-doping eliminates the term 'E_{dep}' and therefore maximizes the free carrier concentration of the heterostructure.

Delta-doped heterostructures have another advantage over their homogeneously doped counterparts: Due to the narrow V-shaped potential well generated by δ-doping in the $Al_x Ga_{1-x} As$, size-quantization occurs also in the $Al_x Ga_{1-x} As$ (Schubert et al., 1987b). The quantization energy, E_0^{δ}, increases the energy of the lowest electron state in the $Al_x Ga_{1-x} As$. As a consequence, the difference in energy between the $Al_x Ga_{1-x} As$ and the GaAs is enhanced which results in a further transfer of electrons into the narrow-gap GaAs layer. Size quantization in the homogeneously doped heterostructure is smaller or may not

(a)

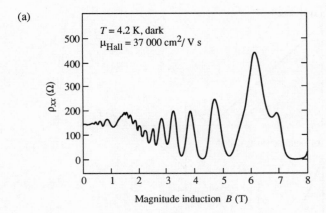

(b)

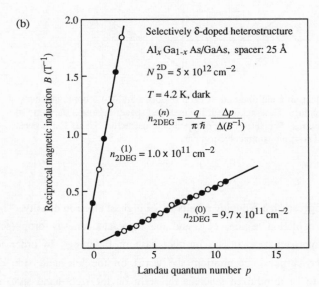

Fig. 10.7. (a) Low-temperature magnetoresistance of a selectively δ-doped heterostructure. (b) Evaluation of the two distinct periods of the Shubnikov–de Haas oscillation yields a concentration of 9.7×10^{11} and $1 \times 10^{11} \, \text{cm}^{-2}$ for the lowest and first excited subband, respectively.

occur at all at low doping concentrations (Vinter, 1983).

The highest electron densities in $Al_x Ga_{1-x} As/GaAs$ are obtained in δ-doped heterostructures. The measured densities exceed $10^{12} \, \text{cm}^{-2}$ (Schubert et al., 1987b) even at low temperatures. Magnetoresistance measurements of a selectively δ-doped heterostructure are shown in Fig. 10.7. The oscillations of the magnetoresistance are called Shubnikov–de Haas oscillations and are due to additional quantization of the energy levels in the magnetic field. The oscillations are periodic in $1/B$. Two clearly distinct periods of the oscillation are observed in Fig. 10.7. The two periods are due to two electronic subbands that are occupied at low temperatures. The concentrations within the two subbands can be evaluated by plotting the Landau quantum numbers of the minima (solid circles) and maxima (open circles) versus reciprocal magnetic field, as shown in Fig. 10.7(b). The slope of this plot yields the concentrations of 9.7×10^{11} cm^{-2} and 1×10^{11} cm^{-2} for the lowest and first excited subbands, respectively. The total concentration is $n_{2DEG} = 1.07 \times 10^{12}$ cm^{-2} at low temperatures. The corresponding mobility is $\mu = 37\ 000$ cm^2/V s. At room temperature, a concentration of $n_{2DEG} = 1.7 \times 10^{12}$ cm^{-2} and a mobility of 8900 cm^2/V s have been obtained from Hall measurements on the same sample.

High densities of the electron gas are advantageous in selectively doped heterostructure transistors. High electron densities result in high transconductances, i.e. high amplification of transistors (Mimura, 1983).

10.2.2 Heterostructures with maximum mobility

The doping distribution influences the carrier mobility via remote impurity scattering. It is an interesting question which doping configuration minimizes remote impurity scattering and results in the highest mobilities. In order to compare different doping configurations it is necessary to compare configurations which result in the *same carrier density*. This restriction is necessary since the carrier mobility depends on the carrier density.

Two different doping distributions which result in the same electron-gas density are shown in Fig. 10.8. The centroid of the two doping distributions has the same distance z_c from the interface. If the thickness of the doped layer, z_d, is reduced and simultaneously the doping concentration, N_D, is increased such that the product is a constant ($z_d N_D = \text{const} = n_{2DEG}$), the free carrier concentration, n_{2DEG}, remains constant as well. It can be easily verified from Poisson's equation that such doping distributions result in the *same* potential and the *same* electric field at the interface, i.e. in the same free carrier density.

In order to estimate the effect of remote ionized impurity scattering, Schubert et al. (1989b) calculated the potential fluctuations occurring at the interface due to random doping impurity distribution. The authors found that potential fluctuations are minimized if the thickness of the doping distribution approaches zero, i.e. for δ-doping in the Al$_x$Ga$_{1-x}$As. The minimization of potential fluctuations is equivalent to a maximization of remote ionized impurity mobility, since the scattering matrix element is proportional to the magnitude of the

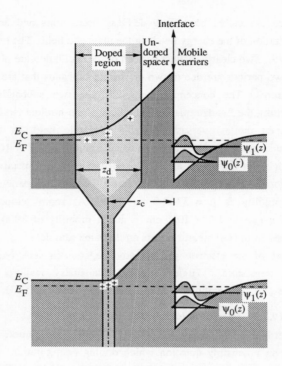

Fig. 10.8. Schematic illustration of the conduction band diagram of a selectively doped heterostructure for different doped layer thicknesses and doping concentrations. The free carrier density remains constant upon changing the doped layer thickness.

scattering potential. The electron mobility monotonically decreases with a decreasing magnitude of the potential fluctuations. This monotonical relationship is valid even in the presence of screening.

Experimental Hall mobilities (left ordinate) and carrier densities (right ordinate) of selectively doped heterostructures with different thicknesses of the doped region are shown in Fig. 10.9. The thickness of the doped region, z_d, is changed, while the distance of the doped region centroid, z_c, remains constant as shown in the inset. The electron mobilities at 4.2 K range between 860 000 cm^2/V s for $z_d = 800$ Å and 4.53×10^6 cm^2/V s for $z_d \to 0$ Å. Mobilities are measured in the dark after illumination. Electron mobilities increase as z_d decreases (see Fig. 10.9) with the highest mobilities obtained as z_d approaches zero. The mobility follows the qualitative trend predicted by the theoretical model.

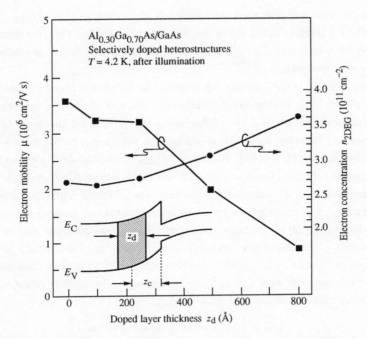

Fig. 10.9. Electron mobility (left ordinate) and electron density (right ordinate) versus doped layer thickness, z_d, of selectively doped heterostructures grown by molecular-beam epitaxy. Samples were grown with $z_d = 0$, 100, 250, 500, and 800 Å, with $z_c = 400$ Å = constant. The densities and mobilities were measured after illumination. The values measured before illumination are typically 15–35% lower but follow the same qualitative trend.

Figure 10.9 further illustrates the carrier density as a function of the doped layer thickness. An increase of n_{2DEG} is observed as z_d increases, which is not expected if only shallow donors are taken into account. A probable cause for this increase is the coexistence of shallow and deep donors in the $Al_x Ga_{1-x} As$: As z_d increases, more deep donors are elevated above the Fermi level in the doped region close to the interface, as illustrated in Fig. 10.8. Consequently, the deep donors nearest the interface become ionized and the mobile carrier density is enhanced. This concentration enhancement is expected for carriers in the dark. It is maintained even after illumination in samples with large z_d, due to photoionization of deep donors close to the interface.

In addition to remote donors in the $Al_x Ga_{1-x} As$, unintentional shallow and deep impurities in the vicinity of the interface reduce the carrier mobility. The influence of any such static imperfection is reduced by screening. The mobility increases with concentration according to the relation $\mu \propto n^{\beta}_{2DEG}$ with $0.5 < \beta < 1.5$ (Pfeiffer et al., 1989a, Price, 1984a and 1984b). Thus, the decrease of mobility with increasing z_d (see Fig. 10.9) would be even more pronounced, if n_{2DEG} were independent of z_d.

Free carriers at the interface are scattered by (i) remote ionized dopants as well as by (ii) unintentional background impurities in the narrow-gap semiconductor. Scattering by remote impurities is reduced for large spacer thicknesses. However, large spacer thicknesses reduce the carrier density *and* therefore the carrier mobility since $\mu \propto n^{\beta}_{2DEG}$. Furthermore, scattering by background impurities becomes more relevant at low carrier densities due to less efficient screening of background impurities. Thus, ionized impurity scattering dominates at low temperatures for small spacer thicknesses (remote intentional impurities) as well as for large spacer thicknesses (unintentional background impurities). The maximum mobility is achieved for intermediate spacer thicknesses (Harris et al., 1986; Pfeiffer et al., 1989b). At low background impurity concentrations of 2×10^{13} cm^{-3}, the maximum mobility was obtained at a spacer thickness of 750 Å (Pfeiffer et al., 1989b).

10.2.3 The inverted interface problem

Inverted heterostructures are known to yield much lower mobilities as compared to regular heterostructures. *Regular heterostructures* are heterostructures which have the doped $Al_x Ga_{1-x} As$ grown on top of the GaAs. *Inverted heterostructures* are heterostructures which have the undoped GaAs grown on top of the doped $Al_x Ga_{1-x} As$. Experimental results of electron mobilities indicated that the mobilities are much lower in inverted heterostructures as compared to regular heterostructures (Drummond et al., 1981b). Despite a wide variety of growth conditions and different designs of inverted heterostructures, the mobilities in such inverted structures are significantly lower than in their regular counterparts (Cho et al., 1988).

The low mobilities obtained in inverted heterostructures have been attributed to the microscopically rough surface morphology of $Al_x Ga_{1-x} As$ (see, for example, Alexandre et al., 1985) and to dopant segregation from the doped $Al_x Ga_{1-x} As$ into the GaAs channel region. The problem of dopant segregation is illustrated in Fig. 10.10, which shows a GaAs quantum well embedded in two $Al_x Ga_{1-x} As$ barriers. The $Al_x Ga_{1-x} As$ barriers are δ-doped 400 Å below and above the quantum well interface. The secondary ion mass spectrometry (SIMS) profile of the Si dopants of the quantum well is also shown in Fig. 10.10. The figure illustrates (i) a pronounced Si tail towards the growth surface and (ii) a shift of the Si peak concentration towards the surface.

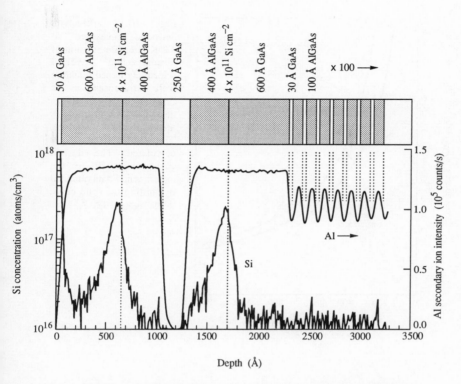

Fig. 10.10. Si and Al secondary ion mass spectrometry (SIMS) profile of an $Al_xGa_{1-x}As/GaAs$ quantum well structure δ-doped on both sides of the GaAs well.

The movement of Si along with the growing surface (see Chap. 8 for further discussion) results in an effectively increased spacer thickness for regular structures and an effectively decreased spacer thickness for inverted structures as compared to the nominal spacer thickness. As a consequence, the densities of the two-dimensional electron gas is expected to be higher for inverted structures as compared to the nominal spacer thickness. The electron concentrations of regular and inverted heterostructures are shown in Fig. 10.11 as a function of the nominal spacer thickness. The inverted heterostructure has a higher electron density as compared to the regular heterostructure for the same nominal spacer thickness (Pfeiffer et al., 1991).

In order to achieve comparable mobilities in inverted heterostructures, a larger spacer thickness must be used in order to compensate for the Si migration towards the surface. The mobilities of regular and inverted $Al_xGa_{1-x}As/GaAs$

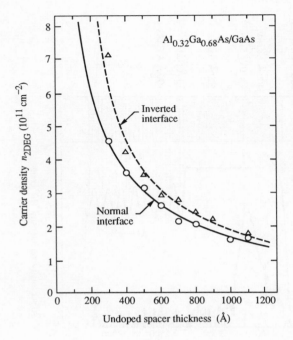

Fig. 10.11. Electron densities measured at 4.2 K after illumination of normal and inverted selectively doped quantum well structures as a function of spacer thickness. Circles and triangles are experimental values. The solid curve is calculated. The dashed curve is a fit to the experimental data obtained by shifting the solid curve to the right by 130 Å.

heterostructures are shown in Fig. 10.12 as a function of the nominal spacer thickness. The electron mobility of the inverted structure reaches its maximum of 2.4×10^6 cm^2/V s at a nominal spacer thickness of 800 Å. The regular heterostructure reaches maximum mobility at a smaller nominal spacer thickness of 700 Å.

The Si migration problem in inverted heterostructures can only be partly solved by the increased spacer thickness. The SIMS profile of Fig. 10.10 also shows an *exponential tail* of the Si concentration towards the semiconductor surface. This tail extends into the undoped GaAs channel region. Any unwanted impurities in the undoped GaAs channel region have a deleterious effect on the electron mobility. Such residual impurities in the GaAs channel region lower the electron mobility in the optimized inverted heterostructure. As a result, the electron mobility in the inverted structure is lower as compared to the regular structure (see Fig. 10.12).

10.3 Doping superlattices

Semiconductor doping superlattices are structures which contain alternately n-type and p-type doped regions. Different from compositional superlattices,

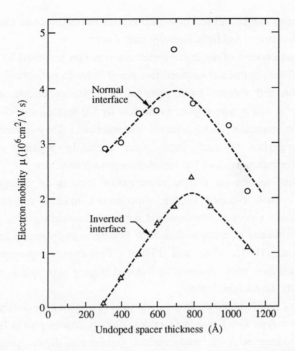

Fig. 10.12. Electron mobilities measured at 4.2 K in the dark after exposure to light for normal and inverted selectively doped quantum well structures whose carrier densities are plotted in Fig. 10.11.

doping superlattices contain only one semiconductor material. They are periodic with typically 10–100 periods. Consider a doping superlattice which has the same doping concentration in the n- and p-type regions. For sufficiently closely spaced doping regions electrons from donors recombine with acceptors; the entire doping superlattice is then depleted of free carriers. The periodic positive and negative charge (due to the ionized donors and acceptors) results in a potential modulation of the conduction and valence band.

The periodic potential, which is called the superlattice potential, gives rise to a number of unique properties not found in any other modulated semiconductor structure. Among the intriguing properties of doping superlattices are (1) a reduced energy gap of the superlattice compared with the (undoped) host semiconductor; (2) an enhanced carrier lifetime for radiative recombination; and (3) the tunability of the superlattice energy gap by the excitation intensity. These unique properties can be utilized in semiconductor devices such as tunable light-emitting diodes, lasers, and light-intensity modulators.

The general concept of doping superlattices was first proposed by Esaki and Tsu (1970). These authors envisioned two possibilities to periodically modulate the conduction and valence band potential of a semiconductor, namely by (i) alternating n- and p-type doping as well as by (ii) periodically changing the semiconductor materials (compositional superlattice). They predicted novel electronic properties of such superlattices including negative differential conductivity for transport along the superlattice growth direction.

The present section on doping superlattices focuses on δ-doped doping superlattices. Such δ-doped doping superlattices have a higher potential modulation, shorter periods, and reduced potential fluctuations (due to random impurity distribution) as compared to their homogeneously doped counterparts (Schubert et al., 1985, 1988a and 1988b). Furthermore, quantum-confined interband transitions were observed in δ-doped doping superlattices but not in homogeneously doped superlattices.

The doping profile of the doping superlattices discussed here consist of a train of alternating n-type and p-type δ-functions. Such a doping profile is shown in Fig. 10.13(a) along with its mathematical expression. If the period of the superlattice, z_p, is sufficiently small, electrons originating from donors recombine with acceptors. If the donors and acceptors have the same density (and all of them are ionized) there are no excess carriers of extrinsic origin and the superlattice is depleted from free carriers.

The dominant charges in a doping superlattice under thermal equilibrium conditions are impurity charges. The schematic band diagram of a doping superlattice is shown in Fig. 10.13(b) along with the electron and hole wave functions of the superlattice. Also shown is a schematic optical absorption transition from the top heavy-hole subband to the lowest conduction subband. The band diagram of δ-doped doping superlattices has the shape of a sawtooth. Such doping superlattices are therefore also called *sawtooth superlattices*. Inspection of the transition energy reveals that it is smaller than the fundamental gap energy of the host semiconductor. That is, doping superlattices allow one to *extend the gap of any semiconductor to lower energies,* which is an intriguing feature. For example, a GaAs doping superlattice can be used as a light-detecting or light-emitting device at $\lambda > 0.87$ μm, i.e. below the fundamental gap of the GaAs host lattice.

This suggests that the energy gap of a doping superlattice be defined as the energy between the lowest conduction subband and the highest valence subband. Assuming a degenerate valence band of the host semiconductor at $k = 0$, heavy

(a)

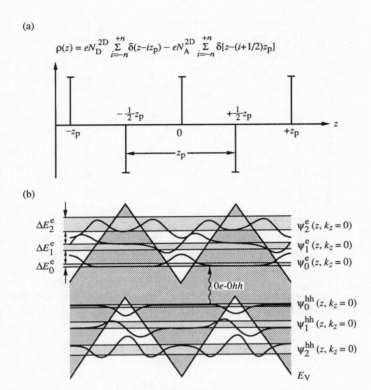

(b)

Fig. 10.13. (a) Doping profile of a δ-doped doping superlattice consisting of a train of alternate n-type and p-type δ-functions . (b) Schematic band diagram and wave functions of the δ-doped doping superlattice. The wave functions $\psi_n(z)$ are shown for $k_z = 0$. Minibands have a width of ΔE_n.

as well as light-hole subbands are formed. We define the superlattice energy gap, E_g^{SL}, as

$$E_g^{SL} = E_g - eV_z + E_0^e + E_0^{hh} \tag{10.18a}$$

where the amplitude of the superlattice modulation is given by

$$V_z = \frac{1}{4} \frac{eN^{2D}}{\varepsilon} \tag{10.18b}$$

and N^{2D} is the (donor and acceptor) impurity density. The lowest subband energies for electrons and heavy holes, E_0^e and E_0^{hh}, are given by (Schubert, 1990)

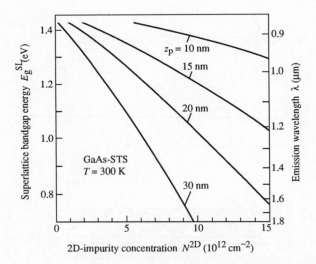

Fig. 10.14. Energy gap and emission wavelength of a sawtooth superlattice as a function of doping densitiy and period. The energy gap has values *smaller* than the gap of the GaAs host lattice.

$$E_0 = \frac{3}{10} \left[\frac{9^2}{2} \right]^{1/3} \left[\frac{e^2 \hbar^2 \, \mathcal{E}^2}{2 m^*} \right]^{1/3} \tag{10.19}$$

where m^* is the electron or heavy-hole effective mass. The superlattice bandgap energy is shown in Fig. 10.14 for a GaAs doping superlattice. The superlattice band gap decreases with increasing doping concentration and period. That is, doping superlattice have an effective energy gap that is smaller as compared to the host semiconductor.

The concept of generating a periodic potential by means of ionized impurities was proposed by Esaki and Tsu (1970). Novel properties such as negative differential conductivity were postulated in such doping superlattices. Shortly after the initial proposal of doping superlattices, published work was exclusively of a theoretical nature (Romanov, 1972; Romanov and Orlov, 1973; Döhler, 1972a and 1972b). Experimental work started in the 1980s and included the observation of the tunability of the energy gap in doping superlattices (Döhler et al., 1981). Extended reviews on homogeneously doped and δ-doped doping superlattices have become available (Ploog and Döhler, 1983; Döhler, 1986; Schubert, 1990).

Quantum-confined interband transitions were readily observed in the compositional $Al_xGa_{1-x}As/GaAs$ superlattices. However, quantum-confined interband transitions were not observed in homogeneously doped superlattices. Doping superlattices which were grown using the δ-doping technique resulted in a dramatic improvement of their optical properties. Quantum-confined optical interband transitions were observed for the first time in absorption spectroscopy (Schubert et al., 1988b) as well as in photoluminescence spectroscopy (Schubert et al., 1989a).

The improved δ-doped doping superlattice structure consists of a train of alternating n-type and p-type δ-doping sheets separated by intrinsic (undoped) layers. A periodic n-i-p-i sequence results, whose band diagram consists of linear sections. The δ-doped superlattice has several important advantages over the homogeneously doped structure including (i) larger superlattice modulation, (ii) the feasibility of shorter periods, and (iii) the minimization of potential fluctuations (Schubert et al., 1988a).

Optical transitions in δ-doped doping superlattices are not governed by the 'conventional' selection rules. The 'conventional' selection rules apply to compositional quantum wells with the quantum states $n = 0, 1, 2, ...$ and allow interband transitions (in the absence of electric fields) only if $\Delta n = 0$. For example, an $n = 0$ conduction subband electron can recombine with a $n = 0$ valence subband hole. In doping superlattices, the selection rules are fundamentally different. That is, the optical dipole matrix elements are finite and non-zero for all transitions. The matrix element involves an initial and final state which have an exponentially decaying part and a spatially oscillating part, yielding a finite, non-zero transition probability for all transitions. This property of doping superlattices is in contrast to compositional superlattices where 'conventional' selection rules do apply.

The GaAs doping superlattice epitaxial layers discussed here were grown by gas-source molecular-beam epitaxy on undoped semi-insulating GaAs substrates. The growth temperature was kept below $T = 550$ °C to avoid diffusion of n-type (Si) and p-type (Be) impurities. The design parameters of the superlattice include a period of $z_p = 150$ Å and a two-dimensional doping density of $N_D^{2D} = N_A^{2D} = 1.25 \times 10^{13}$ cm^{-2}. The samples have 10 periods of 20 dopant sheets separated by $\frac{1}{2}z_p = 75$ Å. The samples have a closely balanced impurity concentration, i.e., $N_D^{2D} \cong N_A^{2D}$. Such a balance is essential, because its absence would blue-shift the absorption edge (Burstein–Moss shift). Absorption measurements were performed on polished, 0.25-cm^2 samples using a dual-beam Perkin–Elmer Model 330 spectrophotometer and a variable-temperature cold-finger cryostat.

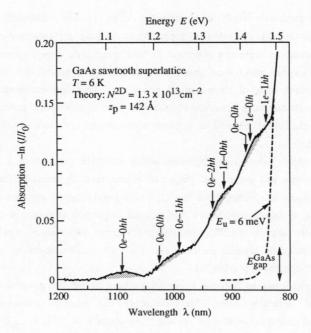

Fig. 10.15. Optical interband absorption spectrum of a GaAs sawtooth superlattice at $T = 6$ K. Theoretical transition energies are indicated by arrows. The lowest electron to lowest heavy hole transition is denoted as $0e$–$0hh$. The parameters used for the calculation are a period of 142 Å and a dopant concentration of 1.3×10^{13} cm^{-2}. The energy gap of the substrate is marked by a double arrow. The absorption tail of the substrate is characterized by an Urbach-tail energy of $E_U = 6$ meV.

Absorption spectra on GaAs sawtooth doping superlattices measured at $T = 6$ K are shown in Fig. 10.15. The gap energy of the undoped GaAs substrate corresponds to a wavelength of $\lambda = 820$ nm and is shown by a double arrow. The substrate absorbs light at energies slightly below the fundamental gap; this absorption of bulk material is known as the *Urbach tail* (Urbach, 1953). The corresponding Urbach-tail energy is $E_U = 6$ meV for undoped GaAs samples. A typical absorption spectrum of an undoped GaAs sample is shown as a dashed curve in Fig. 10.15.

The absorption spectrum shown in Fig. 10.15 shows strong absorption up to 400 meV below the fundamental gap of the GaAs host lattice. The striking aspect of the absorption spectrum are four distinct features: an absorption maximum (peak) at $\lambda = 1090$ nm and three shoulders at wavelengths of $\lambda = 1000, 920,$

and 865 nm. The structure can be attributed to transitions between quantum-confined states in the valence and conduction band. Such quantum-confined interband transitions have not been observed in homogeneously doped doping superlattices. Furthermore, the absorption does not increase monotonically with energy, but has a clear *peak* at $\lambda = 1090$ nm. Unlike the absorption spectrum shown in Fig. 10.15, the joint density of states increases *monotonically* with energy. The occurrence of such an absorption peak indicates excitonic or electron–hole correlation effects.

The formation of excitons necessitates a deeper understanding of the physical properties of doping superlattices. Electron–hole separation naturally hinders the formation of excitons. However, the absorption peak shown in Fig. 10.15 elucidates exciton formation in sawtooth superlattices with appropriate design parameters. Furthermore, excitonic absorption increases the absorption coefficient by several orders of magnitude over nonexcitonic absorption.

Proetto (1990) calculated properties of the ground state exciton in a sawtooth superlattice structure using a variational wave function. He calculated the exciton binding energy as a function of the superlattice period and the dopant density. This calculation revealed that the binding energy is of a similar magnitude (e.g. $E_b = 6$ meV for a period of 100 Å) as compared to bulk GaAs, despite the spatial separation of electrons and holes.

The built-in electric field in the sawtooth superlattice is given by $\mathcal{E} = eN^{2D}/2\varepsilon$ which is equal to $\mathcal{E} \geq 5 \times 10^5$ V/cm. At such high fields, excitonic absorption is not observable in homogeneous semiconductors or square-shaped quantum wells due to a field-induced ionization of excitons. In contrast to homogeneous semiconductors or square-shaped quantum wells, the sawtooth structure, despite the presence of an extremely high field, opposes a field-induced separation of carriers over more than half a superlattice period. Thus, electron–hole correlation effects can be observed even at field strengths exceeding 10^5 V/cm. Excitonic enhancement of the absorption can, however, occur in doping superlattices, only if the spatial electron–hole separation (which is approximately $\frac{1}{2}z_p$ for $n = 0$ states) is smaller than the electron–hole interaction length (excitonic diameter). This condition is indeed satisfied for samples with $\frac{1}{2}z_p = 75$ Å. Excitonic enhancement of the absorption has not been observed in doping superlattices with $z_p \gg a_B^*$.

We next compare the experimental absorption data with theoretical transition energies. The arrows shown in Fig. 10.15 are calculated energies of quantum-confined transitions. The lowest electron ($n = 0$) to lowest heavy-hole ($n = 0$) transition is referred to as the $0e$–$0hh$ transition. There is good agreement between calculated quantum-confined transition energies and experimental ones over a wide range of energies. For the calculation, a period of $z_p = 142$ Å and a doping density of $N_D^{2D} = N_A^{2D} = 1.3 \times 10^{13}$ cm^{-2} were used.

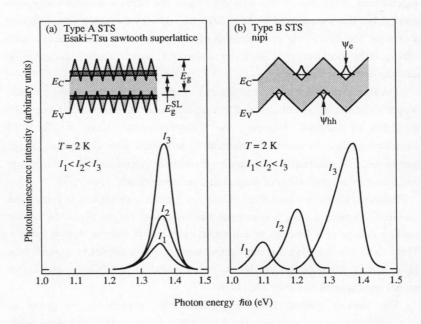

Fig. 10.16. Low-temperature photoluminescence spectra on (a) short-period (Type A) sawtooth superlattices and (b) long-period (Type B) sawtooth superlattices at three different excitation intensities . The emission wavelength is constant for the type A superlattice, whereas it is tunable for the type B superlattice.

The optical properties of doping superlattices strongly depend on the period of the superlattice. Doping superlattices with short and long periods are displayed schematically in the inset of Fig. 10.16(a) and (b), respectively. In short-period (or Type A) sawtooth doping superlattices, the barriers between adjacent quantum wells are thin (e.g. 100 Å) and coupling between adjacent wells becomes significant. Furthermore, the overlap between electron and hole wave function is largest for short-period doping superlattices. Long-period (or Type B) doping superlattices have thick barriers between adjacent quantum wells (e.g. 600 Å) which results in negligible tunneling between adjacent quantum wells. The long period also results in negligible overlap between electron and hole wave functions, i.e. long recombination lifetimes of carriers.

Drastic differences in the optical properties of long-period and short-period doping superlattices were found in low-temperature photoluminescence experiments (Schubert et al., 1987a). The photoluminescence spectra of a short-period and a long-period doping superlattice are shown in Fig. 10.16(a) and (b) respectively. The periods of the superlattices are 150 Å and 600 Å. The doping density is 1×10^{13} cm^{-2} for both types of superlattices. The photoluminescence spectrum shown in Fig. 10.16(a) reveals an emission peak energy at 1.37 eV. Note that the emission energy is smaller than the energy gap of GaAs, which is $E_g = 1.512$ eV at low temperatures. Three different excitation intensities $I_1 < I_2 < I_3$ are used for excitation. The peak energy of the spectra slightly shifts to higher energies with increasing excitation intensity. However, the shift is very small, if compared to the width of the spectra. Thus, an approximately constant peak energy as a function of excitation intensity is found in short-period doping superlattices.

Strong tunability is found in long-period doping superlattices as shown in Fig. 10.16(b). The wavelength is continuously tunable from 1.1 eV to 1.4 eV. The tuning range is larger than the full-width half-maximum of the luminescence line and confirms earlier observations of tunability in doping superlattices (Ploog and Döhler, 1983).

Clearly resolved quantum-confined interband transitions in photoluminescence spectra are shown in Fig. 10.17. The samples have a balanced donor and acceptor density of 1.25×10^{13} cm^{-2} and a period of 150 Å. Three photoluminescence peaks are observed at $\lambda \cong 0.98$, 1.02, and 1.09 μm. Furthermore, a shoulder is observed at the high-energy side of the spectrum at $\lambda \cong 0.95$ μm. The luminescence peaks can be attributed to transitions between quantum-confined conduction and valence band states. The assignment of luminescence peaks is confirmed by calculated transition energies and the comparison with experimentally observed peak energies. A very good fit between experimental and calculated peak energies is obtained by using $N^{2D} = 1.3 \times 10^{13}$ cm^{-2} and $z_p = 142$ Å. Five transitions can be identified, namely due to $0e \rightarrow 0hh$, $0e \rightarrow 0lh$, $0e \rightarrow 1hh$, $0e \rightarrow 2hh$, and $1e \rightarrow 0hh$ transitions.

The photoluminescence spectrum of Fig. 10.17 displays not only the *lowest* transition but also transitions via excited states, e.g., the $0e \rightarrow 1hh$ transition. Furthermore, excited-state transitions (e.g., $0e \rightarrow 0lh$ or $0e \rightarrow 1hh$) are more intense than the ground-state ($0e \rightarrow 0hh$) transition. The light-hole transition is stronger than the heavy-hole transition, even though the density of states of the light-hole subband is much smaller (approximately a factor of $m_{hh}^*/m_{lh}^* \cong 7$). Characteristics of the photoluminescence spectrum can be consistently explained

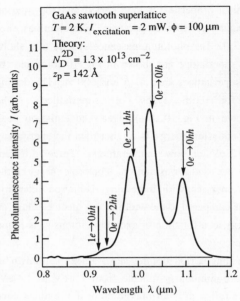

Fig. 10.17. Low-temperature photoluminescence spectrum of a sawtooth doping superlattice. The arrows indicate theoretical transition energies calculated for a superlattice period of $z_p = 142$ Å and a doping density of $N_D^{2D} = N_A^{2D} = 1.3 \times 10^{13}$ cm^{-2}.

in terms of the unique energy dependence of the oscillator strength of the sawtooth structure (Schubert et al., 1989a).

The photoluminescence line shape in semiconductors and semiconductor quantum-well structures can usually be determined by the product of the joined density of states and the thermal distribution of carriers. The latter is frequently modeled in terms of a Boltzmann distribution and a carrier temperature. The carrier temperature depends on the photoluminescence excitation intensity and is typically in the range $10 \leq T_C \leq 50$ K at a lattice temperature of $T_1 = 2$ K. The transition-matrix element depends weakly on energy in homogeneous semiconductors or compositional semiconductor quantum wells. In contrast, the oscillator strength increases exponentially with energy in the sawtooth superlattice. Thus, the oscillator strength has an opposite dependence on energy, as compared to the thermal distribution. Consequently, transitions from excited states can be observed in the sawtooth structure, even though they are sparsely populated.

10.4 Impurities in quantum wells and quantum barriers

Impurities located in *quantum wells* have different characteristics as compared to impurities in the bulk. The change in impurity characteristics is due to the quantum well potential which confines the electron (hole) in addition to the Coulomb potential of the shallow impurity. The effect of the quantum well potential is significant, if the quantum well width is on the order of or smaller than the Bohr radius of the impurity. The change in impurity characteristics include the impurity wave function and the impurity binding energy. In this section, the properties of shallow impurities located in quantum wells are discussed. More recently, structures with impurities located in *quantum barriers* have been demonstrated and these are also included here.

Electrons and holes are bound to hydrogenic impurities by means of the attractive Coulomb potential given by

$$V(r) = \frac{e}{4\pi\varepsilon r} \tag{10.20}$$

where the spherical coordinate r is the distance from the impurity. The wave function of the bound states of the Coulomb potential can be obtained from the Schrödinger equation. In bulk semiconductors, shallow hydrogenic impurities can be adequately described within the effective-mass approximation (see Chap. 1). The effective-mass Hamiltonian operator of a coulombic potential is given by

$$H = -\frac{\hbar^2}{2m^*} \nabla^2 - \frac{e^2}{4\pi\varepsilon r} \ . \tag{10.21a}$$

If the band structure is isotropic and non-degenerate and if the dispersion relation is parabolic, then the Schrödinger equation reduces to the hydrogen problem. In spherical coordinates it is given by (Kohn, 1957)

$$\left[\frac{p^2}{2m^*} - \frac{e^2}{4\pi\varepsilon r} \right] \psi(r) = E\psi(r) \tag{10.21b}$$

where p is the particle momentum, m^* is the carrier effective mass, and ε is the permittivity of the material. The solution of the equation with the lowest energy, i.e. the 1s hydrogenic wave function is given by

$$\psi_{1s}(r) = \frac{1}{\sqrt{\pi a_B^{*3}}} \exp(-r/a_B^*) \tag{10.22}$$

where a_B^* is the effective Bohr radius

$$a_B^* = \frac{4\pi\varepsilon\hbar^2}{m^* e^2} = 0.529 \,\text{Å} \, \frac{\varepsilon}{\varepsilon_0} \, \frac{m_0}{m^*} \tag{10.23}$$

in which 0.529 Å is the hydrogen atom Bohr radius. For GaAs, the effective

Bohr radius for donors is $a_B^* = 103$ Å. The binding energy of the lowest state is given by

$$E_{Ryd} = \frac{m^* e^4}{32 \pi^2 \varepsilon^2 \hbar^2} = 13.6 \text{ eV} \frac{\varepsilon_0^2}{\varepsilon^2} \frac{m^*}{m_0} \tag{10.24}$$

where 13.6 eV is the Rydberg energy. For GaAs, the effective Rydberg energy is $E_{Ryd} = 5.31$ meV.

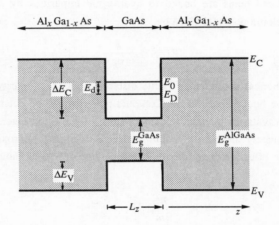

Fig. 10.18. Schematic band diagram of an $Al_x Ga_{1-x}As/GaAs$ quantum well structure. A donor level (E_D) and the lowest subband level (E_0) are separated by the donor ionization energy (E_d).

Impurity characteristics change if the impurity is located in a quantum well. Furthermore, since quantum well structures lack translational invariance the binding energy and the wave function depend on the position of the impurity in the well. A schematic energy level scheme of a quantum well with a donor impurity is shown in Fig. 10.18. The energy of the lowest state in the well is denoted by E_0. An impurity provides an additional confining potential. As a consequence, the donor energy is *below* the lowest conduction subband. The donor ionization energy, E_d, is defined as the difference between the lowest conduction subband energy and the donor energy in analogy to bulk impurities. The evolution of the impurity wave function with quantum well thickness was discussed by Bastard (1981) and is schematically shown in Fig. 10.19 for different well widths and different impurity positions in the well. For thick wells ($L_z \gg a_B$) the Coulomb potential is the dominant confining potential. In this

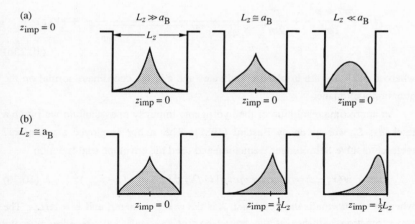

Fig. 10.19. Schematic illustration of the wave function of an electron bound to a hydrogenic impurity for (a) different quantum well width and (b) different positions of the impurity in the quantum well.

case, the impurity wave function is similar to the impurity wave function in the bulk. For thin wells ($L_z \ll a_B$), the quantum well potential is the dominant confining potential. In this case, the impurity wave function is similar to the free quantum-confined state. Figure 10.19 also reveals that the impurity wave function (and therefore also its energy) depends on the impurity position.

In order to calculate the energy and the wave function of the impurity in a quantum well, the quantum well potential must be included in the Hamiltonian operator. The Hamiltonian of Eq. (10.21a) needs to be modified according to

$$H = -\frac{\hbar^2}{2m^*} \nabla^2 - \frac{e^2}{4\pi\varepsilon r} + eV(z) \tag{10.25a}$$

where

$$eV(z) = \begin{cases} 0 & \text{for } z \text{ in the well} \\ \Delta E_c & \text{for } z \text{ in the barrier} . \end{cases}$$

Note that both m^* and ε can in principle be different in the well and the barrier material. For an isotropic and parabolic band, the Hamiltonian can be written as

$$H = \frac{p_x^2 + p_y^2}{2m^*} + \frac{p_z^2}{2m^*} - \frac{e^2}{4\pi\varepsilon\sqrt{x^2 + y^2 + (z - z_{\text{imp}})^2}} + eV(z)$$

$$(10.25b)$$

where x and y are the in-plane coordinates and z is the coordinate normal on the quantum well plane.

An approximate solution of the hydrogenic impurity in a quantum well with a thickness L_z was given by Bastard (1981). The author employed a variational method to solve Schrödinger's equation and used the envelope trial function

$$\psi(x,y,z) = A\cos(kz)\exp[(1/\lambda)\sqrt{x^2 + y^2 + (z - z_{\text{imp}})^2}] \qquad (10.26)$$

where A is a normalization constant, λ is the trial parameter, and $k = \pi/L_z$. The wave function vanishes at the boundaries of the well, i.e. tunneling into the barriers is neglected. Thus, the trial function of Eq. (10.26) is strictly valid only for a quantum well with infinitely high barriers.

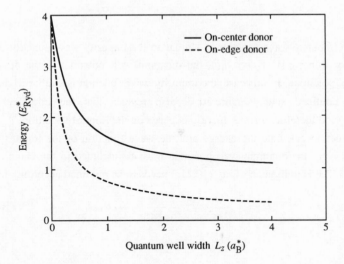

Fig. 10.20. Binding energy for a hydrogenic impurity located at the center or edge of a quantum well as a function of the quantum well thickness. For GaAs, the energy and length units are $E_{\text{Ryd}}^* = 5.31$ meV and $a_B^* = 103$ Å, respectively (after Bastard, 1981).

The energy obtained from the trial function is the energy of the donor ground state relative to the bottom of the quantum well which was defined as $E = 0$. It is instructive to measure the donor ground-state energy relative to the lowest conduction subband. This energy can be obtained by using the same trial function (Eq. (10.26)) but by omitting the Coulomb term in the Hamiltonian operator (Eq. (10.25)) (Masselink et al., 1986). The calculated donor ground-state energy relative to the lowest conduction subband is shown in Fig. 10.20 for donors in the center (solid line) and at the edge (dashed line) of the quantum well (Bastard, 1981). The donor ionization energy equals E_{Ryd} in the bulk limit ($L_z \gg a_{\mathrm{B}}$). With decreasing quantum well thickness, the donor binding energy increases. The increase is due to the fact that the wave function is further localized at the site of the donor atom where the electron experiences a lower potential energy. For infinitely small quantum well thicknesses, the problem reduces to the two-dimensional hydrogen atom. The donor binding energy is then $E_{\mathrm{d}}^{2D} = 4\,E_{\mathrm{Ryd}}$. The Bohr radius of the two-dimensional hydrogen atom equals $a_{\mathrm{B}}^{2D} = \frac{1}{2}\,a_{\mathrm{B}}$ (Bastard et al., 1982). Donors located on the edge of the quantum well have a smaller binding energy except for the two-dimensional limit $L_z \to 0$. The overlap of the electron wave function (Eq. (10.26)) with the on-edge donor is smaller, which results in a smaller binding energy as shown in Fig. 10.20.

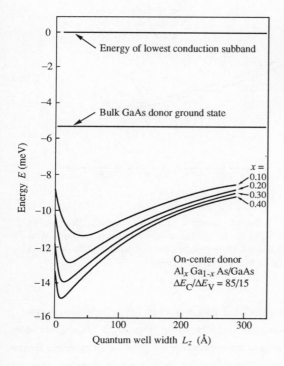

Fig. 10.21. Calculated donor binding energy in $Al_xGa_{1-x}As$/GaAs quantum wells as a function of well widths (after Mailhiot et al., 1982b).

The simple model of Bastard (1981) has been subsequently refined by Mailhiot et al. (1982a and b) and by Greene and Bajaj (1983). The authors took into account (i) the finite barrier height of an $Al_x Ga_{1-x} As/GaAs$ quantum well, (ii) the different effective mass in $Al_x Ga_{1-x} As$ and GaAs, and (iii) the different dielectric constant of the two materials. The calculated on-center donor energy relative to the edge of the lowest conduction subband is shown in Fig. 10.21 (Mailhiot et al., 1982b). The donor binding energy increases with decreasing quantum well thickness, reaches a maximum at a small well thickness and finally decreases again for very small quantum well thicknesses. In the two limits, $L_z \to \infty$ and $L_z \to 0$, the donor binding energy is given by the GaAs and $Al_x Ga_{1-x} As$ donor binding energy, respectively. In the intermediate range $(0 < L_z < a_B)$ the binding energy reaches a maximum because the impurity wave function is compressed at the location of the donor atom by the quantum well potential. The localization of the wave function at the location of the impurity atom results in an increase of the donor binding energy.

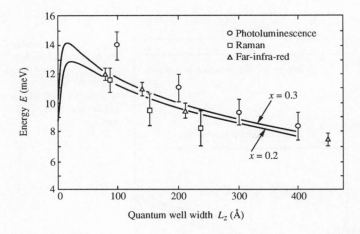

Fig. 10.22. Calculated binding energies for hydrogenic donors in $Al_x Ga_{1-x} As/GaAs$ quantum wells for two different Al mole fractions along with experimental data obtained through photoluminescence, Raman, and far-infra-red absorption measurements (after Masselink et al., 1986).

A comparison between calculated and measured donor binding energies in $Al_x Ga_{1-x} As/GaAs$ quantum wells is shown in Fig. 10.22 (Masselink et al., 1986). The experimental data were obtained from photoluminescence, Raman, and far-infra-red spectroscopy. The calculated donor binding energies agree with nearly all measured data within the experimental error. A comparison of experimental and calculated donor and acceptor binding energies was also made by Greene and Bajaj (1985b). The authors found only minor discrepancies between theory and experiment.

The effect of a non-parabolic dispersion relation in the conduction band on hydrogenic donors in quantum wells has been investigated by Chandhuri and Bajaj (1984). The effect of non-parabolicity was included by using an energy-dependent effective mass obtained from the $\vec{k} \cdot \vec{p}$ approximation. The authors found a small increase in donor binding energies upon inclusion of non-parabolicity. The increase was observed for all values of L_z. Excited states of hydrogenic impurity centers and the effect of magnetic fields were analyzed by Greene and Bajaj (1985a, 1988). The authors used a variational approach in which the trial wave functions were expanded in terms of gaussian basis sets. The hydrogenic binding energies were found to increase with magnetic field.

The characteristics of *acceptors in quantum wells* have been studied in detail by Masselink et al. (1983, 1984, 1985, 1986). Acceptor binding energies were calculated in a multiband effective-mass approximation. The valence band structure of GaAs and other III–V semiconductors is more complicated as compared to the isotropic, non-degenerate conduction band structure. It is well known that the valence band structure consists of the degenerate heavy- and light-hole bands and the split-off band. The kinetic energy term of the Hamiltonian is not a scalar as for donors, but a matrix because of the multiple valence bands. To obtain the acceptor energy, a trial function of the appropriate symmetry must be constructed. The trial function typically contains a large number of variational parameters. The ground-state energy is then calculated using the variational method, i.e. minimizing the energy with respect to all variational parameters. The resulting energy is taken as a good approximation to the true ground state energy. Finally, the binding energy is determined by subtracting the acceptor ground-state energy from the lowest valence subband edge. The latter energy can be calculated as described above, but without the Coulomb term in the Hamiltonian. The acceptor binding energies depend on the quantum well width in the same qualitative way as the donor binding energies. The acceptor binding energy increases with decreasing quantum well thickness, reaches a maximum at small L_z, and finally decreases for very small L_z. This dependence is qualitatively the same as displayed by donors (see Fig. 10.21 and 10.22). The acceptor binding energy was also investigated in the presence of an external magnetic field (Masselink et al., 1985). Different central cell potentials were employed for chemically different acceptors. The calculated acceptor energies were found to agree well with experimental acceptor binding energies.

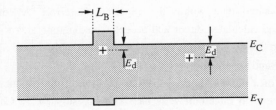

Fig. 10.23. Energy band diagram with donor levels (i) in a quantum barrier and (ii) in the bulk. The donor level in the quantum barrier has a smaller activation energy compared to the donor in the bulk.

Impurities in quantum barriers have different characteristics as compared to impurities in the bulk (Tsang et al., 1992). Consider a semiconductor heterostructure of Type I consisting of a thick well material and a very thin barrier material as shown in Fig. 10.23. Assume further that the heterostructure can be selectively doped either in the thin barriers or in the well. Impurities in the well will have bulk-type properties if the thickness of the well is much larger than the effective Bohr radius of the impurity. However, the characteristics of impurities in the barrier can be significantly changed by the barrier potential. To illustrate the effect of the quantum barrier, we consider hydrogenic donors in a quantum barrier as shown in Fig. 10.23. In addition to the *attractive* Coulomb potential, the *repulsive* barrier potential must be taken into account. For sufficiently thin and sufficiently low barriers, the attractive Coulomb potential dominates. In this case, the barrier thickness is much smaller than the effective Bohr radius of the donor, i.e. $L_B \ll a_B^*$. Hence, donor impurities in the quantum barrier have a bound state. However, the effective activation energy and the impurity wave function will be different as compared to the bulk value. We define the 'effective activation energy' of the donor to be the difference between the lowest continuum state of the semiconductor and the ground-state energy of the donor, as schematically shown in Fig. 10.23. For thin barriers, the effective donor activation energy is expected to be reduced as compared to the activation energy in the bulk. For sufficiently thick and high barriers, the impurity state is shifted above the continuum edge of the well, which corresponds to a *negative* effective activation energy of the impurity. The impurity will then be ionized at all temperatures. The case of negative effective activation energy is reminiscent of selectively doped heterostructures, in which no freeze-out effects are observed even at the lowest temperatures.

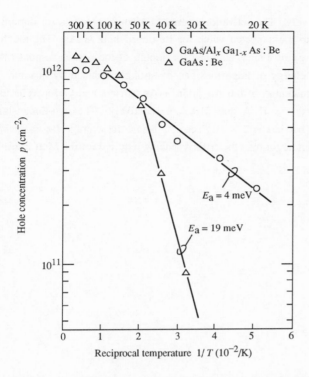

Fig. 10.24. Hole concentration versus reciprocal temperature for a Be-doped $Al_xGa_{1-x}As$/GaAs quantum barrier structure and homogeneously Be-doped GaAs. The activation energy of Be acceptors in the doped quantum barrier structure is significantly lower as compared to homogeneously doped GaAs (after Tsang et al., 1992).

Experimental results on doping of quantum barriers were reported by Tsang et al. (1992) for Be doping of $Al_xGa_{1-x}As$/GaAs heterostructures. The authors showed that the activation energy of impurities can be 'engineered' by appropriate parameters of the quantum barrier. The Hall carrier concentration of the Be-doped heterostructure is shown in Fig. 10.24 as a function of reciprocal temperature for doping in the wide GaAs wells (triangles) and for doping of the thin $Al_xGa_{1-x}As$ barriers (circles). The slope of the carrier concentration versus reciprocal temperature was used to deduce the effective activation energy of the Be impurities using the proportionality $p \propto \exp(-E_a/kT)$. Activation energies of 4 and 19 meV were inferred for doping of the barriers and for doping of the wells, respectively. The activation energy of 19 meV obtained for doping of the well is close to the expected value for Be-doped bulk GaAs. The much lower activation energy of 4 meV obtained for doping of the barrier demonstrates that the activation energy of impurities can be strongly reduced by quantum-barrier doping. The thickness of the $Al_{0.30}Ga_{0.70}As$ barriers and GaAs wells used for the experiments are 15 Å and 210 Å, respectively. The two-dimensional Be density in the barriers is 2×10^{10} cm^{-2}. Thus, the average Be-to-Be impurity spacing is much larger than the critical spacing required for the Mott transition.

11

Delta doping

Epitaxially grown semiconductors can be doped by incorporating impurities during growth. During molecular-beam epitaxy (MBE) and during organometallic vapor-phase epitaxy (OMVPE) the doping process is continuous and the growth processes are not influenced by the doping process. During delta doping the doping process is fundamentally different. In the first step of the doping process, epitaxial growth is suspended. In the second step, doping impurities are deposited on the non-growing semiconductor surface. In the third step, regular crystal growth is resumed.

This doping procedure results in very narrow doping distributions, typically one or a few monolayers wide. Such narrow doping distributions can be considered as δ-function-like. The doping profile can then be written as

$$N(z) = N^{2D}\, \delta(z - z_D) \qquad (11.1)$$

where N^{2D} is the two-dimensional doping density per cm^2. Impurities are confined to the plane $z = z_D$ of a cartesian coordinate system. The first report of such a growth-interrupted dopant deposition is due to Bass (1979), who found that a strong surface adsorption of Si to the non-growing GaAs surface results in sharp doping spikes. The widths of the doping spikes were not explicitly mentioned in the publication. However, a doping profile shown in the publication reveals a width of approximately 250 Å. Adsorption of dopants to the non-growing crystal surface may also be the origin of high dopant concentrations of the substrate–epilayer interface found earlier by Di Lorenzo (1971). The versatility of growth-interrupted doping was realized by Wood et al. (1980), who mentioned that complex doping profiles can be achieved by 'atomic-plane' or δ-doping. The authors further found that Ge doping of GaAs leads to reduced autocompensation. However, the publication does not investigate possible diffusion of dopants from the atomic plane. Clear indications of such diffusion in δ-doped semiconductors were first found by Lee et al. (1985). They concluded that diffusion over 126 Å occurred during crystal growth. It was not until 1988 that Schubert et al. (1988a) showed that highly confined Si doping profiles in GaAs can be achieved at growth temperatures ≤ 550 °C. Si dopants were spatially confined to a layer whose thickness was found to be comparable to the lattice constant. Diffusion over more than two lattice constants was excluded. This chapter summarizes the structural, electronic, and optical properties of δ-doped III–V semiconductors. In addition, device applications of δ-doped semiconductors and the advantages of such applications are discussed.

11.1 Electronic structure

Impurities are confined to one atomic monolayer in δ-doped semiconductors grown under idealized conditions which include the absence of diffusion and other redistribution effects as well as the availability of atomically flat crystal growth. Such an idealized δ-doped III–V semiconductor is shown in Fig. 11.1. Impurities occupy group-III (cation) sites and are located in an atomic plane of the semiconductor. Certainly, such doping distributions are idealized and we refer to Sect. 12.2 for the experimental assessment of spatial impurity distributions in δ-doped semiconductors.

The doping distribution shown in Fig. 11.1 results in a one-dimensional, V-shaped electrostatic potential. To illustrate the V-shaped potential, we consider a sheet of donors with density N_D^{2D} located in the xy-plane at $z = z_D$. The electric field and the potential of a δ-function-like dopant distribution are obtained by integration of Poisson's equation

$$\frac{d^2\phi}{dz^2} = -\frac{d\mathcal{E}}{dz} = -\frac{eN_D(z)}{\varepsilon} \tag{11.2}$$

where e is the elementary charge and $\varepsilon = \varepsilon_r \varepsilon_0$ is the permittivity of the semiconductor. The electric field is obtained by integration as

$$\mathcal{E}(z) = \begin{cases} -\dfrac{1}{2}\dfrac{eN_D^{2D}}{\varepsilon} & \text{for } z < z_D \\[4mm] +\dfrac{1}{2}\dfrac{eN_D^{2D}}{\varepsilon} & \text{for } z > z_D \end{cases} \tag{11.3a}$$

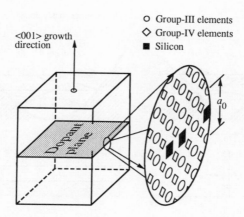

Fig. 11.1. Schematic illustration of a semiconductor grown in the <001> direction containing Si impurities in a group-III plane of a zincblende structure.

or

$$\mathcal{E}(z) = -\frac{1}{2} \frac{eN_{\mathrm{D}}^{2\mathrm{D}}}{\varepsilon} + \frac{eN_{\mathrm{D}}^{2\mathrm{D}}}{\varepsilon} \sigma(z - z_{\mathrm{D}}) \quad \text{for all } z \qquad (11.3b)$$

where $\sigma(z - z_{\mathrm{D}})$ is the step-function. The integration constant required for the solution of Eq. (11.2) is chosen to obtain a symmetric field, i.e. $|\mathcal{E}(z < z_{\mathrm{D}})| = |\mathcal{E}(z > z_{\mathrm{D}})|$, as shown in Fig. 11.2. The second integration of Poisson's equation yields the potential

$$\phi(z) = \begin{cases} -\dfrac{eN_{\mathrm{D}}^{2\mathrm{D}}}{2\varepsilon}(z - z_{\mathrm{D}}) & \text{for } z < z_{\mathrm{D}} \\[2mm] \dfrac{eN_{\mathrm{D}}^{2\mathrm{D}}}{2\varepsilon}(z - z_{\mathrm{D}}) & \text{for } z > z_{\mathrm{D}} \end{cases} \qquad (11.4a)$$

or

$$\phi(z) = -\frac{eN_{\mathrm{D}}^{2\mathrm{D}}}{2\varepsilon}(z - z_{\mathrm{D}}) + \frac{eN_{\mathrm{D}}^{2\mathrm{D}}}{\varepsilon}(z - z_{\mathrm{D}}) \sigma(z - z_{\mathrm{D}}) \quad \text{for all } z . \qquad (11.4b)$$

The potential well given by this equation is V-shaped and is illustrated in Fig. 11.2. (Equation (11.4) is actually the negative potential, i.e. the potential for electrons).

It is worthwhile noting that the singularity of the δ-function is physically

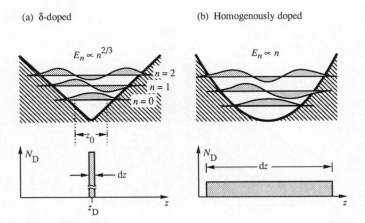

Fig. 11.2. Comparison of a δ-doped and a homogeneously doped potential well. The subband energies are proportional to $n^{2/3}$ and to n for the δ-doped and the homogeneously doped well, respectively.

problematic: According to Eq. (11.1), N_D approaches infinity for $z \to z_D$. Of course, the doping density cannot exceed the host lattice site concentration (e.g. 2.2139×10^{22} cation sites per cm^3 in GaAs) in a real semiconductor. It is useful, however, to recall that mostly the *integral* properties of the doping distribution are relevant. For example, the potential and the electric field are (according to Poisson's equation) *integral* properties of the δ-function. Therefore, the singularity of the δ-function does not represent a problem.

The V-shaped potential of Eq. (11.4) is a result of the impurity charges. In addition to the impurity charges, the charges of free carriers must be taken into account. In the following, we first consider impurity charges only. Free carrier charges can be neglected, if the δ-doped semiconductor is depleted of free carriers, e.g. in the case of doping superlattices (see Sect. 10.3). Subsequently, undepleted δ-doped semiconductors will be considered.

The slopes of the V-shaped potential well, caused by a sheet of charged impurities, depend on the doping concentration. The slopes of the V-shaped potential well (i.e. the electric field) are $> 10^4$ V/cm for typical doping densities. The spatial width of the potential then becomes comparable to the de Broglie wavelengths of carriers. Size quantization (Schrieffer, 1957) occurs and quantum effects must be taken into account via the Schrödinger equation. The V-shaped potential well can then be termed a *quantum* well.

The V-shaped quantum well problem can be solved (i) exactly and analytically using Airy functions, (ii) using the variational method, (iii) the WKB method, and (iv) in a zero-under approximation. These methods were reviewed by Schubert (1990a). All of the methods provide very similar results. The variational method, which will be discussed below, provides excellent trade-off between simplicity and accuracy for the solutions of the V-shaped well problem.

11.1.1 Variational solutions of the V-shaped well

The variational method is a simple approach to obtain the wave functions and eigenstate energies of a one-dimensional quantum well problem. The accuracy of the eigenstate energies and the wave functions certainly depends on the quality of the trial function but is usually very good. The usefulness of the variational approach for quantum wells was first demonstrated for the triangular well of a Si/SiO$_2$ inversion layer (Fang and Howard, 1966).

The calculation of the eigenstate energies and wave functions of the V-shaped quantum well can be conveniently carried out in the effective-mass approximation. The one-dimensional V-shaped potential well problem then leads to text-book type Hamiltonian operators. Using the product method, the wave function in such a one-dimensional quantum well potential can be written as

$$\psi_{n,k_x,k_y}^{c,v}(x,y,z) = e^{ik_x x}\, e^{ik_y y}\, \psi_n^{c,v}(z)\, u^{c,v}(x,y,z) \qquad (11.5)$$

where $u^{c,v}(x,y,z)$ is the periodic Bloch-type eigenfunction close to the zone center of the conduction (c) or valence (v) band (where the effective-mass approximation is valid). The terms $e^{ik_x x}$ and $e^{ik_y y}$ indicate the free movement of carriers in the x- and y-direction. The following calculations are restricted to conduction-band type Bloch-functions, $u^c(x,y,z)$, and to those of the valence band, $u^v(x,y,z)$. The function $\psi_n^{c,v}(z)$ is the eigenfunction of the nth bound state of the Hamiltonian

$$H = -\frac{\hbar^2}{2m^*}\frac{\partial^2}{\partial z^2} + eV(z) \qquad (11.6)$$

where $V(z)$ is the potential of either the conduction or the valence band edge. The function $\psi_n^{c,v}(z)$ is varying slowly on the length scale of the lattice constant ($a_0 \cong 5$ Å), while the Bloch function is periodic in the lattice constant. Therefore, $\psi_n^{c,v}(z)$ is called the envelope function. The following calculation will be carried out within the framework of the envelope-function approximation (Bastard, 1981). For simplicity we consider one band only and omit the superscripts, c,v.

The following trial-function for the ground state is of even symmetry (see Fig. 11.2), and is exponentially decaying for large absolute values of z.

$$\psi_0(z) = \begin{cases} A_0(1 + \alpha_0 z)e^{-\alpha_0 z} & (z \geq 0) \\[2mm] A_0(1 - \alpha_0 z)e^{\alpha_0 z} & (z < 0) \ . \end{cases} \tag{11.7}$$

The constant A_0 is determined by the normalization condition $\langle \psi | \psi \rangle = 1$ to be $A_0^2 = 2\alpha_0/5$. The trial-parameter α_0 determines the exponential decay of the wave function in the classically forbidden region, i.e. beyond the classical turning points. The expectation value of the ground-state energy is given by

$$\langle E_0 \rangle = \langle \psi | H | \psi \rangle = \frac{1}{5} \left[\frac{\hbar^2}{2m^*} \alpha_0^2 + \frac{9}{2} \frac{e\mathcal{E}}{\alpha_0} \right] . \tag{11.8}$$

Minimization of $\langle E_0 \rangle$ with respect to the trial-parameter α_0 yields

$$\alpha_0 = \left[\frac{9}{4} e\mathcal{E} \frac{2m^*}{\hbar^2} \right]^{1/3} . \tag{11.9}$$

Insertion of this result into Eq. (11.7) yields the normalized ground-state wave function. Insertion into Eq. (11.8) yields the ground-state energy according to

$$\boxed{E_0 = \frac{3}{10} \left[\frac{9^2}{2} \right]^{1/3} \left[\frac{e^2 \hbar^2 \mathcal{E}^2}{2m^*} \right]^{1/3}} \tag{11.10}$$

Comparison of this variational result with the mathematically exact result yields that both methods agree to within 1% (Schubert, 1993).

The calculation for the wave function and energy of the first excited state is performed analogously. The trial-parameter α_1 is used in the wave function

$$\psi_1(z) = \begin{cases} A_1 z\, e^{-\alpha_1 z} & (z \geq 0) \\[2mm] A_1 z\, e^{\alpha_1 z} & (z < 0) \ . \end{cases} \tag{11.11}$$

The normalization condition yields $A_1^2 = 2\alpha_1^3$. The expectation value for the energy is given by

$$\langle E_1 \rangle = \frac{\hbar^2}{2m^*} \alpha_1^2 + \frac{3}{2} \frac{e\mathcal{E}}{\alpha_1} . \tag{11.12}$$

The minimization of the energy expectation value yields the trial parameter

$$\alpha_1 = \left[\frac{3}{4} e\mathcal{E} \frac{2m^*}{\hbar^2} \right]^{1/3} \tag{11.13}$$

and finally the eigenstate energy of the first excited state

$$E_1 = \frac{3}{2} \left[\frac{9}{2} \right]^{1/3} \left[\frac{e^2 \hbar^2 \mathcal{E}^2}{2m^*} \right]^{1/3} \tag{11.14}$$

The requirements for the second excited state wave function are (i) even symmetry, (ii) two nodes, and (iii) exponential decay of the wave-function amplitude beyond the classical turning points. The following trial-function satisfies these requirements

$$\psi_2(z) = \begin{cases} A_2 (\alpha_2^2 z^2 - 1) (1 + \alpha_2 z) e^{-\alpha_2 z} & (z \geq 0) \\ A_2 (\alpha_2^2 z^2 - 1) (1 - \alpha_2 z) e^{\alpha_2 z} & (z < 0) \end{cases} \tag{11.15}$$

where the constant $A_2^2 = (4/63) \alpha_2$ is obtained from the normalization condition. The calculation of the energy expectation value is quite tedious and yields

$$\langle E_2 \rangle = \frac{1}{7} \left[3\alpha_2^2 \frac{\hbar^2}{2m^*} + \frac{47}{2} \frac{e\mathcal{E}}{\alpha_2} \right] . \tag{11.16}$$

The minimum of $\langle E_2 \rangle$ is obtained for

$$\alpha_2 = \left[\frac{47}{12} \right]^{1/3} \left[e\mathcal{E} \frac{2m^*}{\hbar^2} \right]^{1/3} \tag{11.17}$$

which results in the eigenstate energy of the second excited state

$$E_2 = \frac{9}{7} \left[\frac{47}{12} \right]^{2/3} \left[\frac{e^2 \hbar^2 \mathcal{E}^2}{2m^*} \right]^{1/3} \tag{11.18}$$

The deviation of this variational result from the exact solution is $< 2\%$ (Schubert,

1993).

It is worthwhile pointing out that the V-shaped well and the triangular well share some of the eigenstate energies and wave functions. The triangular well has one infinitely steep wall and a second wall of slope $\mathcal{E}$. Potential wells of such triangular shape occur in selectively doped $Al_x Ga_{1-x} As/GaAs$ heterostructures and in Si/SiO_2 inversion layers. The first excited state of the V-shaped well coincides with the ground state of the triangular well. Furthermore, the third excited state of the V-shaped well coincides with the first excited state of the triangular well.

11.1.2 Solutions of the parabolic well

A uniform doping distribution results in a potential well of parabolic shape. Consider a semiconductor with a spatially uniform donor impurity concentration N_D. Assume further that the doped layer is depleted of free carriers. Such a situation occurs, for example, in homogeneously doped doping superlattices (see Sect. 10.3). The corresponding parabolic potential well is shown in Fig. 11.2(b). The parabolic potential of the conduction band edge is given by

$$\phi(z) = \frac{e^2}{2\varepsilon} N_D z^2 . \tag{11.19}$$

The exact solutions of the eigenstate (subband) energies are the *harmonic oscillator solutions* given by (Saxon, 1968)

$$E_n = \left[n + \frac{1}{2} \right] \hbar e \left[\frac{N_D}{\varepsilon m^*} \right]^{1/2} \quad \text{for} \quad n = 0,1,2 \cdots . \tag{11.20}$$

The subband separation is constant and independent of the quantum number, i.e. $E_n \propto n$.

11.1.3 Self-consistent solutions of the V-shaped potential

In Sects. 11.1.1 and 11.1.2 the energetic subband structure of a V-shaped and parabolic potential well was calculated without taking into account the charge of free *carriers* in the potential well. The total charge of free carriers equals the total charge of ionized dopants in a δ-doped semiconductor with a single dopant sheet (the n^- or p^--type background concentration, which always occurs in a realistic semiconductor, is neglected). Therefore, the δ-doped carrier system is *neutral*, i.e. the electric field $\mathcal{E} \to 0$ for distances sufficiently far from the doped region.

A *self-consistent* calculation is required if the charge of free carriers is taken into account. Such self-consistent calculations simultaneously satisfy Poisson's and Schrödinger's equation. Poisson's equation allows one to calculate the potential as a function of the spatial coordinates x, y, z from a given charge distribution (both ionized dopant charge and free carrier charge). On the other hand, Schrödinger's equation enables one to calculate the charge distribution in a given potential.

Self-consistent calculations are greatly simplified if the charge distribution is assumed to depend on one spatial coordinate only. Such one-dimensional self-consistent calculations of the energy-space structure of semiconductor quantum systems (see e.g. Ando, 1982a, 1982b; Stern and Das Sarma, 1984; Vinter, 1984) were initially carried out for selectively doped $Al_x Ga_{1-x} As/GaAs$ heterostructures. The solution of Poisson's equation is straightforward whereas the accurate solution of Schrödinger's equation is more elaborate. Therefore, several groups used approximate solutions of the Schrödinger equation, including the variational solution (Stern, 1983; Fang and Howard, 1966) and the WKB approximation (Ando, 1985).

The band diagram obtained from a self-consistent calculation is shown in Fig. 11.3 for δ-doped GaAs. The two-dimensional doping density of the structure is 5×10^{12} cm^{-2} located at $z = 599 - 601$ Å. The doping charge is assumed to be distributed homogeneously over 2 Å. A low temperature of $T = 4$ K was assumed for the calculation. Figure 11.3 reveals that four subbands are populated at the chosen doping density. The electron densities in each subband are as follows: $n_0 = 3.05 \times 10^{12}$ cm^{-2}, $n_1 = 1.18 \times 10^{12}$ cm^{-2}, $n_2 = 0.531 \times 10^{12}$ cm^{-2}, and $n_3 = 0.188 \times 10^{12}$ cm^{-2}.

Comparison of the width of the dopant distribution with the spatial extent of the free carrier distribution allows one to define a δ-doped semiconductor. The spatial extent of the electron distribution of the ground-state subband is approximately 50 Å. Higher excited states have a wider distribution. In δ-doped semiconductors, the dopant distribution is much narrower than the free carrier distribution. In Fig. 11.3, the dopants are assumed to be distributed within 2 Å. It is, however, quite obvious that any impurity distribution with a spatial extent $\ll$ 50 Å will result in a free carrier distribution very similar to the one shown in Fig. 11.3.

As a numerical example, we consider the spatial extent of the ground-state wave function using the variational wave function of Eq. (11.7). We define the spatial extent of the wave function as

$$z_0 = 2(\langle z^2 \rangle - \langle z \rangle^2) \tag{11.21}$$

where $\langle z \rangle$ is the position expectation value of the ground-state wave function. Using Eqs. (11.7)–(11.10), one obtains

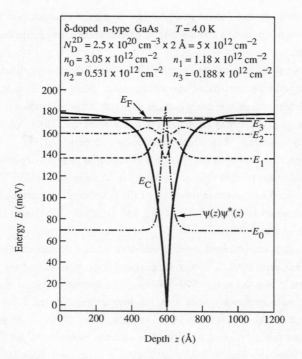

Fig. 11.3. Subband structure and charge distribution in a δ-doped semiconductor obtained from the self- consistent solution of Poisson's and Schrödinger's equation. At a doping density of 5×10^{12} cm^{-2}, four eigenstates are populated.

$$z_0 = 2\sqrt{\frac{7}{5}} \left[\frac{4}{9} \frac{\hbar^2}{e\mathcal{E}2m^*} \right]^{1/3} \tag{11.22}$$

For an n-type doping concentration of $N_D^{2D} = 5 \times 10^{12}$ cm^{-2} one obtains $\mathcal{E} = 3.45 \times 10^5$ V/cm and $z_0 = 46$ Å.

We thus define δ-function-like doping distributions as follows: *one-dimensional doping distributions are δ-function-like, if the spatial extent of the doping distribution is much narrower than the ground-state wave function of the V-shaped quantum well*; the width of the ground-state wave function is given by Eq. (11.22). Application of this definition to the band diagram shown in Fig. 11.3 implies that any doping distribution localized within a length << 50 Å can be considered to be δ-function-like. For example, distribution widths of $dz = 2$ Å or $dz = 20$ Å can both be regarded as δ-function-like. It is clear from this definition, that impurities need not be confined to a single monolayer, in order to result in a δ-function-like doping profile.

11.1.4 Discreteness of charge

The discreteness of charge has not been taken into account in the self-consistent and variational calculations described above. Instead, the charge was assumed to be distributed homogeneously in the dopant plane (jellium model). The coherence length of the wave function in the lateral direction (i.e. the direction of the dopant plane) allows us to estimate the effect of dopant density fluctuations experienced by the free carriers assuming random distribution of dopants. The phase-coherence length in the lateral direction can be estimated from the inelastic scattering time

$$\tau_{inel} = \mu m^*/e \tag{11.23}$$

where μ is the carrier mobility. Certainly, any scattering event, both elastic *and* inelastic, changes the phase of the electron, but since the elastic scattering time is difficult to estimate, only the inelastic time is estimated in Eq. (11.23). The Fermi velocity

$$\tfrac{1}{2} m^* v_F^2 = E_F - E_0 \tag{11.24}$$

allows one to estimate the mean free path between two inelastic scattering events according to

$$l_{inel} = v_F \, \tau_{inel} \, . \tag{11.25}$$

As a numerical example, we consider electrons in GaAs with a density $N_D^{2D} = 1 \times 10^{13}$ cm^{-2}, a mobility of $\mu = 2000$ cm^2/V s, and a Fermi energy of $E_F - E_0 = 100$ meV. The mean free path is estimated to be $l_{inel} = 550$ Å. Within an area $l_{inel} \times l_{inel}$, there are an average of 300 dopant atoms. This number fluctuates depending on location, since the distribution of dopants can be assumed to be random. The average fluctuation of doping atoms within this area is $\sqrt{300} = 17$ assuming Poisson statistics. The relative variation in doping atoms is $\pm\sqrt{N}/N \cong 0.06$. Because the energy of the lowest state is proportional to $\left[N_D^{2D}\right]^{1/3}$ (see Eqs. (11.3) and (11.10)), the state energies fluctuate by approximately ±4%. At lower doping densities the density fluctuations of the doping atom become larger.

The fluctuations in the subband structure of δ-doped semiconductors are due to the fact that dopant atoms are an 'infrequent' event. Only several tens to several hundreds of dopant atoms are within the area defined by the coherence length of the electron wave function. In compositional quantum wells (e.g.

$Al_x Ga_{1-x} As/GaAs$) such uncertainties are exceedingly small (<< 1%). Potential fluctuations due to compositional fluctuations are negligibly small because the potential is defined by a very *large* number of Al and Ga atoms. Consequently, the Poisson statistical fluctuations $\sqrt{N}/N$ are negligible in compositional quantum wells.

Exchange effects, correlation effects, and an energy-dependent effective mass due to a non-parabolic dispersion relation were taken into account in self-consistent calculations on heterostructures (see e.g. Ando, 1982a and 1982b; Stern and Das Sarma, 1984). Exchange, correlation, and non-parabolicity effects are adequate also for δ-doped structures. However, their influence is typically smaller than the uncertainties due to random dopant distribution. Thus, it does not appear useful to consider exchange and correlation effects and, at the same time, ignore effects due to random dopant distribution. Random potential fluctuations can be taken into account by truly three-dimensional self-consistent calculations, in which a random placement of dopants is realized. Unfortunately, such three-dimensional calculations are quite elaborate and have not hitherto been carried out.

11.1.5 Superlattices with V-shaped potentials

Alternating n- and p-type δ-doping spikes equally spaced and of the same density result in a periodic potential with sawtooth-shaped band-edge potentials. Such doping superlattices have a number of unusual optical and electronic properties including a recombination lifetime which strongly depends on the superlattice period. Experimental properties of doping superlattices are discussed in detail in Sect. 10.3 and 11.5. Here, only theoretical aspects of the superlattice band structure are discussed. Complete self-consistent calculations of the band structure and the charge distribution of compensated and non-compensated δ-doped GaAs sawtooth superlattices as a function of period and doping concentration have been reported by Proetto and coworkers (Proetto, 1990; Ferreira and Proetto, 1990, 1991; Reboredo and Proetto, 1992).

The Kronig–Penney model can be used to calculate the dispersion relation and the wave functions in δ-doped doping superlattices (Schubert, 1993). The calculation of the superlattice dispersion relation and the eigenstate energies are carried out in the effective-mass approximation. The sawtooth-shaped periodic potential is given by

$$V(z) = \begin{cases} (z - m z_p) \, \mathcal{E} \\ \\ (-z - m z_p) \, \mathcal{E} \end{cases} \quad \text{for } m = 0, \pm 1, \pm 2, \, \cdots \quad (11.26)$$

where z_p is the superlattice period. The electric field, $\mathcal{E}$, and the amplitude of the superlattice potential, V_{zz}, are related by $V_{zz} = \frac{1}{2} \mathcal{E} z_p$. The wave function in this periodic, one-dimensional potential can be obtained by the product method using the wave function

$$\psi_{n,k_x,k_y}^{c,v} (x,y,z) = e^{ik_x x} \, e^{ik_y y} \, \psi_n^{c,v}(z) \, u^{c,v}(x,y,z) \quad (11.27)$$

where $e^{ik_x x}$ and $e^{ik_y y}$ represent plane electron waves in the lateral directions (x- and y-directions), $\psi_n^{c,v}(z)$ is the envelope wave function of the conduction (c) or valence (v) band, and $u^{c,v}(x,y,z)$ is the conduction- or valence-band Bloch function of the host material. Note that the envelope function $\psi_n(z)$ depends on z only and describes the quantized motion in the z-direction. The subscript n refers to the nth quantized state of the superlattice. The solution of the envelope function in a linear periodic potential are Airy functions (Abromowitz and Stegun, 1972) which can be applied to doping superlattices (Schubert, 1993). Using this approach, the miniband structure, including its wave functions, miniband width, etc., can be calculated.

The broadening of subbands into minibands allows one to estimate the efficiency of vertical transport in doping superlattices using the Kronig–Penney model. The effective mass along the z-direction, i.e. perpendicular to the doping planes, can be calculated using this model. Using the definition of the effective mass $m_z^* = \hbar^2 \left[d^2 E_z(k_z)/dk_z^2 \right]^{-1}$ and the linearized dispersion relation from the Kronig–Penney model, one obtains the effective dispersion mass for motion in perpendicular direction

$$m_{z,n}^* = \frac{\hbar^2}{z_p^2 \Delta E_n} \quad (11.28)$$

where $2\Delta E_n$ is the energetic width of the nth miniband. Large superlattice periods result in small miniband widths, i.e. (according to Eq. (11.28)) into large effective masses. Carrier transport along the z-direction is then strongly reduced. On the other hand, short periods result in large miniband widths, i.e. small effective dispersion masses and more efficient transport along the superlattice direction (Schubert, 1993).

11.2 Growth, localization, and redistribution

Delta-doped semiconductors can be fabricated solely by epitaxial growth techniques such as OMVPE and MBE. Other doping techniques such as implantation or diffusion are not capable of producing sharp doping profiles. The assessment of the spatial localization of impurities requires the use of highly sensitive techniques such as secondary ion mass spectrometry (SIMS), capacitance–voltage profiling (CV), or Auger spectroscopy. These techniques have resolution widths of 10–20 Å, which allows the accurate determination of spatial localization and redistribution effects. Redistribution of impurities in δ-doped semiconductors is due to several effects. Diffusion, drift, and segregation have been found at high growth temperatures. Furthermore, redistribution of impurities can occur during epitaxial growth as well as during post-growth annealing. In this section, the topics of crystal growth, spatial localization, and redistribution of impurities are summarized and discussed. Although the three topics are interrelated, they are discussed separately for the convenience of the reader.

11.2.1 Growth by VPE

The first report on growth-interrupted doping of III–V semiconductors is due to Bass (1979) who reported the occurrence of Si and Ge doping spikes in GaAs grown by OMVPE. It was general practice during sulphur doping experiments to initiate the hydrogen sulphide flow before growth was started. This was done to reduce the sulfur memory effect. However, if this procedure was done for Si doping, a pronounced Si doping spike occurred at the substrate–epitaxial layer interface. Bass (1979) concluded that Si adsorbs strongly on the GaAs surface before growth is initiated. The adsorption persisted at substrate temperatures below or at the growth temperature. Measurements of the Si doping profile after completion of the epitaxial growth revealed that the Si distribution was several hundred ångströms wide.

More recently, Hobson et al. (1989) reported Zn δ-doping in GaAs grown by OMVPE. The narrowest impurity profiles were 70 Å wide and were obtained for Zn concentrations not exceeding 3×10^{18} cm^{-3}. The δ-doped layers were formed as follows. First, the TMGa was switched to the vent line, stopping growth for 10 s. This was followed by introduction of DEZn for a duration of 5–60 s and then switched back to vent, allowing another 10 s before resuming regular epitaxial growth.

Yang et al. (1992) reported Si δ-doping in AlGaAs/GaInAs heterostructures grown by atmospheric pressure OMVPE. Despite the relative high growth temperatures of 650–700 °C, very narrow doping profiles of 32 Å full width at half-maximum (FWHM) were obtained. The authors pointed out that the typical broadening observed in Si δ-doped GaAs grown by MBE at high temperatures was not observed during OMVPE. The δ-doped heterostructures had high sheet concentrations as well as high electron mobilities.

Si δ-doping was also reported by flow-rate modulation epitaxy (FME) which is a growth technique related to OMVPE (Kobayashi et al., 1986; Makimoto, 1986). In FME, organometallics and hydrides are used as group-III and group-V precursors, respectively. The two precursors are provided alternately, resulting in the growth of one GaAs molecular layer per cycle. This unique characteristic allowed the authors to selectively δ-dope a Ga or an As plane with Si. Higher free carrier concentrations (by a factor of 5) were obtained when the Ga plane was doped. The different concentrations were attributed to a higher SiH_4 sticking coefficient on the atomic As layer compared to a Ga layer. Very narrow Si doping profile widths < 30 Å were obtained for Si densities of 1.1×10^{13} cm^{-3}. The excellent spatial localization was attributed to the growth temperature which was approximately 100 °C lower in FME as compared to conventional OMVPE.

11.2.2 Growth by MBE

During the δ-doping deposition process, epitaxial growth of the III–V semiconductor is suspended. The epitaxial growth is interrupted by closing the group–III element shutter of the effusion cell. The group-V element effusion cell is kept open to provide an As-stabilized surface. Since most popular dopant species, such as Si, Be, Zn, etc., occupy cation sites in III–V semiconductors, the anion-stabilized surface makes cation sites readily available. Typically, the growth temperature is kept between 400 °C and 630 °C for the growth of δ-doped GaAs, i.e. below the congruent sublimation temperature. It is well known that the sticking coefficient of excess As on an As-rich GaAs surface is zero at these growth temperatures (see for example Joyce, 1985), i.e. the surface stoichiometry is maintained.

Upon termination of epitaxial growth, the dopant effusion cell is opened and dopants are evaporated on the non-growing semiconductor surface. It was first realized that Ge is strongly adsorbed (chemisorbed) by the GaAs surface and does not reevaporate significantly. Dopants thus accumulate on the crystal surface during the deposition. The evaporation of dopants on an As-stabilized surface was shown to result in reduced autocompensation for Ge-doping (Wood et al., 1980), since virtually all As sites are occupied and Ge is incorporated predominantly on Ga sites. Similar observations were made for Si in GaAs (Schubert et al., 1987a) and for Si in $Al_x Ga_{1-x} As$ (English et al., 1987).

The two-dimensional (2D) density of donors or acceptors, $N_{D,A}^{2D}$, deposited during growth interruption can be calculated from the growth suspension time, τ, and the dopant cell temperature T. The 2D dopant density is given by

$$N^{2D} = N\,v_g\,\tau \tag{11.29}$$

where N is the three-dimensional dopant concentration obtained at the doping cell temperature T and a growth rate v_g. For example, a time $\tau = 18$ s is required to deposit 10^{12} dopants per cm^2 at a dopant cell temperature, which results in a 3D concentration of 2×10^{18} cm^{-3} at a growth rate of 1 μm/hr. It is advisable to choose a *high* dopant cell temperature and keep the growth interruption time minimal and hence reduce the incorporation of unwanted impurities. After dopant deposition, the regular crystal growth is resumed. For atomically flat crystal surfaces and in the absence of diffusion or segregation processes, dopants are confined to a single monolayer of the host semiconductor.

The surface reconstruction changes during growth interruption and during dopant deposition. During growth below the congruent sublimation temperature at Ga/As$_4$ flux ratios close to one, a (1×1) surface reconstruction on GaAs (001) is found by reflection high-energy electron diffraction (RHEED). For Ga/As$_4$ flux ratios < 1, a (2×4) surface reconstruction is found. When the Ga flux is terminated, the surface remains As-stabilized and in a (2×4) As-rich reconstruction. After prolonged deposition of Si on GaAs, a change from the (2×4) to the (1×3) surface reconstruction occurs (Schubert et al., 1986b).

11.2.3 Assessment of localization by SIMS

Secondary ion mass spectrometry (SIMS) is a powerful structural technique with high depth resolution. SIMS has been employed to study the spatial localization of impurities in δ-doped III–V semiconductors by several research groups (Beall et al., 1988, 1989; Lanzillotto et al., 1989; Schubert et al., 1990d). These reports concluded that δ-function-like doping profiles of Si and Be in GaAs can be achieved at sufficiently low growth temperatures. In this case, impurities are confined to one or a few monolayers of the host semiconductor. Detectable broadening of the impurity distribution was found at elevated growth temperatures and at high impurity densities.

The SIMS profile of a Be δ-doped GaAs film is shown in Fig. 11.4. A narrow peak is observed at a depth of ~200 Å which corresponds to the Be δ-doped layer. The full width at half maximum of the Be peak is 29 Å, which is the narrowest SIMS profile measured for δ-doped GaAs grown at typical growth temperatures. Such narrow SIMS profiles can only be obtained under optimized measurement conditions such as a low primary ion energy (1–2 keV), a shallow

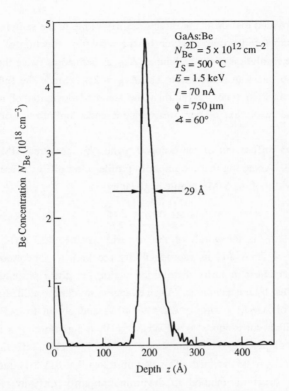

Fig. 11.4. Secondary ion mass spectrometry (SIMS) profile on epitaxial GaAs containing a Be δ-doped layer 200 Å below the surface. The following parameters were used for the MBE growth and the SIMS measurement: growth temperature 500 °C, primary ion energy 1.5 keV, primary ion current 70 nA, raster diameter 750 μm, sputtering angle 60°.

sputtering angle (60°), and a shallow depth of the δ-doped layer (100–300 Å). Several broadening mechanisms determine the width of the SIMS profiles. The two dominant broadening mechanisms are (i) surface roughening and (ii) ion mixing, also called the 'knock-on' effect (Benninghoven et al., 1987). Surface roughening is expected to broaden SIMS profiles symmetrically, while the knock-on effect skews the profile predominantly along the sputter direction. The symmetric shape of the profile shown in Fig. 11.4 indicates that surface roughness is the dominant broadening mechanism rather than the 'knock-on' effect.

The resolution of the SIMS technique is generally obtained by assuming that the shape of the profile is gaussian (Benninghoven et al., 1987). The symmetric gaussian distribution is justified since the profile shown in Fig. 11.4 is indeed symmetric. Provided that diffusion effects are negligible in the sample shown in Fig. 11.4, the measured profile is given by the resolution function of the SIMS technique. The resolution of the technique, Δz_R, is defined as twice the standard deviation of the gaussian distribution, i.e., $\Delta z_R = 2\sigma_r$. Since the full width at half-maximum is 2.36 times larger than the standard deviation of a gaussian distribution, the resolution of our measurement under optimized conditions is $\Delta z_R = 25$ Å.

If impurities diffuse out of the δ-doped plane, the measured SIMS profiles broaden as well. Assuming that the impurity profiles after diffusion are gaussian, the measured width of the SIMS profile is given by

$$\Delta z_M^2 = \Delta z_D^2 + \Delta z_R^2 \tag{11.30}$$

where $\Delta z_D = 2\sigma_D$ is the width of the impurity profile after diffusion. The resolution $\Delta z_R = 25$ Å and its reproducibility are known. Therefore, we can determine the smallest impurity distribution width, i.e. diffusion length, which can be determined by the technique. As an example, we choose a diffusion length for impurities of $\Delta z_D/2 = \sigma_D = \sqrt{2D\tau} = 10$ Å and obtain from Eq. (11.30) $z_M = 32$ Å which corresponds to a FWHM = 38 Å representing a significant broadening as compared to the profile shown in Fig. 11.4. Thus, diffusion lengths of ≤10 Å can clearly be detected. This demonstrates that SIMS is currently the most sensitive structural method to determine impurity diffusion lengths in semiconductors.

The depth dependence of the SIMS resolution is usually expressed by the linear equation

$$\Delta z_R = \alpha + \beta z \tag{11.31}$$

where α is the resolution for $z = 0$ and β is a parameter which represents the deterioration of the resolution with depth. The deterioration of depth resolution increases with sputter depth because of the roughening of the crater surface due to the statistical nature of the sputter process. The parameter β can be obtained from the broadening of the peak width with depth (Schubert et al., 1990d). Evaluation of β yielded $\beta = 5 \times 10^{-3}$ for low-energy sputtering conditions. Similar values were obtained by Clegg and Beall (1989) and Lanzillotto et al. (1989).

11.2.4 Assessment of localization by the CV-technique

The capacitance–voltage profiling technique has been used extensively to assess the spatial localization of impurities in δ-doped semiconductors (Bass, 1979; Lee et al., 1985; Sasa et al., 1985; Schubert et al., 1988a, 1990a). The capacitance–voltage (CV) profiling is a well-known electrical characterization technique to spatially resolve impurity distributions in semiconductors. During the measurement, the capacitance of a metal–semiconductor junction, a p^+n-, or an n^+ p-junction is measured as a function of the applied bias. For such junctions, the CV-concentration, N_{CV}, is inferred from the CV-measurement using the equation

$$N_{CV} = \frac{-2}{e\varepsilon}\,(d\frac{1}{C^2}/dV)^{-1} = \frac{C^3}{e\varepsilon}\frac{dV}{dC} \qquad (11.32)$$

where e and ε are the elementary charge and the permittivity of the semiconductor, respectively, and C is the capacitance of the junction per unit area. The CV-depth, z_{CV}, is obtained from the reciprocal of the capacitance

$$z_{CV} = \varepsilon/C . \qquad (11.33)$$

Equations (11.32) and (11.33) allow one to plot N_{CV} versus z_{CV}, which is called the *CV-profile*.

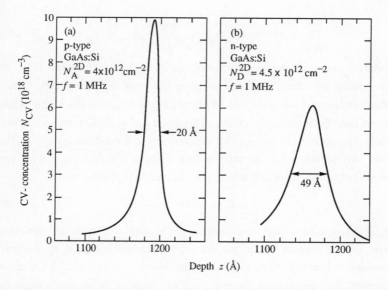

Fig. 11.5. CV-profiles on (a) p-type and (b) n-type δ-doped GaAs. The Be and Si densities are 4.0×10^{12} and 4.5×10^{12} cm^{-2}, respectively.

The CV-profiles measured at 300 K of Be and Si δ-doped GaAs are shown in Fig. 11.5. The Be and Si densities are 4×10^{12} and 4.5×10^{12} cm^{-2}, respectively. The CV-profile widths are 20 Å and 49 Å for the Be and Si profile, respectively. Note that the profile widths are significantly different for the p-type and n-type profile. The difference can thus not be explained by the conventional Debye limitation of the CV-profiling technique.

The spatial resolution of the CV-profiling technique is given by the screening length and the spatial extent of the carrier system for classical and quantum-mechanical carrier distributions, respectively. If the doping concentration varies along the growth direction, the free carrier profile is smeared out as compared to the doping profile. Kennedy et al. (1968) and Kennedy and O'Brien (1969) showed, that the resolution of the CV-technique is limited to the majority carrier screening length. Furthermore, the authors showed that the CV-profile $N_{CV}(z)$, closely follows the free carrier profile, $n(z)$. The majority carrier screening or Debye length in non-degenerately doped semiconductors is given by

$$r_D = \sqrt{\frac{\varepsilon kT}{e^2 n}} \; . \tag{11.34}$$

The Thomas–Fermi screening length for degenerately doped semiconductors is

$$r_{TF} = \pi^{2/3} \sqrt{\frac{\varepsilon \hbar^2}{e^2 m^* (3n)^{1/3}}} \tag{11.35}$$

Note that the two screening lengths have different dependences on n and T.

The spatial resolution of the CV-profiling technique in *semiconductors with quantum confinement* was shown to be limited to the spatial extent of the carrier wave function (Schubert et al., 1990a). This result can be understood as follows. Consider a quantized electron system described by a wave function $\psi(z)$. The position expectation value of the electron system is $\langle z \rangle = \langle \psi | z | \psi \rangle$. The differential capacitance per unit area of such an electron system equals

$$C = \varepsilon / \langle z \rangle \; . \tag{11.36}$$

Upon application of a bias, the position expectation value of the quantized electron system changes due to the potential perturbation caused by the bias. Comparison of Eqs. (11.36) and (11.33) yields that $z_{CV} = \langle z \rangle$, i.e. the CV-depth (at a certain bias) equals the position expectation value of the electron system (at a certain bias). Thus, the *resolution* of the CV-technique, Δz_{CV}, is given by the change of the position expectation value, $\Delta \langle z \rangle$, upon sweeping the bias, i.e.

$$\Delta z_{CV} = \Delta \langle z \rangle \; . \tag{11.37}$$

Schubert et al. (1990a) showed that the change in position expectation value is approximately equal to the spatial width of the wave function, z_0, i.e. approximately equal to the width of the confining potential. For δ-doped semiconductors, the spatial extent of the ground-state wave function has been calculated using a variational method. The resolution, i.e. CV-profile width, was obtained as

$$\Delta z_{CV} = z_0 = 2\sqrt{\frac{7}{5}} \left[\frac{4}{9} \frac{\varepsilon \hbar^2}{e^2 N^{2D} m^*} \right]^{1/3} \qquad (11.38)$$

Using this result, the calculated resolution for the two profiles shown in Fig. 11.5 is given by $\Delta z_{CV} = 26$ Å and 48 Å, respectively, which agrees very well with the experimental result. Thus, the CV-profiles shown in Fig. 11.5 are resolution limited, i.e. broadening of the impurity distribution due to redistribution effects is negligible.

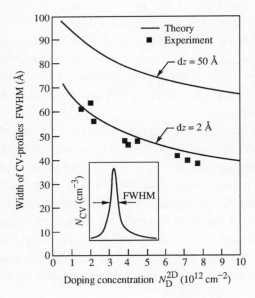

Fig.11.6. Comparison of theoretical (solid lines) and experimental (squares) full-widths at half-maximum (FWHM) of CV-profiles obtained from n-type δ-doped GaAs at different doping densities. The two theoretical curves are obtained from self-consistent calculations assuming doping distribution widths of 2 Å (δ-doped) and 50 Å (pulse-doped). Good agreement is obtained, only if diffusion is assumed to be negligible.

Self-consistently calculated CV-profile widths are shown in Fig. 11.6 as a function of n-type doping density in GaAs (Schubert et al., 1988a). Size quantization of the carrier system was taken into account via the Schrödinger equation. The impurity distribution assumed for the calculation is top-hat-shaped with a width of $dz = 2$ Å (δ-doping) and $dz = 50$ Å (pulse doping). Comparison of the calculated width with experimentally measured widths (squares) indicates that the profile widths are again resolution limited. Furthermore, an impurity distribution width of $dz = 50$ Å is too wide to account for the experimental results.

11.2.5 Assessment of localization by Auger technique

Auger electron spectroscopy is usually not considered to be a useful technique to study impurity distributions in semiconductors. This is due to the low surface sensitivity of Auger spectroscopy of about 0.1%–1%. For Si in GaAs, 10^{19} cm^{-3} is considered a high doping concentration, but in terms of bulk atomic concentration, 10^{19} cm^{-3} equals about 0.02%, i.e. below the typical Auger sensitivity. Nevertheless, Webb (1989) showed that Auger electron spectroscopy can be very useful for studying the segregation of impurities in δ-doped semiconductors. Doping densities of 10^{13} cm^{-2} exceed 1% in atomic surface concentration which can be detected by Auger analysis.

In order to investigate Si segregation, Webb (1989) grew a buffer layer of undoped GaAs by MBE, then deposited 3×10^{13} cm^{-2} Si atoms, and finally grew a 10 Å GaAs cap layer. Without any Si segregation, a decrease in the Si-Auger-electron intensity by a factor of $\exp(10 \text{ Å}/l_{\text{electron}}) \cong 7$ is expected for an electron mean free path of $l_{\text{electron}} = 5$ Å. If segregation does occur, the attenuation factor will be smaller. Thus, Auger electron spectroscopy allows one to study impurity redistribution towards the surface in δ-doped semiconductors with a sufficiently high impurity density.

Webb (1989) investigated segregation of Si in δ-doped GaAs as a function of growth temperature using Auger spectroscopy. The results of the study are summarized as follows. No segregation was detectable at growth temperatures of 520 °C, indicating good spatial confinement of Si δ-doped GaAs even at Si densities as high as 3×10^{13} cm^{-2}. Considering the high spatial resolution of the experiment, it can be concluded that the Si impurities are confined to a layer of thickness < 10 Å. At elevated growth temperatures of approximately 540–590 °C, Si segregation did occur. The redistribution of Si increased at higher growth temperatures. An activation energy of 2.4 eV was estimated for the thermally activated redistribution of Si impurities.

11.2.6 Redistribution of impurities

Spatial redistribution of impurities can occur during epitaxial growth at elevated growth temperatures as well as during post-growth annealing. The magnitude of the impurity profile broadening depends on a number of parameters such as the temperature, impurity species, growth conditions, defect concentration in the semiconductor, etc. Despite the large number of parameters, there are two general tendencies for impurity redistribution. First, the magnitude of redistribution increases at high temperatures. Second, for some impurity species (e.g. Be or Zn), the magnitude of redistribution increases strongly with impurity concentration.

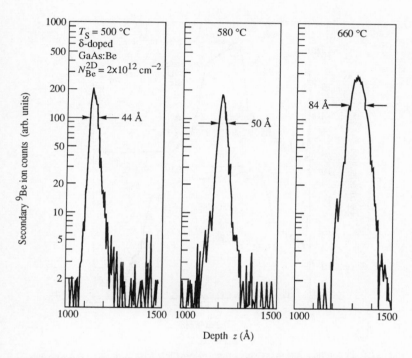

Fig. 11.7. Secondary ion mass spectrometry (SIMS) profiles on three Be δ-doped GaAs samples grown at 500, 580, and 660 °C. The SIMS profiles broaden systematically at high growth temperatures.

An example of impurity redistribution during growth is shown in Fig. 11.7 for Be δ-doped GaAs grown by MBE at temperatures of 500 °C, 580 °C, and 660 °C. The Be impurity profile in Fig. 11.7 was measured by SIMS. The nominal Be density of the three samples is 2×10^{12} cm^{-2}. At low growth temperatures of 500 °C, the width of the Be profile is 44 Å which is a resolution-limited profile for the measurement conditions employed. As the growth temperature is increased to 580 °C, the profile width increases to 50 Å. The profile broadening is more pronounced for a growth temperature of 660 °C where the profile width is 84 Å. The basic trend of impurity profile broadening at elevated growth temperatures has been observed for many different impurities such as Be, Zn, C, and Si, as well as different III–V semiconductors such as

GaAs, $Al_x Ga_{1-x} As$, $Ga_{0.47} In_{0.53} As$, $Al_{0.48} In_{0.52} As$ and other materials (Schubert, 1990b; Schubert et al., 1989b, 1990c). The first indication of impurity redistribution in δ-doped semiconductors was reported by Lee et al. (1985).

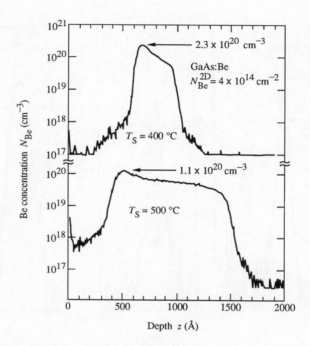

Fig. 11.8. Secondary ion mass spectrometry profile of Be δ-doped GaAs grown at (top) 400 °C and (bottom) 500 °C. The lower growth temperature results in reduced broadening and a higher Be peak concentration.

The occurrence of *impurity redistribution during growth* cannot be explained by a single physical mechanism. Instead, several redistribution mechanisms are required to fully account for the multitude of redistribution effects occurring during growth at elevated growth temperatures. The mechanisms and their physical driving forces were discussed in detail in Chap. 8. Here, we restrict ourselves to summarizing briefly the mechanisms. *Diffusion* of impurities results in a symmetric (i.e. gaussian) impurity distribution which increases in width with temperature. The diffusion coefficient can depend on the impurity concentration. For example, the substitutional–interstitial model (see Chap. 8) predicts an increase of the diffusion coefficient with the concentration according to a power law. Redistribution of impurities toward the semiconductor surface is termed *impurity segregation.* The driving force for surface segregation can be (i) an electric field due to a surface dipole (Schubert et al., 1990b) or (ii) a lower energy of the impurity atom on the surface as compared to the bulk (Harris et al., 1984). The lower surface energy can be due to a different bonding configuration and strength of the impurity on the surface as opposed to the bulk. Some impurities exhibit a *rigid upper limit* of their concentration (e.g. N_{Be}, max $\cong 10^{20}$ cm^{-3} in GaAs). The physical mechanism limiting the maximum concentration can be a chemical solubility limit. In addition, *repulsive interaction* between impurities can also explain the maximum concentration limit observed for some impurities in semiconductors. According to this model (see Chap. 8), the maximum achievable doping concentration increases with decreasing growth temperature. Experimental results confirming this trend are shown in Fig. 11.8. The illustration shows the Be profiles of two GaAs samples grown by MBE at different temperatures. The Be impurities of density 4×10^{14} cm^{-2} were deposited during growth interruption. Subsequently, a 1000 Å thick top layer was grown by MBE. The entire growth and the Be deposition were performed at the two temperatures 500 °C and 400 °C. The profiles of Fig. 11.8 show (i) that significant broadening occurs for both growth temperatures, and (ii) that the profile of the sample grown at 400 °C is narrower and has a higher maximum concentration as compared to the sample grown at 500 °C. The change in width and maximum concentration is approximately a factor of two for the decrease in temperature by 100 °C. Thus, the results illustrated in Fig. 11.8 are consistent with the model of repulsive interaction between impurities.

The redistribution of impurities in III–V semiconductors follows some general tendencies which depend primarily on the impurity element rather than on the host semiconductor. The general characteristics of impurities used in δ-doping are summarized as follows.

Beryllium: Be is a relatively stable p-type impurity in III–V semiconductors. Excellent spatial localization can be achieved for concentrations below 10^{13} cm^{-2} (see also Chap. 8). However, when bulk concentrations are in the high 10^{19} cm^{-3} range, strong redistribution of Be in GaAs occurs (Schubert et al., 1990c).

Zinc: The characteristics of Zn are qualitatively similar to Be. Good spatial localization can be achieved at low growth temperatures and low Zn densities. However, strong spatial redistribution occurs for high Zn concentrations, typically in the $10^{18} - 10^{19}$ cm^{-3} range. Quantitatively, Zn is a faster diffuser than Be in many III–V semiconductors.

Carbon: C diffuses very slowly in III–V semiconductors and is characterized by exceptional stability (see also Chap. 8). However, the electrical activity of C was found to decrease for C densities exceeding 10^{12} cm^{-2} in GaAs (Nagle et al., 1991).

Silicon: Excellent spatial localization can be achieved with Si at low growth temperatures (see also Chap. 8). The electrical activity decreases strongly for Si densities exceeding 10^{13} cm^{-2} in GaAs (see also Sect. 11.3).

11.3 Transport properties

The electronic properties of δ-doped semiconductors are fundamentally different from homogeneously doped semiconductors. Impurities are randomly distributed in such homogeneously doped semiconductors. In contrast, impurities are confined to a plane in δ-doped semiconductors. This confinement results in changes of the electronic and transport properties of δ-doped as compared to homogeneously doped semiconductors which will be discussed below.

The free carrier concentrations of Be and Si δ-doped GaAs at $T = 77$ and 300 K are shown in Fig. 11.9 as a function of the impurity concentration (Schubert et al., 1986c). The 2D impurity density was inferred from the deposition time, growth rate, and the impurity effusion cell flux using Eq. (11.29). The 2D Hall carrier density was then measured for the n-type and p-type samples shown in Fig. 11.9 at $T = 77$ and 300 K. The solid line represents full activation of the impurities, i.e. $n^{2D} = N_D^{2D}$ and $p^{2D} = N_A^{2D}$. The n- as well as the p-type carrier concentration closely follow the solid line for impurity densities $\leq 10^{13}$ cm^{-2} indicating full electrical activation of the impurities. Full electrical activity of Be δ-doped GaAs is maintained at concentrations exceeding 10^{13} cm^{-2}. However, broadening of the Be distribution may occur at such high densities (Schubert et al., 1990c).

For Si doping densities exceeding 10^{13} cm^{-2}, the free carrier concentration saturates at approximately 10^{13} cm^{-2}. The saturation of the free carrier concentration was investigated by several groups. Beall et al. (1989) concluded that the free carrier saturation is caused by non-substitutional Si impurities in GaAs. The authors demonstrated the presence of electrically inactive Si complexes, Si_{As} acceptors, and $Si_{Ga} - Si_{As}$ pairs in heavily δ-doped GaAs. Similar results were previously reported by Maguire et al. (1987) for homogeneously doped GaAs. Schubert et al. (1990a) confirmed the conclusion of electrically inactive impurities by a comparison of CV and Hall measurements. Koenraad et al. (1990a) concluded from measurements under hydrostatic pressure that the saturation is *not* caused by the population of DX-centers. The exact

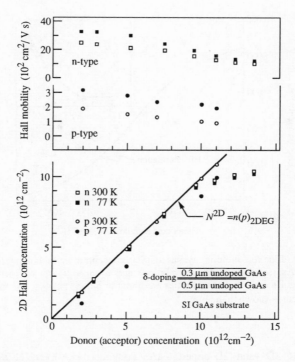

Fig. 11.9. Mobility (top) and free carrier Hall concentration (bottom) of Si and Be δ-doped GaAs versus dopant concentration.

density at which the saturation occurs, depends on the crystal growth conditions. The saturation density varies between 8×10^{12} cm^{-2} and 2.5×10^{13} cm^{-2} (Gillman et al., 1988).

The carrier mobility in δ-doped semiconductors is significantly different from homogeneously doped semiconductors. Schubert et al. (1987a) showed that ionized impurity scattering in δ-doped GaAs is reduced compared to homogeneously doped GaAs. Experimental electron mobilities of δ-doped GaAs at $T = 300$ K are shown in Fig. 11.10 as a function of the equivalent 3D doping concentration. This equivalent concentration is inferred from the 2D density according to $N_D = \left[N_D^{2D} \right]^{3/2}$. The equivalent 2D and 3D doping densities have the same mean distance between impurity atoms.

The experimental points shown in Fig. 11.10 are next compared with theoretical electron mobilities of n-type GaAs obtained from the Hilsum relation (Hilsum, 1974) shown by the solid line. At low doping concentrations, the

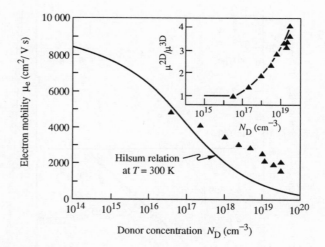

Fig. 11.10. Electron mobility versus doping concentration in homogeneously doped GaAs (solid line, Hilsum relation) and δ-doped GaAs (experimental points). Significantly higher mobilities are found in δ-doped GaAs. The mobility enhancement is shown in the inset.

mobilities of 3D- and 2D-doped GaAs coincide. However, at high doping concentrations, the mobility of 2D-doped GaAs significantly exceeds the mobility of homogeneously doped material (see Fig. 11.10 for illustration). At a concentration of $N_D^{3D} = 3.5 \times 10^{19}$ cm^{-3}, the mobility-enhancement μ^{2D}/μ^{3D} is more than a factor of 4. The enhancement is shown in the inset of Fig. 11.10. The distinguishing feature is the *increasing enhancement* of the electron mobility *at high doping concentrations*.

Temperature-dependent Hall measurements were performed to clarify the origins of the mobility enhancement (Schubert et al., 1987a). The electron mobility depends weakly on temperature in the entire temperature range, indicating a reduced role of impurity scattering. Ionized impurity scattering in conventionally, *non-degenerately*, 3D-doped samples depends on the sample temperature according to (Brooks–Herring)

$$\mu_{II} = \frac{128(2\pi)^{1/2} \, \varepsilon^2 \, (kT)^{3/2}}{(m^*)^{1/2} \, N_D^{3D} \, e^3} \left[\ln \frac{24m^* \, \varepsilon(kT)^2}{n \, e^2 \, \hbar^2} \right]^{-1} . \qquad (11.39)$$

If impurity scattering were important, a dependence of $\mu \propto T^{3/2}$ would be expected at low temperatures. Ionized impurity scattering in *degenerately* doped semiconductors is beset with a number of difficulties (Chattopadhyay and Queisser, 1981), and cannot be expressed in closed form. However, a weaker temperature dependence is expected for degenerate doping, since the *Fermi* velocity rather than the *thermal* velocity determines the average kinetic energy of carriers. The Fermi velocity depends weakly on temperature. A temperature dependence of $\mu \propto T^{0.29}$ is found in low temperatures for a heavily δ-doped sample. The weak $T^{0.29}$ dependence is significantly smaller than the $T^{1.5}$ dependence inferred from Eq. (11.39), indicating the strong degeneracy of the electron gas.

Two theoretical origins will be discussed for the mobility enhancement, i.e. (i) high degeneracy of the Fermi gas, and (ii) spatial electron–donor separation in the $n = 1, 3, ...$ subbands. The 2DEG of δ-doped epitaxial layers is highly degenerate (i.e. high kinetic energy parallel to the donor plane). Carriers energetically close to the Fermi surface, which are in turn most relevant to elastic and inelastic scattering, have a kinetic energy of $E_F - E_n$ where E_n is the eigenstate energy. The kinetic energy can exceed the thermal energy kT considerably. Consequently, the temperature-dependent elastic impurity scattering is reduced as long as the kinetic energy term, $E_F - E_n$, dominates, which occurs at high concentrations. The latter condition is fulfilled in the entire temperature range investigated. It is worthwhile noting that the eigenstate energy itself is partly potential and partly kinetic (in the perpendicular direction to the well plane). The kinetic part can be calculated from

$$\langle E_{kin} \rangle = E_n - \langle \psi(x) | E_c(x) | \psi(x) \rangle \qquad (11.40)$$

and is usually a considerable fraction of the total eigenstate energy. The kinetic part of the eigenstate energy does not, however, enter the calculation of ionized impurity scattering in Eq. (11.39).

Spatial separation of free electrons from their parent ionized donor impurities occur in the odd-labeled ($n = 1, 3, ...$) subbands. Carriers in the $n = 1$ state therefore have zero probability of being in the donor plane, resulting in a mobility enhancement of electrons in odd-numbered eigenstates. This electron–donor separation in real space in δ-doped structures is reminiscent of the carrier–impurity separation in selectively doped heterostructures.

A mobility enhancement in δ-doped GaAs was also found by Koenraad et al. (1990b) using cyclotron resonance measurements. They attributed the enhanced mobility to the small overlap between the ionized donors and the electron wave function, especially for states with high subband index. Subsequently, an enhancement of electron mobility was also found for Sb δ-doped Si (Gossmann and Schubert, 1993).

The electronic wave function in δ-doped semiconductors can be influenced further by compositional quantum wells. Masselink (1991a) showed that the overlap of the electronic ground-state wave function of a square-shaped well is maximized if the impurities are confined to the center of the well. In this case, the electron mobility of the ground state is lower as compared to bulk GaAs doped at the same level. High field transport measurements in $Al_x Ga_{1-x} As/GaAs$ quantum wells revealed an increasing differential mobility with increasing electric field (Masselink, 1991b). The large increase in differential mobility was attributed to heating of electrons out of the symmetric ground state into the antisymmetric first excited state. Because the excited state has a node at the δ-doped plane of impurities, these hot electrons have a much smaller overlap with ionized impurities in the well center and therefore have a higher mobility.

Delta-doping in n-type $Al_x Ga_{1-x} As$ was shown to strongly influence the formation of DX-centers (Etienne and Thierry-Mieg, 1988). The number of DX-centers in $Al_{0.32} Ga_{0.68} As$ was found to decrease with increasing doping concentration. At sufficiently high doping densities, the effect of persistent photoconductivity, which is known to be due to DX-centers, vanishes. A physical mechanism causing the reduction of DX-centers was proposed by Koenraad (1990a), who argued that the energy of the DX-level depends on the doping concentration. The DX-level shifts away from the Γ conduction band minimum in heavily δ-doped GaAs and, most likely, $Al_x Ga_{1-x} As$. Since the DX-level shifts to higher energies, it becomes unpopulated. As a result, persistent photoconductivity and free carrier freeze out, which are associated with the DX-center, are not observed in heavily δ-doped $Al_x Ga_{1-x} As$.

Elastic electron scattering of carriers by ionized impurities in δ-doped semiconductors has been considered by Levi et al. (1989) for *random* and *ordered* distributions of impurities in the δ-doped plane. It is a common assumption that impurities in homogeneously doped semiconductors are distributed randomly, i.e. are distributed according to the Poisson distribution (Shockley, 1961). In δ-doped semiconductors, the distribution of impurities in the δ-doped plane is usually also assumed to be poissonian. However, Levi et al. (1989) proposed that, under suitable growth conditions, high-density δ-doping may result in partial spatial ordering of the impurity atoms. A reduction of the elastic scattering rate by up to a factor of four was predicted. To date, experimental evidence for ordered doping distributions in III–V semiconductors has not been reported. However, Headrick et al. (1990 and 1991) reported on δ-doping with a full one monolayer coverage of boron (B) impurities in Si. The mobilities measured were comparable to highly B-doped bulk Si.

11.4 Electronic devices

The employment of δ-doping opens up new possibilities for device structures. First, the δ-doping technique allows one to improve and optimize known semiconductor device structures. As will be shown below, a δ-function-like doping profile represents the optimum doping distribution for many well-known device structures. For example, δ-doping in high-mobility heterostructures minimizes remote ionized impurity scattering. Thus, modulation-doped heterostructures are *optimized* by the employment of the δ-doping technique. Furthermore, δ-doping profiles in semiconductors are the ultimate spatial distribution of dopants. It is well-known that the performance and speed of semiconductor devices increase as the spatial dimensions of the device structure decrease. Narrower doping distributions are a natural consequence of the spatial scaling process. Delta-doped semiconductors therefore play an increasingly important role in semiconductor devices scaled down to their ultimate spatial limit.

Second, δ-doped semiconductors lead to entirely novel concepts that cannot be realized in conventional, homogeneously doped structures. Delta-doped structures exhibit a quantized free carrier distribution at sufficiently high doping densities. Such a quantized free carrier system can only be achieved in δ-doped structures but not in homogeneously doped structures. Examples of structures that can be realized only by the δ-doping technique will be discussed below.

11.4.1 Non-alloyed ohmic contacts

Ohmic contacts play a central role in semiconductor devices. Ohmic contact metallizations usually consist of a dopant such as Be, Zn, or Ge, and a low-resistance metal such as Au. Upon alloying, dopants diffuse into the semiconductor and form a highly doped semiconductor–metal junction. Ohmic metal–semiconductor contacts are tunnel contacts and their contact resistance is determined by the barrier height and the width of the depletion region. While the barrier height is a constant for a given metal–semiconductor junction, the width of the depletion region depends on the doping concentration. High doping concentrations decrease the width of the depletion region and lower the contact resistance. The high doping concentrations achievable by the δ-doping technique are therefore advantageous for low-resistance ohmic contact formation.

The band diagram of a δ-doped metal–semiconductor contact is shown in Fig. 11.11 (Schubert et al., 1986b). The δ-doped layer with density N^{2D} is located at a distance z_D from the metal–semiconductor junction. The figure also shows that the effective tunnel barrier thickness depends on the voltage applied to the junction. The doping density must be sufficiently high in order to lower the

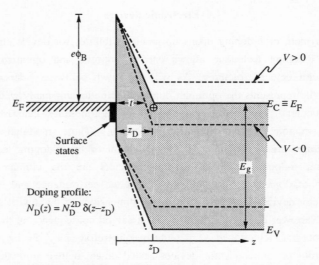

Fig. 11.11. Band diagram of a δ-doped non-alloyed ohmic contact. Since the thickness t is small (~20 Å), tunneling is the dominant transport mechanism across the Schottky barrier.

conduction band at the doped region to or below the Fermi level. The minimum doping density is then given by

$$N^{2D} \geq \frac{\varepsilon \phi_B}{e z_D} \tag{11.41}$$

where ε, e, and ϕ_B are the permittivity, elementary charge and the semiconductor barrier height, respectively.

The specific contact resistance, ρ_c, of a δ-doped ohmic contact can be calculated as a function of the distance z_D, based on an isotropic, parabolic semiconductor band structure and the well-known Fowler–Nordheim tunneling current equation. The result of the calculation is (Schubert et al., 1986b)

$$(\rho_c)^{-1} = \left[\frac{e}{2\pi\hbar z_D} \right]^2 (\hbar + z_D \sqrt{em^* \phi_B}) \exp\left[-\frac{2 z_D}{\hbar} \sqrt{em^* \phi_B} \right]. \tag{11.42}$$

Here the electron effective mass, m^*, is the mass associated with motion perpendicular to the metal–semiconductor interfacial plane. At high doping concentrations, contact resistances $< 10^{-8}$ Ω cm^2 can be inferred from Eq. (11.42).

The current–voltage characteristic of a non-alloyed ohmic contact is shown in

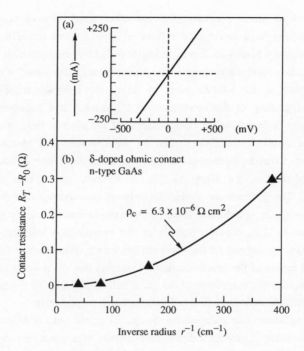

Fig. 11.12. (a) Current–voltage characteristic of the non-alloyed δ-doped ohmic contact to n-type GaAs. Evaluation of the specific contact resistance of typical δ-doped contacts reveals a specific contact resistance of $6.3 \times 10^{-6} \ \Omega \ cm^2$. The lowest contact resistances measured are in the 1×10^{-6} to $2 \times 10^{-6} \ \Omega \ cm^2$ range.

Fig. 11.12(a) for a GaAs:Si sample grown by MBE. The characteristic is strictly linear. The evaluation of the specific contact resistance is shown in Fig. 11.12(b) for circular contacts of various diameters. Evaluation of the contact resistance yields $\rho_c = 6.3 \times 10^{-6} \ \Omega \ cm^2$. Further experimental details were reported by Schubert et al. (1986b).

Non-alloyed ohmic contacts have a smooth surface morphology due to the lack of any thermal processing. The smooth surface morphology is advantageous for photolithographic processing of small patterns. Furthermore, the smooth surface morphology of the metal permits use of the metal as a reflector for optical radiation. Such reflectors are required in many optoelectronic devices.

11.4.2 Elimination of heterojunction band discontinuities

Heterojunction band discontinuities will be an important aspect of future high-speed devices. In these devices, the employment of band discontinuities allows the realization of novel concepts. On the other hand, heterojunction band discontinuities can have detrimental effects in some devices. Specifically, the current flow across a heterojunction is clearly impeded by heterojunction barriers.

Heterojunction band discontinuities can be completely eliminated in either the conduction band or the valence band by using the δ-doping technique and compositional grading of the heterojunction (Gossmann and Schubert, 1993). The elimination is based on the compensation of the electric field of a doping dipole by the quasi-electric field caused by the compositional grading of the heterojunction. Consider a junction between two semiconductors 'A' and 'B'. In the interfacial region, the alloy $A_{1-x}B_x$ is formed, where x is the alloy composition. The junction is graded *linearly*, if the composition x depends *linearly* on the spatial coordinate. If we further assume that the energy gap of the semiconductor $A_{1-x}B_x$ depends linearly on the composition and that the ratio $\Delta E_C / \Delta E_V$ does not depend on the composition, then a quasi-electric field exists in the graded region of the heterojunction. Assuming that Δz is the length of the graded region, then the magnitude of the quasi-field is given by $\mathcal{E} = (\Delta E_C / e \Delta z)$ and $(\Delta E_V / e \Delta z)$ for the conduction and valence band, respectively.

A δ-doping dipole can compensate the quasi-electric field at heterojunction band discontinuities. Consider a dipole that consists of a donor and an acceptor sheet separated by Δz. The donor and acceptor doping densities $N_D^{2D} = N_A^{2D} = N^{2D}$ are assumed to be equal. Then the structure will be 'compensated', i.e. will not contain free carriers originating from impurities. The electric field of the dipole can be calculated from Poisson's equation; the magnitude of the field is given by

$$\mathcal{E} = (e/\varepsilon)\, N^{2D} \ . \tag{11.43}$$

Equating the quasi-field caused by compositional grading and the field caused by the doping dipole allows one to determine the required impurity density according to

$$N^{2D} = \frac{\varepsilon \Delta E_C}{e^2\, \Delta z} \ . \tag{11.44}$$

At this doping density, the built-in field due to the graded composition exactly compensates the electric field due to the doping dipole. Thus, the band discontinuity is eliminated.

Heterojunction band discontinuities can also be eliminated by using modulation doping and parabolic grading (Schubert et al., 1992). Another possibility of modifying the properties of heterojunctions constitutes the concept of 'tuning' the magnitude of band discontinuities by doping dipoles (Capasso et al., 1985a and 1985b). The doping dipoles are closely spaced δ-doped donor and acceptor sheets. For appropriately selected donor and acceptor concentrations, the 'effective' discontinuity can be tuned over a wide range of energies. Thus, the two concepts of *tuning* (doping without grading) and *elimination* (doping and compositional grading) of heterojunction band discontinuities allow significant modifications of such junctions.

11.4.3 Schottky–Gate field effect transistors

The properties of field-effect transistors improve as the spatial dimensions of the transistor are scaled down in size. The specific method of improvement depends on the type of field-effect transistor considered. In this section, δ-doped *homostructure* field-effect transistors (FETs) and selectively δ-doped *heterostructure* FETs are discussed. (Board et al., 1981; Schubert and Ploog 1985; Schubert et al., 1986c).

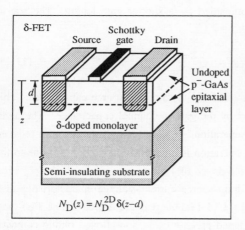

Fig. 11.13. Delta-doped GaAs metal–semiconductor field-effect transistor (δ-FET). The conductive channel is a Si δ-doped layer located at a distance d below the Scottky gate.

A schematic illustration of a δ-doped FET is shown in Fig. 11.13. The simplest structure of such an FET consists of a semi-insulating substrate and an epitaxial layer containing a δ-doped spike. A highly doped region near the semiconductor surface improves ohmic contact resistance as well as the source resistance of the device (Schubert et al., 1986a).

The advantages of the δ-doped Schottky-gate FET are (i) reduced short-channel effects, (ii) higher transconductance, and (iii) higher speed, as compared to a homogeneously doped FET. Reduced short-channel effects are due to the proximity of the channel to the gate metal and the minimized spatial width of the electron channel. The δ-doping technique allows one to place the doped channel in close proximity to the gate. Furthermore, the spatial width of the electron channel is reduced as compared to a homogeneously doped channel. Short channel effects become important if the gate-to-channel distance and the spatial extent of the channel become comparable to the gate length. Since the spatial dimensions of the doping profile are minimized for δ-doped FETs, short-channel effects are less prominent in δ-doped FETs as compared to regular FETs with the same gate length.

The high transconductance of δ-doped FETs is a result of the proximity of the Schottky gate to the electron channel. It is well known that the transconductance of an FET increases with decreasing gate-to-channel distance. The increased transconductance can be derived from either the Shockley transistor model (graded channel approximation) or the saturated velocity transistor model. From a planar capacitor point of view, the increased transconductance is a result of the increased amount of induced charge per unit gate voltage.

An increase in the transit frequency of δ-doped FETs results, if the decreased gate-to-channel distance is accompanied by a shorter gate length. The transit frequency of an FET is given by

$$f_t = \frac{g_m}{2\pi C_{GS}} \tag{11.45}$$

where g_m is the transconductance and C_{GS} is the gate-source capacitance. This capacitance can be approximated by the gate-to-channel capacitance. The latter increases if the gate-to-channel separation is reduced and decreases with reduced gate length. Thus, higher speed operation is inferred from Eq. (11.45) due to the increased transconductance of the δ-doped FET.

The gate-to-source and the drain-source current–voltage characteristics of a δ-doped FET are shown in Fig. 11.14 (Schubert et al., 1986a). The FET was fabricated in a two-mask self-aligned process using non-alloyed ohmic contacts. The transconductance of the 1 μm gate length device is 240 mS/mm.

Delta doping has also been applied to selectively doped heterostructure transistors (Schubert et al., 1987c). The devices are characterized by a high density of the electron gas at the hetero interface. High free carrier densities result in improved transconductance as well as in lower source-to-channel series resistance. Electron densities exceeding 1.5×10^{12} cm^{-2} at 300 K were demonstrated with δ-doped heterostructures. Low-temperature measurements revealed that two electronic subbands are occupied with a total electron density exceeding 10^{12} cm^{-2}. High transconductances of 350 mS/mm were obtained in such selectively δ-doped heterostructures with a gate length of 1 μm.

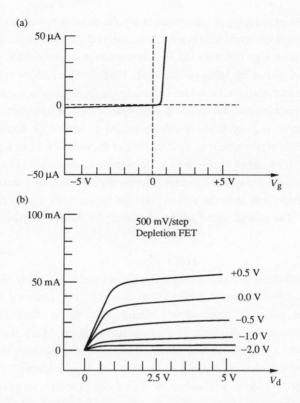

Fig. 11.14. (a) Gate-source and (b) drain-source current–voltage characteristics of a δ-doped depletion-mode metal–semiconductor field-effect transistor (MESFET) with a gate length of 1.3 μm, a width of 150 μm, and a transconductance of 240 mS/mm.

11.4.4 Bipolar transistors

The employment of the δ-doping technique in bipolar transistors is motivated by the decrease in achievable base layer thicknesses. The thickness of the base layer is of special importance. Majority carriers from the emitter traverse the base by either a diffusive motion (thick base layers) or by ballistic flight (thin base layers). For both transport models, i.e. diffusive and ballistic transport, a thin base is advantageous for minimum transit time. A reduction of transit time simultaneously reduces the probability of recombination in the base. In other words, the transport factor of the base, which is a well-known quantity in bipolar transistor analysis, increases with decreasing base thickness.

The concept of δ-doping in homojunction bipolar transistors was motivated by the potentially greater speed achievable (Schubert et al., 1984). Delta-doping in bipolar transistors was first realized in heterostructure devices made of III–V semiconductors grown by MBE (Malik et al., 1988b). The authors pointed out that the base transit time can be reduced by confining the dopants to a thin sheet.

Another application of the δ-doping technique in bipolar transistors is in the collector layer. A doping dipole of closely spaced p- and n-type doping sheets was employed to create a built-in electric field in the collector layer adjacent to the base layer (Tokumitsu et al., 1990). The purpose of the field is to sweep out majority carriers from the base-collector junction region which have traversed the base. The carriers will enter the neutral collector region more rapidly due to the electric field. The overall speed of the transistor can thus be improved by the dipole layer.

11.4.5 Diodes

The field distribution in semiconductor diodes is critical for many diodes, for example planar barrier diodes, IMPATT diodes, etc. The δ-doping technique allows one to tailor the potential and electric field of the diode leading to improved electrical characteristics. The planar barrier diode (Malik et al., 1980a, 1980b, 1982) is a majority carrier diode. It has a low capacitance due to the absence of the diffusion capacitance of conventional bipolar diodes. Therefore, the planar barrier diode is well suited for high-speed operation of mixer diodes and other applications. Furthermore, the planar barrier diode allows one to 'tailor' the current–voltage characteristic to a much higher degree than conventional pn-junction diodes. The planar barrier diode consists of an n-type region, an intrinsic region, a p-type δ-doped region, an intrinsic region, and finally an n-type region. Two optional n-type δ-doped sheets may be included at the boundary between the n-type and the intrinsic region. The p-type δ-doped region in the center of the diode is completely depleted of holes. The diode is thus a unipolar device. Symmetric as well as asymmetric current–voltage characteristics can be achieved by adjusting the thicknesses of the two intrinsic regions.

IMPATT or impact avalanche transit time diodes are active devices used for microwave generation in the frequency range 50–150 GHz. The diodes have an avalanche multiplication region and a transit time region. The multiplication region has a high electric field while the transit time region has a constant lower electric field. It is desirable to control the field distribution accurately in order to obtain devices with well-controlled characteristics. Owing to the stringent requirements of the IMPATT diodes, it was proposed to employ the δ-doping technique for improved characteristics (Schubert et al. 1984). IMPATT diodes incorporating δ-doped layers were grown by Si MBE (Luy, 1990). The author reported efficiencies exceeding 10% at a frequency of 100 GHz. The improvements were attributed to layer design, packaging, and resonator circuit design. Furthermore, a δ-doped spike effectively restricted the ionization region. Delta-doped IMPATT diodes made from III–V semiconductors have not been reported.

11.5 Optical devices

The initial proposal for superlattices comprised not only compositional superlattices, e.g. of the type $Al_x Ga_{1-x} As/GaAs$, but also *doping superlattices* (Esaki and Tsu, 1970). Such doping superlattices consist of a periodic sequence of alternating n-type and p-type regions in a semiconductor. A periodic potential results along the semiconductor growth direction due to positive and negative impurity charges. The periodic superlattice potential results in novel electronic properties not observed in conventional semiconductors. Shortly after the initial proposal of doping superlattices, published work was exclusively of a theoretical nature (Ovsyannikov, 1971; Romanov, 1972; Döhler, 1972a and 1972b; Romanov and Orlov, 1973). Experimental work started in the 1980s and included the observation of the tunability of the energy gap in doping superlattices (Döhler et al., 1981) and the first observation of quantum-confined interband transitions (Schubert et al., 1988b and 1989a). Extended reviews on doping superlattices have now become available (Schubert, 1990a; Döhler, 1986; Ploog and Döhler, 1983).

The band diagram of a symmetric δ-doped doping superlattice is shown in Fig. 11.15(a). The symmetric superlattice consists of alternating n-type and p-type doping sheets separated by $z_p/2$. Using the same n- and p-type doping densities ($N_D^{2D} = N_A^{2D}$) results in a depleted doping superlattice with a sawtooth-shaped band diagram. This is called a sawtooth superlattice (Schubert et al., 1985c). Using different thicknesses between the doping sheets results in an asymmetric band structure as shown in Fig. 11.15 (Glass et al., 1989). Note that the conduction band and valence band edges vary *parallel* in doping superlattices. This characteristic is different from compositional $Al_x Ga_{1-x} A/GaAs$ superlattices, where the band edges vary *antiparallel* as shown in Fig. 11.15(b).

The band diagram of a symmetric sawtooth doping superlattice including its quantized electronic states is shown in greater detail in Fig. 11.16. For symmetric doping superlattices with the same donor and acceptor densities, the total band modulation can be calculated from Poisson's equation and is given by

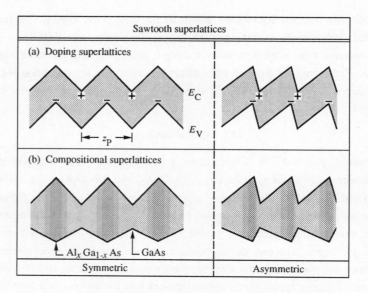

Fig. 11.15. Symmetric and asymmetric sawtooth superlattices. In doping superlattices, the conduction- and valence-band edges are parallel whereas they are antiparallel in compositional superlattices.

$$V_{zz} = eN^{2D}z_p/4\varepsilon \,. \tag{11.46}$$

The effective energy gap of the doping superlattice, E_g^{SL}, is defined as the lowest transition energy between quantum-confined conduction and valence band states, i.e.

$$\boxed{E_g^{SL} = E_g - eV_{zz} + E_0^e + E_0^{hh}} \tag{11.47}$$

where E_g is the original gap energy of the host lattice. It is thus possible to extend the energy gap of any semiconductor to lower energies by using the doping superlattice concept.

A second important property of doping superlattices is the relatively weak oscillator strength of optical transitions. The location of maximum electron concentration is shifted with respect to the location of maximum hole concentration by half a superlattice period, as shown in Fig. 11.16. As a result, the transition matrix element, which is proportional to the overlap of the electron and hole wave functions, is reduced as compared to the bulk case. Due to the

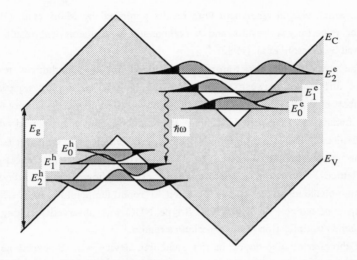

Fig. 11.16. Schematic band diagram of a sawtooth superlattice. The overlap of electron and hole wave functions increases exponentially with subband index.

weaker oscillator strength in doping superlattices, optical measurements are frequently performed at low temperatures (Ploog and Döhler, 1983).

Delta-doped superlattices are the optimum doping superlattice structure. It was shown that the potential modulation of the band edges is maximized by δ-doping. Furthermore, potential fluctuations are minimized in such structures due to the confinement of impurities to a sheet (Schubert, 1990a). As a result, smaller periods can be achieved in the sawtooth superlattice as compared to the conventional doping superlattice structure.

The remainder of this section emphasizes device applications of doping superlattices, including superlattice light-emitting diodes (LEDs) and lasers. The spectroscopic properties of doping superlattices, including quantum-confined interband transitions, were discussed in Chap. 10.

11.5.1 Perpendicular transport

Transport in semiconductor structures in which the dopant plane is perpendicular to the transport direction were reported by Malik et al. (1980a, 1980b). They used n-i-p-i-n doping profiles which result in a triangular barrier structure and obtained a rectifying majority carrier diode. Subsequently, Kazarinov and Luryi (1981) calculated the charge injection over such a triangular barrier. They obtained an exact expression for the current–voltage characteristic of the triangular barrier diode, which was in agreement with results published by Malik et al. (1980b). Further experimental results and a comparison with theoretical results were reported by Gossard et al. (1982).

The study of electron transport in δ-doped doping superlattices revealed drastic differences between short-period and long-period doping superlattices (Schubert et al., 1987b). A new type of negative differential conductivity (NDC) was found in long-period doping superlattices. In short-period doping superlattices, NDC is not observed. Instead, efficient electron transport is found in the perpendicular direction to the superlattice planes in short-period superlattices. Furthermore, Sun and Liu (1991) showed that a diode-like current–voltage characteristic is obtained at room temperature in a sawtooth doping superlattice. At $T = 77$ K, S-type NDC was observed resulting from avalanche multiplication in the superlattice region.

A three-terminal δ-doped barrier switching device with S-shaped negative differential resistance was proposed and realized by Baillargeon et al. (1989). The authors showed that the S-type NDC can be controlled by the third terminal. The barrier height was increased and decreased according to the voltage applied to that terminal. The results showed that the device could be used for oscillators at frequencies exceeding 20 GHz.

Resonant tunneling in δ-doped structures was studied by Wang et al. (1991) and Houng et al. (1992). The authors used p^+-type δ-doping embedded in an n^+ region and n^+-type δ-doping embedded between two metal–semiconductor contacts. The resonant tunneling process was evidenced by structure in the current–voltage curves. A peak-to-valley ratio of 1.5 was reported. Finally, transmission resonances and transmission coefficients were calculated using Airy functions.

11.5.2 Light-emitting diodes

Light-emitting diodes (LEDs), with an active region consisting of a doping superlattice, emit light of energy *below* the band gap of the host semiconductor. As evidenced by Eq. (11.47), the superlattice energy gap can be smaller than the gap of the host semiconductor. Thus, doping superlattice LEDs offer a unique characteristic not found in other semiconductor structures. A sketch of an edge-emitting sawtooth doping superlattice diode is shown in Fig. 11.17 along with the δ-function-like doping profile and the sawtooth-shaped band diagram (Schubert et al., 1985b). The edge-emitting diode has $Al_x Ga_{1-x} As$ confinement layers and a GaAs active region. The p-n junction is located at the superlattice region, which results in carrier recombination in the superlattice region.

Electroluminescent spectra of the LEDs operating at 300 K are shown in

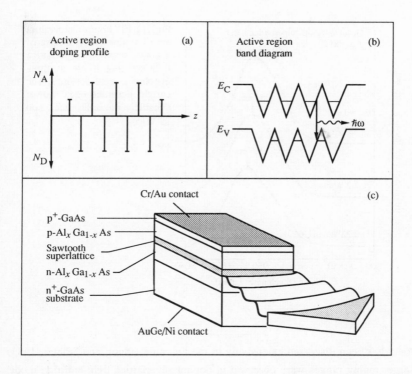

Fig. 11.17. (a) Doping profile, (b) band diagram, and (c) basic structure of the edge-emitting sawtooth superlattice light-emitting diode. The doping superlattice in the active region is sandwiched between $Al_xGa_{1-x}As$ confinement layers.

Fig. 11.18 for three different excitation currents. The period of the doping superlattice active region is $z_p = 200$ Å and the δ-doping densities are $N_D^{2D} = N_A^{2D} = 5 \times 10^{12}$ cm^{-2}. The peak wavelength of the LED is $\lambda \cong 960$ nm, i.e. well below the band gap of GaAs. These wavelengths were inaccessible to conventional GaAs LEDs. At the wavelength corresponding to the bandgap energy of the GaAs host material, no luminescence signal is detected, demonstrating the superlattice character of the active region of the LED.

The three electroluminescence spectra exhibit a small shift of the peak wavelength to shorter wavelengths with increasing excitation intensity, shifting from 966 nm at low excitation intensity to 959 nm at high excitation intensity. The total shift in energy is small if compared to the total linewidth of the luminescence line ($\Delta\lambda > 50$ nm). The energy upshift is probably due to (i) band filling effects and (ii) screening of ionized impurities by free carriers. The latter

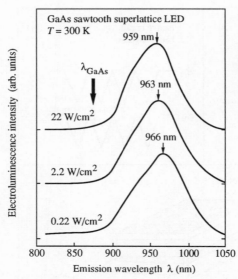

Fig. 11.18. Room-temperature electroluminescence spectra of a GaAs sawtooth superlattice light-emitting diode at three injection current densities. The current densities equal power densities of 0.22, 2.2, and 22 W/cm^2.

effect has been discussed extensively in the literature (Ploog and Döhler, 1983). Larger tuning ranges were observed in doping superlattice light-emitting diodes with longer periods (Hasnain et al., 1986). The authors used selective contacts for the n-type and p-type regions. The period of the superlattice reported by the authors was 600 Å and 900 Å.

11.5.3 Lasers

Lasers are one of the most interesting device applications for doping superlattices. Tunable lasers are desirable for optical wavelength multiplex communication systems. On the other hand, optical transitions in doping superlattices are inherently of weaker oscillator strength as compared to bulk material. Thus, their realization requires high quality epitaxial material. First attempts to realize a doping superlattice laser revealed optical amplification, i.e. gain (Jung et al., 1983). However, laser emission was not achieved.

The first doping superlattice current injection laser (Schubert et al., 1985a) was realized in GaAs using the δ-doping technique with no evidence for tunability. Laser emission in doping superlattices was also reported by Vojak et al. (1986). The photo-pumped lasers emitted below the bandgap of GaAs, also with no evidence for tunability. The first tunable doping superlattice laser was realized by inhomogeneous excitation of the Fabry–Perot cavity (Schubert et al. 1989c). A schematic sketch of the layer sequence, the active region doping

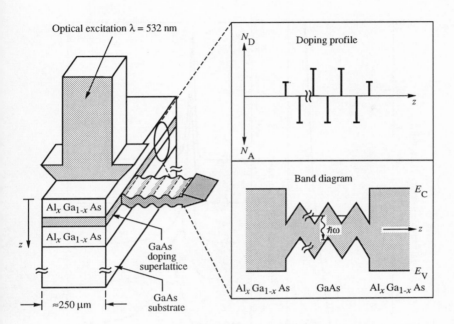

Fig. 11.19. Geometry used for optical excitation of the optically excited tunable sawtooth superlattice laser. The active region doping profile and the band diagram are shown on the right.

profile, and the corresponding band diagram are shown in Fig. 11.19. After epitaxial growth, the layers were cleaved into stripes of nominal 250 μm length and 1 cm width. A frequency-doubled Nd-doped Q-switched YAG laser (λ = 532 nm) was used for optical excitation. Light emission from the sample was detected with a Si detector using gated detection. The laser samples were cooled in a variable-temperature He cryostat.

The spontaneous and stimulated emission spectra at T = 5K are shown in Fig. 11.20 together with the light output versus excitation intensity curve. In the spontaneous emission regime at low excitation intensity, the peak wavelength moves to shorter wavelengths with increasing excitation intensity. At higher excitation intensities, stimulated emission occurs, which is accompanied by the characteristic kink in the light output curve of Fig. 11.20 and a narrowing of the emission spectrum to values much below the thermal energy kT. However, the peak wavelength does not change in the stimulated emission regime. Such a constant emission energy is not unexpected, since upon reaching the laser threshold, the Fermi level remains constant and additional carriers undergo

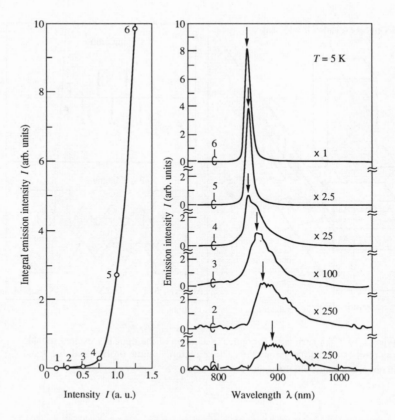

Fig. 11.20. Low-temperature (left) emission intensity versus excitation intensity of the tunable sawtooth doping superlattice laser, and (right) optical spectra below, at, and above threshold intensity. The wavelength remains constant after the laser threshold is reached.

stimulated recombination with a correspondingly very short lifetime. Thus, the emission energy remains constant for different excitation intensities in the stimulated emission regime.

The emission energy can be tuned continuously by inhomogeneous excitation of the Fabry–Perot cavity. Such inhomogeneous excitation is achieved by displacing the exciting beam from its centered position, as shown in the top part of Fig. 11.21. The inhomogeneous excitation results in a higher laser emission energy as compared to the symmetric excitation. Figure 11.21 reveals that the tuning range of the laser is approximately 35 Å. This tuning range does not represent a fundamental limit. The tuning range is limited by the intensity

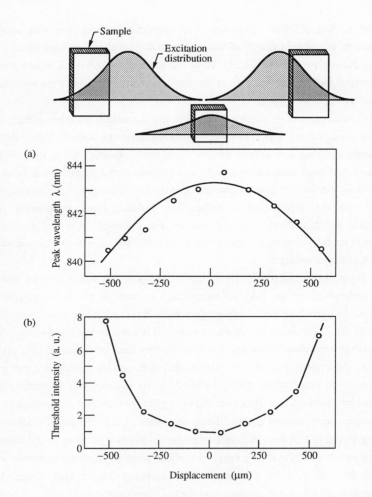

Fig. 11.21. (a) Peak wavelength of the tunable sawtooth doping superlattice laser versus displacement of the optical excitation beam. Tunability of the laser emission is achieved over $\Delta\lambda \cong 35$ Å. (b) Threshold intensity of the laser versus displacement of the exciting beam. The top part of the figure schematically shows the displacement of the exciting beam with respect to the laser bar.

distribution of the exciting source. A wider tuning range can be expected for a more inhomogeneous excitation, which could be achieved in a two- or three-section current-injection laser.

Simultaneously, as the peak of the stimulated emission shifts to shorter wavelengths, the excitation intensity required to reach threshold increases, as illustrated in Fig. 11.21(b). However, it is important to visualize that upon displacement, the sample is excited only by a small part of the exciting beam as shown in the top part of Fig. 11.21. Thus, the increase of threshold intensity is overestimated and the true increase in threshold intensity is not as pronounced as suggested by Fig. 11.21(b).

Under homogeneous excitation conditions, the stimulated emission energy is below the band gap of GaAs. However, the emission energy is much higher than the spontaneous emission energy at low excitation intensities ($E \cong 1.35$ eV). Thus, the superlattice modulation is reduced by photoexcited electrons and holes, which screen the ionized dopant charges of donors and acceptors, respectively. Even though the modulation is reduced, a residual *band modulation is maintained,* as demonstrated by the low emission energy. Thus, stimulated emission is achieved before the bands are completely flat, that is for incomplete screening of dopant charges.

The physical mechanism leading to the tunability of the semiconductor laser can be understood on the basis of increased loss induced by inhomogeneous excitation. As a result of the inhomogeneous excitation, i.e., reduced excitation in one part of the Fabry–Perot cavity, the optical loss is enhanced in this section. In order to obtain stimulated emission, the other section must be subjected to higher excitation. As a result, the band modulation decreases and the superlattice energy gap increases in this section. Once the intentionally induced loss is overcome, stimulated emission occurs. However, the lasing *energy* increases as compared to the homogeneously excited cavity. Thus, tunability of the stimulated emission wavelength is achieved by a different excitation in the two sections of the laser. The principal limit of the tuning range is reached when the flatband condition is achieved in one part of the laser. The corresponding tuning range is approximately 250 Å at low temperatures for the samples studied.

11.5.4 Modulators

Modulation and switching of near-infra-red radiation by means of electrical control is an important functional device characteristic in photonic switching systems. It is desirable that such photonic switching devices (i) have a broad wavelength range in which the light intensity can be modulated, (ii) have a large contrast ratio in the transparent and opaque states, (iii) can operate at voltages compatible with electronic integrated circuits, and (iv) have high-speed capability.

Modulators with such characteristics can be fabricated from doping superlattices with long periods (Schubert and Cunningham, 1988). Contacting the n-type and p-type regions of the superlattice allows one to change the modulation, i.e. the internal electric field of the superlattice, by means of an external bias. Tunneling-assisted absorption (Franz–Keldysh absorption), which occurs at energies below the band gap of the semiconductor, depends exponentially on the electric field. Thus, doping superlattices can be used to modulate the intensity of transmitted light by tunneling-assisted absorption.

Modulation experiments with the conventional doping superlattice structure yielded a transmission change of 22% using selective side-contacts to the superlattice region (Chang-Hasnain et al., 1987). Absorption modulators were also realized with a δ-doped structure (Schubert and Cunningham, 1988). The maximum contrast ratio between the opaque and the transparent state achieved with the δ-doped modulator was 70% at a wavelength of $\lambda \cong 950$ nm.

12

Characterization techniques

Doping and other materials parameters influence the properties of III–V devices in a profound manner. Examples of device parameters which are strongly influenced by the defect and doping concentration are the radiative efficiency of a laser, the minority carrier lifetime in the base of a bipolar transistor, the carrier mobility in the channel of a field-effect transistor, or the quantum efficiency of a pin photo-diode. In this chapter, characterization techniques are discussed that relate directly to shallow impurities as well as deep centers. The characterization techniques are categorized as (i) electronic (ii) optical, and (iii) chemical and structural techniques. Fundamental aspects of characterization techniques as well as practical 'hints' for the experimentalist are emphasized.

12.1 Electronic characterization techniques

Many properties of semiconductors that relate directly to impurities or defects can be assessed by electrical measurements. Such measurements include current–voltage, capacitance–voltage, resistivity, magnetoresistance, and impedance measurements. Frequently, temporal transients of such measurements are of interest, for example the capacitance transient after a semiconductor has been subjected to an electrical pulse. In this section, the Hall effect, capacitance–voltage (CV) profiling technique, deep level transient spectroscopy (DLTS), thermally stimulated capacitance (TSCAP), thermally stimulated current (TSC), and admittance spectroscopy are discussed.

12.1.1 Hall effect measurements

Hall effect measurements (Hall, 1879) allow one to determine the (majority) **Hall carrier concentration** of unipolar semiconductors in which the minority carrier concentration can be neglected. The *Hall carrier concentration* differs from the **true carrier concentration** by a numerical factor, r_H, which is called the **Hall factor** and which is on the order of unity.

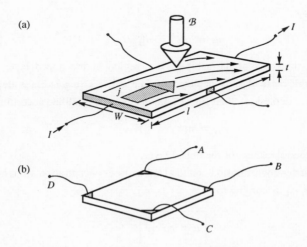

Fig. 12.1. (a) Hall bar geometry and (b) van der Pauw geometry of a semiconductor for Hall effect measurements.

A simple Hall bar geometry is shown in Fig. 12.1(a). The sample is subjected to a current I and to a magnetic field $\mathcal{B}$ perpendicular to the sample surface. Charge carriers flowing through the sample experience the Lorentz force

$$\mathcal{F}_L = -ev \times \mathcal{B} \tag{12.1}$$

where $-e$ is the charge of electrons and v is the carrier velocity. Carriers are deflected by the Lorentz force from their flow pattern without a magnetic field as indicated in Fig. 12.1(a). As a result, an electric field evolves due to the displacement of electron charges. The electric field is called the **Hall field**. Carriers experience a force in the Hall field of magnitude

$$\mathcal{F}_H = -e\mathcal{E}_H . \tag{12.2}$$

The deflection of carriers persists until the Lorentz force is balanced by the Hall force. Equating Eqs. (12.1) and (12.2) yields the Hall carrier concentration according to

$$n_H^{2D} = \frac{I\mathcal{B}}{eV_H} \tag{12.3}$$

where $V_H = \mathcal{E}_H W$ (see Fig. 12.1a) is the Hall voltage, and n_H^{2D} is the two-dimensional (2D) Hall carrier concentration, i.e. the Hall concentration per unit area. The three-dimensional (3D) carrier concentration is obtained from the 2D concentration by

$$n_H = n_H^{3D} = n_H^{2D}/t \tag{12.4a}$$

where t is the thickness of the sample. If a region of the doped layer of the sample, with thickness W_D, is depleted of free carriers (e.g. surface depletion layer or interface depletion), then Eq. (12.4a) needs to be modified according to

$$n_H = n_H^{2D}/(t - W_D) \tag{12.4b}$$

where t is the total thickness of the doped layer.

The *true* carrier concentration and the Hall carrier concentration are related by the **Hall factor**, r_H, according to

$$n = r_H n_H \tag{12.5}$$

where r_H is given by

$$r_H = \langle \tau_m^2 \rangle / \langle \tau_m \rangle^2 \tag{12.6}$$

and τ_m is the energy-dependent momentum-relaxation time and $\langle \tau_m \rangle$ is the expectation value of τ_m at a given temperature. The magnitude of the Hall factor is $r_H = 1.18$ for phonon scattering, $r_H = 1.93$ for ionized impurity scattering and $r_H = 1.0$ for degenerately doped semiconductors (Smith, 1978). Thus, the Hall carrier concentration is always equal or smaller than the true carrier concentration, $n_H \leq n$. For simplicity it is frequently assumed that the Hall factor is unity, i.e. $n_H = n$. We finally note that the quantity

$$R_H = \frac{1}{en_H} = \frac{r_H}{en} \tag{12.7}$$

is called the **Hall coefficient** and is frequently used in the literature.

The Hall effect measurement and a conductivity measurement allows one to determine the **mobility** of carriers. The conductivity of an n-type semiconductor is given by

$$\sigma = en\mu \qquad (12.8)$$

where μ is the **drift mobility**. The drift mobility can be determined, if the conductivity and the carrier concentration are known, that is

$$\mu = \sigma/en . \qquad (12.9a)$$

If, however, the *true* carrier concentration is unknown but instead the *Hall* carrier concentration is known, then Eq. (12.9a) can be written as

$$\boxed{\mu_H = \sigma/en_H} \qquad (12.9b)$$

where μ_H is called the **Hall mobility**. The Hall mobility deviates from the drift mobility by the Hall factor, and is always larger or equal to the drift mobility, because $\mu_H = \sigma/en_H = \sigma r_H/en = \mu r_H$ and $r_H \geq 1$. In other words, the Hall mobility overestimates the true (drift) mobility.

The **van der Pauw** technique (van der Pauw, 1958) is a convenient means of performing Hall and conductivity measurements. A sample with van der Pauw-type geometry is shown in Fig. 12.1(b). The sample is square-shaped and has four ohmic contacts in the corners. The contacts must be much smaller than the sample area. Furthermore, the contacts must be at the very edges of the sample. As an improvement, van der Pauw proposed a 'clover leaf' geometry which reduces the influence of the contacts. In the van der Pauw measurement, a current source is applied to two of the four contacts, e.g. the current enters through contact A and leaves through contact B. The voltage between the contacts C and D is then measured. For convenience, we define the resistance $R_{AB,CD} = V_{CD}/I_{AB}$. Then the Hall coefficient is given by

$$R_H\,(\text{cm}^2/\text{C}) = \frac{2.5 \times 10^4}{\mathcal{B}}[(R^+_{AC,DB} - R^-_{AC,DB}) + (R^+_{DB,AC} - R^-_{DB,AC})]$$

$$(12.10a)$$

where R^+ and R^- indicate the forward and reverse polarization of the magnetic induction $\mathcal{B}$, and the magnitude of $\mathcal{B}$ is measured in units of 10^3 gauss. The two-dimensional Hall carrier concentration is then given by

$$n_H^{2D}\,(\text{cm}^{-2}) = 6.25 \times 10^{18}/R_H \qquad (12.10b)$$

where the Hall density is obtained in cm^{-2}. The two-dimensional resistivity of a sample with van der Pauw geometry is given by

$$\rho^{2D}(\Omega) = \frac{\pi}{4 \ln 2} \left(R_{AD,CB} + R_{DA,CB} + R_{DC,BA} + R_{CD,BA} \right) . \tag{12.10c}$$

The 2D resistivity is frequently measured in ohms or ohms/square. Finally, the Hall mobility can be obtained from the Hall coefficient and the 2D resistivity

$$\mu_H (\text{cm}^2/\text{V s}) = R_H / \rho^{2D} \tag{12.10d}$$

where R_H and ρ^{2D} are in the units of Eqs. (12.10a) and (12.10c).

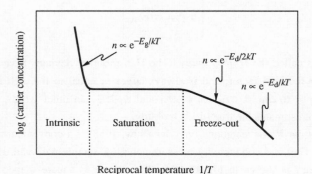

Fig. 12.2. Schematic temperature dependence of the carrier concentration in a doped semiconductor. All impurities are ionized in the saturation regime. Freeze-out of carriers occurs at low temperatures. At high temperatures, the carrier concentration increases due to thermal band-to-band excitation.

Temperature-dependent Hall measurements allow one to evaluate the thermal ionization energy of impurities. The schematic temperature-dependent carrier concentration of an n-type semiconductor is shown in Fig. 12.2. The carrier concentration displays three regions which are the *freeze-out*, the *saturation*, and the *intrinsic* regimes. A detailed discussion of the temperature-dependent carrier concentration was given in Chap. 3. The dependence of the carrier concentration on temperature of an n-type semiconductor with a donor activation energy E_d is

$$n \propto e^{-E_d/2kT} \tag{12.11}$$

which is valid for $N_D \gg \frac{1}{2} N_c e^{-E_d/kT} \gg N_A$ where N_c is the effective density of states and N_A is the residual acceptor background concentration. At even lower temperatures one may observe a regime

$$n \propto e^{-E_d/kT} \tag{12.12}$$

which is valid for $\frac{1}{2} N_c e^{-E_d/kT} \ll N_A$, i.e. for very low temperatures or higher acceptor background concentrations. Thus, Eq. (12.11) rather than Eq. (12.12) applies for most practical purposes, that is high-quality, uncompensated semiconductors. The thermal activation energy of donor impurities can be evaluated from the slope of n versus $(1/T)$ using Eq. (12.11)

$$E_d = -2k \frac{d(\ln n)}{d(1/T)} .$$
(12.13)

The numerical factor before the slope has a value of $-2k = -0.172$ meV.

12.1.2 Capacitance–voltage profiling technique

The capacitance–voltage (CV) profiling technique (Thomas et al., 1962; Meyer and Guldbrandsen, 1963; Amron, 1964, 1967) allows one to measure the CV-concentration, N_{CV}, as a function of depth. In the simplest case, that is for a non-compensated, homogeneously doped semiconductor, the CV-concentration equals the electrically active doping concentration. The CV-measurement can be performed on Schottky junctions, p^+ n-junctions, and n^+ p-junctions. The CV-measurement is based on the *depletion-approximation* which assumes that the depletion region of a junction contains no free carriers and that the semiconductor is neutral outside the depletion region. The transition between the neutral and the depletion region is assumed to be abrupt. In the following, a metal-to-n-type semiconductor junction will be considered. However, the results are also valid for p^+ n- and n^+p-junctions with $p^+ \gg$ n and $n^+ \gg$ p, respectively.

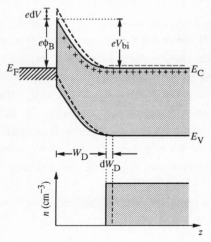

Fig. 12.3. Band diagram and free carrier profile of a semiconductor with the near-surface region depleted of free carriers. Application of a small bias dV changes the depletion width by dW_D.

The band diagram of a semiconductor–metal junction is shown in Fig. 12.3. The barrier height of the junction is ϕ_B. The semiconductor is assumed to be homogeneously n-type doped and uncompensated. The width of the surface depletion region can be obtained from Poisson's equation and is given by

$$W_D = \sqrt{\frac{2\varepsilon}{eN_D} (V_{bi} - V)} \cong \sqrt{\frac{2\varepsilon}{eN_D}(\phi_B - V)} \qquad (12.14)$$

where V_{bi} is the 'built-in' voltage (see Fig. 12.3), which is $V_{bi} \cong \phi_B$ for typical doping concentrations; V is the external voltage applied to the Schottky diode. Upon application of a voltage increment, the depletion width increases by

$$dW_D = -\frac{1}{2} \sqrt{\frac{2\varepsilon}{eN_D(\phi_B - V)}} \, dV . \qquad (12.15)$$

The change in depletion *charge per unit area*, dQ, for a voltage increment, dV, is given by $dQ = eN_D \, dW_D$. The differential depletion capacitance per unit area is defined as $C = dQ/dV$ and is given by

$$C = \frac{dQ}{dV} = \frac{1}{2}\sqrt{\frac{2e\varepsilon\, N_D}{\phi_B - V}} = \frac{\varepsilon}{W_D} . \qquad (12.16)$$

The *CV-concentration* is then defined as

$$\boxed{N_{CV} = \frac{-2}{e\varepsilon}\left[\frac{dC^{-2}}{dV}\right]^{-1} = \frac{C^3}{e\varepsilon}\frac{dV}{dC}} \qquad (12.17)$$

Insertion of Eq. (12.16) into Eq. (12.17) shows that the CV-concentration equals the shallow donor concentration, i.e. $N_{CV} = N_D$. This result applies to homogeneously doped, *uncompensated* semiconductors only. For *compensated* semiconductors with $N_D > N_A$, it is easy to show that $N_{CV} = N_D - N_A$. This equation also applies to semiconductors with *amphoteric* impurities, e.g. Si in GaAs, where $N_{CV} = N_D - N_A$. Similarly, *electrically inactive* impurities in semiconductors are not measured by the CV-technique, i.e. $N_{CV} = N_D - N_D^0$, where N_D is the total donor concentration and N_D^0 is the concentration of electrically inactive donors.

The concentration N_{CV} occurs at the depth, z_{CV}, which is defined as

$$z_{CV} = \varepsilon/C \qquad (12.18)$$

where C is the differential capacitance per unit area. Using Eqs. (12.14) and (12.16) one can show that $z_{CV} = W_D$ for homogeneously doped semiconductors. Finally, the *CV-profile* is obtained by plotting N_{CV} as a function z_{CV}.

There are several limitations to the CV-technique such as diode breakdown at large reverse voltages, leakage currents of the diode, and deep traps, which contribute to the capacitance signal. Several of these limitations were discussed by Blood and Orton (1978). Large series resistances of junctions used for CV-profiling can have a drastic effect on the measurement (Wiley and Miller, 1975). The current–voltage phase angle is generally a good indicator of the reliability of the CV-measurement. A phase-angle near 90° is desirable for all bias voltages in order to obtain reliable CV-data. Although corrections are possible for phase angles $<90°$, the corrections are not always advisable due to the ambiguity of their origin. A variation of the CV-profiling technique has been demonstrated by Miller (1972). Rather than constant voltage increments, the Miller technique employs constant distance increments which results in a better signal-to-noise ratio.

The **spatial resolution** of CV-profiles on semiconductors with a *classical* (non-quantum-mechanical) carrier system is given by the majority carrier screening length (Kennedy et al., 1968, 1969). If the doping impurity concentration varies along a spatial axis, the CV-profile does not coincide with the impurity profile. Furthermore, for abrupt doping profiles, the free carrier concentration does not coincide with the doping concentration. While abrupt changes in the impurity concentration can occur, changes in the free carrier concentration are smeared out with the screening length being the characteristic length. Kennedy et al. (1968, 1969) further showed that the CV-profile, $N_{CV}(z_{CV})$, very closely follows the free carrier profile, $n(z)$. In a non-degenerately doped semiconductor, the majority carrier screening length is the Debye length (see Chap. 1)

$$r_D = \sqrt{\frac{\varepsilon kT}{e^2 n}} \ . \qquad (12.19)$$

The Debye length increases with temperature and does not depend on the carrier effective mass. In degenerately doped semiconductors, the majority carrier screening length is the Thomas–Fermi length (see Chap. 1)

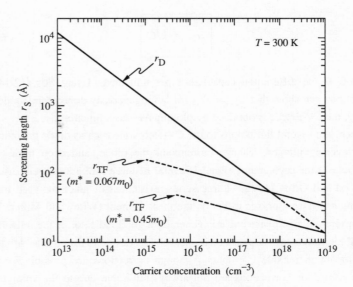

Fig. 12.4. Majority carrier screening length for non-degenerately doped (Debye length, r_D) and degenerately doped (Thomas–Fermi length, r_{TF}) semiconductors. The Thomas–Fermi length is calculated for two different masses.

$$r_{TF} = \pi^{2/3}\sqrt{\frac{\varepsilon\hbar^2}{e^2 m^*(3n)^{1/3}}} \tag{12.20}$$

The Thomas–Fermi screening length is independent of temperature and depends on the carrier effective mass. The calculated Debye length and the Thomas–Fermi screening length are shown in Fig. 12.4 for a temperature of $T = 300$ K. The mass-dependent Thomas–Fermi length is shown for $m^* = 0.067 m_0$ and $m^* = 0.45 m_0$ which correspond to the electron and heavy-hole effective masses in GaAs.

The **spatial resolution** of the CV-profiling technique in semiconductors with **quantum confinement** was shown to be limited to the spatial extent of the carrier wave function (Schubert et al., 1990a). This resolution limit is explained as follows. Consider a quantized electron system described by a wave function ψ. The position expectation value of the electron system is $\langle z \rangle = \langle \psi | z | \psi \rangle$. The differential capacitance per unit area of such an electron system equals

$$C = \varepsilon/\langle z \rangle . \tag{12.21}$$

Upon application of a bias, the position expectation value of the quantized

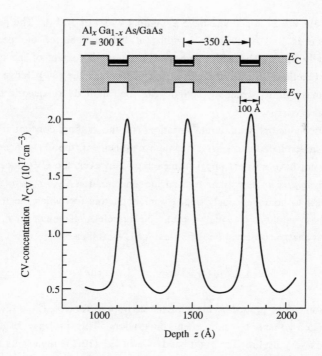

Fig. 12.5. Capacitance–voltage profile of an $Al_xGa_{1-x}As$/GaAs quantum well structure doped in the GaAs quantum wells.

electron system moves due to the potential perturbation caused by the bias. Recall now, that the depth scale of the CV-measurement is calculated from

$$z_{CV} = \varepsilon/C . \qquad (12.22)$$

Comparison of Eqs. (12.21) and (12.22) yields $z_{CV} = \langle z \rangle$, i.e. the CV-depth (at a certain bias) equals the position expectation value of the electron system (at a certain bias). Thus, the *resolution* of the CV-technique, Δz_{CV}, is given by the change of the position expectation values, $\Delta \langle z \rangle$, upon sweeping the depleting bias, i.e.

$$\Delta z_{CV} = \Delta \langle z \rangle . \qquad (12.23)$$

The authors showed that the change in position expectation value is approximately equal to the spatial width of the wave function, i.e. approximately equal to the width of the confining potential. As an example, the CV-profile of an $Al_xGa_{1-x}As$/GaAs multiquantum well sample is shown in Fig. 12.5. The quantum wells are Si-doped and have a nominal width of 100 Å. The period of the structure is 350 Å. The half-widths at half-maxima of the peaks are approximately 50 Å. The width agrees with the spatial extent of the quantum well wave function. It is worthwhile noting that neither the (3D) Debye nor the (3D) Thomas–Fermi screening length can be applied to quantized, two-dimensional carrier systems.

In semiconductors with large variations of the doping concentration and especially in semiconductors with quantum confinement, the CV-concentration, N_{CV}, does not have a direct physical meaning. However, the CV-concentration corresponds *approximately* to the free carrier concentration $N_{CV}(z) \cong n(z)$. This is in contrast to homogeneously doped semiconductors for which the following equation has validity: $N_{CV}(z) = n(z)$. Nevertheless, Kroemer et al. (1980) showed that charge conservation is fulfilled for CV-profiles, that is

$$\int_{-\infty}^{\infty} N_{CV}(z_{CV}) \, dz_{CV} = \int_{-\infty}^{\infty} n(z) \, dz \, . \tag{12.24}$$

The equation can be proved by inserting the definitions for N_{CV} (Eq. (12.17)) and z_{CV} (Eq. (12.18)) and by subsequent integration. This property of the CV-technique is very useful. The right-hand side of Eq. (12.24) represents the two-dimensional (2D) carrier density of an electron system. Thus, the 2D carrier density of doped quantum well structures, selectively doped heterostructures, and similar structures can be precisely measured by the CV-profiling technique.

We consider finally the *influence of deep traps* on the CV-profiling technique. As an example, we consider a semiconductor with shallow donors and deep donor-like impurities with a constant concentration. The band diagram of a metal–semiconductor junction containing shallow and deep levels is shown in Fig. 12.6. It is useful to differentiate between (i) a neutral region, (ii) a shallow donor depletion region of thickness W_D, (iii) a deep donor depletion region, and (iv) a transition region of thickness λ. The potential drops and widths of the respective regions can be calculated from Poisson's equation for the general case $\lambda \leq W_D$:

$$V_\lambda = \frac{1}{e} \, (E_F - E_T) \, , \tag{12.25a}$$

$$\lambda = \sqrt{\frac{2\varepsilon}{e^2 N_D} \, (E_F - E_T)} \, , \tag{12.25b}$$

and

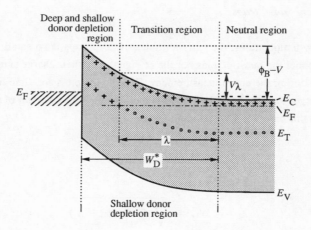

Fig. 12.6. Band diagram of a metal–semiconductor junction with deep and shallow donor impurities. The shallow and deep donor depletion regions are W_D^* and $W_D^* - \lambda$, respectively.

$$\phi_B - V = \frac{e}{\varepsilon} W_D^* [N_D \lambda + (N_D + N_T)(\tfrac{1}{2} W_D^* - \lambda)] , \qquad (12.26a)$$

$$W_D^* = \frac{N_T \lambda + \sqrt{N_T^2 \lambda^2 + 2(N_D + N_T) \dfrac{\varepsilon}{e}(\phi_B - V)}}{N_D + N_T} . \qquad (12.26b)$$

For $N_T = 0$ or for $\lambda = 0$, Eq. (12.26) reduces to the well-known depletion approximation for shallow impurities. The above equations are valid under equilibrium conditions, i.e. when no transient emission and capture processes of the deep traps occur.

In order to estimate the influence of deep traps, recall that the CV-measurement uses an ac small signal voltage [the frequency of the ac small signal voltage is in the radio-frequency (rf) regime and it is, therefore, frequently called the rf signal] to measure the capacitance at a particular dc voltage. The quasi-dc voltage is swept in a quasi-static (point-by-point) mode in order to obtain the C versus V characteristic. Kimerling (1974) differentiated between three cases to estimate the influence of deep traps. As examples, we consider donor-like traps of concentration N_T in an n-type semiconductor with shallow donor concentration N_D.

Case 1: Very 'slow' traps

At very low temperatures, the emission rates of deep traps is so small, that they 'follow' neither the quasi-dc bias nor the ac signal, i.e. their charge state remains unchanged during the entire measurement. That is, the emission time of the trap is much longer than the period of the ac signal and the time constant of the quasi-dc bias. In this case, the CV-technique gives

$$\boxed{N_{CV} = N_D}$$ (12.27a)

$$z_{CV} = W_D = \sqrt{\frac{2\varepsilon}{2N_D} (\phi_B - V)}$$ (12.27b)

which is also obtained if no traps were present.

Case 2: 'Slow' traps

At higher temperatures, the trap charge state follows the quasi-dc bias but does *not* follow the ac signal. The entire ac oscillating charge (per unit area) is located at $x = W_D^*$ and has a magnitude of

$$dQ = eN_D \, dW_D^*$$ (12.28a)

with a corresponding ac voltage of

$$dV = dQ/C = dQ \, W_D^*/\varepsilon = (e/\varepsilon) \, N_D W_D^* dW_D^* .$$ (12.28)

Using the definition of the CV-concentration (Eq. (12.17)) and CV-depth (Eq. (12.18)) along with Eq. (12.26), one obtains

$$\boxed{N_{CV} = N_D + N_T \left[1 - \frac{\lambda}{W_D^*} \right]}$$ (12.29a)

$$z_{CV} = W_D^* .$$ (12.29b)

The concentration N_{CV} approaches $N_D + N_T$ for deep depletion.

Case 3: 'Fast' traps

At sufficiently high temperatures, the trap charge state follows both the quasi-dc bias and the ac signal. The oscillating charge per unit area consists of the shallow donor and the trap contribution

$$dQ = eN_D dW_D^* + eN_T dW_D^* . \tag{12.30a}$$

The corresponding ac voltage change is given by

$$dV = dQ/C = \frac{e}{\varepsilon} [N_D W_D^* dW_D^* + N_T (W_D^* - \lambda) dW_D^*] . \tag{12.30b}$$

Using Eqs. (12.17) and (12.18) one obtains

$$\boxed{N_{CV} = N_D + N_T} \tag{12.31a}$$

$$z_{CV} = \frac{N_D W_D^* + N_T (W_D^* - \lambda)}{N_D + N_T} . \tag{12.31b}$$

The CV-result contains the entire charge, since all impurities follow the ac signal. The CV-depth is between the ionization region of shallow donors and traps, i.e. $(W_D^* - \lambda) < z_{CV} < W_D^*$. More specifically, z_{CV} is the gravity point of the charge distribution.

In a similar way, equations for the CV-concentration and the CV-depth can be derived for acceptor-like traps (Kimerling, 1974) or for p-type semiconductors. Results for non-uniform shallow impurity or trap concentration can be much more complicated and are typically not available in analytic form. Major errors in N_{CV} can occur for such a non-uniform impurity and trap concentration. Caution is also required if the trap emission time constant is comparable to either the ac signal or the quasi-dc bias. Major errors can result if the measurement is performed under such conditions.

A modification of the CV-technique is the *electrochemical CV-profiling technique* which has been reviewed by Blood (1986). The maximum depth of the regular CV-profiling technique is given by electrical avalanche or tunnel breakdown of the junction. This limitation is avoided in electrochemical CV-profiling which can profile many μm deep into the semiconductor. The electrochemical technique employs an electrolyte as a contact which acts as a rectifying barrier Schottky contact. The electrolyte allows one to measure the capacitance at zero bias. Subsequently, the semiconductor is etched by a photochemical etching process that is carried out in the same apparatus (Faktor et al., 1980). The alternative processes of capacitance measurements and etching allows one to measure the CV-concentration versus depths for very large depths.

12.1.3 Deep level transient spectroscopy (DLTS)

Deep level transient spectroscopy (DLTS) is a widely used, versatile capacitance-technique to characterize deep centers in semiconductors including the trap concentration, activation energy, emission properties, electron- and hole-capture cross sections and other properties (Lang, 1974). DLTS is based on temporal capacitance transients that occur after a rapid bias change of a Schottky diode, a p^+n-, or an n^+p-junction. After the rapid bias change, a net emission or a net capture of electrons or holes occurs from deep levels until equilibrium is re-established. The charge redistribution in the depletion layer results in a change in capacitance, i.e. a capacitance transient. Using appropriate models, the capacitance transients allow one to determine the characteristics of the deep levels in the depletion region of the junction.

In order to estimate the influence of deep traps on the depletion capacitance of a Schottky diode, let us consider an n-type semiconductor which, in addition to shallow donors, also contains deep donor impurities with a concentration N_T and an ionization energy E_t. The band diagram of such a semiconductor under various bias conditions is shown in Fig. 12.7 where only the deep but not the shallow states are shown for the sake of clarity. The width of the shallow impurity depletion region is denoted as W_D^*. Within the width λ, shallow donors are depleted while deep donors remain neutral (see Eqs. (12.25) and (12.26)).

Upon a rapid bias change, the band diagram of the metal–semiconductor junction changes as illustrated in Fig. 12.7. Immediately after the junction is subjected to a negative bias, deep traps are not in thermal equilibrium, as shown in Fig. 12.7(b). The width of the depletion region increases to a non-equilibrium value of $W_D^* + \Delta W_D$. Subsequently, the traps located above the Fermi level emit carriers until equilibrium conditions are established as shown in Fig. 12.7(c). During the transient emission of carriers, the charge in the depletion region increases, resulting in a decrease of the depletion region from $W_D^* + \Delta W$ to W_D^* and a corresponding *increase* in capacitance.

The charge modulation induced by the ac small signal voltage of the capacitance measurement occurs at the edge of the depletion layer of width W_D^*. It is assumed that the emission rate of deep traps is so low that the charge state of the traps cannot follow the ac signal. The differential capacitance per unit area is then given by

$$C = \varepsilon/W_D^* . \tag{12.32}$$

The **magnitude of the capacitance transient** is therefore given by the change in depletion layer width during the transient, that is

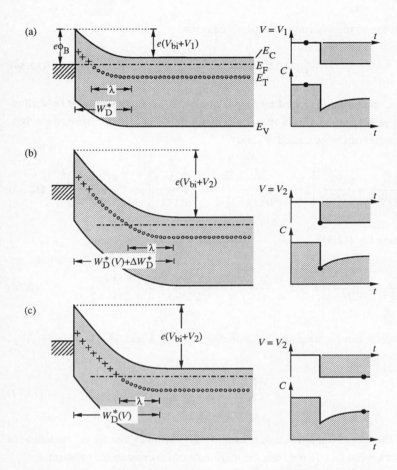

Fig. 12.7. Schematic band diagram of an n-type semiconductor with deep traps (a) under zero bias, (b) immediately following the application of a negative voltage, and (c) in thermal equilibrium. Also shown are the diode voltage pulse and the capacitance transient.

$$\frac{\Delta C}{C} = - \frac{\Delta W_D^*}{W_D^*} . \tag{12.33}$$

In order to calculate the magnitude of the capacitance transient, we consider the voltage drop across the junction and use the approximation $\phi_B \cong V_{bi}$. The voltage drop across the junction remains *constant* during the capacitance transient (see diagram in Fig. 12.7c). Thus, the *increase* in the deep donor ionization region is compensated by a *decrease* in the shallow donor depletion region. This can be expressed in terms of Poisson's equation by

$$\frac{e}{\varepsilon} \, W_D^*(V_2) N_D \Delta W_D^* = \frac{e}{\varepsilon} \int_{W_D^*(V_1) - \lambda}^{W_D^*(V_2) - \lambda} z N_T \, dz \, . \tag{12.34}$$

That is, the potential created by deep donor ionization is compensated by shallow donor neutralization. For a spatially constant trap concentration along the z-axis, the integral can be evaluated to yield

$$\frac{\Delta W_D^*}{W_D^*(V_2)} = \frac{N_T}{2 N_D} \left\{ 1 - \frac{2\lambda}{W_D^*(V_2)} \left[1 - \frac{W_D^*(V_1)}{W_D^*(V_2)} \right] - \left[\frac{W_D^*(V_1)}{W_D^*(V_2)} \right]^2 \right\} \tag{12.35}$$

or, using Eq. (12.33),

$$\frac{\Delta C}{C(V_2)} = \frac{N_T}{2 N_D} \left\{ 1 - 2 \frac{\lambda}{W_D^*(V_2)} \left[1 - \frac{C(V_2)}{C(V_1)} \right] - \left[\frac{C(V_2)}{C(V_1)} \right]^2 \right\} . \tag{12.36}$$

The result can be simplified by assuming $W_D^* \gg \lambda$ and $W_D^*(V_2) \gg W_D^*(V_1)$ which yields

$$\boxed{\frac{\Delta C}{C} = \frac{N_T}{2 N_D}} \tag{12.37}$$

This simple and important result (Lang, 1979) illustrates that the concentration of deep traps can be inferred from the magnitude of the capacitance transient.

A quantitative signature of deep level emission and capture transients can be obtained by measuring the capacitance transient as a function of temperature. Capacitance transients of a deep level as measured at different temperatures are shown schematically in Fig. 12.8. In the DLTS measurement the capacitance is measured at two times, t_1 and t_2, and the DLTS signal is defined as

$$S_{DLTS}(T) = \frac{C(t_1) - C(t_2)}{|\Delta C(0)|} \tag{12.38}$$

where $\Delta C(0)$ is the capacitance change due to the voltage pulse at $t = 0$. At very low temperatures, the DLTS signal is small since thermally stimulated emission of carriers from deep levels is negligible. At intermediate temperatures, the DLTS signal reaches a maximum. At high temperatures, the DLTS signal decreases again, because the emission of carriers from deep levels is faster than the measurement time window. Assuming that the capacitance transient is

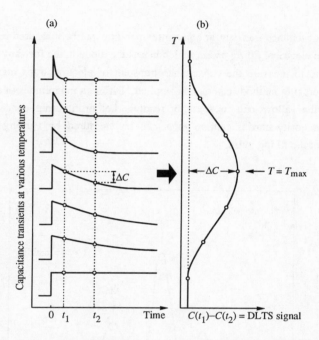

Fig. 12.8. Schematic illustration of (a) capacitance transients at different temperatures and (b) the DLTS signal deduced from the transients (after Lang, 1974).

exponential (Lang, 1974), the DLTS signal can be written as

$$S_{\text{DLTS}}(T) = e^{-t_1/\tau} - e^{-t_2/\tau} .$$ (12.39)

At a certain temperature, $T_{\max}$, the DLTS signal reaches a maximum. The deep level time constant, τ, at the temperature $T_{\max}$ is denoted as $\tau_{\max}$. The relationship between $\tau_{\max}$ and t_1 and t_2 is determined by differentiating $S_{\text{DLTS}}(T)$ with respect to τ and setting the result equal to zero which yields

$$\tau_{\max} = \frac{t_1 - t_2}{\ln(t_1/t_2)} .$$ (12.40)

An alternative method to determine the time constant is to measure the entire transient using many sampling points and fitting an exponential curve to the transient. Lock-in amplifiers and exponential correlates have also been used for this purpose (Miller et al., 1977). The extraction of an exponential capacitance transient can be further facilitated by using a baseline restorer (Miller et al.,

1975).

The capacitance transient at a given temperature can be measured repeatedly. A positive electrical *filling pulse*, which biases the diode in the forward direction, allows one to measure the capacitance transient of electron traps many times. Signal averaging methods can then be applied. Different magnitudes of the filling pulse further allow one to spatially resolve, i.e. profile trap concentrations. Finally, majority carrier capture rates can be measured by varying the time duration of the filling pulse.

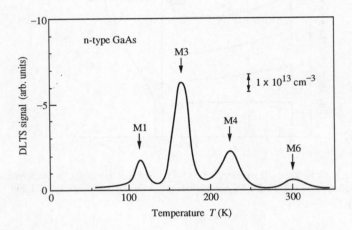

Fig. 12.9. DLTS spectrum of electron traps in n-type GaAs grown by molecular-beam epitaxy.

The experimental DLTS spectrum of n-type GaAs is shown in Fig. 12.9. Four distinct peaks are observed which are labeled M1, M3, M4, and M6 following the assignment of Lang et al. (1976). The four peaks in Fig. 12.9 are attributed to electron traps. Such electron traps emit electrons during the depletion process, i.e. the capacitance *increases* during the transient. Using Eq. (12.38), the DLTS signal is therefore a *negative* quantity for electron traps. Alternatively, hole traps in n-type semiconductors can be measured by a negative filling pulse and a subsequent rapid increase in bias. Hole traps moving below the Fermi level during the rapid bias change emit holes into the valence band. The negatively charged hole traps reduce the shallow donor space charge. As a result, the capacitance decreases during the transient. Thus, hole traps in n-type semiconductors have a *positive* DLTS signal (see Eq. (12.38)).

Activation energies of deep traps can be obtained from DLTS measurements by analyzing the emission kinetics of traps. As an example, we consider again electron traps in an n-type semiconductor. Recall (see Chap. 2) that the *capture probability* per unit time of an electron trap is given by

$$c_n = n v_{th} \sigma_n \qquad (12.41a)$$

where v_{th} is the thermal velocity of free carriers and σ_n is the *electron capture cross section* of the trap. For some deep levels, the capture cross section is independent of temperature, e.g. $\sigma_n = 10^{-15}$ cm^2. However, the capture process may be thermally activated for other deep levels. The capture cross section is then expressed in terms of a capture barrier,

$$\sigma_n = \sigma_{n\infty} e^{-E_B/kT} \qquad (12.41b)$$

where E_B is the height of the capture barrier and $\sigma_{n\infty}$ is the capture cross section at $T \to \infty$. Non-radiative, thermally activated capture was analyzed by Henry and Lang (1977).

The mean time for an impurity to capture an electron is

$$\tau_c = 1/c_n . \qquad (12.42)$$

Note that the capture probability per unit time is frequently called the *capture rate*. The electron *emission probability* per unit time is given by (see Chap. 2)

$$e_n = N_c \, v_{th} \sigma_n \frac{1}{g} \, e^{-(E_C - E_T)/kT} \qquad (12.43)$$

where N_c is the effective density of states at the conduction band edge and g is the degeneracy of the trap level. The mean time to emit an electron from a trap is

$$\tau_e = 1/e_n . \qquad (12.44)$$

Note that the emission probability per unit time is frequently called *emission rate*. Neglecting interactions of the trap with the valence band, the rate equation for electron emission can be written as

$$\frac{dN_T^0}{dt} = -e_n N_T^0 \qquad (12.45)$$

where N_T^0 is the concentration of traps occupied by an electron. The solution of the equation is

$$N_T^0(t) = N_T^0(t = 0) \, e^{-e_n t} . \qquad (12.46)$$

Equations (12.43) and (12.44) allow one to determine the activation energy of a trap by evaluating the emission probability e_n as a function of temperature. In practice, the emission probability is determined by using different time windows (t_1, t_2), which shift the DLTS peak on the temperature scale and allows one to evaluate $\tau_e(T)$ using Eq. (12.40). Finally, the slope of $\tau_e = 1/e_n$ in an Arrhenius plot yields the trap activation energy. Note that the prefactors of Eq. (12.43) have a temperature dependence of T^2, since $N_c \propto T^{3/2}$ and $v_{th} \propto T^{1/2}$. Therefore, Arrhenius plots of T^2/e_n versus $1/T$ can be used to evaluate the activation energy of traps.

The activation energies of electron traps and hole traps in GaAs have been evaluated by Martin et al., (1977) and Mitonneau et al., (1977). The authors used GaAs grown by MBE, VPE, LPE, and by bulk growth techniques and characterized the levels in terms of activation energy and capture cross section (see also Chap. 2). The study revealed that some of the deep levels are characteristic for certain growth technologies and cannot be detected in GaAs grown with different growth methods. Analysis of deep traps in III–V semiconductors by DLTS have also been reported in GaInAs (Mircea et al., 1977), GaPAs (Calleja et al., 1983), and for GaInPAs (Levinson and Temkin, 1983).

Some caution should be exercised when analyzing DLTS results. First, the activation energy is composed of the reaction *enthalpy* and *entropy*. Frequently, the entropy term is neglected and the ionization enthalpy is assumed to be a good approximation for the ionization energy. Second, the measured ionization energy can be lowered by the **Poole–Frenkel effect** (see Chap. 2). This effect lowers the effective ionization energy of an impurity in a strong electric field. A clear electric field dependence of the ionization energy has indeed been observed experimentally (Tasch and Sah, 1970; Lang, 1974; Vincent et al., 1979) and has been explained by the lowering of the emission barrier as well as by tunnel-assisted emission through the barrier (see Chap. 2).

Occasionally, **non-exponential capacitance transients** are observed in DLTS measurements especially in ternary and quaternary alloy semiconductors. Such non-exponential capacitance transients in alloy semiconductors were explained by broadening of the defect levels caused by microscopic statistical variations of the alloy composition (Omling et al., 1983). Broadening of defect levels or fluctuations of the band edges caused by microscopic variations of the alloy composition result in a distribution of the ionization energies. Consequently, the DLTS signal is a superposition of a distribution of exponential transients. The resulting DLTS signal is non-exponential which was observed experimentally in III–V alloy semiconductors (Omling et al., 1983).

12.1.4 Thermally stimulated capacitance (TSCAP) and thermally stimulated current (TSC)

Thermally stimulated capacitance (TSCAP) and thermally stimulated current (TSC) measurements are deep level characterization techniques which can be employed if the sample under investigation has a pn-junction or a metal–semiconductor junction. The methods allow one to identify approximate activation energies and the type of traps, i.e. minority or majority carrier traps. TSCAP and TSC measurements require only simple equipment such as a capacitance bridge, a current meter, and a variable temperature cryostat. In order to fill *majority* carrier traps, the semiconductor junction is cooled to low temperatures with no bias applied to the junction. Subsequently, the junction is reverse biased and heated to room temperature. In order to fill *minority* carrier traps, a pn-junction can be forward biased or optically excited. Again, the junction is subsequently reverse biased and heated to room temperature. At temperatures at which majority carriers are emitted from traps, a step-like increase in the capacitance and a peak in the current occur. Emission of minority carriers from traps results in a step-like decrease of the capacitance and a current peak. The capacitance and current features characteristic for majority carrier traps and minority carrier traps are shown in Fig. 12.10 (Lang, 1979). In the absence of traps, the capacitance does not change with temperature, since it is assumed that the shallow impurities are ionized in the entire temperature range considered, i.e. the depletion width remains constant, as shown in Fig. 12.10(a). The current increases at higher temperatures due to a thermally activated leakage current.

TSC was first proposed by Driver and Wright (1963). TSCAP was proposed and applied to GaAs by Carballes and Lebailly (1968). Sah and coworkers (1967, 1970, 1972, 1976) also contributed significantly to the understanding and analysis of the technique. A quantitative relationship between the trap activation energy, E_t, and the temperature of the maximum-slope capacitance, T_{max}, was given by Buehler and Phillips (1976) for TSCAP measurements

$$E_t = kT_{max} \ln \frac{\alpha kT_{max}^4}{\beta(E_t + 2kT_{max})} \tag{12.47}$$

where α is a constant and β is the heating rate used in the experiment. The authors showed that the equation also applies to TSC measurements, where T_{max} is the temperature of the maximum current (see Fig. 12.10). Exact values for E_t can be obtained from TSC and TSCAP measurements using Eq. (12.47). The constant α in Eq. (12.47) contains the capture cross section of the trap, σ, which cannot be obtained from the TSC and TSCAP measurements. Therefore,

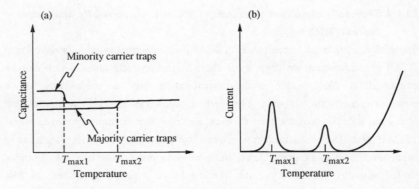

Fig. 12.10. Schematic illustration of (a) a thermally stimulated capacitance (TSCAP) and (b) a thermally stimulated current (TSC) signal of a semiconductor junction. The semiconductor is assumed to have minority and majority carrier traps. The minority carrier traps have a higher concentration and a lower activation energy than the majority carrier traps (after Lang, 1979).

approximate values for the capture cross section must be estimated in order to obtain the desired activation energies.

The total trap concentration can be obtained from the area under a current peak of the TSC curve. The total charge per contact area A is given by

$$Q = \frac{1}{A} \int_t I(t)\,dt = \frac{1}{A} \int_T I(T)\frac{1}{\beta}\,dT \qquad (12.48a)$$

which allows one to estimate the trap concentration

$$\boxed{N_T = \frac{1}{e}\,Q/\Delta W_D} \qquad (12.48b)$$

where ΔW_D is the change in depletion layer width caused by the reverse bias. Assume that the Schottky diode is cooled under flat band conditions, i.e. under forward bias. The initial depletion layer width is then zero. At the lowest temperature the diode is subjected to a reverse bias V. After heating the diode, the trap depletion layer width is (see Eq. (12.26b))

$$\Delta W_D = W_D^* - \lambda = \frac{N_T\lambda + \sqrt{N_T^2\lambda^2 + 2(N_D + N_T)\dfrac{\varepsilon}{e}(\phi_B - V)}}{N_D + N_T} - \lambda$$

$$(12.49)$$

which provides, together with Eq. (12.48b), an implicit solution for N_T. The total trap concentration can also be obtained from the step-height of the capacitance in the TSCAP measurement. The change in depletion layer width caused by free carrier emission from traps can be obtained from Eqs. (12.14) and (12.26b) which represent the depletion layer width for ionized shallow and for ionized shallow plus deep impurities, respectively. The change in capacitance (per unit area) is thus given by

$$\Delta C = \varepsilon \left[\frac{1}{W_D^*} - \frac{1}{W_D} \right] \qquad (12.50)$$

which allows one to calculate the trap concentration, N_T. Furthermore, the *sign* of the capacitance change allows one to distinguish between minority and majority carrier traps. The TSC and TSCAP techniques work best for trap concentrations $N_T \geq 0.1 \, N_D$ and $\Delta E \geq 0.25$ eV.

TSC and TSCAP measurements were reported for several III–V semiconductors including InP (Macrander et al., 1984).

12.1.5 Admittance spectroscopy

Admittance spectroscopy is a deep level characterization technique which can be used to analyze deep levels in the depletion region of a metal–semiconductor junction or a pn-junction. Admittance spectroscopy (Losee, 1972 and 1975) is a powerful equilibrium method which provides activation energies of relatively shallow (< 100 meV) and deep (>> 100 meV) levels as well as their concentrations. Admittance spectroscopy is a further advancement over the TSC method. Rather than measuring the dc signal from a junction, admittance spectroscopy measures the ac admittance. The possibility of measuring the ac admittance at different frequencies allows one to extract more information from the traps as compared to the TSC and TSCAP techniques.

The basic physical concept of admittance spectroscopy is illustrated in Fig. 12.11. The band diagram of a metal–semiconductor junction shown in Fig. 12.11(a) contains deep majority carrier traps in an n-type semiconductor. Upon application of a small-signal voltage of frequency f to the junction, the depletion layer width is modulated. At low temperatures, where the emission rate from the traps is small, the charge state of the deep traps does not 'follow' the ac modulation. At higher temperatures, i.e. for sufficiently large impurity emission rates, the charge modulation due to the ac voltage signal is shown in Fig. 12.11(b). Losee (1975) proposed an electrical circuit shown in Fig. 12.11(c) that can describe the electrical properties of the junction. The circuit describes the traps in terms of a capacitor C_t and a resistor R_t with a time constant

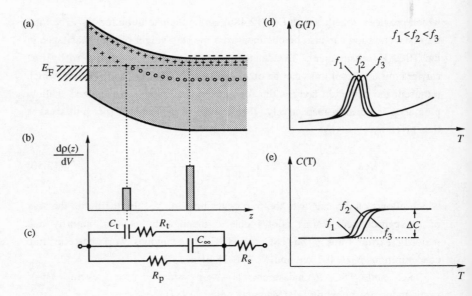

Fig. 12.11. Illustration of the principles of admittance spectroscopy. (a) Band diagram of an n-type semiconductor containing majority carrier traps. (b) Differential small signal charge density as a function of distance. The first and second charge peak are due to deep level emission and free carrier depletion, respectively. (c) Equivalent circuit diagram of the junction (after Losee, 1975). (d) Conductance and (e) capacitance of the junction as a function of temperature for three different measurement frequencies. The peaks of the conductance and the steps in the capacitance move to higher temperature with increasing measurement frequency.

$\tau = R_t C_t = 1/e_n$, where e_n is the emission probability of traps. The capacitor C_∞ describes the junction capacitance for very high frequencies, at which deep traps cannot follow the ac signal. In addition, the two resistors R_p and R_s model the leakage current and series resistance of the diode.

Peaks in the conductance and steps in the capacitance of the diode occur if the measurement frequency is resonant with the reciprocal emission probability of the trap. Schematic conductance and capacitance traces are shown in Fig. 12.11(d) and (e). The emission probability of an electron from a trap is given by (see Chap. 2)

$$e_n = \tau_n^{-1} = N_c v_{th} \sigma_n \frac{1}{g} e^{-E_t/kT} \tag{12.51}$$

where g is the degeneracy of the trap level and E_t the activation energy of the trap. In Eq. (12.51), a simple capture probability $c_n = n v_{th} \sigma_n$ is assumed. Peaks in the conductance and steps in the capacitance occur at temperatures at which the measurement frequency coincides with the emission probability, that is

$$f \cong e_n . \tag{12.52}$$

Since e_n strongly depends on temperature, the peaks in conductance and steps in capacitance depend on the measurement frequency. Inspection of the temperature dependence reveals that the peaks and steps shift to higher temperatures for increasing measurement frequencies. This dependence is schematically shown in Fig. 12.11(d) and (e). A qualitative physical explanation of the conductance peaks and capacitance steps is as follows. At low temperatures or high frequencies, the emission rate is small compared to the measurement time constant, i.e. the traps can *not* follow the ac signal. The capacitance is then given by

$$C_l = \varepsilon / W_D^* \tag{12.53}$$

where W_D^* is given by Eq. (12.26b). At higher temperatures, the traps follow the ac signal. The capacitance is then given by

$$C_h = \varepsilon \left[\frac{W_D^* N_D + (W_D^* - \lambda) N_T}{N_D + N_T} \right]^{-1} \tag{12.54}$$

where the term in the parenthesis is the gravity point of the charge induced by the ac signal. The change in capacitance is then given by

$$\Delta C = C_h - C_l \tag{12.55}$$

which represents an implicit solution for the trap density. At intermediate temperatures or frequencies, the deep level emission follows but lags the ac small signal. A charge oscillation which lags the ac signal implies a current that is in phase with the ac-signal, since $j = dQ/dT$. Thus, at intermediate temperatures or frequencies the diode has a finite conductance. Since the conductance above and below the intermediate regime is very small, a peak in the *real part* of the admittance (i.e. conductance) occurs in the intermediate regime. The peak in the conductance occurs approximately when the emission probability of traps at the Fermi level coincides with the oscillation frequency of the ac signal, $f \cong 1/\tau_e \cong e_n$. Losee (1975), in a more rigorous treatment of the problem, calculated the frequency at which the conductance is maximum as

$$\omega = 2\pi f = 1/\tau_n = e_n . \tag{12.56}$$

The activation energy of the trap can be obtained from an Arrhenius plot. Rewriting of Eqs. (12.51) and (12.56) yields

$$E_t = -k\frac{d}{d(1/T_{max})} \ln \frac{\omega}{N_c v_{th} \sigma_n g^{-1}} \tag{12.57}$$

where T_{max} is the temperature of maximum conductance at an angular measurement frequency of ω. Thus, the slope of the Arrhenius plot of $\ln(\omega)$ versus $(1/T_{max})$ is given by $-E_t/k$ and the value of the ordinate intercept at $(1/T_{max}) = 0$ is $(N_c v_{th} \sigma_n g^{-1})$. Rather than Arrhenius plots of $\ln(\omega)$ vs $1/T_{max}$, plots of $\ln(\omega T_{max}^{-3/2})$ vs. $1/T_{max}$ and $\ln(\omega T_{max}^{-2})$ vs. $1/T_{max}$ can be found in the literature. The additional temperature dependences $T_{max}^{-3/2}$ and T_{max}^{-2} arise from the temperature dependence of N_c and $N_c v_{th}$, respectively.

Admittance spectroscopy has been used in the characterization of a number of III–V semiconductors including GaInPAs (Takanohashi et al., 1984), InP (Takanohashi et al., 1988), AlGaAs (Chakravarty et al., 1989; Duenas et al., 1991), and GaAs (Hickmott, 1991).

12.2 Optical characterization techniques

Optical characterization in III–V semiconductors is a wide field which includes luminescence spectroscopy, absorption spectroscopy, photocurrent spectroscopy, excitation spectroscopy, and Raman spectroscopy. Furthermore, these spectroscopic techniques can be performed under the influence of external perturbations such as electric fields, magnetic fields, or mechanical stress. Optical spectroscopy has been important for both fundamental semiconductor research and technological materials evaluation. Optical spectroscopy always involves majority and minority carriers. The techniques are therefore sensitive to both the majority and minority carrier properties of semiconductors. Since minority carrier properties of semiconductors such as carrier lifetime or radiative efficiency depend sensitively on material quality (i.e. defect concentration), optical spectroscopic techniques are well suited to determine the quality of materials.

The signature of both shallow and deep impurities can be observed in optical techniques. Radiative transitions can involve shallow impurities as the initial or final state of the electronic transitions. The signature of shallow impurities can therefore be seen in luminescence or absorption techniques. Transitions involving deep impurities are mostly non-radiative. Deep impurities are therefore rarely observed in optical spectroscopic techniques. However, deep levels have a profound indirect influence on optical measurements in terms of reducing the radiative efficiency via Shockley–Read recombination processes. This section summarizes and discusses optical spectroscopic techniques with emphasis on extrinsic, i.e. impurity related aspects. The techniques discussed include photoluminescence, electroluminescence, cathodoluminescence, absorption, far-infra-red absorption, photocurrent, excitation, modulation, Raman, and local vibrational mode spectroscopy.

12.2.1 Luminescence spectroscopies

Luminescence spectroscopies probe the radiative intrinsic and extrinsic transitions of semiconductors. High concentrations of non-equilibrium electron–hole pairs are excited by an external source. The detection system typically consists of a monochromator (spectrometer) and a single or multi-channel detector. Photomultiplier tubes, pn-junction diodes, and charge-coupled devices (CCDs) are popular detector types for luminescence spectroscopy. The phase-sensitive 'Lock-In' technique is commonly used to enhance the signal-to-noise ratio of the detection system.

Different excitation sources are used for luminescence experiments. In *photoluminescence*, an optical source, typically a laser, is used as a means of excitation. The excitation energy is greater than the fundamental gap of the semiconductor in order to generate electron–hole pairs. The lateral excitation spot size depends on the excitation optics and can vary between 1 and 100 μm. The excitation depth is $l \cong \alpha^{-1}$, where α is the absorption coefficient of the semiconductor at the excitation wavelength. Additional diffusion of free carriers can enlarge the effective excitation volume. In *electroluminescence*, electrons and holes are injected into a pn-junction by means of electrodes. The requirement of a pn-junction restricts the use of electroluminescence spectroscopy to diode structures. That is, electroluminescence experiments cannot be performed on unipolar materials. In *cathodoluminescence*, a high-energy electron beam is used for excitation. The energy of a single electron of the cathode beam (>> 1 keV) is much larger than the energy of the fundamental gap. Thus, many electron–hole pairs are excited per incident electron. The probing depth of cathodoluminescence is on the order of a few μm. For example, an electron range of approximately 1 μm has been estimated for a 10 keV electron beam in GaAs (Yacobi and Holt, 1986). The lateral extent of the cathode beam can be several 100 Å for sufficiently high electron energies. However, such values cannot be reached in cathodoluminescence experiments due to free carrier diffusion. Improvement of the lateral resolution can be achieved by means of time-resolved cathodoluminescence in which the incident electron beam and the detected luminescence is gated. Detection of the luminescence for times shortly after the incidence of an electron pulse allows one to improve the lateral resolution of cathodoluminescence.

A schematic band diagram with several extrinsic and intrinsic luminescence

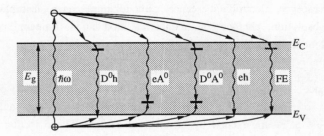

Fig. 12.12. Schematic radiative transitions in direct-gap semiconductors including donor-to-hole (D^0h), electron-to acceptor (eA^0), donor-to-acceptor pair (D^0A^0), band-to-band (eh), and free exciton (FE) transitions.

transitions is shown in Fig. 12.12. Electrons are 'pumped' by the exciting radiation of energy $\hbar\omega$ from the valence to the conduction band. For $\hbar\omega \gg E_g$, carriers initially have a high kinetic energy and rapidly thermalize via optical and acoustic phonon scattering. Thermalized carriers can recombine via different recombination channels. In the following, we assume that the semiconductor has a *direct gap*. Optical transitions in a direct-gap semiconductors can occur without the assistance of phonons. We further assume that the wave functions of shallow donors and acceptors are conduction-band-like and valence-band-like, respectively, i.e. optical transitions via shallow impurities are allowed as well. Prominent impurity transitions are ***neutral-donor-to-hole*** (D^0h), and ***electron-to-neutral-acceptor*** (eA^0). The energy of these transitions for $kT \cong 0$ is given by

$$(D^0h): \quad E = E_g - E_d \, , \tag{12.58a}$$

$$(eA^0): \quad E = E_g - E_a \tag{12.58b}$$

where E_d and E_a are the donor and acceptor ionization energies, respectively. At finite temperatures, the peak transition energies increase by a factor of $\frac{1}{2}kT$ (Eagles, 1960)

$$(D^0h): \quad E = E_g - E_d + \frac{1}{2}kT \tag{12.59a}$$

$$(eA^0): \quad E = E_g - E_a + \frac{1}{2}kT \tag{12.59b}$$

The intensity of the (D^0h) or the (eA^0) transition is given by (Eagles, 1960; Pickering et al., 1983).

$$I_{D^0 h, eA^0}(\hbar\omega) \propto (\hbar\omega - E_g + E_{d,a})^{1/2} \exp\left[-\frac{\hbar\omega - E_g + E_{d,a}}{kT}\right] \qquad (12.60)$$

where the square-root dependence of the prefactor originates in the density of states and the exponential function originates from a thermal Boltzmann distribution.

The *donor–acceptor pair* $(D^0 A^0)$ transition is an extrinsic transition occurring in semiconductors with high concentrations of residual donors and acceptors. The energy of donor-to-acceptor transitions is given by

$$(D^0 A^0): \quad E = E_g - E_a - E_d + \frac{e^2}{4\pi\varepsilon r} \qquad (12.61)$$

where $e^2/4\pi\varepsilon r$ is a Coulomb term and r is the spatial separation between donors and acceptors. The Coulomb term accounts for the electrostatic energy of the $(D^+ A^-)$ pair in the ionized final state of the $(D^0 A^0)$ transition. A further van der Waals correction to the energy in the $(D^0 A^0)$ transition has been discussed by Dean (1973). The distance r between the constituents of the pair can only assume discrete values due to the discreteness of lattice sites. As a consequence, the Coulomb term can assume only discrete values. Therefore, a set of discrete lines is expected for $(D^0 A^0)$ transitions. Such discrete lines have indeed been observed in GaP (Thomas et al., 1964). The spacing of the discrete $(D^0 A^0)$ lines is not constant in energy. For sufficiently large r, the energy separation between the $(D^0 A^0)$ lines decreases, and discrete $(D^0 A^0)$ lines cannot be resolved.

Donor–acceptor transitions can only occur, if the separation between the constituents of the pair is larger than a critical radius, r_c (Hopfield, 1964). A $(D^+ A^-)$ pair has the electrostatic energy $e^2/4\pi\varepsilon r$. If the pair is neutralized by an electron and a hole, the electrostatic pair interaction energy is zero and an energy $E_d + E_a$ is transferred to the lattice. Thus, the neutralization process can only occur if

$$E_d + E_a \geq \frac{e^2}{4\pi\varepsilon r_c} \qquad (12.62)$$

which allows one to evaluate the critical radius r_c. In GaAs and InP, the shallow impurity ionization energies are small which results in a large r_c. Discrete $(D^0 A^0)$ transitions are therefore not observed in GaAs, InP, or other semiconductors with small impurity ionization energies. The intensity is proportional to the overlap integral between the electron and hole wave functions, that is

$$I_{D^0A^0}(\hbar\omega) \propto \exp\left[-\frac{2r}{a_B}\right] \tag{12.63}$$

where a_B is the Bohr radius of the impurity with the smaller binding energy.

Band-to-band (eh) recombination involves free electrons in the conduction band that recombine with free holes in the valence band. Band-to-band recombination occurs predominantly at high temperatures, for example room temperature. The energy of (eh) transitions at finite temperatures occurs at

$$(\text{eh}): \quad E = E_g + \tfrac{1}{2}kT . \tag{12.64}$$

The intensity distribution of (eh) recombination is obtained from the density of states and the Fermi–Dirac distribution function and is given by

$$I_{eh}(\hbar\omega) \propto \int_{E_c}^{\hbar\omega - E_g} \rho_{DOS,C}(E) F_e(E) \rho_{DOS,V}[1 - F_e(E - \hbar\omega)]\, dE \tag{12.65}$$

where ρ_{DOS} and F_e are the density of states and the Fermi–Dirac distribution function and the subscripts C and V refer to the conduction and valence band, respectively. Evaluation of the integral yields

$$I_{eh}(\hbar\omega) \propto (\hbar\omega)^2 (\hbar\omega - E_g)^{1/2} \exp\left[\frac{-(\hbar\omega - E_g)}{kT}\right] . \tag{12.66}$$

The equation can be further simplified for $E_g \gg kT$ which yields

$$\boxed{I_{eh}(\hbar\omega) \propto (\hbar\omega - E_g)^{1/2} \exp\left[\frac{-(\hbar\omega - E_g)}{kT}\right]} \tag{12.67}$$

The (eh) recombination line has a maximum at $\hbar\omega = E_g + \tfrac{1}{2}kT$, which is consistent with Eq. (12.64). The full width at half-maximum of the emission line is $\Delta E \cong 1.8kT$ which can be calculated from Eq. (12.67).

Excitonic luminescence is observed at low temperatures such as liquid helium temperature. **Excitons** are electron–hole pairs that are bound by coulombic interaction. The coulombic interaction is similar to that of a hydrogen atom. Electron–hole pairs bound by coulombic interaction are called **free excitons** if they do not interact with any other center. The free exciton binding energy can be obtained from the analogy of the hydrogen atom and is given by

$$E_{\text{FE}} = E_{\text{g}} - \frac{m_r^* e^4}{32 \pi^2 \varepsilon^2 \hbar^2} \tag{12.68}$$

where ε is the permittivity of the semiconductor and m_r^* is the reduced effective mass

$$\frac{1}{m_r^*} = \frac{1}{m_e^*} + \frac{1}{m_h^*} , \tag{12.69}$$

m_e^* and m_h^* being the effective electron and hole mass, respectively. The effective Bohr radius of the exciton is given by

$$a_{\text{B}}^* = \frac{4 \pi \varepsilon \hbar^2}{m_r^* e^2} . \tag{12.70}$$

Equations (12.68) and (12.70) can be expressed as

$$E_{\text{FE}} = E_{\text{g}} - \left(\frac{\varepsilon_0}{\varepsilon} \right)^2 \frac{m_r^*}{m_0} E_{\text{Ryd}} , \tag{12.71}$$

$$a_{\text{B}}^* = \frac{\varepsilon}{\varepsilon_0} \frac{m_0}{m_r^*} a_{\text{B}} \tag{12.72}$$

where $E_{\text{Ryd}} = 13.6$ eV and $a_{\text{B}} = 0.53$ Å are the Rydberg energy and Bohr radius, respectively. As an example, we consider GaAs with $m_e^* = 0.067 m_0$, $m_{\text{hh}}^* = 0.45 m_0$, and $m_r^* = 0.058 m_0$. With $\varepsilon_r = \varepsilon/\varepsilon_0 = 13.1$, one obtains $E_{\text{FE}} = E_{\text{g}} - 4.6$ meV and $a_{\text{B}}^* = 120$ Å.

Finally, *bound excitons* (Haynes, 1960, 1966) are excitons that are bound to an attractive center such as a shallow neutral or ionized impurity, a deep impurity, or some other defect of the semiconductor lattice. The transition energy of bound excitons obeys the relation

$$E_{\text{BE}} < E_{\text{FE}} . \tag{12.73}$$

That is, the energy of the bound exciton is lowered by the binding center. The bound exciton does not possess kinetic energy since it is bound to a center. The linewidth is the natural linewidth given by the uncertainty principle

$$\Delta E = \hbar \tau \tag{12.74}$$

where τ is the spontaneous lifetime of the exciton. The lineshape of the exciton is therefore lorentzian

$$E_{\text{BE}} \propto \frac{\Delta E/2\pi}{(E - E_{\text{BE}})^2 + \frac{1}{4}\Delta E^2} . \tag{12.75}$$

Bound excitons in high-purity binary III–V semiconductors can have linewidths as narrow as 0.1 meV (Harris, 1992). Excellent reviews on excitons were given by Rashba and Sturge (1982) and by Dean and Herbert (1979).

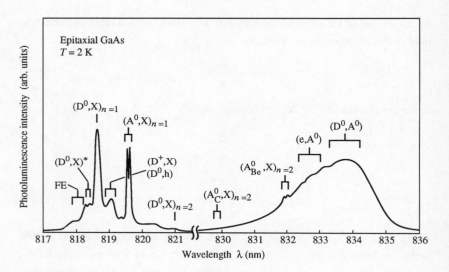

Fig. 12.13. Low-temperature photoluminescence spectrum of epitaxial GaAs. The entire luminescence occurs below the band gap ($\lambda_g = 816$ nm) (courtesy of T. D. Harris).

The near-band-edge photoluminescence spectrum of GaAs measured at 2 K is shown in Fig. 12.13. The fundamental gap of GaAs at 2 K is at $E_g = 1.519$ eV which corresponds to a wavelength of $\lambda_g = 816$ nm. No luminescence is observed at that wavelength on the intensity scale of Fig. 12.13 which indicates that (eh) recombination is very weak at low temperatures. The luminescence spectrum displays a wealth of transitions in the near-bandgap regime. Three impurity-assisted transitions can be identified which are the $(D^0 h)$, the (eA^0), and the $(D^0 A^0)$ transitions. The (eA^0) transition occurs at approximately 832.7 nm which corresponds to an energy of 30.5 meV below the band gap. Beryllium ($E_a = 28$ meV), Mg (28 meV), or Zn (31 meV) are candidates for the acceptors involved in this transition. The $(D^0 h)$ transition at approximately 819 nm corresponds to a donor energy of 5.5 meV. Since many chemically different shallow donors have very similar ionization energies, it is hardly possible to identify the chemical nature of shallow donors by the energy of the $(D^0 h)$ transition in GaAs.

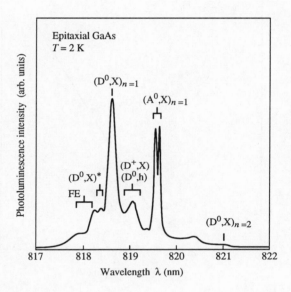

Fig. 12.14. Exitonic regime of a low-temperarure photoluminescence spectrum of epitaxial GaAs (courtesy of T. D. Harris).

The excitonic regime of a low-temperature photoluminescence spectrum of epitaxial GaAs is shown in Fig. 12.14. The major features of the spectrum are the free exciton (FE) and several bound excitons including the neutral-donor-bound exciton $(D^0,X)_{n=1}$, the ionized-donor-bound exciton (D^+,X), and the neutral-acceptor-bound exciton $(A^0,X)_{n=1}$. Note that the (D^+,X) transition and the $(D^0 h)$ transition have similar energies and cannot be differentiated spectroscopically. A transition that leaves the neutral donor in an excited state is the $(D^0,X)_{n=2}$ transition occurring at 821 nm. The $(D^0,X)_{n=2}$ transition has a *lower* energy than the $(D^0,X)_{n=1}$ transition since the neutral donor is left in an excited state in the former transition. Analogously, the $(A^0,X)_{n=2}$ transition has a lower energy than the $(A^0,X)_{n=1}$ transition. For a detailed discussion of excitonic and impurity transitions in III–V semiconductors, the reader is referred to Williams and Bebb (1972).

A quantitative relation between the bound exciton (BE) localization energy and the ionization energy of the binding center was found by Haynes (1960).

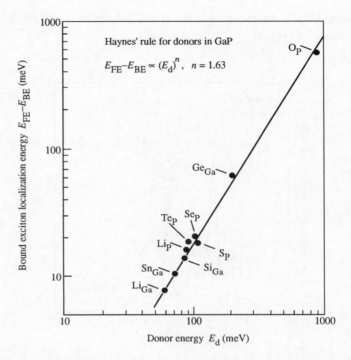

Fig. 12.15. Variation of the bound exciton (BE) localization energy for different donors in GaP (after Dean, 1983).

This relation allows one to identify impurity elements from their BE energy. Let us define the BE localization energy as the energy required to remove the BE from its binding center to obtain a FE and an isolated, non-interacting center. Haynes found experimentally that the BE localization energy increases with the impurity ionization energy. This rule is known as ***Haynes' rule*** and can be written as

$$E_{FE} - E_{BE} = \alpha(E_d)^n \tag{12.76}$$

where E_d is the ionization energy of the binding center, α is a constant with typical values of $0.05 - 0.5$, and n is a constant with typical values of $1.0 - 2.0$. (For acceptors in Si, Haynes found $\alpha = 0.10$ and $n = 1.0$). The variation of the BE localization energy for nine different donors in GaP is shown in Fig. 12.15 (Dean, 1983). The author modeled the binding energy in terms of Eq. (12.76) and found that an exponent of $n = 1.63$ provides a good fit to the experimental data.

In GaAs and InP, the shallow donor and acceptor energies do not vary much for different chemical impurities. It is therefore difficult to evaluate the parameters α and n in Eq. (12.76). Assuming $n = 1.0$, the parameter $\alpha \cong 0.2$ and $\alpha \cong 0.1$ for neutral donor and acceptor transitions in GaAs, respectively, as inferred from Fig. 12.14.

The sharpness of the BE lines allows one to resolve many different lines and thereby to identify chemical impurities in III–V semiconductors. A systematic study of acceptor energies in GaAs using the bound exciton, electron-to-acceptor, and donor-to-acceptor lines has been reported by Ashen et al. (1975). Donors in GaAs were identified by Harris et al. (1988), Almassy et al. (1981), and Reynolds et al. (1983). Donor identification in InP using photoluminescence techniques was reported by Dean et al. (1984a, 1984b) and Skolnick et al. (1984).

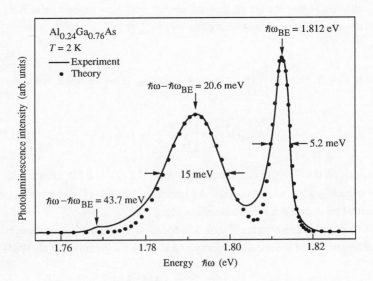

Fig. 12.16. Low-temperature photoluminescence spectrum of $Al_{0.24}Ga_{0.76}As$. Experimental data (solid line) compares well with theoretical data (dots) calculated from alloy broadening.

Spectral features of luminescence spectra are significantly broadened in ternary and quaternary semiconductor alloys by *alloy broadening* (Schubert et al., 1984). Consider a ternary semiconductor alloy such as $Al_x Ga_{1-x} As$. Assume further that the composition x varies on a microscopic scale and is caused by the random distribution of Al and Ga on the cation sites of the zincblende lattice. Since the bandgap energy depends on the local alloy composition, the gap energy varies spatially as well. As a result, luminescence of alloys reflects the spatial variation of the bandgap energy. The low-temperature photoluminescence spectrum of $Al_{0.24}Ga_{0.76}As$ shown in Fig. 12.16 illustrates the effect of alloy broadening. The BE line, which is typically 0.1 meV wide in binary III–V semiconductors, has a full width at half-maximum of 5.2 meV. The (eA^0) transition is 15 meV wide. The alloy broadening of the BE and (eA^0) lines can be calculated assuming random distribution of cations or anions. For a ternary alloy semiconductor with composition x (e.g. $Al_xGa_{1-x}As$), the luminescence linewidth is given by (Schubert et al., 1984)

$$\Delta E_{BE} = 2\sqrt{2\ln 2}\, \frac{dE_g}{dx} \left[\frac{x(1-x)}{KV_{exc}} \right]^{1/2} \tag{12.77}$$

where dE_g/dx is the change of the gap energy with composition, and K is the cation density of the semiconductor given by

$$K = 4a_0^{-3} \tag{12.78}$$

which is $K = 2.21 \times 10^{22}$ cm^{-3} in GaAs and V_{exc} is the excitonic volume given by

$$V_{exc} = \frac{4}{3}\pi \left[\frac{\varepsilon\, m_0}{\varepsilon_0\, m_r^*}\, a_B \right]^3 . \tag{12.79}$$

The width of the BE line calculated from Eqs. (12.77)–(12.79) agrees with the experimental width as illustrated by the comparison of experimental and theoretical luminescence spectra shown in Fig. 12.16.

The spectral width of the eA^0 line was found to be predominantly determined by the fluctuation of the acceptor energy and is given by (Schubert et al. 1984)

$$\Delta E_{eA^0} = \frac{\Delta E_v}{\Delta E_g}\, \frac{dE_g}{dx} \left[\frac{x(1-x)}{KV_{acc}} \right]^{1/2} \tag{12.80}$$

where $\Delta E_V/\Delta E_g$ is the relative variation of the valence band edge with composition, also called valence band offset; V_{acc} is the volume of the acceptor wave function

$$V_{acc} = \frac{4}{3}\pi \left[\frac{\varepsilon\, m_0}{\varepsilon_0\, m_{hh}^*} \right] a_B \tag{12.81}$$

where m_{hh}^* is the heavy-hole effective mass. The experimental (eA^0) transition-line-width agrees with the calculated linewidth as shown in Fig. 12.16. A

$\Delta E_V / \Delta E_g$ ratio of 0.15 was assumed in the calculation. Alloy broadening also prevails in luminescence spectra of other ternary alloy semiconductors such as $Ga_x In_{1-x} As$, $Al_x In_{1-x} As$, or $Ga_x In_{1-x} P$. Luminescence spectra of quaternary III–V semiconductors (see, for example, Göbel, 1982) have even broader luminescence features due to random concentration fluctuations of cations *and* anions.

For completeness, we summarize other broadening mechanisms of luminescence lines in semiconductors. We assume that a radiative electronic transition occurs between an initial and a final state. Each of the two quantum states can be broadened. Two types of broadening mechanisms, namely *homogeneous* and *inhomogeneous broadening* are differentiated. *Homogeneous* broadening is due to the finite lifetime of an electron occupying a quantum state. This lifetime can be limited by, for example, (i) radiative transitions, (ii) tunneling of the electron to another state (e.g. tunneling of an electron in a multi-quantum well structure from one well to another well), or (iii) scattering of the electron by a phonon to another state. The broadening due to a finite lifetime, τ, can be obtained from the uncertainty relation

$$\Delta E = \hbar / \tau \qquad (12.82a)$$

where ΔE is called the *natural linewidth*. If several physical mechanisms, e.g. recombination, tunneling, and phonon scattering, limit the state lifetime, then the total broadening is given by

$$\Delta E^2 = \Delta E^2_{\text{recombination}} + \Delta E^2_{\text{tunneling}} + \Delta E^2_{\text{phonon}} . \qquad (12.82b)$$

The lineshape of a homogeneously broadened luminescence transition is lorentzian (Davydov, 1965)

$$I(E) = \frac{\Delta E / 2\pi}{(E - E_{\text{ave}})^2 + \tfrac{1}{4} \Delta E^2} \qquad (12.82c)$$

where E_{ave} is the average luminescence energy.

The *inhomogeneous* broadening mechanisms of a quantum state include any statistical fluctuations which perturb the energy of the state. Examples of such mechanisms are (i) compositional fluctuations in semiconductor alloys, (ii) quantum well width fluctuations in semiconductor heterostructures, or (iii) impurity density fluctuations due to the random distribution of impurities. The lineshape of inhomogeneously broadened transitions depends on the specific broadening mechanism, but, frequently, a gaussian lineshape is assumed.

Finally, the *thermal broadening* of radiative transitions is due to the thermal distribution of carriers among a set of quantum states. Thermal broadening of the

(eA^0), (D^0h), and (eh) transitions results in a linewidth of

$$\Delta E \cong 1.8 \ kT \tag{12.83}$$

as inferred from Eqs. (12.60) and (12.67). Note that thermal broadening is neither a homogeneous nor an inhomogeneous broadening mechanism, since it is *not* due to the broadening of a quantum state. Instead, thermal broadening reflects the distribution of carriers among a set of initial and final states.

12.2.2 Absorption characterization

Light traversing a semiconductor decreases in intensity along the propagation direction due to absorption processes. Consider a light beam propagating along the z-direction which enters an absorbing semiconductor at $z = 0$. The light intensity can then be written as

$$I = I_0 \ e^{-\alpha z} \tag{12.84}$$

where α is the *absorption coefficient* and I_0 is the light intensity at $z = 0$. Different physical processes give rise to the absorption of photons in semiconductors.

Intrinsic absorption processes are (i) band-to-band (ii) excitonic, (iii) free carrier absorption, (iv) inner-shell electron absorption, and (v) phonon absorption processes. In the near-bandgap wavelength range, band-to-band absorption is the most important process at room temperature. The lowest energy of band-to-band transitions is given by the gap energy of the semiconductor which is also called the *fundamental absorption edge*.

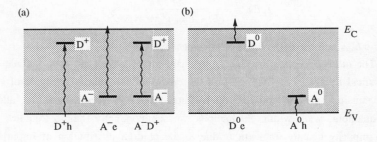

Fig. 12.17. Schematic illustration of (a) the near-band-edge absorption processes D^+h, A^-e, and (b) the far-infra-red absorption processes D^0e and A^0h.

Extrinsic absorption processes are illustrated in Fig. 12.17 and are either in the near-bandgap regime or in the far-infra-red regime. Near-gap transitions (Fig. 12.17a) are the valence-band-to-donor (D^+h), the acceptor-to-conduction-band (A^-e), or the acceptor-to-donor (A^-D^+) transition. The far-infra-red (FIR) transitions (Fig. 12.17b) for shallow impurities are the donor-to-conduction-band (D^0e) and the valence-band-to-acceptor (A^0h) transition. The FIR excitation wavelength can be as long as $\lambda = 250$ μm and requires very different experimental methods as compared to the near-band-edge transitions. A further extrinsic FIR absorption process is due to local vibrational modes of impurities or of impurity pairs. Such local vibrational mode absorption will be discussed later in this section.

Near-bandgap absorption spectra of GaAs at two different temperatures are shown in Fig. 12.18 (Sturge, 1962). The magnitude of the below-bandgap absorption coefficient (Fig. 12.18a) is several orders of magnitude smaller than that for above-bandgap absorption. The absorption spectra taken at low temperatures ($T = 21$ K) display well-resolved features as compared to the room temperature spectra (T = 294 K). Sturge (1962) proposed that the shoulders observed in Fig. 12.18(a) are associated with impurity levels. The impurity levels are at energies 0.30 eV, 0.47 eV, and 0.70 eV above the valence band or below the conduction band. Comparison of the absorption data with photoconductivity experiments (Bube, 1960) led Sturge (1962) to the conclusion that the 0.47 eV and the 0.70 eV levels are associated with donor levels. Sturge (1962) also showed that the initial rise of the absorption can be described with the empirical formula $\alpha \propto (E - E_i)^2$ where E_i was defined as the *threshold* of an impurity. The threshold, E_i, varied with temperature; however ($E_g - E_i$) was shown to be constant with temperature. This can be explained if the impurity level is 'attached' to either the valence or the conduction band.

Band-edge and above-band-edge absorption is shown in Fig. 12.18(b) for the two temperatures $T = 21$ K and 294 K. The shift of the onset of absorption reflects the shift of the band edge from $E_g \cong 1.51$ eV to 1.42 eV for low and high temperatures, respectively. The pronounced enhancement of the band-edge absorption at low temperatures is attributed to free excitonic absorption. Such excitonic enhancement is not observed at room temperature since excitons with a binding energy on the order of 5 meV dissociate rapidly at room temperature.

Several aspects of near-band-edge absorption need to be mentioned. At high doping concentrations, e.g. donor concentrations, the Fermi level moves into the conduction band. Since all states below E_F are occupied, the absorption edge shifts from $E_C - E_V = E_g$ to $E_F - E_V > E_g$. Analogous arguments apply to high acceptor concentrations. The shift of the absorption edge to higher energies

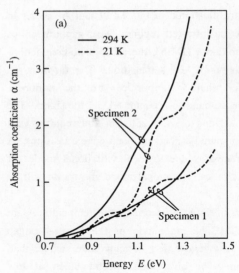

Fig. 12.18. (a) Below-bandgap and (b) above-bandgap absorption spectra in GaAs at $T =$ 21 K and 294 K. Absorption processes below the band gap shown in (a) are due to impurities (Sturge, 1962).

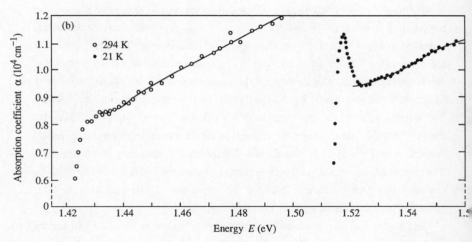

is called the **Burstein–Moss shift** (Burstein, 1954; Moss, 1954) and is due to band filling effects. On the other hand, the formation of tail states and bandgap renormalization occur at high doping density which counteracts the band filling. Impurity **tail states** are formed when impurity states form an impurity band and finally merge with the valence and conduction band. Impurity tail states were correlated with internal electric fields in the semiconductor caused by random impurity distribution (Redfield, 1963; Pankove, 1965). The authors correlated the magnitude of the band tail with the impurity concentration. In addition to tail states, **bandgap renormalization** is a many-body effect at high carrier densities which reduces the band gap. Bandgap renormalization was discussed by Wolf (1962) and others. A detailed discussion of such impurity effects is found in Chap. 1. Even at low impurity concentrations, absorption occurs below the fundamental gap. Urbach (1953) showed that the absorption in the tail follows the relationship $d(\ln\alpha)/d(\hbar\omega) = 1/kT$. Such band tails are referred to as **Urbach tails**. The temperature dependence of Urbach tails led Knox (1963) to the conclusion that the tails are caused by phonon assisted transitions. For a quantitative discussion of band tails, the reader is referred to Chap. 1.

Photocurrent absorption spectroscopy is a simple alternative to regular absorption spectroscopy. In photocurrent spectroscopy, absorption processes can be detected that create free carriers in the valence or conduction band. These carriers are detected in terms of a current. Two major processes complicate the interpretation of photocurrent measurements. First, *surface recombination* of photogenerated carriers reduces the photocurrent signal. Because of surface recombination, the photocurrent signal is smaller than expected. Surface recombination is especially important for high photon energies, i.e. $\hbar\omega \gg E_g$. The absorption coefficient for such photon energies is large, resulting in a small penetration depth $l = \alpha^{-1}$, i.e. free carriers are generated in the near-surface region, in which they are subject to surface recombination. Second, *bulk recombination* of photogenerated carriers reduces the photocurrent signal. This problem can be alleviated, if pin-junctions are used for the photocurrent measurement. Electrons and holes created in the i-region of the diode will be spatially separated by the built-in electric field. The spatially separated electrons and holes do not recombine and thus fully contribute to the photocurrent. Nevertheless, the employment of pin-junctions does not solve the problem of surface recombination. The two problem areas of surface and bulk recombination make photocurrent absorption spectroscopy a qualitative method whose quantitative interpretation is difficult.

For absorption in the far-infra-red (FIR), photocurrent spectroscopy is the preferred method due to its sensitivity and due to the lack of efficient detectors that work in a broad wavelength range, e.g. $\lambda = 100 - 400$ μm. In addition to photocurrent, photovoltage measurements were used in the FIR region (Stillman et al., 1977), and these are preferred for highly resistive samples.

The FIR conductivity spectrum of high-purity GaAs measured at 4.2 K is shown in Fig. 12.19 (Wolfe and Stillman, 1971). The background impurity concentration of the GaAs sample is in the 10^{13} cm^{-3} range. The photoconductivity spectrum shows clearly resolved peaks at energies ranging

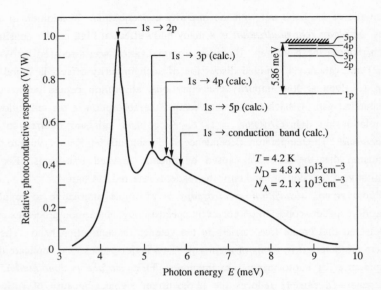

Fig. 12.19. Far-infra-red photoconductivity spectrum of high-purity GaAs. Also shown are theoretical transition energies (arrows) calculated from the hydrogenic model using the experimental (1s → 2p) energy as a fit parameter. The hydrogenic energy level diagram is shown in the inset (after Wolfe and Stillman, 1971).

from 4 meV to 6 meV. The major line at 4.40 meV was identified as the 1s→2p transition of a hydrogenic donor impurity, i.e. a transition from the ground state ($n = 1$) to the first excited state ($n = 2$). In zero magnetic field, the 2s and 3s states are degenerate with the 2p and 3p states. However, transitions between the ground state (1s) and the s-type excited states are not allowed. The matrix elements for the transitions, which are proportional to $\langle \psi_{1s} | r | \psi_{2s} \rangle$ and $\langle \psi_{1s} | r | \psi_{3s} \rangle$, are zero, since all of the wave functions ψ_{1s}, ψ_{2s}, and ψ_{3s} are functions of even symmetry with respect to the radius r (Bohm, 1951). Using the energy of the 1s → 2p transition, the authors calculated the 1s → 3p, 1s → 4p, 1s → 5p, and 1s-to-continuum transitions. The energies of the calculated transitions are shown in Fig. 12.19 and agree with the experimental peaks. Using the 1s → 2p transition energy, the shallow donor ionization energy was calculated to be $E_d = 5.86$ meV in GaAs.

If the excited states (2p, 3p, 4p, ···) were really bound states, electrons in these states could not contribute to the conductivity of the sample and the excited-state absorption would not result in peaks in the photoconductivity spectrum. Stillman et al. (1977) proposed several mechanisms by which excited hydrogenic impurity atoms become ionized, that is contribute to the photoconductivity. First, excited-state wave functions are more likely to overlap with neighboring impurities as compared to the spatially localized 1s wave function. Conduction via an excited-state impurity band is then more likely. Second, excited donor atoms can be ionized by field-induced tunneling from the excited state into the conduction band. Third, energetic free electrons can ionize excited-state donors by impact ionization. Fourth, excited donors can be thermally ionized, that is by absorption of one or several phonons. Fifth, an excited donor can be photoionized, that is by absorption of a second photon. Although it is difficult to estimate the relative importance of the processes, it is evident that excited donors can become fully ionized and contribute to the conductivity of the semiconductor.

Finally, FIR radiation can also be absorbed by localized vibrational modes of impurities or impurity pairs. Such absorption will be discussed later in this section.

12.2.3 Raman characterization

Raman characterization is a spectroscopic technique in which elementary excitations are induced by incident light. In the process of an elementary excitation, the incident photon is scattered such that the scattered photon has a different energy than the incident photon. Raman spectroscopy is therefore also called *inelastic light scattering* spectroscopy. The energy and momentum balance of a Raman scattering event can be written as

$$E_i \pm E_{exc} = E_s \, , \tag{12.85a}$$

$$k_i \pm q_{exc} = k_s \tag{12.85b}$$

where E_{exc} and q_{exc} are the energy and momentum of the elementary excitation and the subscript i and s refer to the incident and the scattered photon, respectively. The minus sign in Eq. (12.85) is valid if a quantum process is *excited* in the semiconductor. As a result, the energy of the scattered photon is *lower*, i.e. *Stokes shifted*. The plus sign in Eq. (12.85) is valid if an elementary process is *de-excited*. As a result, the energy of the scattered photon is *higher*, i.e. *anti-Stokes*-shifted. The intensity of the Stokes-shifted scattered light is usually much stronger than the intensity of the anti-Stokes-shifted light.

A Raman backscattering measurement configuration is shown in Fig. 12.20(a). The excitation beam has near normal incidence with respect to the sample surface. Also shown is the momentum conservation condition (Fig. 12.20b) for an elementary excitation, i.e. Stokes scattering. The average

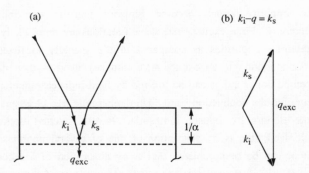

Fig. 12.20. (a) Illustration of the Raman backscattering geometry with wavevector of incident (k_i) and scattered (k_s) light. The penetration depth of the exciting light into the sample is $1/\alpha$. (b) Momentum conservation for Stokes scattering.

probing depth of the incident light beam depends on the absorption coefficient of the semiconductor, $\alpha(\lambda)$, at the wavelength λ of the incident light. The average probing depth of the measurement is $[\alpha(\lambda)]^{-1}$. Thus, the probing depth of Raman scattering can be chosen by proper selection of the excitation wavelength.

The most widespread use of Raman spectroscopy is for lattice vibrations, that is if the elementary excitations are phonons. Due to the small momentum of photons as compared to the range of momenta of phonons, only *optical* (transverse and longitudinal modes) photons can be excited in Raman experiments. Such optical phonons can have zero momentum and a finite energy. Several comprehensive reviews have been published on Raman spectroscopy for optical phonon excitation (see, for example, Cardona and Güntherodt, 1984, 1989).

There are two major areas where Raman spectroscopy is used to characterize impurities. These are *electronic Raman spectroscopy* and **local vibrational mode (LVM) Raman spectroscopy**. In electronic Raman spectroscopy, carriers are excited from the ground state of an impurity to an excited impurity state. In LVM Raman spectroscopy, local vibrational modes of impurities or impurity complexes are excited by the incident light.

Electronic Raman spectroscopy was first employed for acceptors in GaAs (Wright and Mooradian, 1968) and GaP (Henry et al., 1966; Chase et al., 1977). The first well-resolved electronic Raman spectra were reported for acceptors in semi-insulating GaAs (Wan and Bray, 1985). These spectra allowed the authors to identify C and Zn in GaAs grown by the liquid encapsulated Czochralski (LEC) technique. Wagner et al. (1986a, 1986b, 1986c, 1988a, 1988b) extensively studied shallow acceptors in semi-insulating GaAs using electronic Raman spectroscopy. The studies revealed that electronic Raman spectroscopy allows one to evaluate quantitatively acceptor concentrations in semi-insulating GaAs (Wagner et al., 1986a). A comparison of electronic Raman and luminescence experiments revealed that quantitative assessment of acceptor concentrations is difficult with either technique. Electronic Raman spectroscopy has also been used for relatively deep acceptors in GaAs (Shanabrook et al., 1986). The concentration of shallow acceptors compensated by the EL2 center in semi-insulating GaAs was studied by Wagner et al. (1988b). Electronic Raman scattering was also reported for donors in $GaAs/Al_x Ga_{1-x}As$ quantum wells (Perry et al., 1985a, 1985b). They observed a line which they attributed to $1s \rightarrow 2s$ transitions of the donor and additional peaks attributed to donor-resonance excitations. Gammon et al. (1986) reported electronic Raman transitions for Be acceptor transitions in $GaAs/Al_x Ga_{1-x}As$ quantum wells. Experimental line positions were found to agree with theoretical predictions. Although LVM will be discussed in detail in the next section, some specific characteristics of LVM Raman are listed here. As mentioned above, the probing depth of Raman is $(1/\alpha)$ which can be of the order of one or several 1000 Å. It is therefore possible to probe thin layers by LVM Raman, e.g. epitaxially grown layers. Furthermore, LVM Raman allows measurements at relatively high free carrier concentrations. Free carrier absorption, which interferes with the LVM signal (see next section), does not affect the LVM-Raman signal in an excessive way due to the high energy of incident and scattered photons.

12.2.4 Local vibrational mode characterization

Acoustic and optical phonons are thermal vibrations of atoms in a perfect lattice. Such lattice vibrations are propagating plane waves and are thus spatially delocalized. The frequency of optical phonons is approximately given by (Moll, 1964)

$$\omega_{LO} \cong \omega_{TO} \cong \sqrt{\frac{k}{M}} \qquad (12.86)$$

where k is the force constant of the lattice and M is the mass of a lattice atom. The subscripts LO and TO refer to longitudinal and transverse optical phonon modes, respectively. In compound semiconductors, Eq. (12.86) needs to be modified according to

$$\omega_{LO} \cong \omega_{TO} \cong \sqrt{\frac{2k}{M_1 + M_2}} \qquad (12.87)$$

where M_1 and M_2 are the two atomic masses of the binary compound semiconductor.

The characteristics of lattice vibrations are changed drastically if an impurity atom is added to the lattice. Consider an impurity atom with a mass M_i such that $M_i < M$. The angular frequency of the vibration of an impurity atom is then given in analogy to Eq. (12.86) by

$$\omega_i \cong \sqrt{\frac{k}{M_i}} \ . \tag{12.88}$$

Since $M_i < M$, the frequency of the impurity vibration is higher as compared to regular lattice vibrations, i.e. the energy of the impurity vibrations, $\hbar\omega_i$, is larger than the (maximum) energy of optical phonons. Hence, the energy of an impurity vibration cannot be transferred to the lattice by the emission of a single phonon. The amplitude of the vibration of the impurity and its neighbors is exponentially attenuated with distance from the impurity (Spitzer, 1971). Such impurity vibrations are called spatially *localized vibrational modes* (LVM).

The site of a substitutional impurity atom has tetrahedral symmetry and the potential for small displacements of the impurity is that of a spherical harmonic oscillator (Spitzer, 1971). Thus vibrations in the x-, y-, and z-directions are equivalent and the oscillation mode is triply degenerate. Furthermore, due to the isotropic, spherical potential, only *one* (triply degenerate) fundamental oscillation is expected for a substitutional impurity atom. The calculated and experimental wavenumbers of substitutional impurities in GaAs and InP are shown in Table 12.1.

Several modes of local vibrations with different frequencies may be possible for impurity pairs, impurity clusters, and impurity–native defect complexes. As an example, a Si_{Ga}–Si_{As} pair in GaAs is schematically shown in Fig. 12.21 (Spitzer, 1971). The Si_{Ga} – Si_{As} pair has four vibrational modes which are indicated by arrows. The approximate values for the vibrational angular frequencies are $\omega_1 = 327 \text{ cm}^{-1}$, $\omega_2 = 369 \text{ cm}^{-1}$, $\omega_3 = 390 \text{ cm}^{-1}$, and $\omega_4 = 419 \text{ cm}^{-1}$. Not all vibrational modes of impurity complexes may be *infra-red active*, but only those modes in which a dipole moment is changed during the impurity displacement. For the case of the Si–Si pair, all four modes are infra-red active (Spitzer, 1971).

Vibrational modes can have fine structures with several resonances at slightly different frequencies due to isotopes of the semiconductor lattice. For example, Ga has two isotopes with atomic masses 69 and 71 and relative abundances of 60.4% and 39.6%. A substitutional impurity occupying an As site in GaAs has four Ga next neighbors. As a result, five different

Table 12.1. *Fundamental local mode wavenumbers in* cm^{-1} (8.065 cm^{-1} ≅ 1 meV) *for substitutional impurities in GaAs and InP (Newman, 1974).*

	Theory	Experiment
$^{31}P_{As}$	351	355.4
$^{27}Al_{Ga}$	369	362.0
$^{28}Si_{Ga} - {}^{28}Si_{As}$	369	367.2
$^{28}Si_{Ga} - V_{Ga}$		368.5
$^{30}Si_{Ga}$	352	373.4
$^{29}Si_{Ga}$	357.5	378.5
$^{28}Si_{Ga}$	363	383.7
$^{28}Si_{Ga} - {}^{28}Si_{As}$	390	393.0
$^{28}Si_{Ga}$	365	398.2
$^{28}Si_{Ga} - {}^{28}Si_{As}$	419	464.0
$^{11}B_{Ga}$	542	517.0
$^{10}B_{Ga}$	570	540.2
$^{13}C_{As}$	500	561.2
$^{12}C_{As}$	519	582.4
$^{10}B_{In}$	558	543.5
$^{11}B_{In}$	536	522.8

combinations of Ga atoms with different masses are possible which are (1) 4 × ^{69}Ga, (2) 1 × ^{71}Ga & 3 × ^{69}Ga, (3) 2 × ^{71}Ga & 2 × ^{69}Ga, (4) 3 × ^{71}Ga & 1 × ^{69}Ga, and (5) 4 × ^{71}Ga. Due to the strong spatial localization of the vibrational mode at the impurity atom, next neighbor isotopes do not shift the frequency dramatically. A high-resolution detection system is required to resolve the splitting due to isotopes. Furthermore, if the four nearest neighbors consist of two different isotopes, the degeneracy of local mode vibrations along the x-, y-, and z-axis is lifted. Sinai and Wu (1989) calculated the frequencies for such a mixed nearest neighbour environment.

The localized vibrational mode absorption band of ^{12}C in GaAs is shown in Fig. 12.22 (Theis et al., 1982). The measurement is taken at 80 K. The four lines shown in Fig. 12.22 merge to a single line for lower resolutions of the detection system. The single line peaks at 582.4 cm^{-1}. A higher resolution of 0.06 cm^{-1} reveals a series of at least four peaks occurring at 582.18, 582.31, 582.41, and

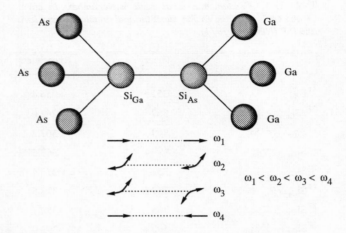

Fig. 12.21. Schematic illustration of the four local vibrational modes of a Si impurity pair in GaAs (after Spitzer, 1971).

582.65 cm^{-1}. The four lines of ^{12}C in GaAs were first observed by Newman et al. (1972) at the same wavenumbers. The overlapping bands were explained by the different isotopic arrangements of the four nearest Ga neighbors of the ^{12}C impurity, and the resulting different mode frequencies. Theis et al. (1982) further concluded that C only rarely occupies Ga sites and estimated the compensation ratio $[C_{Ga}]/[C_{As}] \leq 0.01$. Even at high concentrations, C was shown to mostly occupy As sites for OMMBE grown $Al_x Ga_{1-x} As$ (Abernathy et al. 1990).

LVM can be measured by either far-infra-red (FIR) absorption or by Raman spectroscopy. The two techniques are usually referred to as *LVM-FIR* and *LVM-Raman*. A severe limitation of LVM-FIR is free carrier absorption which occurs in the same spectral range as the local vibrational modes. In order to reduce this absorption, the free carrier concentration can be reduced by electron irradiation, by diffusion of a compensating impurities, or by implantation of H, He, or O. However, the species involved in these processes and the impurity under study may interact and change the experimental results as compared to the undamaged semiconductor. The absorption due to free carriers is less significant in LVM-Raman. First, the probing depth of the exciting laser can be chosen to be very small by choosing high-energy excitation ($\hbar \omega \gg E_g$). Second, the incident and scattered photons are not subject to free carrier absorption due to the high energy of the photons.

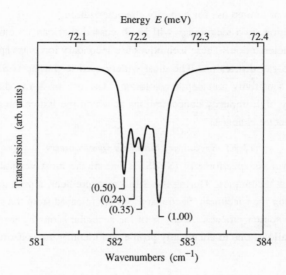

Fig. 12.22. Localized vibrational mode absorption for ^{12}C in GaAs measured at 80 K. The values in parentheses are estimated band stengths (after Theis et al., 1982).

LVM spectroscopy has been widely used for the analysis of Si in GaAs (Ramsteiner et al., 1988; Maguire et al., 1987; Murray et al., 1989; Wagner and Ramsteiner, 1989; Nakamura and Katoda, 1985; Holtz et al., 1986; Leigh and Newman, 1982) which includes Si doping during growth and during post-growth implantation. LVM spectroscopy has also been employed for p-type impurities such as Be (Wagner and Ramsteiner, 1989) and C (Leigh and Newman, 1982; Theis et al., 1982; Kitagawara et al., 1986; Abernathy et al., 1990) in GaAs. The sensitivity of LVM can be very high. Kitagawara et al. (1986) demonstrated a C detection limit of 5×10^{14} cm^{-3} in GaAs. The authors also related the magnitude of the LVM absorption to the C concentration.

12.3 Chemical and structural characterization techniques

Many chemical and structural characterization techniques do not have sufficient sensitivity to be useful for characterization of impurities. Sensitivities of 10^{-9} are required in order to detect impurities at low concentrations in III–V semiconductors. Auger spectroscopy, for example, has a sensitivity of typically $0.1\% = 10^{-3}$. Hence, only very high impurity concentrations can be detected. Although Auger spectroscopy is a widely used semiconductor characterization technique, it is of limited use for impurity characterization.

In this section, two techniques will be discussed that can be employed for impurity characterization. These techniques are secondary ion mass spectrometry (SIMS), and X-ray diffraction. The most widely used technique is SIMS which has excellent sensitivity and depth resolution. The use of X-ray diffraction is limited to very high impurity concentrations at which the lattice constant of the host semiconductor changes.

12.3.1 Secondary ion mass spectrometry

Secondary ion mass spectrometry (SIMS) is one of the most versatile chemical characterization techniques. During the SIMS measurement, a primary ion beam is used to sputter the specimen. Secondary ions are released from the specimen as a result of the sputter process. Analysis of the secondary ions by means of mass spectrometry allows one to chemically profile and identify the constituents of the specimen.

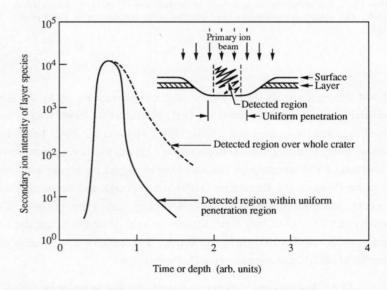

Fig. 12.23. Relationship of depth profile shape to rastered and detected areas during secondary ion mass spectrometry (SIMS) (after Wilson et al., 1989).

The **primary ion beam** has typical energies of 1–50 keV and is rastered across a squared area of typically $100 \times 100~\mu m^2$ to $500 \times 500~\mu m^2$. The ion beam and the specimen is schematically shown in the inset of Fig. 12.23. The collisions between primary ions and the semiconductor specimen are inelastic. As a result of these collisions, atoms or molecules of the semiconductor are released into the surrounding vacuum. The primary ion beam may have an off-normal incidence with respect to the sample surface. Off-normal incidences $\leq 60°$ are common. At off-normal incidence, the transfer of primary ion momentum occurs primarily in a lateral direction and scattering processes are close to the specimen surface. As a consequence, improved depth resolution is attainable. Typical chemical species of the primary ion beam are O_2^+, Cs^+, Ar^+, Kr^+, and Xe^+. When chemical reactions between the primary ions and the specimen are undesired, noble gas atoms are frequently used as primary ion species. However, the use of O_2^+ and Cs^+ enhance the yields of negative and positive secondary ions, respectively, and thus are most commonly used for the analysis of impurities.

The *sputter rate* depends on the primary ion energy, the area profiled, the angle of the incident primary ions, the primary ion species, the primary ion flux, and the semiconductor material. The primary ion flux (ions per unit time), also called the primary beam current, can be easily adjusted. One μA of primary ion current corresponds to 6.25×10^{12} ions/s. Assuming that each primary ion creates one secondary atom and recalling that one monolayer of III–V semiconductors contains about 6×10^{14} atoms/cm^2, a primary ion current of 100 μA is required for a sputter rate of one monolayer per second (for a sputter area of 1 cm^2). In practice, the sputter yield, i.e. the number of specimen atoms sputtered by one primary ion, varies and can assume values of 1–100. Most of the sputtered atoms and molecules are neutral. A small fraction of the particles are ionized, typically $10^{-3} - 10^{-7}$, and this fraction is called the *secondary ion yield*. Secondary ions are extracted by an electric field and analyzed by mass spectrometry. The spectrometric analysis yields the charge-to-mass ratio (e/m) of the secondary ions. Note that isotopes of impurities can be detected by SIMS due to the high mass sensitivity of spectrometers.

The sputtered area and the detected area are schematically shown in Fig. 12.23 (Wilson et al., 1989). During the measurement, the detected element is measured as a function of *time*. The time is later converted to *depth* by measuring the total depth of the sputtered crater. The area of the detected region affects the depth resolution of the SIMS profile. It is desirable to limit the detected region to the area of uniform sputtering. An improved depth resolution is obtained by limiting the area of detection as illustrated in Fig. 12.23. The *concentration* of a measured species is obtained by comparison with a measured standard (e.g. an ion-implanted standard) and by assuming that the secondary ion count varies linearly with the concentration of the detected element.

Historically, the SIMS technique was developed independently by Castaing et al. (1960, 1962) and by Herzog and Liebl (1963). Bennighoven (1970) demonstrated that planar surfaces can be maintained during the sputter process. Several comprehensive reviews on SIMS have been published (Benninghoven et al., 1987; Wilson et al., 1989). A newer development of SIMS is the secondary neutral mass spectrometry (SNMS) which has been reviewed by Oechsner (1984). In SNMS, secondary atoms are ionized after the sputter process (post-ionization), e.g. by means of optical excitation or electron beam bombardment. In conventional SIMS, the secondary ion yield is on the order of $10^{-7} - 10^{-3}$. Due to post-ionization, the secondary ion yield can be increased to assume values close to unity. To date, however, the sensitivity of SNMS was not shown to be superior on a routine basis as compared to conventional SIMS.

Impurities in the mid and high doping regime ($> 10^{17}$ cm^{-3}) can be easily detected by SIMS. For some impurities, the sensitivity is higher allowing detection of impurities in the $10^{13} - 10^{15}$ cm^{-3} concentration range. It is important to note that the concentrations measured by SIMS are the impurity atom concentrations regardless of their *electrical activity*. In some cases, such as highly Si doped GaAs, the *electrically active* Si concentration can be much lower than the Si atom concentration (see, for example, Greiner and Gibbons, 1984).

The depth resolution of the SIMS technique for impurity profiles was investigated by Schubert et al. (1990b) using δ-doped semiconductors. The SIMS profiles of the Be impurity spikes in GaAs had a full width at half-maximum of 29 Å. The resolution of the profiling technique is defined as twice the standard deviation of a gaussian distribution fitted to the experimental profile (Benninghoven et al., 1987). Using this procedure, Schubert et al. (1990b) obtained a resolution of 25 Å. Low-energy sputtering (1.5 keV), a shallow sputtering angle (60° off-normal), and a Be doping spike in the proximity of the surface (200 Å) were required in order to obtain this resolution. Clegg and Beall (1989) used δ-doped semiconductors to analyze the leading and trailing slope of the SIMS profiles over several orders of magnitude of the SIMS signal.

The interpretation of impurity distributions assessed by the SIMS technique is complicated by **matrix effects**. If an impurity species is detected in a homogeneous semiconductor, the secondary ion yield is constant and does not change with time. This situation can change in semiconductor multilayer structures, since the secondary ion yield of an impurity species depends on the host semiconductor (i.e. matrix). Assume a semiconductor heterostructure comprising two different semiconductors and both containing the same impurity species and the same impurity concentration. Then changes in the semiconductor composition (matrix) result in changes of the secondary ion yield. As a consequence, the SIMS signal of an impurity changes, even if the impurity concentration is the same in the two matrices. This effect is known as the *matrix* *effect*. The physical origin of the matrix effect is due to the interdependence of secondary ion yield of an impurity and the host semiconductor composition (Wilson et al., 1989). The presence of electronegative elements such as O, N, or F, can increase the yield of positive secondary ions. As an example, we consider Be impurities in $Al_x Ga_{1-x} As/GaAs$ heterostructures subjected to O_2^+-sputtering during SIMS (Downey and Hozack, 1989). The Be^+ secondary ion yield increases with the presence of O and O_2 due to its strong electronegativity. The O surface coverage of $Al_x Ga_{1-x} As$ is larger compared to GaAs due to the strong Al–O affinity. As a result, the Be^+ secondary ion yield is higher in $Al_x Ga_{1-x} As$ and the measured SIMS Be concentrations are higher in $Al_x Ga_{1-x} As$. To avoid this artificial reading, a separate calibration standard must be run for each matrix. Even though a separate standard is helpful for bulk material analysis, it is nevertheless problematic for thin multilayer structures and concentration profiles near interfaces (Downey et al., 1990). Matrix effects can be alleviated by using SNMS rather than SIMS. In SNMS, the secondary ion yield approaches unity under idealized conditions and does not depend on the matrix as long as parasitic reactions between the measured species and other constituents of the semiconductor can be neglected. SNMS impurity profiles indeed exhibit a reduced matrix effect as compared to SIMS (Downey et al., 1990).

12.3.2 X-ray diffraction

X-ray diffraction is a versatile characterization technique which is predominantly used to determine lattice constants of ternary and quaternary III–V semiconductors. For example, the lattice constant of epitaxial $Ga_x In_{1-x} As$ on InP substrates is measured to determine if the epitaxial crystal is lattice-matched to the substrate. X-ray diffraction can also be used to determine the lattice distortion caused by impurities at very high doping concentrations. For example, C impurities cause lattice contraction in GaAs or InP at high C concentrations. The change in lattice constant can be determined by x-ray diffraction. Conversely, impurity concentrations can be determined by x-ray diffraction, assuming that the relation between impurity concentration and lattice constant is known. In this section, the x-ray diffraction technique is discussed with respect to impurity characterization.

Electromagnetic waves are elastically scattered by periodic structures such as semiconductor lattices. The magnitude of the incident and the scattered wavevector are identical due to the elastic nature of the scattering event, i.e.

$$k_{incident} = k_{scattered} = \frac{2\pi}{\lambda} \tag{12.89}$$

where k is the magnitude of the wave vector and λ is the wavelength of the electromagnetic wave. The total reflection of electromagnetic radiation in periodic media is called **Bragg reflection** and occurs at specific wavelengths. The nth Bragg reflection condition is given by

$$n\lambda = 2d\sin\theta \qquad\qquad (12.90)$$

where θ is the angle of incidence with respect to the crystal surface and d is the spacing of the lattice planes of interest, e.g. the (001), (004), (011), or (111) plane. Frequently, the (004) Bragg reflection is used to determine the lattice constant. This reflection originates from xy-planes separated by a quarter of the lattice constant ($a_0/4$). The lattice plane spacing, d, is on the order of the lattice constant, a_0. It is therefore desirable to use electromagnetic wavelengths on the order of the lattice constant, i.e. in the x-ray regime. For practical x-ray generators, the wavelength is a fixed quantity given by atomic core transitions of the x-ray source. In order to observe the Bragg reflections, the angle θ (see Eq. (12.90)) must be varied by 'rocking' the specimen. X-ray rocking curves are thus plots of the diffracted x-ray intensity as a function of the rocking angle θ.

In very highly doped semiconductors, the lattice constant changes and these changes can be detected by x-ray diffraction. Lattice **contraction** and lattice **expansion** results from impurities which are smaller and larger than the host semiconductor atoms, respectively. For example, C atoms are smaller than either Ga or As resulting in lattice contraction of highly C-doped GaAs. If the highly doped layer is an epitaxial layer grown on a substrate, **mechanical strain** occurs between the two layers due to the mismatch in lattice constant. For sufficiently thin mismatched epitaxial layers, the layers are strained *elastically*. Epilayers which exceed a certain *critical layer thickness* relax by means of misfit dislocations. The lattice constant of such relaxed layers is the equilibrium lattice constant which is different from that of the substrate.

An elastically strained epitaxial layer grown on a substrate with a larger lattice constant is shown schematically in Fig. 12.24. The equilibrium lattice constant of the epitaxial layer is assumed to be smaller than that of the substrate. As a result of the lattice mismatch, the epitaxial layer is tetragonally distorted (Hornstra and Bartels, 1978). Tetragonal distortion occurs for symmetry planes such as the (001), (011), or the (111) plane and manifests itself in two different lattice constants in the growth plane, $a_\parallel$, and along the growth direction, $a_\perp$. For the *lattice contraction* illustrated in Fig. 12.24, the in-plane lattice constant, $a_\parallel$, is the same as the lattice constant of the substrate, a_0. The normal lattice constant, $a_\perp$, is smaller or larger than a_0 for lattice contraction and expansion, respectively.

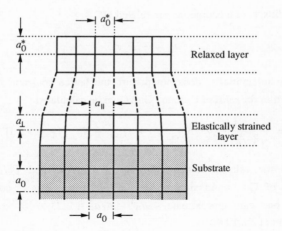

Fig. 12.24. Schematic illustration of a cubic lattice (substrate) with lattice constant a_0, a tetragonally distorted, elastically strained layer with lattice parameters $a_\parallel$ and $a_\perp$, and a relaxed layer with equilibrium lattice constant a_0^*.

Note that the above considerations apply only to elastically strained epitaxial layers. Epitaxial layers exceeding a certain thickness relax and assume an equilibrium lattice constant a_0^* as schematically shown in Fig. 12.24. The transition from a strained layer to a relaxed layer occurs if the elastic strain energy stored in the epitaxial layer exceeds the energy required to create misfit dislocations. Since the elastic strain energy is proportional to the thickness of the epitaxial layer, the two energies coincide at a certain thickness which is called the **critical thickness**.

Lattice contraction was found in highly C-doped GaAs (Akatsuka et al. 1990, 1991; de Lyon et al., 1990; George et al., 1991; Hoke et al., 1991) and to a smaller extent in Be-doped GaAs (Hoke et al., 1991). The splitting of the x-ray diffraction peaks originating from the GaAs substrate with lattice constant a_0 and the tetragonally distorted GaAs film with the perpendicular lattice constant $a_\perp$ is give by

$$\frac{a_\perp - a_0}{a_0} = -\Delta\theta \, \cot\theta \qquad (12.91)$$

where θ is the diffraction angle (33.024°) for the (004) substrate reflection (Hoke et al., 1991). The lattice constant of GaAs:C can be calculated from the covalent tetrahedral radii of the constituents. The lattice constant, a_0, and the covalent tetrahedral radius, r, of a compound are related by

$$a_0 = \frac{8}{\sqrt{3}}\, r\,. \tag{12.92}$$

Assuming that interstitial C does not perturb the lattice constant, the expected lattice contraction for *relaxed* C-doped GaAs can be calculated

$$a_0^* - a_0 = \frac{4}{\sqrt{3}\ (4/a_0^{*3})}\ [N_{C_{Ga}}\,(r_C - r_{Ga}) + N_{C_{As}}(r_C - r_{As})] \tag{12.93a}$$

where a_0^* is the relaxed lattice constant of GaAs:C and $(4/a_0^{*3})$ is the concentration of Ga or As sites of the GaAs lattice. For C concentrations $N_C \ll 10^{22}$ one can approximate $4/a_0^{*3} \cong 4/a_0^3 \cong 2.21 \times 10^{22}$ cm^{-3} which yields (de Lyon et al., 1990)

$$a_0^* - a_0 = \frac{4}{\sqrt{3}\ (2.21 \times 10^{22}\ \text{cm}^{-3})}\ [N_{C_{Ga}}(r_C - r_{Ga}) + N_{C_{As}}\,(r_C - r_{As})]$$

$$\tag{12.93b}$$

where $N_{C_{Ga}}$ and $N_{C_{As}}$ are the concentrations of C atoms on Ga and As sites, respectively, and r_C, r_{Ga}, and r_{As} are the respective covalent tetrahedral radii. The covalent radii of Ga, As, C and Be are $r_{Ga} = 1.26$ Å, $r_{As} = 1.20$ Å, $r_C = 0.77$ Å, and $r_{Be} = 0.90$ Å, respectively. Due to $r_C < r_{Be}$, a larger lattice contraction is expected for C-doped GaAs as compared to Be-doped GaAs. While de Lyon et al. (1990) used the radii listed above, Hoke et al. (1991) used $r_{Ga} = r_{As} = r_{GaAs} = 1.225$ Å. The quantitative difference between the two approaches is very small. The lattice constant of *elastically strained*, tetragonally distorted C-doped GaAs is given by (de Lyon et al., 1990)

$$(a_\perp - a_0) = \frac{(1 + \nu)}{(1-\nu)}\ (a_0^* - a_0) = 1.90\ (a_0^* - a_0) \tag{12.94}$$

where a value of $\nu = 0.31$ has been assumed as the Poisson ratio of the GaAs epilayer. Note that $a_0^* = a_\perp^{\text{relaxed}} = a_\parallel^{\text{relaxed}}$ for the relaxed layer. Comparison of the lattice constant of highly C-doped GaAs measured by x-ray diffraction with the calculated lattice constants (Eqs. (12.93) and (12.94)) yielded agreement between experiment and theory. Quantitatively, the lattice mismatch between highly C-doped GaAs and the substrate can be quite significant. At a C-doping concentration of 10^{21} cm^{-3}, a lattice mismatch of $0.5 - 0.8\%$ was determined (George et al., 1991). For sufficiently thick films, misfit dislocations are created which deteriorate the minority carrier properties of the material. For example, a decrease in photoluminescence efficiency is expected in films with dislocations.

For completeness, the relations between the substrate lattice constant a_0, the elastically strained constants $a_\perp$ and $a_\parallel$, and the relaxed constant a_0^* are summarized. The in-plane element of the strain tensor, ε, is defined as

$$\varepsilon_\parallel = \frac{a_\parallel - a_0^*}{a_0^*} . \tag{12.95a}$$

For elastically strained epilayers, the lattice constants are

$$a_\parallel = a_0 , \tag{12.95b}$$

$$a_\perp = \left[1 + \frac{2v}{1-v} \, \varepsilon_\parallel \right] a_0 \tag{12.95c}$$

where v is again the Poisson number which is 0.31 for GaAs and $v \cong 0.3$ for most semiconductors. Thus the knowledge of a_0 and a_0^* allows one to calculate $a_\perp$.

The change in lattice constant predicted by Eqs. (12.93) and (12.94) are *linear* with the C concentration, if the compensation ratio $N_{C_{As}}/N_{C_{Ga}}$ does not depend on the C concentration. Such a linear dependence agrees with **Vegard's law**. Vegard's law predicts that the lattice constant changes linearly with the composition of a multicomponent system. Consider, for example, a binary system AB consisting of the components A and B. Assuming that A and B have the same crystal structure and that the solid solution $A_x B_{1-x}$ is miscible over the entire composition range, the lattice constant of the binary compound is given by (Vegard's law)

$$\boxed{a_{AB} = a_A x + a_B (1-x)} \tag{12.96}$$

where a_A and a_B are the lattice constants of the material A and B, respectively. Vegard's law in the form of Eq. 12.96 can be applied to compounds such as $Al_x Ga_{1-x} As$ (A = AlAs, B = GaAs, $A_x B_{1-x}$ = $Al_x Ga_{1-x} As$). However, Eq. 12.96 cannot be applied to systems such as GaAs:C, since a component 'GaC' does not exist.

Note that deviations from Vegards law may occur at very high doping concentrations. Equation (12.93) predicts a *linear* dependence of the lattice constant on the C concentration as long as the C compensation ratio remains constant. However, if the ratio $N_{C_{Ga}}/N_{C_{As}}$ changes at high C doping concentrations, Eq. (12.93) predicts a *non-linear* dependence between the lattice constant and the C concentration. Furthermore, the concentration of interstitial C is expected to increase at high doping concentrations. Such interstitials will perturb the lattice in a different way than substitutional C atoms, which is also expected to result in a deviation from Vegard's law. To date, however, deviations from Vegard's law in highly C-doped GaAs have not been reported.

Appendix A: Properties of III–V semiconductors

Table A.1 *Characteristics of GaAs, InP, and InAs*

Quantity	Unit	GaAs value	InP value	InAs value
Crystal structure	—	zincblende	zincblende	zincblende
Lattice constant	Å	5.6533	5.8690	6.0583
Band structure	—	direct	direct	direct
Bandgap energy				
(300 K)	eV	1.424	1.350	0.354
(2 K)	eV	1.519	1.425	0.418
Absorption edge (300 K)	μm	0.870	0.918	3.50
Electron affinity	eV	4.07	4.5	4.9
Electron effective mass	m_0	0.067	0.077	0.022
Electron mobility (300 K)	cm^2/V s	8500	5000	33 000
Heavy hole effective mass	m_0	0.45	0.60	0.40
Light hole effective mass	m_0	0.082	0.12	0.026
Hole mobility (300 K)	cm^2/V s	400	150	450
Dielectric constant				
ε_r	—	13.1	12.6	15.1
ε_∞	—	10.9	9.61	12.3
Refractive index				
$\lambda = 1.24$ μm	—	3.45	3.22	3.5
Optical phonon energy				
LO	meV	36	43	30
TO	meV	33	37	27
Thermal conductivity (300 K)	W/K cm	0.46	0.68	0.30
Thermal expansion	10^6 K^{-1}	6.4	4.56	4.52
Density	g/cm^3	5.32	4.79	5.67
Melting point	°C	1238	1335	942

(sources are listed in references)

Table A.2 *Characteristics of $Al_x Ga_{1-x} As$*

Quantity	Unit	$Al_x Ga_{1-x}$ As value
Crystal structure	—	zincblende
Lattice constant	$\overset{\circ}{A}$	$5.6533 + 0.0072x$
Band structure	—	$x < 0.45$: direct
		$x > 0.45$: indirect
Bandgap energy		
(300 K)	eV	$x < 0.45$: $1.424 + 1.247x$
		$x > 0.45$: $1.424 + 1.247x + 1.147(x - 0.45)^2$
Electron affinity	eV	$x < 0.45$: $4.07 - 1.1x$
		$x > 0.45$: $3.64 - 0.14x$
Electron effective mass		
Density of states:		
m_Γ^*	m_0	$0.067 + 0.083x$
m_L^*	m_0	$0.55 + 0.12x$
m_X^*	m_0	$0.85 - 0.07x$
Conductivity:		
m_Γ^*	m_0	$0.067 + 0.83x$
m_L^*	m_0	$0.11 + 0.03x$
m_X^*	m_0	$0.32 - 0.06x$
Electron mobility (300 K)	cm^2/V s	$x < 0.45$: $8000 - 22000x + 10000x^2$
		$x > 0.45$: $255 + 1160x - 720x^2$
Heavy hole effective mass	m_0	$0.45 + 0.30x$
Light hole effective mass	m_0	$0.08 + 0.057x$
Hole mobility (300 K)	cm^2/V s	$370 - 970x + 740x^2$
Dielectric constant		
ε_r	—	$13.1 - 3.0x$
ε_∞	—	$10.9 - 2.7x$
Refractive index		
$\lambda = 1.239$ μm	—	$3.45 - 0.54x$
Optical phonon energy		
LO (GaAs like; $x < 1$)	meV	$36.2 - 6.5x + 1.8x^2$
LO (AlAs like; $x > 0$)	meV	$44.6 + 8.8x - 3.3x^2$
TO (GaAs like; $x < 1$)	meV	$33.3 - 0.6x - 1.2x^2$
TO (AlAs like; $x > 0$)	meV	$44.6 + 0.5x - 0.3x^2$
Thermal resistivity (300 K)	K cm/W	$2.27 + 28.83x - 30x^2$
Thermal expansion	10^6 K^{-1}	$6.4 - 1.2x$
Density	g/cm^3	$5.36 - 1.6x$
Melting point	°C	$1238 - 58x + 560x^2$

Table A.3 *Characteristics of Ga$_{0.47}$In$_{0.53}$As and Al$_{0.48}$In$_{0.52}$As*

Quantity	Unit	Ga$_{0.47}$In$_{0.53}$As value	Al$_{0.48}$In$_{0.52}$As value
Crystal structure	—	zincblende	zincblende
Lattice constant	Å	5.8690	5.8690
Band structure	—	direct	direct
Bandgap energy			
(300 K)	eV	0.75	1.450
(2 K)	eV	0.812	1.508
Absorption edge (300 K)	μm	1.65	0.855
Electron effective mass	m_0	0.041	0.083
Electron mobility (300 K)	cm^2/V s	13 800	4600
Heavy hole effective mass	m_0	0.5	—
Light hole effective mass	m_0	0.051	—
Hole mobility (300 K)	cm^2/V s	350	—
Dielectric constant			
ε_r	—	14.1	—
ε_∞	—	11.6	—
Refractive index			
$\lambda = 1.65$ μm	—	3.56	—
$\lambda = 1.24$ μm	—	—	3.2
Optical phonon energy			
LO	meV	33	—
TO	meV	28	—
Thermal conductivity (300 K)	W/K cm	0.063	—
Thermal expansion	10^{-6}K^{-1}	5.55	—
Density	g/cm^3	5.54	4.85

Table A.4 *Characteristics of GaInPAs (λ = 1.3 and 1.55 μm) lattice matched to InP*

Quantity	Unit	Ga$_{.27}$In$_{.73}$P$_{.40}$As$_{.60}$ value	Ga$_{.40}$In$_{.60}$P$_{.15}$As$_{.85}$ value
Crystal structure	—	zincblende	zincblende
Lattice constant	Å	5.8690	5.8690
Band structure	—	direct	direct
Bandgap energy			
(300 K)	eV	0.953	0.800
(2 K)	eV	1.018	0.881
Absorption edge (300 K)	μm	1.3	1.55
Electron effective mass	m_0	0.053	0.045
Electron mobility			
(300 K, $N_D = 1\times10^{16)}$	cm^2/V s	6300	7700
Heavy hole effective mass	m_0	0.69	0.64
Light hole effective mass	m_0	0.072	0.062
Hole mobility			
(300 K, $N_A = 1\times10^{17}$)	cm^2/V s	80	150
Dielectric constant			
ε_r	—	13.3	13.7
ε_∞	—	10.9	11.4
Refractive index			
$\lambda = 1.30$ μm	—	3.52	—
$\lambda = 1.55$ μm	—	3.40	3.55
$\lambda = 1.65$ μm	—	3.38	3.49
Optical phonon energy			
LO$_1$	meV	41	—
LO$_2$	meV	32	33
TO$_1$	meV	40	—
TO$_2$	meV	28	28
Thermal conductivity	W/K cm	0.044	0.043
(300 K)			
Thermal expansion	10^{-6}K^{-1}	5.32	5.65
Density	g/cm^3	5.26	5.42

Appendix B: Constants and Conversions

Table B.1 *Constants*

α	7.2974×10^{-3}	Sommerfeld's fine structure constant
a_B	0.52918 Å	Bohr radius
ε_0	8.8542×10^{-12} A s/V m	dielectric permittivity in vacuum
e	1.6022×10^{-19} C	elementary charge
c	2.9979×10^{8} m/s	velocity of light in vacuum
E_{Ryd}	13.606 eV	Rydberg energy
h	6.6261×10^{-34} J s	Planck's constant
$\hbar$	1.0546×10^{-34} J s	Planck's constant divided by 2π
k	1.3807×10^{-23} J/K	Boltzmann's constant
μ_0	1.2566×10^{-6} V s/A m	magnetic permeability in vacuum
m_e	9.1094×10^{-31} kg	free electron mass
N_{Avo}	6.0221×10^{23} mol^{-1}	Avogadro's number
R	8.3145 J K^{-1} mol^{-1}	ideal gas constant

Table B.2 *Useful conversions*

1 eV	=	1.6022×10^{-19} J
kT	=	25.86 meV (at 300 K)
E	=	$1239.8 / \lambda$ (E: photon energy in eV; λ: wavelength in nm)
E	=	$(1/8065.5) \times \lambda^{-1}$ (E: photon energy in eV; λ^{-1}: inverse wavelength in cm^{-1})

References

Preface

Seidman, A. H. and Marshall S. L. (1963) *Semiconductor Fundamentals* **VI** (Wiley, New York) p. 86.

Shockley, W. (1950) *Electrons and Holes in Semiconductors* (Van Nostrand, New York) p. 12.

Wilson, A. H. (1931) *Proc. Roy. Soc. A.* **134**, 279.

Chapter 1

Abarenkov, I. V. and Heine, V. (1965) *Phil. Mag.* **12**, 529.

Abram, R. A., Rees, G. J., and Wilson, B. L. H. (1978) *Adv. Phys.* **27**, 799.

Adler, D. (1982) in *Handbook on Semiconductors* **1**, eds. Moss, T. S. and Paul, W. (North-Holland, Amsterdam).

Adler, P. N. (1969) *J. Appl. Phys.* **40**, 3554.

Altarelli, M. and Bassani, F. (1982) in *Handbook on Semiconductors* **1**, eds. Moss, T. S. and Paul W. (North Holland, Amsterdam).

Ando, T., Fowler, A. B., and Stern, F. (1982) *Rev. Mod. Phys.* **54**, 437.

Austin, I. G. and Mott, N. F. (1969) *Adv. Phys.* **18**, 41.

Baldareschi, A. and Lipari, N. O. (1973) *Phys. Rev. B***8**, 2697.

Baldareschi, A. and Lipari, N. O. (1974) *Phys. Rev. B9*, 1525.

Bassani, F., Iadonisi, G., and Preziosi, B. (1969) *Phys. Rev.* **186**, 735.

Bennet, H. S. and Lowney, J. R. (1987) *J. Appl. Phys.* **62**, 521.

Bethe, H. (1942) unpublished.

Böer, K. (1990) *Survey of Semiconductor Physics* (van Nostrand Reinhold, New York).

Bohm, D. (1951) *Quantum Theory* (Prentice-Hall, Englewood Cliffs).

Borowitz, S. (1967) *Fundamentals of Quantum Mechanics* (Benjamin, New York).

Brinkman, W. F. and Rice, T. M. (1973) *Phys. Rev. B7*, 1508.

Burstein, E. (1954) *Phys. Rev.* **93**, 632.

Casey, H. C., Jr., Sell, D. D., and Wecht, K. W. (1975) *J. Appl. Phys.* **46**, 250.

Casey, H. C., Jr and Stern, F. (1976) *J. Appl. Phys.* **47**, 631.

Chelikowsky, J. R. and Cohen, M. L. (1976) *Phys. Rev. B14*, 556.

Dapkus, P. D., Holonyak, N. Jr., Rossi, J. A., Williams, L. V., and High, D. A. (1969) *J. Appl. Phys.* **40**, 3300.

Davydov, A. S. (1965) *Quantum Mechanics* (Pergamon, Oxford).

Debye, O. P. and Hückel, E. (1923) *Zeit Phys.* **24**, 185, 305.

Debye, P. and Conwell, E. M. (1954) *Phys. Rev.* **93**, 693.

Deppe, D. G., Gerrard, N. D., Pinzone, C. J., Dupuis, R. D., and Schubert, E. F. (1990) *Appl. Phys. Lett.* **56**, 315.

Efros, A. L. and Shklovskii, B. I. (1975) *J. Phys. C8*, L49.

Finkelnburg, W. (1958) *Introduction to Atomic Physics* (in German) (Springer, Berlin) p. 97.

Flügge, S. (1971) *Practical Quantum Mechanics* **1** (Springer, Berlin).

Halperin, B. I. and Lax, M. (1966) *Phys. Rev.* **148**, 722.

Halperin, B. I. and Lax, M. (1967) *Phys. Rev.* **153**, 802.

Haug, H. and Schmitt-Rink, S. (1984) *Prog. Quantum Electron*, **9**, 3.

Haug, H. and Schmitt-Rink, S. (1985) *J. Opt. Soc. Am. B***2**, 1135.

Hubbard, J. (1963) *Proc. Roy. Soc. (London) Ser. A***276**, 238.

Jain, S. C., McGregor, J. M., and Roulston, D. J. (1990) *J. Appl. Phys.* **68**, 3747.

Kane, E. O. (1963) *Phys. Rev.* **131**, 79.

Kohn, W. (1957a) in *Solid State Physics* **5**, eds. Seitz, F. and Turnbull, D. (Academic, New York).

Kohn, W. (1957b) *Phys. Rev.* **105**, 509.

Landsberg, P. T., Neutroschel, A., Lindholm, F. A., and Sah, C. T. (1985) *Phys. Stat. Sol. B***130**, 255.

Lax, M. and Halperin, B. I. (1966) *J. Phys. Soc. Japan* **21** (Supplement), 218.

Luttinger, J. M. (1956) *Phys. Rev.* **102**, 1030.

Madelung, O. (1978) *Introduction to Solid-State Theory* (Springer, Berlin).

Mahan, G. D. (1980) *J. Appl. Phys.* **51**, 2634.

Mahan, G. D. (1990) *Many-Particle Physics* (Plenum, New York).

Merzbacher, E. (1970) *Quantum Mechanics* (John Wiley, New York).

Morgan, T. N. (1965) *Phys. Rev.* **139**, A343.

Moss, T. S. (1961) *Optical Properties of Semiconductors* (Academic, New York).

Mott, N. F. (1987) *Conduction in Non-Crystalline Materials* (Clarendon, Oxford).

Mott, N. F. (1990) *Metal–Insulator Transitions* (Taylor & Francis, London).

Olego, D. and Cardona, M. (1980) *Phys. Rev. B***22**, 886.

Onton, A., Yacoby, Y., Chicotka, R. J. (1972) *Phys. Rev. Lett.* **28**, 966.

Pantelides, S. T. (1975) in *Advances in Solid State Physics Vol. 15*, ed. Queisser, H. J. (Pergamon-Vieweg, Braunschweig) p. 149.

Pantelides, S. T. (1978) *Rev. Mod. Phys.* **50**, 797.

Pearsall, T. P. (ed.) (1982) *GaInAsP Alloy Semiconductors* (Wiley, New York).

Phillips, J. C. (1970a) *Phys. Rev. B***1**, 1540.

Phillips, J. C. (1970b) *Phys. Rev. B1*, 1545.

Phillips, J. C. (1973) *Bonds and Bands in Semiconductors* (Academic, New York).

Pollak, M. and Knotek, M. L. (1974) *Phys. Rev. B9*, 644.

Saxon, D. S. (1968) *Elementary Quantum Mechanics* (Holden-Day, San Francisco,).

Sernelius, B. E. (1986) *Phys. Rev. B33*, 8582.

Sherwin, C. W. (1959) *Introduction to Quantum Mechanics* (Henry Holt, New York).

Shklovskii, B. I. and Efros, A. L. (1984) *Electronic Properties of Doped Semiconductors* (Springer, Berlin).

Shockley, W. (1961) *Solid-State Electron.* **2**, 35.

Sommerfeld, A. (1920), see Finkelnburg (1958).

Stern, F. (1967) *Phys. Rev. Lett.* **18**, 546.

Stillman, G. E., Cook, L. W., Roth, T. J., Low, T. S., and Skromme, B. J. (1982) in *GaInAsP Alloy Semiconductors*, ed. Pearsall, T. P. (Wiley, New York).

Stoneham, A. M. (1975) *Theory of Defects in Solids* (Clarendon, Oxford).

Stoneham, A. M. (1986) in *Defects in Semiconductors*, ed. Bardeleben, H. J. (Trans. Tech., Switzerland).

Thomas, D. G. and Hopfield, J. J. (1966) *Phys. Rev.* **150**, 680.

Verie, C. (1967) *Proc. Int. Conf. on II–VI Compounds* (Benjamin, New York) p. 1124.

White, H. E. (1934) *Introduction to Atomic Spectra* (McGraw-Hill, New York).

Wolf, P. A. (1962) *Phys. Rev.* **126**, 405.

Yao, H. and Compaan, A. (1990) *Appl. Phys. Lett.* **57**, 147.

Chapter 2

Banks, P. W., Brand, S., and Jaros, M. (1980) *J. Phys. C.: Solid State Phys.* **13**, 6167.

Baraff, G. A. and Schlüter, M. (1985) *Phys. Rev. Lett.* **55**, 1327.

Bardeen, J. (1947) *Phys. Rev.* **71**, 717.

Bethe, H. A. and Salpeter, E. E. (1957) *Quantum Mechanics of One- and Two-Electron Atoms* (Academic, New York).

Bishop, S. G. (1986) in *Deep Centers in Semiconductors*, ed. Pantelides, S. T. (Gordon and Breach, New York).

Blakemore, J. S. (1967) *Phys. Rev.* **163**, 809.

Blakemore, J. S. (1987) *Semiconductor Statistics* (Dover, New York).

Bourgoin, J. and Lannoo, M. (1983) *Point Defects in Semiconductors II: Experimental Aspects* (Springer, Berlin).

Davydov, A. S. (1976) *Quantum Mechanics*, 2nd ed. (Pergamon, Oxford).

Finkelnburg, W. (1958) *Introduction to Atomic Physics* (in German) (Springer, Berlin).

Frenkel, J. (1938a) *Phys. Rev.* **54**, 647.

Frenkel, J. (1938b) *Tech. Phys. USSR* **5**, 685.

Hall, R. N. (1951) *Phys. Rev.* **83**, 228.

Hall, R. N. (1952) *Phys. Rev.* **87**, 387.

Hartke, J. L. (1968) *J. Appl. Phys.* **39**, 4871.

Henry, C. H. and Lang, D. V. (1977) *Phys. Rev.* **B15**, 989.

Hjalmarson, H. P., Vogl, P., Wolford, D. J. and Dow, J. D. (1980) *Phys. Rev. Lett.* **44**, 810.

Huang, K. and Rhys, A. (1950) *Proc. Roy. Soc. (London)* **A204**, 406.

Jaros, M. (1982) *Deep Levels in Semiconductors* (Adam Hilger, Bristol).

Jaros, M. and Band, S. (1976) *Phys. Rev.* **B14**, 4494.

Korol, E. N. (1977) *Sov. Phys.-Solid State* **19**, 1327.

Kovarskii, V. A. (1962) *Sov. Phys.-Solid State* **4**, 1200.

Landsberg, P. T. (1970) *Phys. Stat. Solidi* **41**, 457.

Lang, D. V. (1974a) *J. Appl. Phys.* **45**, 3014.

Lang, D. V. (1974b) *J. Appl. Phys.* **45**, 3023.

Lang, D. V. and Logan, R. A. (1977) *Phys. Rev. Lett.* **39**, 635.

Lang, D. V., Logan, R. A. and Jaros, M. (1979) *Phys. Rev. B***19**, 1015.

Lannoo, M. and Bourgoin, J. (1981) *Point Defects in Semiconductors I: Theoretical Aspects* (Springer, Berlin).

Lax, M. (1959) *J. Phys. Chem. Sol.* **8**, 66.

Lax, M. (1960) *Phys. Rev.* **119**, 1502.

Lax, M. (1992) private communication.

Madelung, O. (1978) *Introduction to Solid State Theory* (Springer, Berlin).

Martin, G. M., Mitonneau, A., and Mircea, A. (1977) *Electron. Lett.* **13**, 191.

Milnes, A. G. (1973) *Deep Impurities in Semiconductors* (Wiley, New York).

Mitonneau, A., Martin, G. M., and Mircea, A. (1977) *Electron. Lett.* **13**, 666.

Nelson, R. J. (1977) *Appl. Phys. Lett.* **31**, 351.

Pantelides, S. T. (1986) *Deep Centers in Semiconductors: A State of the Art Approach* (Gordon and Breach, New York).

Pons, D. and Makram-Ebeid, S. (1979) *J. Physique* **40**, 1161.

Poole, H. H. (1914) Philos. Mag. **27**, 58.

Queisser, H. J. (1971) in *Advances in Solid State Physics Vol. 11*, (Vieweg, Braunschweig).

Riddoch, F. A. and Jaros, M. (1980) *J. Phys. C: Solid State Phys.* **13**, 6181.

Sah, C. T. (1967) *Proc. IEEE* **55**, 654.

Schubert, E. F. and Ploog, K. (1984) *Phys. Rev. B***29**, 4562.

Seeger, K. (1982) *Semiconductor Physics* (Springer, Berlin).

Sheinkman, M. K. (1965) *Sov. Physics-Solid State* **7**, 18.

Shockley, W. and Read, W. T. (1952) *Phys. Rev.* **87**, 835.

Solar, N. and Burstein, E. (1955) *Phys. Rev.* **98**, 1757.

van Roosbroeck, W. and Shockley, W. (1954) *Phys. Rev.* **94**, 1558.

Vincent, G., Chantre, A., and Bois, D. (1979) *J. Appl. Phys.* **50**, 5484.

Vogl, P. (1981) in *Advances in Solid State Physics* **21** (Vieweg, Braunschweig).

Vogl, P. (1985) in *Advances in Solid State Physics* **25** (Vieweg, Braunschweig).

Zunger, A. (1986) in *Solid State Physics* **39**, eds. Ehrenreich, H. and Turnbull, D. (Academic, New York).

Chapter 3

Blakemore, J. S. (1982) *Sol. State Electron.* **25**, 1067.

Chang, T. Y. and Izabelle, A. (1989) *J. Appl. Phys.* **65**, 2162.

de Boer, J. H. and van Geel, W. C. (1935) *Phys.* **2**, 186.

Ehrenberg, W. (1950) *Proc. Phys. Soc. London.* A**63**, 75.

Frank (1931) Rev. Mod. Phys. **3**, 1.

Joyce, W. B. (1978) *Appl. Phys. Lett.* **32**, 680.

Joyce, W. B. and Dixon, R. W. (1977) *Appl. Phys. Lett.* **31**, 354.

Kittel, C. and Kroemer, H. (1980) *Thermal Physics*, 2nd ed. (Freeman, San Francisco).

Lang, D. V., Grimmeiss, H. G., Meijer, E., and Jaros. M. (1980) *Phys. Rev.* B**22**, 3917.

Laufer, P. M., Pollak, F. H., Nahory, R. E., and Pollak, M. A. (1980) *Solid State Commun.* **36**, 978.

Madelung, O. (1970) *Fundamentals of Semiconductor Physics*, (in German) (Springer, Berlin).

Morin, F. F. (1959) in *Semiconductors*, ed. Hannay, N. B. (Reinhold, New York) p. 31.

Mott, N. F. and Gurney, R. W. (1940) *Electronic Processes in Ionic Crystals* (Oxford University Press, Oxford) p. 156.

Nilsson, N. G. (1973) *Phys. Stat. Sol.* A**19**, K75.

Reif, F. (1965) *Fundamental of Statistical and Thermal Physics* (McGraw-Hill, New York).

Schubert, E. F. and Ploog, K. (1984) *Phys. Rev. B30*, 7021.

Smith, R. A. (1986) *Semiconductors* (Cambridge University Press, Cambridge).

Sommerfeld, A. (1928) *Zeitschrift Physik* **47**, 1; see also A. Sommerfeld and N. H. Frank
     (1931) *Rev. Mod. Phys.* **3**, 1.

Thurmond, C. D. (1975) *J. Electrochem. Soc.* **112**, 1133.

Van Vechten, J. A. and Thurmond, C. D. (1976a) *Phys. Rev. B14*, 3539.

Van Vechten, J. A. and Thurmond, C. D. (1976b) *Phys. Rev. B14*, 3551.

Chapter 4

Arthur, J. R., Jr. (1968) *Proc. Conf. Struct. Chem. Solid Surfaces*, ed. G. A. Somorjai
     (Wiley, New York), p. 46–1.

Arthur, J. R., Jr. (1968) *J. Appl. Phys.* **39**, 4032.

Calawa, A. R. (1981) *Appl. Phys. Lett.* **38**, 701.

Chiu, T. H. (1991) et al., unpublished.

Chiu, T. H., Tsang, W. T., Schubert, E. F., and Agyekum, E. (1987) *Appl. Phys. Lett.* **51**,
     1109.

Cho, A. Y. (1985) in *The Technology and Physics of Molecular Beam Epitaxy*, ed. Parker
     E.H.C. (Plenum, New York).

Cho, A. Y. (1971) *J. Vac. Sci. Technol.* **8**, S31.

Cho, A. Y. (1977) *Jpn. J. Appl. Phys.* **16**, Suppl. 16-1, 435.

Cho, A. Y. and Arthur, J. R. (1975) *Prog. Solid-State Chem.* **10**, 157.

Cho, A. Y. and Casey, H. C. (1974) *Jr. Appl. Phys. Lett.* **35**, 288.

Chow, R. and Chai, Y. G. (1983) *Appl. Phys. Lett.* **42**, 383.

Davey, J. E. and Pankey, T. (1968) *J. Appl. Phys.* **39**, 1941.

Dupuis, R. D. and Dapkus, P. D. (1978) *Appl. Phys. Lett.* **32**, 406.

Foxon, C. T. and Joyce, B. A. (1975) *Surf. Sci.* **50**, 434.

Foxon, C. T. and Joyce, B. A. (1977) *Surf. Sci.* **64**, 293.

Foxon, C. T., Boudry, M. R., and Joyce, B. A. (1974) *Surf. Sci.* **44**, 69.

Günther, K. G. (1958) *Zeit. Naturforsch.* **13a**, 1081.

Hellman E. S. and Harris, J. S., Jr. (1987) *J of Cryst. Growth* **81**, 38.

Horiguchi, S., Kimura, K., Kamon, K., Mashita, M., Shimatzu, M., Mihara, M., and Ishii, M. (1986) *Jpn. J. Appl. Phys.* **25**, L979.

Johnston, W. D., Jr., DiGuiseppe, M. A., and Wilt, D. P. (1989) *AT&T Technical Journal* **68**, 53.

Knudsen, M. (1909) *Ann. Phys. (Leipzig)* **4**, 999.

Ludowise, M. I. (1985) *J. Appl. Phys.* **58**, R31.

Manasevit, H. M. (1968) *Appl. Phys. Lett.* **12**, 156.

Manasevit, H. M. and Simpson, W. I. (1969) *J. Electrochem. Soc.* **116**, 1725.

Panish, M. B. (1980) *J. Electrochem. Soc.* **127**, 2729.

Panish, M. B. (1987) *J. Cryst. Growth* **81**, 249.

Panish, M. B. and Temkin, H. (1989) *Annu. Rev. Mater. Sci.,* **19**, 209.

Panish, M. B., Temkin, H., and Sumski, S. (1985) *J. Vac. Sci. Technol.* **B3**, 657.

Pütz, N., Heinecke, H., Heyen, M., Balk, P., Weyers, M., and Lüth, H. (1986) *J. Cryst. Growth* **74**, 292.

Razeghi, M., (1989) *The MOCVD Challenge* (Adam Hilger, Bristol).

Schubert, E. F., Cunningham, J. E., Tsang, W. T., and Timp, G. L. (1987) *Appl. Phys. Lett.* **51**, 1170.

Schubert, E. F., Tsang, W. T., Feuer, M. D., and Mankiewich, P. M. (1988) *IEEE Electron Device Letters* **9**, 145.

Stringfellow, G. B. (1989) *Organometallic Vapor-Phase Epitaxy: Theory and Practice* (Academic, Boston).

Swaminathan, V. and Macrander, A. T. (1991) *Materials Aspects of GaAs and InP Based Structures* (Prentice Hall, Englewood Cliffs).

Tanaka, H., Itoh, H., O'Hori, T., Takikawa, M., Kasai, K., Takechi, M., Suzuki, M., and Komeno., J. (1987) *Jpn. J. Appl. Phys.* **26**, L1456.

Tietjen, J. J. and Amick, J. A. (1966) *J. Electrochem. Soc.* **113**, 724.

Tsang, W. T. (1985a) *Appl. Phys. Lett.* **45**, 1234.

Tsang, W. T. (1985b) in *Molecular Beam Epitaxy and Heterostructures*, eds. Chang, L. L. and Ploog, K. (Martinus Nijhoff, Dordrecht).

Tsang, W. T. (1989) in *VLSI Electronics: Microstructure Science*, **21**, ed. Einspruch, N. G. (Academic, New York) p. 255.

Tsang, W. T. (1990) private communication.

Tsang, W. T. and Schubert, E. F. (1986) *Appl. Phys. Lett.* **49**, 220.

Tsang, W. T., Dayem, A. H., Chiu, T. H., Cunningham, J. E., Schubert, E. F., Ditzenberger, J. A., Shah, J., Zyskind, J. L., and Tabatabaie, N. (1986) *Appl. Phys. Lett.* **49**, 170.

Tsang, W. T., Levi, A. F. J., and Burkhardt, E. G. (1988) *Appl. Phys. Lett.* **53**, 983.

Tsang, W. T., Schubert, E. F., Chiu, T. H., Cunningham, J. E., Burkhardt, E. G., Ditzenberger, J. A., and Agyekum, E. (1987) *Appl. Phys. Lett.* **51**, 761.

Veuhoff, E., Pletschen, W., Balk, P., and Lüth, H. (1981) *J. Cryst. Growth* **55**, 30.

Chapter 5

Ambridge, T. and Faktor, M. M. (1975a) *J. Appl. Electrochem.* **5**, 319.

Ambridge, T. and Faktor, M. M. (1975b) *Gallium Arsenide and Related Compounds* ed. Bok, J., *Inst. Phys. Conf. Ser.* **24**, 320.

Arthur, J. R. (1973) *Surface Science* **38**, 394.

Ashen, D. J., Dean, P. J., Hurle, D. T. J., Mullin, J. B., White, A. M., and Greene, P. D. (1975) *J. Phys. Chem. Sol.* **36**, 1041.

Beall, R. B., Clegg, J. B., and Harris, J. J. (1988) *Semicond. Sci. Technol.* **3**, 612.

Blakemore, J. (1982) *J. Appl. Phys.* **53**, R123.

Blood, P. (1986) *Semicond. Sci. Technol.* **1**, 7.

Braslau, N. (1981) *J. Vac. Sci. Technol.* **19**, 803.

Calderon, L., Lu, Y., Shen, H., Pamulapati, J., Dutta, M., Chang, W. H., Yang, L. W., and Wright, P. D. (1992) *Appl. Phys. Lett.* **60**, 1597.

Chai, Y. G., Chow, R., and Wood, C. E. C. (1981) *Appl. Phys. Lett.* **39**, 800.

Chang, C. A., Ludeke, R., Chang, L. L., and Esaki, L. (1977) *Appl. Phys. Lett.* **31**, 759.

Chiu, T. H., Ditzenberger, J. A., Luftman, H. S., Tsang, W. T., and Ha, N. T. (1990) *Appl. Phys. Lett.* **56**, 1688.

Cho, A. Y. (1975) *J. Appl. Phys.* **46**, 1733.

Cho, A. Y. and Hayashi, I. (1971) *J. Appl. Phys.* **42**, 4422.

Cho, A. Y. and Panish, M. B. (1972) *J. Appl. Phys.* **43**, 5118.

Collins, D. M., Miller, J. N., Chai, Y. G., and Chow, R. (1982) *J. Appl. Phys.* **53**, 3010.

Davies, G. J., Andrews, D. A., and Heckingbottom, R. (1981) *J. Appl. Phys.* **52**, 7214.

Duhamel, N., Henoc, P., Alexandre, F., and Rao, E. V. K. (1981) *Appl. Phys. Lett.* **39**, 49.

Enquist, P., Hutchby, J. A., and de Lyon, T. J. (1988) *J. Appl. Phys.* **63**, 4485.

Greiner, M. E. and Gibbons, J. F. (1984) *Appl. Phys. Lett.* **44**, 750.

Hamm, R. A., Panish, M. B., Nottenburg, R. N., Chen, Y. K., and Humphrey, D. A. (1989) *Appl. Phys. Lett.* **54**, 2586.

Harris, J. J., Ashenford, D. E., Foxon, C. T., Dobson, P. J. and Joyce, B. A. (1984) *Appl. Phys. A***33**, 87.

Hartley, F. and Patai, S. (1982) *The Chemistry of the Metal-Carbon Bond* (Wiley, New York) p. 53-71.

Heckingbottom, R. (1985) in *Molecular Beam Epitaxy and Heterostructures* ed. Chang, L. L. and Ploog, K. (Martinus Nijhoff, Dordrecht).

Heiblum, M., Wang, W. I., Osterling, L. E., and Deline V. (1983) *Appl. Phys. Lett.* **54**, 6751.

Hess, K., Stath, N., and Benz, K. W. (1974) *J. Electrochem Soc.: Solid State Sci. Technol.* **121**, 1208.

Hirama, V. (1964) *J. Chem. Eng. Data* **9**, 65.

Hobson, W. S., Pearton, S. J., Schubert, E. F., and Cabaniss, G. (1989) *Appl. Phys. Lett.* **55**, 1546.

Hoke, W. E., Lemonias, P. J., Weir, D. G., Hendricks, H. T., and Jackson, G. S. (1991a) *J. Appl. Phys.* **69**, 511.

Hoke, W. E., Lemonias, P. J., Lyman, P. S., Hendricks, H. T., Weir, D. G., and Colombo, P. (1991b) *J. Cryst. Growth* **111**, 269.

Ilegems, M. (1977) *J. Appl. Phys.* **48**, 1278.

Ilegems, M. and O'Mara, W. C. (1972) *J. Appl. Phys.* **43**, 1190.

Ito, H. and Ishibashi, T. (1991) *Jpn. J. Appl. Phys.* **30**, L944.

Joyce, B. A. and Foxon (1977) *Jpn. J. Appl. Phys.* **16**, Suppl. 16-1, 17.

Kadhim, M. A. and Tuck, B. (1972) *J. Mater. Sci.* **7**, 68.

Kirchner, P. D., Woodall, J. M., Freeouf, J. L., Wolford, D. J., Pettit, G. D. (1981) *J. Vac. Sci. Technol.* **19**, 604.

Kittel, C. and Kroemer, H. (1980) *Thermal Physics* 2nd ed. (Freeman, New York) p. 393.

Knudsen, M. (1909) *Ann. Phys. (Leipzig)* **4**, 999.

Kobayashi, N., Makimoto, T., and Horikoshi, Y. (1987) *Appl. Phys. Lett.* **50**, 1435.

Kopf, R. F. and Schubert, E. F. (1990) unpublished.

Künzel, H., Knecht, J., Jung, H., Wünstel, K., and Ploog, K. (1982) *Appl. Phys.* **A28**, 167.

Lanzillotto, A. M., Santos, M., and Shayegan, M. (1989) *Appl. Phys. Lett.* **55**, 1445.

Maguire, J., Murray, R., Newman, R. C., Beall, R. B., and Harris, J. J. (1987) *Appl. Phys. Lett.* **50**, 516.

Malik, R. J., Nottenburg, R. N., Schubert, E. F., Walker, J. F., and Ryan, R. W. (1988) *Appl. Phys. Lett.* **53**, 2661.

Malik, R. J., Nagle, J., Micovic, M., Harris, T. D., Ryan, R. W., and Hopkins, L. C. (1992) *J. Vac. Sci. Technol.* **B10**, 850.

Marton, L. and Marton C. (1979) (ed.), *Methods of Experimental Physics and Technology* **14**, (Academic, New York) p. 350.

McLevige, W. V., Vaidyanathan, K. V., Streetman, B. G., Ilegems, M., Comas, J., and Plew, L. (1978) *Appl. Phys. Lett.* **33**, 127.

Nagle, J., Malik, R. J., Gershoni, D. (1991) *J. Cryst. Growth* **111**, 264.

Ogawa, M. and Baba, T. (1985) *Jpn. J. Appl. Phys.* **24**, L572.

Pfeiffer, L. N. and West, K. W. (1987) unpublished.

Ploog, K. and Fischer, A. (1978) *J. Vac. Sci. Technol.* **15**, 255.

Poole, I., Lee, M. E., Singer, K. E., Frost, J. E. F., Kerr, T. M., Wood, C. E. C., Andrews, D. A., Rothwell, W. J. M., and Davies, G. J. (1988) *J. Appl. Phys.* **63**, 395.

Putley, E. H. (1960) *The Hall-Effect and Related Phenomena* (Butterworths, London).

Reynolds, W. N. (1968) *Physical Properties of Graphite* (Elsevier, Amsterdam).

Rupprecht, H., Woodall, J. M., Konnerth, K. and Pettit, D. G. (1966) *Appl. Phys. Lett.* **9**, 221.

Schubert, E. F. (1990) *J. Vac. Sci. Technol.* A**8**, 2980.

Schubert, E. F. and Kopf, R. F. (1990) unpublished.

Schubert, E. F., Gilmer, G. H., Kopf, R. F., and Luftman, H. S. (1992) *Phys. Rev. B***46**, 15 078.

Schubert, E. F., Kuo, J. M., Kopf, R. F., Luftman, H. S., Hopkins, L. C., and Sauer, N. J. (1990a) *J. Appl. Phy.* **67**, 1969.

Schubert, E. F., Kuo, J. M., Kopf, R. F., Jordan, A. S., Luftman, H. S., and Hopkins L. C. (1990b) *Phys. Rev. B***42**, 1364.

Schubert, E. F., Stark, J. B., Chiu, T. H., and Tell, B. (1988) *Appl. Phys. Lett.* **53**, 293.

Schubert, E. F., Tu, C. W., Kopf, R. F., Kuo, J. M., Lunardi, L. M. (1989) *Appl. Phys. Lett.* **54**, 2592.

Skromme, B. J., Stillman, G. E., Obsevstan, J. D., and Chan, S. S. (1984) *J. Electron Mater.* **13**, 463.

Spitzer, W. G. and Panish, M. B. (1969) *J. Appl. Phys.* **40**, 4200.

Stillman, G. E., Cook, L. W., Roth, T. J., Low, T. S., and Skromme, B. J. (1982) in *GaInAsP Alloy Semiconductors*, ed. Pearsall, T. P. (Wiley, New York) p. 121.

Subbanna, S., Tuttle, G., and Kroemer, H. (1988) *J. Electron. Mater.* **17**, 297.

Theis, T. N., Mooney, P. M., and Wright, S. L. (1988) *Phys. Rev. Lett.* **60**, 361.

Tuck, B. (1988) *Atomic Diffusion in III–V Semiconductors* (Adam Hilger, Bristol).

Tuck, B. and Kadhim, M. A. (1972) *J. Mater. Sci.* **7**, 585.

Vescan, L., Selders, J., Maier, M., Kräutle, H., and Beneking., H. (1984) *J. Cryst. Growth* **67**, 353.

White, A. M., Dean, P. J., Joyce, B. D., Clarke, R. C., and Taylor, L. L. (1972) *Proc. Int. Conf. Phys. Semicond., Warsaw, 1972* (Polish Scientific Publishers, Warsaw) p. 190.

Williams, E. W., Elder, W., Astles, M. G., Webb, M., Mullin, J. B., Straughan, B., and Tufton, P. J. (1973) *J. Electrochem. Soc.: Solid State Sci. Technol.* **120**, 1741.

Wood, C. E. C. (1978) *Appl. Phys. Lett.* **33**, 770.

Wood, C. E. C. (1985a) in *The Physics and Technology of MBE* ed. Parker, E. H. C. (Plenum, New York) p. 61.

Wood, C. E. C. (1985b) in *Molecular Beam Epitaxy and Heterostructures*, ed. Chang, L. L. and Ploog, K. (Martinus Nijhoff, Dordrecht).

Wood, C. E. C. and Joyce, B. A. (1978) *J. Appl. Phys.* **49**, 4854.

Wood, C. E. C., Desimone, D., Singer, K., and Wicks, G. W. (1982) *J. Appl. Phys.* **53**, 4230.

Yano, M., Suzuki, Y., Ishii, T., Matsushima, Y. and Kimata, M. (1978) *Jpn. J. Appl. Phys.* **17**, 2091.

Chapter 6

Aebi, V., Cooper, C. B., Moon, R. L. and Saxena, R. R. (1981) *J. Crystal Growth* **55** 517.

André, J. P., Gallais, A. and Hallais, J. (1977) in: *Gallium Arsenide and Related Compounds 1976, Inst. Phys. Conf. Ser.* **33a** (Inst. Phys. London) p. 1.

Bass, S. J. (1979) *J. Crystal Growth* **47**, 613.

Bass, S. J. and Oliver, P. E. (1977) in *Gallium Arsenide and Related Compounds 1976, Inst. Phys. Conf. Ser.* **33b** (Inst. Phys., London) p. 1.

Blaauw, C., Emmerstosfer, B., and Springthrope, A. J. (1987) *J. Crystal Growth* **84**, 431.

Bottka, N., Sillmon, R. S., and Tseng, W. F. (1984) *J. Crystal Growth* **68**, 54.

Chang, C. Y., Lee, M. K., Su, Y. K., and Hsu, W. C. (1983) *J. Appl. Phys.* **54**, 5464.

Field, R. J. and Ghandhi, S. K. (1986a) *J. Crystal Growth* **74**, 543.

Field, R. J. and Ghandhi, S. K. (1986b) *J. Crystal Growth* **74**, 551.

Fraas, L. M., McLeod, P. S., Cape, J. A., and Partain, L. D. (1984) *J. Crystal Growth* **68**, 490.

Gersten, S. W., Vendura, G. J., and Yeh, Y. C. M. (1986) *J. Crystal Growth* **77**, 286.

Giling, L. J. and van de Ven, J. (1987) in *Advanced Crystal Growth*, ed. Dryburth, P. M., Cockayne, B., and Barraclough, K. G. (Prentice Hall, New York) p. 309.

Glade, M., Grünzmacher, D., Meyer. R., Woelk, E. G., and Balk, P. (1989) *Appl. Phys. Lett.* **54**, 2411.

Glew, R. W. (1982) in *Gallium Arsenide and Related Compounds*, ed. Sugano, T. (Inst. Phys., Bristol) p. 581.

Glew, R. W. (1984) *J. Crystal Growth* **68**, 44.

Hallais, J. P. (1978) *Acta Electron.* **21** 129.

Hayafuji, N., Mizuguchik, Ochi, S., and Murotani, T. (1986) *J. Crystal Growth* **77**, 281.

Heyen, M., Bruch, H., Bachen, K.-H., and Balk, P. (1977) *J. Crystal Growth* **42**, 127.

Houng, Y.-M. and Low, T. S. (1986) *J. Crystal Growth* **77**, 272.

Hsu, C. C., Yuan, J. S., Cohen, R. M., and Stringfellow, G. B. (1986) *J. Crystal Growth* **74**, 535.

Keil, G., Le Metayer, M., Cuquel, A. and LePollotec, D. (1982) *Rev. Phys. Appl.* **17** 405.

Konagai, M., Yamada, T., Akatsuka, T., Nozaki, S., Miyake, R., Saito, K., Fukamachi, T., Tokumitsu, E., and Takahashi, K. (1990) *J. Crystal Growth* **105** 359.

Konagai, M., Yamada, T., Akatsuka, T., Saito, K., Tokumitsu, E., and Takahashi, K. (1989) *J. Crystal Growth* **98**, 167.

Kozen, A., Najima, S., Tenmyo, J., and Asahi, H. (1986) *J. Appl. Phys.* **59**, 1156.

Kuech, T. F., Tischler, M. A., Wang, P.-J., Scilla, G., Polenski, R., and Cardone. F. (1988) *Appl. Phys. Lett.* **53**, 1317.

Kuech, T. F., Veuhoff E., and Meyerson, B. S. (1984) *J. Crystal Growth* **68**, 48.

Lewis, C. R., Dietze, W. T., and Ludowise, M. J. (1983) *J. Electron. Mater.* **12**, 507.

Lewis, C. R., Ludowise, M. J., and Dietze, W. T. (1984) *J. Electron. Mater.* **13**, 447.

Manasevit, H. M. and Thorsen, A. C. (1972) *J. Electrochem. Soc.* **119**, 99.

Matsumoto, K., Hidaka, J., and Uchida, K. (1989) *J. Appl. Phys.* **65**, 3849.

Matsumoto, K., Hidaka, J., and Uchida, K. (1990) *J. Crystal Growth* **99**, 329.

Nelson, A. W. and Westbrook, L. D. (1984) *J. Crystal Growth* **68**, 102.

Nishizawa, J. and Kurabayashi, T. (1983) *J. Electrochem Soc.* **130** 413.

Ohno, H., Ikeda, E., and Hasegawa, H. (1984) *J. Crystal Growth* **68**, 15.

Parsons, J. D. and Krajehbrink, F. G. (1983) *J. Electrochem. Soc.* **130**, 1782.

Parsons, J. D. and Krajenbrink, F. V. (1984) *J. Crystal Growth* **68**, 60.

Parsons, J. D., Lichtmann, L. S., Krajenbrink, F. G., and Brown, D. W. (1986) *J. Crystal Growth* **77**, 32.

Pütz, N., Heinecke, H., Heyen, M., Balk, P., Weyers, M., and Lüth, H. (1986) *J. Crystal Growth* **74**, 292.

Razeghi, M. (1989) *The MOCVD Challenge* (Adam Hilger, Bristol).

Roberts, J. S., Mason, N. J., and Robinson, M. (1984) *J. Crystal Growth* **68**, 422.

Roth, A. P., Yakimova, R., and Sundaram, V. S. (1984) *J. Crystal Growth* **68**, 65.

Seki, Y., Tanno, K., Iida, K., and Ichiki, E. (1975) *J. Electrochem. Soc.* **122**, 1108.

Shapira, Y. and Lichtman, D. (1979) in *Methods of Experimental Physics* **14**, eds. Marton, L. and Marton, C. (Academic, New York).

Skevington, P. J., Andrews, D. A., and Davies, G. J. (1990) *J. Cryst. Growth* **105**, 371.

Stringfellow, G. B. (1989) in *Organometallic Vapor-Phase Epitaxy* (Academic, San Diego).

Tanaka, H., Itoh, H., O'Hori, T., Takikawa, M., Kosai, K., Takehi, M., Suzuki, M., and Komeno, J. (1987) *Jpn. J. Appl. Phys.* **26**, L1456.

Timmons, M. L., Chiang, P. K., and Hattangady, S. V. (1986) *J. Crystal Growth* **77**, 37.

Tsang, W. T. (1989) in *VLSI Electronics: Microstructure Science* **21**, ed. Einspruch, N. G. (Academic, New York) p. 255.

Tuck, D. G. (1982) in *Comprehensive Organometallic Chemistry*, eds. Wilkinson, G., Stone, G. A., and Abel, F. W. (Pergamon, Oxford) p. 684.

van de Ven, J., Rutten, G. M. J., Raaijmakers, M. J., and Giling, L. J. (1986) *J. Crystal Growth* **76**, 352.

Veuhoff, E., Kuech, T. F., and Meyerson, B. S. (1985) *J. Electrochem. Soc.* **132**, 1958.

Veuhoff, E., Maier, M., Bachen, K.-H., and Balk, P. (1981) *J. Crystal Growth* **53**, 598.

Weyers, M., Musoff, J., Marx, D., Kohl, A., and Balk, P. (1990) *J. Crystal Growth* **105**, 383.

Weyers, M., Pütz, N., Heinecke, H., Heyer, M., Lüth, H., and Balk, P. (1986) *J. Electron. Mater.* **15**, 57.

Wiley, J. D. (1975) in *Semiconductors and Semimetals* **10**, eds. Willardson, R. K. and Beer, A. C. (Academic, New York) p. 154.

Wood, C. E. C., DeSimeone, D., Singer, K., and Wicks, G. W. (1982) *J. Appl. Phys.* **53**, 4230.

Yakimova, R., Roth, A. P., Williams, D. F., and Sundaran, V. S. (1984) *J. Crystal Growth* **68**, 71.

Yoshida, M., Watanabe, H., and Uesugi, F. (1985) *J. Electrochem. Soc.* **132**, 677.

Zilko, J. L. and Schubert, E. F. (1991) unpublished.

Chapter 7

Aliev, M. I., Safaraliev, G. I. and Abdinova, S. G. (1967) *Phys. Stat. Sol.,* **22**, 741.

Andrews, D. A., Heckingbottom, R., and Davies, G. J. (1983) *Appl. Phys.* **54**, 4421.

Anholt, R., Balasingam, P., Chou, S. Y., Sigmon, T. W., and Deal, M. (1988) *J. Appl.*

Arnold, N., Dämbkes, H., and Heime, K. (1980) *Jpn. J. Appl. Phys.* **19**, Suppl. 19-1, 361.

Arnold, N., Schmitt, R., and Heime, K. (1984) *J. Phys. D.: Appl. Phys.* **17**, 443.

Arthur, J. R. (1967) *J. Phys. Chem. Solids* **28**, 2257.

Astles, M. G., Smith, F. G. H., and Williams, E. W. (1973) *J. Electrochem. Soc.* **120**, 1750.

Bahir, G. and Merz, J. L. (1988) *Inst. Phys. Conf. Ser.* **91**, 467.

Bahir, G., Merz, J. L., Abelson, J. R., and Sigmon, T. W. (1989) *J. Appl. Phys.* **65**, 1009.

Baillargeon, J. N., Cheng, K. Y., Hsieh, K. C., and Wei, C. L. (1990) *J. Cryst. Growth* **105**, 106.

Ballingall, J. M. and Wood, C. E. C. (1982) *Appl. Phys. Lett.* **41**, 947.

Baraff, G. A. and Schlüter, M. (1985) *Phys. Rev. Lett.* **55**, 1327.

Bass, S. J. and Oliver, P. E. (1977) in *Gallium Arsenide and Related Compounds,* ed. Eastman, L. F. (Inst. Phys., Bristol) p. 1.

Beadle, W. E., Tsai, J. C. C., and Plummen, R. D. (1985) *Quick Reference Manual (QRM),* (Wiley, New York) pp. 2–71.

Beall, R. B., Clegg, J. B., Castagné, J., Harris, J. J., Murray, R., and Newman, R. C. (1989) *Semicond. Sci. Technol.* **4**, 1171.

Benchimol, J. L., Alaoui, F., Le Roux, G., Gao, Y. and Alexandre, F. (1989) *2nd Int. Conf. on Chemical Beam Epitaxy, Houston.*

Biersack, J. P. and Haggmark, L. G. (1980) *Nucl. Instrum. Methods* **174**, 257.

Biersack, J. P. and Ziegler, J. F. (1982) *Ion Implantation Techniques* (Springer, New York) p. 122.

Bose, S. S., Lee, B., Kim, M. H., Stillman, G. E., and Wang, W. I. (1988) *J. Appl. Phys.* **63**, 743.

Braslau, N. (1981) *J. Vac. Sci. Technol.* **19**, 803.

Camlibel, I., Chin, A. K., Guggenheim, H., Singh, S., van Uitert, L. G., and Zydzik, G. J. (1984) *J. Electrochem. Soc.* **131**, 1687.

Chadi, D. J. and Chang, K. J. (1988) *Phys. Rev. Lett.* **61**, 873.

Chai, Y. G., Chow, R., and Wood, C. E. C. (1981) *Appl. Phys. Lett.* **39**, 800.

Cheng, K. Y., Cho, A. Y. and Wagner, W. R. (1981) *J. Appl. Phys.*, **52**, 6328.

Chiang, T.-C., Ludeke, R., Aono, M., Landgren, G., Himpsel, F. J., and Eastman, E. E. (1983) *Phys. Rev. B27*, 4770.

Chin, A. K. and Bonner, W. A. (1982) *Appl. Phys. Lett.*, **40**, 248.

Cho, A. Y. and Arthur, J. R. (1975) *Prog. Solid State Chem.* **10**, 157.

Cristel, L. A., Gibbons, J. F., and Mylroic, S. (1980) *J. Appl. Phys.* **51**, 6176.

De Souza, J. P., Sadana, D. K., Baratte, H., and Cardone, F. (1990) *Appl. Phys. Lett.* **57**, 1129.

Donnelly, J. P. and Hurwitz, C. E. (1980) *Solid State Electron.* **23**, 943.

Drummond, T. J., Lyons, W. G., Fisher, R., Thorne, R. E., Morkoc, H., Hopkins, C. G. and Evans, C. A., J. (1982) *J. Vac. Sci. Technol.* **21**, 957.

Etienne, B. and Thierry-Mieg, V. (1988) *Appl. Phys. Lett.* **52**, 1237.

Feldman, L. C. and Mayer, J. W. (1986) *Fundamentals of Surface and Thin Film Analysis* (North-Holland, New York).

Ferris, S. D., Leamy, H. J., and Poate, J. M. eds. (1979) *Laser-Solid Interactions and Laser Processing* (American Institute of Physics, New York).

Fujita, S., Bedair, S. M., Littlejohn, M. A. and Hauser, J. R. (1980) *J. Appl. Phys.* **51**, 5438.

Gibbons, J. F., Johnson, W. S., and Mylroic, S. W. (1975) *Projected Range Statistics* (Dowolan, Hutchinson, and Ross, Stroudsburg, PA).

Gill, S. S. and Sealy, B. J. (1984) *J. Appl. Phys.* **56**, 1189.

Gill, S. S. and Sealy, B. J. (1986) *J. Electrochem. Soc.: Solid-State Sci. Technol.* **133**, 2590.

Gilmer, G. H. and Gossmann, H. J. (1993) to be published.

Gilmer, G. H. (1992) personal communication.

Glew, R. W. (1984) *J. Crystal Growth* **68**, 44.

Goldstein, B. (1984) *Phys. Rev.* **121**, 1305.

Greiner, M. E. and Gibbons, J. F. (1984) *Appl. Phys. Lett.* **44**, 750.

Hamm, R. A., Panish, M. B., Nottenburg, R. N., Chen, Y. K., and Humphrey, D. A. (1989) *Appl. Phys. Lett.* **54**, 2586.

Harris, T. D. and Schubert, E. F. (1990) unpublished results.

Harrison, R. J. and Houston, P. A. (1986) *J. Crystal Growth* **78**, 257.

Hartley, F. and Patai. S. (1982) *The Chemistry of the Metal-Carbon Bond* (Wiley, New York) pp. 53–71.

Heckingbottom (1985) in *Molecular Beam Epitaxy and Heterostructures*, eds. L. Chang, L. and Ploog, K. (Martinus Nijhoff, Dordrecht) p. 71.

Heime, K. (1967) *Solid-State Electron.* **10**, 732.

Heime, K. (1989) *InGaAs Field-Effect Transistors* (Research Studies Press LTD., Taunton, England) p. 136.

Hoke, W. E., Lemonias, P. J., Weir, D. G., Hendriks, H. T., and Jackson, G. S. (1991a) *J. Appl. Phys.* **69**, 511.

Hoke, W. E., Lemonias, P. J., Lyman, P. S., Hendriks, H. T., Weir, D., and Colombo, P. (1991b) *J. Cryst. Growth* **111**, 269.

Inada, T., Taka, S., Yamamato, Y. (1981) *J. Appl. Phys.* **52**, 6623.

Iwamoto, T., Mori, K., Mizuta, M. and Kukimoto, H. (1984) *J. Crystal Growth,* **68**, 27.

Kadoya, Y., Sato, A., Kano, H., and Sakaki, H. (1991) *J. Cryst. Growth,* **111**, 280.

Kamiya, T. and Wagner, E. (1977) *J. Appl. Phys.* **48**, 1928.

Kanber, H., Cipolli, R. J., Henderson, W. B., and Whelan, J. M. (1985) *J. Appl. Phys.* **57**, 4732.

Kawaguchi, Y. and Nakashima, K. (1989) *J. Cryst. Growth* **95**, 181.

Kim, S. J. et al. (1987) *IEEE Electron Dev. Lett.* **8**, 518.

Kisker, D. W. and Zawadzki, A. G. (1988) *J. Crystal Growth* **89**, 378.

Kittel, C. and Kroemer, H. (1980) *Thermal Physics*, 2nd ed. (Freeman, New York).

Kohzu, H., Kuzuhara, M., and Takayama, Y. (1983) *J. Appl. Phys.* **54**, 4998.

Krautle, H. (1988) *J. Appl. Phys.* **63**, 4418.

Kreuzer, H. J. (1991) in *Low Temperature Epitaxial Growth of Semiconductors*, ed. Harris, T. (World Scientific, Singapore) p. 1.

Kreyszig, E. (1972) *Advanced Engineering Mathematics*, 3rd ed. (Wiley, New York) p. 726.

Kroger, F. A. (1973) *Chemistry of Imperfect Crystals* 2nd ed. (Elsevier, Amsterdam).

Künzel, H., Knecht, J., Jung, H., Wünstel, K., and Ploog, K. (1982) *Appl. Phys.* **A28**, 167.

Kuphal, E. (1981) *Solid State Electron.* **24**, 69.

Lannoo, M. and Bourgoin, J. (1981) *Point Defects in Semiconductors I*, (Springer, Berlin) p. 203.

Larsen, K. K. and Larsen, A. N. (1990) *Materials Science Forum*, **65-66**, 293.

Lee, B., Bose, S. S., Kim, M. H., Reed, A. D., Stillman, G. E., Wang, W. I., Vina, L., and Colter, P. C. (1989a) *J. Cryst. Growth* **96**, 27.

Lee, B., Bose, S. S., Kim, M. H., Stillman, G. E., and Wang, W. I. (1990) *J. Appl. Phys.* **67**, 255 (1990).

Lee, B., Szafranek, I., Stillman, G. E., Arai, K., Nashimoto, Y., Shimizu, K., Iwata, N., and Sakuma, I. (1989b) *Surf. Interf. Analysis* **14**, 619.

Lievin, J. L., Alexandre, F. and Dubon-Chevaillier, C. (1988) in *Properties of Impurity States in Superlattice Semiconductors* eds., Fong, C. Y., Batra, I. P. and Ciraci, S., (Plenum, New York), p. 19.

Lu, Z. H., Hanna, M. C., Szmyd, D. M., Oh, E. G., and Majerfeld, A. (1990) *Appl. Phys. Lett.* **56**, 177.

Maguire, J., Murray, R., Newman, R. C., Beall, R. B., and Harris, J. J. (1987) *Appl. Phys. Lett.* **50**, 516.

Maier, M. and Selders, J. (1986) *J. Appl. Phys.* **60**, 2783.

Malik, R. J., Nagle, J., Micovic, M., Harris, T. D., Ryan, R. W., and Hopkins, L. C. (1992) *J. Vac. Sci. Technol.* **B10**, 850.

Mathiot, D. and Pfister, J. C. (1982) *J. Physique Letters* **43**, L453.

Mathiot, D. and Pfister, J. C. (1984) *J. Appl. Phys.* **55**, 351.

Mathiot, D. and Pfister, J. C. (1989) *J. Appl. Phys.* **66** 1970.

Miller, D. L. and Asbeck, P. M. (1987) *J. Crystal Growth* **81**, 368.

Morgan, T. N. (1972) in *Proc. 11th Int. Conf. on the Physics of Semiconductors*, **2**, ed. Miasek, M. (Elsevier, Amsterdam) p. 989.

Mottram, A., Peaker, A. R. and Sudlow, P. D. (1971) *J. Electrochem. Soc.*, **118**, 318.

Munoz-Yague, A. and Baceinedo, S. (1982) *J. Electrochem. Soc.: Solid-State Sci. and Technol.* **129**, 2108.

Nakajima, M., Takomori, A., Yokosuka, T., Uchiyama, K., and Abe, T. (1989) *2nd Int. Conf. on Chemical Beam Epitaxy, Houston.*

Nanichi, Y. and Pearson, G. L. (1969) *Solid State Electron.* **12**, 341.

Neave, J. H. and Joyce B. A. (1983) *Appl. Phys. A***31**, 190.

Ogawa, M. and Baba, T. (1985) *Jpn. J. Appl. Phys.* **24**, L572.

Panish, M. B., Hamm, R. A., Hopkins, L. C., and Chu, S. N. G. (1990a) *Appl. Phys. Lett.* **56**, 1137.

Panish, M. B., Hamm, R. A., and Hopkins, L. C. (1990b) *Appl. Phys. Lett.* **56**, 2301.

Panish, M. B. (1990) personal communication.

Pearton, S. J. (1988) *Solid State Phenomena* **1&2**, 247.

Pickar, K. A. (1975), in *Applied Solid State Science Vol. 5*, ed. Wolfe, R. (Academic, New York).

Poole, I., Lee, M. E., Singer, K. E., Frost, J. E. F., Kerr, T. M., Wood, C. E. C., Andrews, D. A., Rothwell, W. J. M., and Davies, G. J. (1988) *J. Appl. Phys.* **63**, 395.

Razeghi, M. (1989) *The MOCVD Challenge* (Adam Hilger, Bristol) p. 10.

Ryssel, H. and Biersack, J. P. (1986) in *Process and Device Modeling*, ed. Enggl, W. L. (North-Holland, Amsterdam).

Ryssel, H. and Ruge, I. (1986) *Ion Implantation*, (Wiley, New York).

Schmitt, F., Su, L. M., Franke, D., and Kaumanns, R. (1984) *IEEE Trans. Electron Dev.* **31**, 1083.

Schubert, E. F., Gilmer, G. H., Monroe, D. P., Kopf, R. F., and Luftman, H. S. (1992) *Phys. Rev. B***46**, 15 078.

Schubert, E. F., Kopf, R. F., Kuo, J. M., Luftman, H. S., and Garbinski, P. A. (1990a) *Appl. Phys. Lett.* **57**, 497.

Schubert, E. F., Kuo, J. M., Kopf, R. F., Jordan, A. S., Luftman, H. S., and Hopkins, L. C. (1990b) *Phys. Rev.* **B42**, 1364.

Schubert, E. F., Kuo, J. M., Kopf, R. F., Luftman, H. S., Hopkins, L. C., and Sauer, N. J. (1990c) *J. Appl. Phys.* **67**, 1969.

Selberherr, S. (1984) *Analysis and Simulation of Semiconductor Devices* (Springer, Wien) Chap. 3.

Sharma, B. L. (1989) *Solid State Technology*, November issue, 113.

Shockley, W. (1961) *Solid State Electron.* **2**, 35.

Singh, S., Zydzik, G. J., and van Uitert, L. G. (1984) unpublished.

Singh, S., Baiocchi, F., Butherus, A. D., Grodkiewicz, W. H., Schwartz, B., Van Uitert, L. G., Yesis, L., and Zydzik, G. J. (1988) *J. Appl. Phys.* **64**, 4194.

Stringfellow, G. B. (1985) in *Semiconductors and Semimetals* **22**, ed. Tsang, W. T., (Academic, New York) p. 209.

Stringfellow, G. B. (1986) *J. Crystal Growth* **75**, 91.

Stringfellow, G. B. (1989) *Organometallic Vapor-Phase Epitaxy* (Academic, New York).

Tan, T. Y. and Gösele, U. (1988) *Appl. Phys. Lett.* **52**, 1240.

Tell, B. (1991) personal communication.

Teramoto, I. (1972) *J. Phys. Chem. Solids* **33**, 2089.

Theis, T. N., Mooney, P. M., and Wright, S. L. (1988) *Phys. Rev. Lett.* **60**, 361.

Tokumitsu, E. (1990) *Jpn. J. Appl. Phys.* **29**, L698.

Tsang, W. T. (1989) in *VLSI Electronics: Microstructure Science* **21**, ed. Einspruch, N. G. (Academic, New York) p. 255.

Tsang, W. T., Choa, F. S., and Ha, N. T. (1991) *J. Electron. Mater.* **20**, 541.

Valco, G. J., Kapoor, V. J., Biedenbender, M. D., and Williams, W. D. (1989) *J. Electrochem. Soc.* **136**, 175.

Vescan, L., Selders, J., Kräutle, H., Kütt, W., and Beneking, H. (1982) *Electron. Lett.* **18**, 534.

Walukiewicz, W. (1989) *Appl. Phys. Lett.* **54**, 2094.

Wang, K. W. (1987) *Appl. Phys. Lett.* **51**, 2127.

Wang, K. W. (1989) US Patent 4,818,721 April 4.

Weimer, P. K. (1962) *Proc. IRE* **50**, 1462.

Weisberg, L. R. and Blanc, J. (1963) *Phys. Rev.* **131**, 1548.

Wilson, R. G., Novak, S. W. and Zavada, J. M. (1988) *Inst. Phys. Conf. Ser.* **91**, 479.

Wood, C. E. C., Desimone, D., Singer, K. E., and Wicks (1982) *J. Appl. Phys.* **53**, 4230.

Woodhouse, J. D., Donnelly, J. P., Nitishiu, P. M., Owens, E. B., and Ryan, J. L. (1984) *Solid State Electron.* **27**, 677.

Wu, M. C., Su, Y. K., Cheng, K. Y. and Chang, C. Y. (1988) *Solid State Electron.* **31**, 251.

Yamada, T., Tokumitsu, E., Saito, K., Akatsuka, T., Miyauchi, M., Konagai, M. and Takahashi, K. (1989) *J. Crystal Growth* **95**, 145.

Ziegler, J. F., Biersack, J. P. and Cuomo, G., TRIM computer code, Copyright 1991, IBM Corporation.

Ziegler, J. F., Biersack, J. P., and Littmark, U. (1985) *The Stopping Range of Ions in Solids* **1**, (Pergamon, New York).

Chapter 8

Arnold, N., Schmitt, R., and Heime, K. (1984) *J. Phys. D: Appl. Phys.* **17**, 443.

Baraff, G. A. and Schlüter, M. (1985) *Phys. Rev. Lett.* **55**, 1327.

Bardeen, J. (1947) *J. Phys. Rev.* **71**, 717.

Beall, R. B., Clegg, J. B., Castagné, J., Harris, J. J., Murray, R. and Newman, R. C. (1989) *Semicond. Sci., Technol.* **4**, 1171.

Boltaks, B. I., Ksendzov, S., and Rembeza, S. I. (1969) *Sov. Phys. Solid State* **10**, 2186.

Brozel, M. R., Foulkes, E. J., and Tuck, B. (1982) *Phys. Stat. Sol. (a)* **72**, K159.

Brozel, M. R., Tuck, B., and Foulkes, E. J. (1981) *Electron. Lett.* **17**, 532.

Casey, H. C., Jr., Panish, M. B., and Wolfstirn, K. B., (1971) *J. Phys. Chem. Solids* **32**, 571.

Chang, L. L. and Casey, H. C. (1964) *Solid State Electron.* **7**, 481.

Chang, L. L. and Pearson, G. L. (1964a) *J. Appl. Phys.* **35**, 374.

Chang, L. L. and Pearson, G. L. (1964b) *J. Appl. Phys.* **35**, 1960.

Chevrier, J., Armand, M., Huber, A. M., and Linh, N. T. (1980) *J. Electron. Mater.* **9**, 745.

Chiang, T.-C., Ludeke, R., Aono, M., Landgren, G., Himpsel, F. J., and Eastman, D. E. (1983) *Phys. Rev. B***27**, 4770.

Chin, A. K., Camlibel, I., Dutt, B. V., Swaminathan, V., Bonner, W. A., and Ballman, A. A. (1983a) *Appl. Phys. Lett.* **42**, 901.

Chin, A. K., Camlibel, I., Sheng, T. T., and Bonner, W. A. (1983b) *Appl. Phys. Lett.* **43**, 495.

Cunnell, F. A. and Gooch, C. H. (1960) *J. Phys. Chem. Sol.* **15**, 127.

Deal, M. D. and Stevenson, D. A. (1986) *J. Appl. Phys.* **59**, 2398.

Deppe, D. G. and Holonyak, Jr., N. (1988) *J Appl. Phys.* **64**, R93.

Deppe, D. G., Holonyak, Jr. N., Kish, F. A., and Baker, J. E. (1987) *Appl. Phys. Lett.* **50**, 998.

Devine, R. L. S., Foxon, C. T., Joyce, B. A., Clegg, J. B., and Govers, J. P. (1987) *Appl. Phys. A***44**, 195.

Dutt, B. V., Chin, A. K., Camlibel, I., and Bonner, W. A. (1984) *J. Appl. Phys.* **56**, 1630.

Einstein, A. (1905) *Ann. Phys.* **17**, 549.

Einstein, A. (1906) *Ann. Phys.* **19**, 371.

Fane, R. W. and Goss, A. J. (1963) *Solid State Electron.* **6**, 383.

Frank, F. C. and Turnbull, D. (1956) *Phys. Rev.* **104**, 617.

Frieser, R. G. (1965) *J. Electrochem. Soc.* **112**, 697.

Gibbon, C. F. and Ketchow, D. R. (1971) *J. Electrochem. Soc.* **118**, 975.

Goldstein, B. (1961) *Phys. Rev.* **121**, 1305.

Goldstein, B. and Keller, H. (1961) *J. Appl. Phys.* **32**, 1180.

Greiner, M. E. and Gibbons, J. F. (1984) *Appl. Phys. Lett.* **44**, 750.

Hamm, R. A., Panish, M. B., Nottenburg, R. N., Chen, Y. K., and Humphrey, D. A. (1989) *Appl. Phys. Lett.* **54**, 2586.

Harris, J. J., Ashenford, D. E., Foxon, C. T., Dobson, P. J., and Joyce, B. A. (1984) *Appl. Phys. A***33**, 87.

Harris, J. J., Joyce, B. A., Gowers, J. P., and Neave, J. H. (1982) *Appl. Phys. A***28**, 63.

Heine, V. (1965) *Phys. Rev.* **138**, A1689.

Hofmann, S. and Erlewein, J. (1979) *Surf. Sci.* **77**, 591.

Holmes, D. E., Wilson, R. G., and Yu, P. W. (1981) *J. Appl. Phys.* **52**, 3396.

Huber, A. M., Morillot, G., Bonnet, M., Merenda, P., and Bessoneau, G. (1982) *Appl. Phys. Lett.* **41**, 638.

Huijser, A., van Laar, J., and van Rooy, T. L. (1977) *Surf. Sci.* **62**, 472.

Ihm, J., Chadi, D. J., and Joannopoulos, J. D. (1983) *Phys. Rev. B***27**, 5119.

Ilegems, M. (1977) *J. Appl. Phys.* **48**, 1278.

Ilegems, M. and O'Mara, W. C. (1972) *J. Appl. Phys.* **43**, 1190.

Kadhim, M. A. and Tuck, B. (1972) *J. Mater. Sci.* **7**, 68.

Kahng, D. (1964) *Bell. Syst. Techn. J.* **42**, 215.

Kaliski, R. W., Nam, D. W., Deppe, D. G., Holonyak, Jr. N., Hsieh, K. C., and Burnham, R. D. (1987) *J. Appl. Phys.* **62**, 998.

Karelina, T. A., Lavrishchev, T. T., Prikhod'ko, G. L., and Khludov, S. S. (1974) *Inorganic Mater.* **10**, 194.

Kato, H., Yokozawa, M., Kohara, R., Okabayashi, Y., and Takayanagi, S. (1969) *Solid State Electron.* **12**, 137.

Kendall, D. L. (1968) in *Semiconductors and Semimetals*, eds. Willardson, R. K. and Beer, A. C. (Academic, New York) p. 163.

Kendall, D. L. and Huggins, R. A. (1969) *J. Appl. Phys.* **40**, 2750.

Khludkov, S. S. and Lavrishchev (1976) *Inorganic Mater.* **12**, 972.

Kobayashi, N., Makimoto, T., and Horikoshi, Y. (1987) *Appl. Phys. Lett.* **50**, 1435.

Kundukhov, R. M., Metreveli, S. G., and Siukaev, N. v. (1967) *Sov. Phys.-Semicond.* **1**, 765.

Laidig, W. D., Holonyak, Jr., N., Camvas, M. D., Hess, K., Coleman, J. J., Dapkus, P. D., and Bardeen, J. (1981) *Appl. Phys. Lett.* **38**, 776.

Laidig, W. D., Holonyak, Jr., N., Coleman, J. J. and Dapkus, P. D. (1982) *J. Electron. Mater.* **11**, 1.

Linh, N. T., Huber, A. M., Etienne, P., Morillot, G., Duchemin, P., and Bonnet, M. (1980) in *Semi-Insulating III–V Materials*, ed. Rees, G. J. (Shiva, Nantwich) p. 206.

Longini, R. L. and Greene (1956) *Phys. Rev.* **102**, 992.

Luther, L. C. and Wolfstirn, K. B. (1973) *J. Electron. Mater.* **2**, 375.

Manning, J. R. (1968) *Diffusion Kinetics for Atoms in Crystals* (van Nostrand, Princeton) pp. 95, 166.

Matino, H. (1974) *Solid State Electron.* **17**, 35.

McLevige, W. V., Vaidyanathan, K. V., Streetman, B. G., Ilegems, M., Comas, J., and Plew, L. (1978) *Appl. Phys. Lett.* **33**, 127.

Meehan, K., Holonyak, Jr., N., Brown, J. M., Nixon, M. A., Gavrilovic, P., and Burnham, R. D. (1984) *Appl. Phys. Lett.* **45**, 549.

Mei, P., Yoon, H. W., Venkatesan, T., Schwarz, S. A., and Harbison, J. P. (1987) *Appl. Phys. Lett.* **50**, 1823.

Mott, N. F. (1938) *Proc. Cambridge, Philos. Soc.* **34**, 568.

Nanichi, Y. and Pearson, G. L. (1969) *Solid State Electron.* **12**, 341.

Oberstar, J. D., Streetman, B. G., Baker, J. E., and Williams, P. (1981) *J. Electrochem. Soc.* **128**, 1814.

Palfrey, H. D., Brown, M., and Willoughby, A. F. W. (1981) *J. Electrochem. Soc.* **128**, 2224.

Palfrey, H. D., Brown, M., and Willoughby, A. F. W. (1983) *J. Electron. Mater.* **12**, 863.

Panish, M. B. (1973) *J. Phys.* **44**, 2659.

Panish, M. B., Hamm, R. A., and Hopkins, L. C. (1990) *Appl. Phys. Lett.* **56**, 2301.

Panish, M. B., Hamm, R. A., Ritter, D., Luftman, H. S., and Cotell, C. M. (1991) *J. Crystal Growth* **112**, 343.

Pankove, J. I. and Johnson, N. M. (1991) ed., *Hydrogen in Semiconductors* (Academic, New York).

Phillips, J. C. (1973) *Bonds and Bands in Semiconductors* (Academic, New York) p. 222.

Prince, F. C., Oren, M., and Lam, M. (1986) *Appl. Phys. Lett.* **48**, 546.

Rekalova, G. I., Kebe, U., Persiyanov, T. V., Krymov, V. M., and Krymova, E. D. (1971) *Sov. Phys.-Semicond.* **5**, 134.

Rekalova, G. I., Shakov, A. A., and Gavrushko, A. A. (1969) *Sov. Phys.-Semicond.* **2**, 1452.

Risken, H. (1984) *The Fokker–Planck Equation* (Springer, Berlin).

Schillmann, E. (1962) in *Compound Semiconductors*, eds. Willardson, R. K. and Goering, H. L. (Reinhold, New York) p. 358.

Schneider, M. and Nebauer, E. (1975) *Phys. Stat. Sol. (a)* **32**, 333.

Schottky, W. (1940) *Naturwissenschaften* **26**, 843.

Schottky, W. (1940) *Z. Phys.* **41**, 570.

Schubert, E. F., Kuo, J. M., Kopf, R. F., Jordan, A. S., Luftman, H. S., and Hopkins, L. C. (1990a) *Phys. Rev. B* **42**, 1364.

Schubert, E. F., Kuo, J. M., Kopf, R. F., Luftman, H. S., Hopkins, L. C., and Sauer, N. J. (1990b) *J. Appl. Phys.* **67**, 1969.

Schubert, E. F., Stark, J. B., Chiu, T. H., and Tell, B. (1988) *Appl. Phys. Lett.* **53**, 293.

Schubert, E. F., Tu, C. W., Kopf, R. F., Kuo, J. M., and Lunardi, L. M. (1989) *Appl. Phys. Lett.* **54**, 2592.

Shaw, D. (1975) *Phys. Stat. Sol. (B)* **72**, 11.

Shishiyanu, F. S. and Boltaks, B. I. (1966) *Sov. Phys.-Solid State* **8**, 1053.

Shishiyanu, F. S., Gheorghiu, V. G., and Palazov, S. K. (1977) *Phys. Stat. Sol. A***40**, 29.

Shockley, W. and Last, J. T. (1957) *Phys. Rev.* **107**, 392.

Showan, S. R. and Shaw, D. (1969) *Phys. Stat. Sol.* **32**, 97.

Small, M. B., Potemski, R. M., Reuter, W., and Ghez, R. (1982) *Appl. Phys. Lett.* **41**, 1068.

Spicer, W. E., Gregory, P. E., Chye, P. W., Babalola, I. A., and Sukegawa, T. (1975) *Appl. Phys. Lett.* **27**, 617.

Spicer, W. E., Lindau, I., Skeath, P., Su, C. Y., and Chye, P. W. (1980) *Phys. Rev. Lett.* **44**, 420.

Sze, S. M. and Wei, L. Y. (1961) *Phys. Rev.* **124**, 84.

Tan, T. Y. and Gösele, U. (1987) *J. Appl. Phys.* **61**, 1841.

Tan, T. Y. and Gösele, U. (1988) *Appl. Phys. Lett.* **52**, 1240.

Tuck, B. (1988) *Atomic Diffusion in III–V Semiconductors* (Adam Hilger, Bristol).

Tuck, B. and Adegboyega, G. A. (1979) *J. Phys. D: Appl. Phys.* **12**, 1895.

Tuck, B. and Badawi, M. H. (1978) *J. Phys. D: Appl. Phys.* **11**, 2541.

Tuck, B. and Hooper, A. (1975) *J. Phys. D.: Appl. Phys.* **8**, 1806.

Tuck, B. and Jay, P. R. (1977a) *J. Phys. D: Appl. Phys.* **10**, 1315.

Tuck, B. and Jay, P. R. (1977b) *J. Phys. D.: Appl. Phys.* **10**, 2089.

Tuck, B. and Zahari, M. D. (1977a) in *Gallium Arsenide and Related Compounds 1976*, *Inst. Phys. Conf. Ser.* **33a**, p. 177.

Tuck, B. and Zahari, M. D. (1977b) *J. Phys. D.: Appl. Phys.* **10**, 2473.

Uskov, V. A. (1974) *Sov. Phys.-Semicond.* **8**, 1573.

Vasudev, P. K., Wilson, R. G., and Evans, C. A. (1980) *Appl. Phys. Lett.* **37**, 837.

Vieland, L. J. (1961) *J. Phys. Chem. Solids* **21**, 318.

Watanabe, K., Matsuoka, Y., Imamura, Y., and Ito, K. (1982) in *Gallium Arsenide and Related Compounds 1981*, *Inst. Phys. Conf. Ser.* **63** p. 1.

Weisberg, L. R. and Blanc, J. (1963) *Phys. Rev.* **131**, 1548.

Willoughby, A. F. W. (1983) in *Defects in Semiconductors II*, eds. Mahajan, S. and Corbett, J. W. (Elsevier, New York) p. 237.

Wilson, R. B. (1982) in *Gallium Arsenide and Related Compounds 1981, Inst. Phys. Conf. Ser. 63*, p. 1.

Wolfe, C. M. and Nichols, K. H. (1977) *Appl. Phys. Lett.* **31**, 356.

Wolfe, C. M. and Stillman, G. E. (1973) *Solid State Commun.* **12**, 283.

Wood, C. E. C. and Joyce, B. A. (1978) *J. Appl. Phys.* **49**, 4854.

Yamada, M., Tien, P. K., Martin, R. J., Nahony, R. E., and Ballman, A. A. (1983) *Appl. Phys. Lett.* **43**, 594.

Yamazaki, H., Kawasaki, Y., Fujimoto, M., and Kudo, K. (1975) *Jpn. J. Appl. Phys.* **14**, 717.

Young, A. B. and Pearson, G. L. (1970) *J. Phys. Chem. Sol.* **31**, 517.

Zschauer, K.-H. and Vogel, A. (1970) in *Proc. 3rd Int. Symp. GaAs*, (Institute of Physics, London), p. 100.

Chapter 9

Ainslie, N. G., Blum, S. E., and Woods, J. F. (1962) *J. Appl. Phys.* **33**, 2391.

Allen, J. W. (1960) *Nature* **187**, 403.

Baba, T., Mizuta, M., Fujisawa, T., Yoshino, J., and Kukimoto, H. (1989) *Jpn. J. Appl. Phys.* **28**, L891.

Baba, T., Mizutani, T., and Ogawa, M. (1983) *Jpn. J. Appl. Phys.* **22**, L627.

Bachelet, G. B., Schlüter, M., and Baraff, G. A. (1983) *Phys. Rev.* **B27**, 2545.

Bachrach, R. Z. and Bringans, R. D. (1983) *J. Vac. Sci. Technol.* **B1**, 142.

Baraff, G. A. (1992) in *Deep Centers in Semiconductors*, 2nd ed., ed. Pantelides, S. T. (Gordon and Breach, Philadelphia) p. 547.

Baraff, G. A., Lannoo, M., and Schlüter, M. (1988) *Phys. Rev.* **B38**, 6003.

Bergman, K., Omling, P., Samuelson, L., and Grimmeiss, H. G. (1988) in *Semi-Insulating III-V Materials*, eds. Grossmann, G. and Ledebo, L. (Adam Hilger,

Bristol).

Boenig, H. V. (1982) *Plasma Science and Technology* (Cornell University Press, Ithaca).

Briddon, P. and Jones, R. (1989) *Inst. Phys. Conf. Ser.* **95**, 459.

Calawa, A. R. (1978) *Appl. Phys. Lett.* **33**, 1020.

Chadi, D. J. (1992) *Phys. Rev. B***46**, 6777.

Chadi, D. J. and Chang, K. J. (1988a) *Phys. Rev. Lett.* **60**, 2187.

Chadi, D. J. and Chang, K. J. (1988b) *Phys. Rev. Lett.* **61**, 873.

Chadi, D. J. and Chang, K. J. (1989) *Phys. Rev. B***39**, 10063.

Chadi, D. J. and Zhang, S. B. (1991) *J. Electron Mater.* **20**, 55.

Chand, N., Henderson, T., Clem, J., Masselink, W. T., Fischer, R., Chang, Y. C., and Morkoc, H. (1984) *Phys. Rev. B***30**, 4481.

Chen, Y. K., Temkin, H., Tanbun-Ek, T., Logan, R. A., and Nottenburg, R. A. (1989) *IEEE Electron Dev. Lett.* **10**, 162.

Chevallier, J., Clerjaud, B., and Pajot, B. (1991) in *Hydrogen in Semiconductors*, eds. Pankove, J. I. and Johnson, N. M. (Academic, New York).

Chevallier, J., Jalil, A., Azoulay, R., and Mircea, A. (1986) *Materials Science Forum* **10-12**, 591.

Chevallier, J., Pajot, B., Jalil, A., Mostefaoui, R., Rahbi, R., and Boissy, M. C. (1988) *Mat. Res. Soc. Symp. Proc.* **104**, 337.

Corbett, J. W. (1971) in *Ion Implantation*, eds. Eisen, F. H. and Chadderton, L. T. (Gordon and Breach, London).

Cotton, F. A. (1971) *Chemical Applications of Group Theory* 2nd ed. (Wiley, New York).

Dabrowski, J. and Scheffler, M. (1988) *Phys. Rev. Lett.* **60**, 2183.

Dabrowski, J. and Scheffler, M. (1989) *Phys. Rev. B***40**, 10391.

Dean, P. (1986) in *Deep Centers in Semiconductors*, ed Pantelides, S. T. (Gordon and Breach, New York) p. 185.

Devenaud, B. and Favennec (1979) *Inst. Phys. Conf. Ser.* **45**, 292.

Dischler, B., Fuchs, F., and Kaufmann, U. (1986) *Appl. Phys. Lett.* **48**, 1282.

Dutta, N. K. (1989) *AT&T Techn. J.* **68**, 5.

Ennen, H. and Schneider, J. (1985) in *Defects in Semiconductors*, eds. Kimerling, L. C. and Parsey, J. M., Jr. (TMS-AIME) p. 115.

Ennen, H., Kaufmann, U., Pomrenke, G., Schneider, J., Windscheif, J. and Axmann, A. (1983a) *J. Cryst. Growth* **64**, 165.

Ennen, H., Schneider, J., Pomrenke, G., and Axmann, A. (1983b) *Appl. Phys. Lett.* **43**, 943.

Favennec, P. N. (1976) *J. Appl. Phys.* **47**, 2532.

Fertin, J. L., Lebailly, J., and Deyris, E. (1967) *Gallium Arsenide and Related Compounds 1966, Inst. Phys. Conf. Ser.* **3**, 46.

Foyt, A. G., Lindley, W. T., Wolfe, C. M., and Donnelly, J. P. (1969) *Solid State Electron.* **12**, 209.

Fujisawa, T., Yoshino, J., and Kukimoto, H. (1990) *Jpn. J. Appl. Phys.* **29**, L388.

Garmire, E., Stoll, H., Yariv, A., and Hunsperger, R. G. (1972) *Appl. Phys. Lett.* **21**, 87.

Gershenzon, M. and Mikulyak, R. M. (1962) *Solid State Electron.* **5**, 313.

Gooch, C. H., Hilsum, C., and Holeman, B. R. (1961) *J. Appl. Phys.* **32**, 2069.

Guenter, K. G. (1958) *Z. Naturforschung* **13a**, 1081.

Haisty, R. W., Mehal, E. W., and Stratton, R. (1962) *J. Phys. Chem. Solids* **23**, 829.

Hofmann, D., Müller, G., and Streckfuss, N. (1989) *Appl. Phys.* **A48**, 315.

Holmes, D. E., Chen, R. T., Elliot, K. R. and Kirkpatrick, C. G. (1982) *Appl. Phys. Lett.* **40**, 46.

Huber, A. M., Linh, N. T., Valladon, M., Debrun, J. L., Martin, G. M., Mitonneau, A., and Mircea, A. (1979) *J. Appl. Phys.* **50**, 4022.

Iseler, G. W. (1979) *Inst. Phys. Conf. Ser.* **45**, 144.

Ishibashi, T., Tarucha, S., and Okamoto, H. (1982) *Jpn. J. Appl. Phys.* **21**, L476.

Jackson, G. S., Hall, H. O., Guido, L. J., Plano, W. E., Pan, N., Holonyak, Jr. N., and Stillman, G. E. (1988) *Appl. Phys. Lett.* **52**, 691.

Jaros, M. (1982) *Deep Levels in Semiconductors* (Adam Hilger, Bristol).

Kainosho, K., Shimakura, H., Kanazawa, T., Inoue, T., and Oda, O. (1990) *Inst. Phys. Conf. Ser.* **106**, 25.

Kaminska, M. (1984) *Proc. 17th Int. Conf. Phys. Semicond.* (Springer, New York) p. 741.

Kaminska, M., Skowronski, M., Lagowski, J., Parsey, J. M., and Gatos, H. C. (1983) *Appl. Phys. Lett.* **43**, 302.

Kaminska, M., Skowronski, M., and Kuszko, W. (1985) *Phys. Rev. Lett.* **55**, 2204.

Kaniewska, M. and Kaniewski, J. (1985) *Solid State Commun.* **53**, 485.

Kasatkin, V. A., Kesamanly, F. P., Makarenko, V. K., Masterov, V. F., and Samorukov, B. E. (1981a) *Sov. Phys. Semicond.* **14**, 1092.

Kasatkin, V. A., Kesamanly, F. P., and Samorukov, B. E. (1981b) *Sov. Phys. Semicond.* **15**, 352.

Kaufman, U. (1989) in *Advances in Solid State Physics,* **29**, (Vieweg, Braunschweig).

Kazmierski, C., Theys, B., Rose, B., Mircea, A., Jalil, A., and Chevallier (1989) *J. Electron. Lett.* **25**, 1933.

Khachaturyan, K. A., Awshalom, D. D., Rozen, J. R., and Weber, E. R. (1989) *Phys. Rev. Lett.* **63**, 1311.

Kim, H. K., Hwang, J. S., Noh, S. K., and Chung, C. H. (1986) *Jpn. J. Appl. Phys.* **25**, L888.

Kinchin, G. H. and Pease, R. S. (1955) *Rept. Progr. Phys.* **18**, 1.

Klein, P. B., Herry, R. L., Kennedy, T. A., and Wilsey, N. D. (1986) in *Defects in Semicond.*, ed. von Bardeleben, H. J., *Materials Science Forum* **10-12**, 1259.

Klein, P. B., Moore, F. G., and Dietrich, H. B. (1991) *Appl. Phys. Lett.* **58**, 502.

Kopev, P. S., Ivanov, S. V., Yegorov, A. Yu., and Uglov, D. Yu (1989) *J. Cryst. Growth* **96**, 533.

Künzel, H., Fischer, A., Knecht, J., and Ploog, K. (1983) *Appl. Phys.* A**32**, 69.

Künzel, H., Ploog, K., Wünstel, K., and Zhou, B. L. (1984) *J. Electron. Mater.* **13**, 281.

Lang, D. V. (1986) in *Deep Centers in Semiconductors*, ed. Pantelides, S. T. (Gordon and Breach, New York).

Lang, D. V. and Logan, R. A. (1975) *J. Electron. Mat.* **4**, 1053.

Lang, D. V. and Logan, R. A. (1977) *Phys. Rev. Lett.* **39**, 635.

Lang, D. V., Logan, R. A., and Jaros, M. (1979) *Phys. Rev. B***19**, 1015.

Lindquist, P. F. (1977) *J. Appl. Phys.* **48**, 1262.

Long, J. A., Riggs, V. G., and Johnston, Jr. W. D. (1984) *J. Cryst. Growth* **69**, 10.

Makram-Ebeid, S., Gautard, D., Devillard, P., and Martin, G. M. (1982) *Appl. Phys. Lett.* **40**, 161.

Martin, G. M. and Makram-Ebeid, S. (1986) in *Deep Centers in Semiconductors*, ed. Pantelides, S. T. (Gordon and Breach, New York).

Martin, G. M., Mitonneau, A., and Mircea, A. (1977) *Electron. Letters* **13**, 191.

Matsumura, T., Emori, H., Terashima, K., and Fukuda, T. (1983) *Jpn. J. Appl. Phys.* **22**, L154.

McCluskey, F. P., Pfeiffer, L., West, K. W., Lopata, J., Lamont–Schnoes, M., Harris, T. D., Pearton, S. J., and Dautremont–Smith, W. C. (1989) *Appl. Phys. Lett.* **54**, 1769.

Miller, M. D., Olsen, G. H., and Ettenberg (1977) *Appl. Phys. Lett.* **31**, 538.

Miniscalco, W. J. (1991) *J. Lightwave Technol.* **9**, 234.

Miyazawa, S. and Nanishi, Y. (1983) *Jpn. J. Appl. Phys.* **22**, Supplement 22-1, 419.

Mizuta, M., and Kitano, T. (1987) *Appl. Phys. Lett.* **52**, 126.

Mizuta, M. and Mori, K. (1988) *Phys. Rev. B***37**, 1043.

Mizuta, M., Tachikawa, M., Kukimoto, H., and Minomura, S. (1985) *Jpn. J. Appl. Phys.* **24**, L143.

Monberg, E. M., Brown, H., Chu, S. N. G., and Parsey, J. M. (1988) *Proc. 5th Conf. on Semi-insulating III–V Materials* (Malmö, Sweden).

Monberg, E. M., Gault, W. A., Dominguez, F., and Simchock, F. (1987) *J. Cryst. Growth* **83**, 174.

Mooney, P. M. (1990) *J. Appl. Phys.* **67**, R1.

Mooney, P. M., Theis, T. N., and Calleja, E. (1991) *J. Electron. Mater.* **20**, 23.

Müller, G. (1987) in *Advanced Crystal Growth*, eds. Dryburgh, P. M., Cockayne, B., and Barraclough, K. G. (Prentice Hall, Englewood Cliffs) p. 144.

Mullin, J. B., Royle, A., and Straughan, B. W. (1971) *Inst. Phys. Conf. Ser.* **9**, 41.

Muszalski, J., Babinski, A., Korona, K. P., Kaminska, E., Piotrowska, A., Kaminska, M., Weber, E. R. (1991) *Acta Physica Polonica A***80**, 413.

Nandhra, P. S., Newman, R. C., Murray, R., Pajot, B., Chevallier, J., Beall, R. B., and Harris, J. J. (1988) *Semicond. Sci. Technol.* **3**, 356.

Nelson, R. J. (1977) *Appl. Phys. Lett.* **31**, 351.

Nissen, M. K., Steiner, T., Beckett, D. J. S., and Thewalt, M. L. W. (1990) *Phys. Rev. Lett.* **65**, 2282.

Nissen, M. K., Villemaire, A., and Thewalt, M. L. W. (1991) *Phys. Rev. Lett.* **67**, 112.

Oberstar, J. D., Streetman, B. G., Baker J. E. and Williams, P. (1981) *J. Electrochem. Soc.* **128**, 1814.

Oda, O., Kainosho, K., Kohiro, K., Hirano, R., Shimakura, H., Inoue, T., Yamamoto, H., and Fukui, T. (1991) *Proc. 3rd Int. Conf. InP and Related Materials* (Cardiff, United Kingdom).

Oda, O., Katagiri, K., Shinohara, K., Katsura, S., Takahashi, Y., Kainosho, K., Kohiro, K., and Hiranoi, R. (1990) in *Semiconductors and Semimetals*, eds. Willardson, R. K., and Beer, A. C. (Academic Press, Boston).

Orito, F., Tsujikawa, Y., and Tajima, M. (1984) *J. Appl. Phys.* **55**, 1119.

Otsubo, M., Segawa, K., and Miki, H. (1973) *Jpn. J. Appl. Phys.* **12**, 797.

Ozeki, M., Komeno, J., and Ohkawa, S. (1979) *Fujitsu Sci. Tech. J.* **15**, 83.

Pajot, B., Chevallier, J., Chaumont, J., and Azoulay, R. (1988a) *Mat. Res. Soc. Symp. Proc.* **104**, 345.

Pajot, B., Newman, R. C., Murray, R., Jalil, A., Chevallier, J., and Azoulay, R. (1988b) *Phys. Rev. B***37**, 4188.

Pankove, J. I., Zanzucchi, P. J., Magee, C. W., and Lucovsky, G. (1985) *Appl. Phys. Lett.* **46**, 421.

Pao, Y.-C., Liu, D., Lee, W. S., and Harris, J. S. (1986) *Appl. Phys. Lett.* **48**, 1291.

Peale, R. E., Mochizuki, Y., Sun, H., and Watkins, G. D. (1992a) *Phys. Rev. B***45**, 5933.

Peale, R. E., Sun, H., and Watkins, G. D. (1992b) *Phys. Rev. B***45**, 3353.

Pearton, S. J. (1990) *Materials Science Reports* **4**, 313.

Pearton, S. J., Corbett, J. W., and Shi, T. S. (1987a) *Appl. Phys. A***43**, 153.

Pearton, S. J., Dautremont-Smith, W. C., Chevallier, J., Tu, C. W., and Cummings, K. D. (1986) *J. Appl. Phys.* **59**, 2821.

Pearton, S. J., Dautremont-Smith, W. C., Lapato, J., Tu, C. W., and Abernathy, C. R. (1987b) *Phys. Rev. B***36**, 4260.

Samuelson, L., Omling, P., Titze, H., and Grimmeiss, H. G. (1981) *J. Cryst. Growth* **55**, 164.

Schubert, E. F. and Ploog, K. (1984) *Phys. Rev. B***30**, 7021.

Schubert, E. F., Fischer, A., and Ploog, K. (1985) *Phys. Rev. B***31**, 7937.

Seager, C. H. (1991) in *Hydrogen in Semiconductors*, eds. Pankove, J. I. and Johnson, N. M. (Academic, New York) p. 17.

Seager, C. H. and Ginley, D. S., *Sandia Report, SAND82-1701*, p. 19.

Shimakura, H., Kainosho, K., Inoue, T., Yamamoto, H., and Oda, O. (1989) *Proc. Int. Conf. Sci. Technol. Defect Control in Semiconductors* (Yokohoma, Japan).

Spaeth, J.-M., Krambrock, K. and Hofmann, D. M. (1990) in *20th Int. Conf. Phys. Semicon. Vol. 1*, ed. Anastassakis, E. M. and Joannopoulus, J. D. (World Scientific, Singapore) p. 441.

Speier, P., Schemmel, G., and Kuebart, W. (1986) *Electron. Lett.* **22**, 1217.

Stavola, M., Pearton, S. J., Lopata, J., Abernathy, C. R., and Bergman, K., (1989) *Phys. Rev. B***39**, 8051.

Stavola, M., Kozuch, D. M., Abernathy, C. R., and Hobson, W. S. (1991) *Proc. E. MRS Fall Meeting* (MRS, Pittsburgh).

Steeples, K., Dearnaley, G., and Stoneham A. M. (1980) *Appl. Phys. Lett.* **36**, 981.

Street, R. A. and Mott, N. F. (1975) *Phys. Rev. Lett.* **35**, 1293.

Sugiura, H., Kawashima, M., and Horikoshi, Y. (1986) *Jpn. J. Appl. Phys.* **25**, 950.

Tachikawa, M., Mizuta, M., and Kukimoto, H. (1984) *Jpn J. Appl. Phys.* **23**, 1594.

Taguchi, A., Taniguchi, M., and Takahei, K. (1992) *Appl. Phys. Lett.* **60**, 965.

Tajima, M. (1987) *Jpn. J. Appl. Phys.* **26**, L885.

Tajima, M., Iino, T., and Ishida, K. (1987) *Jpn. J. Appl. Phys.* **26**, L1060.

Takikawa, M. and Ozeki, O. (1985) *Jpn. J. Appl. Phys.* **24**, 303.

Tell, B., Brown Goebler K. F., Cunningham, J. E., Chiu, T. H., and Jan, W. Y. (1990) *Appl. Phys. Lett.* **56**, 2657.

Temkin, H., Chen, Y. K., Garbinski, P., Tanbun-Ek, T., and Logan, R. A. (1988) *Appl. Phys. Lett.* **53**, 2534.

Theis, T. N (1990) *Proceedings of the MRS* (Boston).

Tu, L. W., Wang, Y. H., Schubert, E. F., Weir, B. E., Zydzik, G. J., and Cho, A. Y. (1992) unpublished.

Van der Ziel, J. P., Tsang, W. T., Logan, R. A., and Augustyniak, W. M. (1981) *Appl. Phys. Lett.* **39**, 376.

Vincent, G. and Bois, D. (1978) *Solid State Commun.* **27**, 431.

Von Neida, A. E., Pearton, S. J., Hobson, W. S., and Abernathy, C. R. (1989) *Appl. Phys. Lett.* **54**, 1540.

Wagner, E. E., Mars, D. E., Hom, G., and Stringfellow, G. B. (1980) *J. Appl. Phys.* **51**, 5434.

Watanabe, M. O. and Maeda, H. (1984) *Jpn. J. Appl. Phys.* **23**, L734.

Watanabe, M. O., Morizuka, K., Mashita, M., Ashizawa, Y., and Zohta, Y. (1984) *Jpn. J. Appl. Phys.* **23**, L103.

Weber, E. R. (1985) in *Microscopic Identification of Electronic Defects in Semiconductors, Proc. Mat. Res. Soc.*, (MRS, Pittsburgh).

Weber, E. R. (1991) personal communication.

Wesch, W., Wendler, E., Götz, G., and Kekelidse, N. P. (1989) *J. Appl. Phys.* **65**, 519.

Wolford, D. J. Editor (1991) *J. Electron. Mater.* **20** pp. 1-70.

Zakharenkov, L. F., Kasatkin, V. A., Kesamanly, F. P., Samorukov, B. E., and Sokolova, M. A. (1981) *Sov. Phys. Semicond.* **15**, 946.

Zavada, J. M., Jenkinson, H. A., Sarkis, R. G., and Wilson, R. G. (1985) *J. Appl. Phys.* **58**, 3731.

Zavada, J. M., Wilson, R. G., Novak, S. W., Von Neida, A. R., and Pearton, S. J. (1988) *Mat. Res. Soc. Symp. Proc.* **104**, 331.

Zhang, S. B. and Chadi, D. J. (1990) *Phys. Rev. B***42**, 7174.

Ziegler, J. F., Biersack, J. P., and Cuomo, G. (1991) TRIM computer code, copyright, IBM Corporation.

Chapter 10

Alexandre, F., J., Goldstein, L., Leroux, G., Joncour, M. C., Thibierge, H., and Rao, E. V. K. (1985) *Vac. Sci. Technol. B***3**, 950.

Ando, T. (1982a) *J. Phys. Soc. Jpn.* **51**, 3893.

Ando, T. (1982b) *J. Phys. Soc. Jpn.* **51**, 3900.

Ando, T. (1985) *J. Phys. Soc. Jpn.* **54**, 2676.

Ando, T. and Mori, S. (1979) *J. Phys. Soc. Japan* **47**, 1518.

Bastard, G. (1981) *Phys. Rev. B***24**, 4714.

Bastard, G., Mendez, E. E., Chang, L. L., and Esaki, L. (1982) *Phys. Rev. B***26**, 1974.

Brooks, H. in *Advance in Electronics and Electron Physics*, ed. Marton, L. (Academic, New York, 1955) p. 85, p. 156.

Casey, H. C. and Panish, M. B. (1978) in *Heterostructure Lasers* (Academic, New York) Part B, p. 16.

Chattopadhyay, D. and Queisser, H. J. (1981) *Rev. Mod. Phys.* **53**, 745.

Chaudhuri, S. and Bajaj, K. K. (1984) *Phys. Rev. B***29**, 1803.

Cho, N. M., Kim, D. J., Madhukar, A., Newman, P. G., Smith, D. D., Aucoin, T., and Iafrate, G. J. (1988) *Appl. Phys. Lett.* **52**, 2037.

Conwell, E. and Weisskopf, V. F. (1950) *Phys. Rev.* **77**, 388.

Dingle, R., Stormer, H. L., Gossard, A. C., and Wiegmann, W. (1978) *Appl. Phys. Lett.* **33**, 665.

Döhler, G. H. (1972a) *Phys. Stat. Sol.* **52**, 79.

Döhler, G. H. (1972b) *Phys. Stat. Sol.* **52**, 533.

Döhler, G. H. (1986) *IEEE J. Quant. Electron.* **22**, 1682.

Döhler, G. H., Künzel, H., Olego, D., Ploog, K., Ruden, P., Stolz, H. J., and Abstreiter, G. (1981) *Phys. Rev. Lett.* **47**, 864.

Drummond, T. J., Kopp, W., Fischer, R. J., and Morkoc, H. (1982) *J. Appl. Phys.* **53**, 1028.

Drummond, T. J., Kopp, W., Morkoc, H., Hess, K., Cho, A. Y., and Streetman, B. G. (1981a) *J. Appl. Phys.* **52**, 5689.

Drummond, T. J., Morkoc, H., Sn, S. L., Fischer, R. J., and Cho, A. Y. (1981b) *Electron. Lett.* **17**, 870.

Esaki, L. and Tsu, R. (1970) *IBM J. Res. Develop.* **14**, 61.

Fang, F. F. and Howard, W. E. (1966) *Phys. Rev. Lett.* **16**, 797.

Greene, R. L. and Bajaj, K. K. (1983) *Solid State Commun.* **45**, 825.

Greene, R. L. and Bajaj, K. K. (1985a) *Phys. Rev.* **B31**, 913.

Greene, R. L. and Bajaj, K. K. (1985b) *Solid State Commun.* **53**, 1103.

Greene, R. L. and Bajaj, K. K. (1988) *Phys. Rev.* **B37**, 4604.

Harris, J. J., Foxon, C. T., Lacklison, D. E., and Barnham, K. W. J. (1986) *Superlattices and Microstructures* **2**, 563.

Hess, K. (1979) *Appl. Phys. Lett.* **35**, 484.

Hess, K., Morkoc, H., Shichijo, H., and Streetman, B. G. (1979) *Appl. Phys. Lett.* **35**, 469.

Kohn, W. (1957) *Solid State Phys.* **5**, 257.

Kopf, R. F., Schubert, E. F., and Harris, T. D. (1991) *Appl. Phys. Lett.* **58**, 631 and references therein.

Mailhiot, C., Chang, Y.-C., and McGill, T. C. (1982a) *J. Vac. Sci. Technol.* **21**, 519.

Mailhiot, C., Chang, Y.-C., and McGill, T. C. (1982b) *Phys. Rev.* **B26**, 4449.

Masselink, W. T., Chang, Y. C., and Morkoc, H. (1984) *J. Vac. Sci. Technol.* **B2**, 376.

Masselink, W. T., Chang, Y.-C., Morkoc, H., Litton, C. W., Bajaj, K. K., and Yu, P. W. (1986) *Solid State Electron.* **29**, 205.

Masselink, W. T., Chang, Y.-C., and Morkoc, H. (1983) *Phys. Rev.* **B28**, 7373.

Masselink, W. T., Chang, Y.-C., and Morkoc, H. (1985) *Phys. Rev.* **B32**, 5190.

Mimura, T. (1983) *Jpn. Annu. Rev. Electron., Comput., & Telecommun: Semicond. Technology* **8**, 277.

Mori, S. and Ando, T. (1980) *J. Phys. Soc. Japan* **48**, 865.

Noda, T., Tanaka, M. and Sakaki, H. (1991) *J. Cryst. Growth* **111**, 348.

Pfeiffer, L. (1990) personal communication.

Pfeiffer, L., Schubert, E. F., West, K. W., and Magee, C. W. (1991) *Appl. Phys. Lett.* **58**, 2258.

Pfeiffer, L., West, K. W., Stormer, H. L., and Baldwin, K. W. (1989a) *Appl. Phys. Lett.* **55**, 1888.

Pfeiffer, L., West, K. W., Stormer, H. L., and Baldwin, K. W. (1989b) *Mat. Res. Symp. Proc.* **145**, 3.

Ploog, K. and Döhler, G. H. (1983) *Adv. Phys.* **32**, 285.

Price, P. J. (1981) *Ann. Phys.* **133**, 217.

Price, P. J. (1984a) *Solid State Commun.* **51**, 607.

Price, P. J. (1984b) *Surf. Sci.* **143**, 145.

Proetto, C. R. (1990) *Phys. Rev.* **B41**, 6036.

Romanov, Y. A. (1972) *Sov. Phys. Semicond.* **5**, 1256.

Romanov, Y. A. and Orlov, L. K. (1973) *Sov. Phys. Semicond.* **7**, 182.

Schubert, E. F. (1990) *Optical and Quantum Electronics* **22**, S141.

Schubert, E. F., Cunningham, J. E. and Tsang, W. T. (1987a) *Phys. Rev.* **B36**, 1348.

Schubert, E. F., Cunningham, J. E., Tsang, W. T., and Timp, G. L. (1987b) *Appl. Phys. Lett.* **51**, 1170.

Schubert, E. F., Harris, T. D. and Cunningham, J. E. (1988a) *Appl. Phys. Lett.* **53**, 2208.

Schubert, E. F., Harris, T. D., Cunningham, J. E. and Jan W. (1989a) *Phys. Rev. B* **39**, 11011.

Schubert, E. F., Horikoshi, Y. and Ploog, K. (1985) *Phys. Rev. B* **32**, 1085.

Schubert, E. F., Pfeiffer, L., West, K. W., and Izabelle, A. (1989b) *Appl. Phys. Lett.* **54**, 1350.

Schubert, E. F., Ullrich, B., Harris, T. D. and Cunningham, J. E. (1988b) *Phys. Rev. B* **38**, 8305.

Seeger, K. (1982) *Semiconductor Physics* (Springer, Berlin).

Stern, F. (1983) *Appl. Phys. Lett.* **43**, 974.

Stern, F. and Das Sarma, S. (1984) *Phys. Rev. B* **30**, 840.

Stormer, H. L. (1983) *Surface Science* **132**, 519.

Stormer, H. L., Dingle, R., Gossard, A. C., and Wiegmann, W. (1978) *Inst. Phys. Conf. Ser.* **43**, ed. Wilson, B. L. (Inst. Physics, London) p. 557.

Stormer, H. L., Dingle, R., Gossard, A. C., Wiegmann, W., and Sturge, M. D. (1979) *Solid State Commun.* **29**, 705.

Stormer, H. L., Gossard, A. C., Wiegmann, W., and Baldwin, K. (1981a) *Appl. Phys. Lett.* **39**, 912.

Stormer, H. L., Gossard, A. C., and Wiegmann, W. (1981b) *Appl. Phys. Lett.* **39**, 712.

Stormer, H. L., Pinczuk, A., Gossard, A. C., and Wiegmann, W. (1981c) *Appl. Phys. Lett.* **38**, 691.

Tsang, W. T., Schubert, E. E., and Cunningham, J. E. (1992) *Appl. Phys. Lett.* **60**, 115.

Urbach, F. (1953) *Phys. Rev.* **92**, 1324.

Vinter, B. (1983) *Solid State Commun.* **48**, 151.

Vinter, B. (1984) *Appl. Phys. Lett.* **44**, 307.

Walukiewicz, W., Ruda, H. E., Lagowski, J., and Gatos, H. C. (1984) *Phys. Rev. B* **30**, 4571.

Wolfe, C. M., Stillman, G. E., and Lindely, W. T. (1970) *J. Appl. Phys.* **41**, 3088.

Chapter 11

Abramowitz, M. and Stegun, L. A. (eds.) (1972) *Handbook of Mathematical Functions*, (National Bureau of Standards, Washington D.C.).

Ando, T. (1982a) *J. Phys. Soc. Jpn.*, **51**, 3893.

Ando, T. (1982b) *J. Phys. Soc. Jpn.*, **51**, 3900.

Ando, T. (1985) *J. Phys. Soc. Jpn.*, **54**, 2676.

Baillargeon, J. N., Cheng, K. Y., Laskar, J., and Kolodzey, J. (1989) *Appl. Phys. Lett.* **55**, 663.

Bass, S. J. (1979) *J. Cryst. Growth* **47**, 613.

Bastard, G. (1981) *Phys. Rev. B24*, 5693.

Beall, R. B., Clegg, J. B., and Harris, J. J. (1988) *Semicond. Sci. Technol.* **3**, 612.

Beall, R. B., Harris, J. J., Clegg, R. J., Gowers, J. P., Joyce, B. A., Castagnè, J., and Welch, V. (1989) in *Gallium Arsenide and Related Compounds 1988* ed. Harris, J. S. (IOP Publishing Ltd, Bristol).

Benninghoven, A., Rüdenauer, F. G., and Werner, H. W. (1987) *Secondary Ion Mass Spectrometry* (Wiley, New York).

Board, K., Chandra, A., Wood, C. E. C., Judaprawira, S., and Eastman, L. F. (1981) *IEEE Trans. Electron Dev.* **28**, 505.

Capasso, F., Cho, A. Y., Mohammed, K., and Foy, P. W. (1985a) *Appl. Phys. Lett.* **46**, 664.

Capasso, F., Mohammed, K., and Cho, A. Y. (1985b) *Vac. Sci. Technol. B3*, 1245.

Chang-Hasnain, C. J., Hasnain, G., Johnson, N. M., Döhler, G. H., Miller, J. N., Whinnery, J. R., and Dienes, A. (1987) *Appl. Phys. Lett.* **50**, 915.

Chattopadhyay, D. and Queisser, H. J. (1981) *Rev. Mod. Phys.* **53**, 745.

Clegg, J. B. and Beall, R. B. (1989) *Surf. Interface Anal.* **14**, 307.

Di Lorenzo, J. V. (1971) *J. Electrochem. Soc.* **118**, 1645.

Döhler, G. H. (1972a) *Phys. Stat. Sol.* **52**, 79.

Döhler, G. H. (1972b) *Phys. Stat. Sol.* **52**, 533.

Döhler, G. H. (1986) *IEEE J. Quant. Electron.* **22**, 1682.

Döhler, G. H., Künzel, H., Olego, D., Ploog, K., Ruden, P., Stolz, H. J., and Abstreiter, G. (1981) *Phys. Rev. Lett.* **47**, 864.

English, J. H., Gossard, A. C., Störmer, H. L., and Baldwin, K. W. (1987) *Appl. Phys. Lett.* **50**, 1826.

Esaki, L. and Tsu, R. (1970) *IBM J. Res. Develop.* **14**, 61.

Etienne, B. and Thierry-Mieg, V. (1988) *Appl. Phys. Lett.* **52**, 1237.

Fang, F. F. and Howard, W. E. (1966) *Phys. Rev. Lett.* **16**, 797.

Ferreira, J. M. and Proetto, C. R. (1990) *Phys. Rev. B***42**, 5657.

Ferreira, J. M. and Proetto, C. R. (1991) *Phys. Rev. B***44**, 11231.

Gillman, G., Vinter, B., Barbier, E., and Tardella, A. (1988) *Appl. Phys. Lett.* **52**, 972.

Glass, A. M., Schubert, E. F., Wilson, B. A., Bonner, C. E., Cunningham, J. E., Olson, D. H., and Jan, W. (1989) *Appl. Phys. Lett.* **54**, 2247.

Gossard, A. C., Kazarinov, R. F., Luryi, S., and Wiegmann, W. (1982) *Appl. Phys. Lett.* **40**, 832.

Gossmann, H. J. and Schubert, E. F. (1993) *CRC Crit. Rev. Solid State Mater. Sci.* **18**, 1.

Harris, J. J., Ashenford, D. E., Foxon, C. T., Dobson, P. J., and Joyce, B. A. (1984) *Appl. Phys. A***33**, 87.

Hasnain, G., Döhler, G. H., Whinnery, J. R., Miller, J. N., and Dienes, A. (1986) *Appl. Phys. Lett.* **49**, 1357.

Headrick, R. L., Weir, B. E., Levi, A. F. J., Eaglesham, D. J., and Feldman, L. C. (1990) *Appl. Phys. Lett.* **57**, 2779.

Headrick, R. L., Weir, B. E., Levi, A. F. J., Freer, B., Bevk, J., and Feldman, L. C. (1991) *J. Vac. Sci. Technol. A***9**, 2269.

Hilsum, C. (1974) *Electron Lett.* **10**, 259.

Hobson, W. S., Pearton, S. J., Schubert, E. F., and Cabaniss, G. (1989) *Appl. Phys. Lett.* **55**, 1546.

Houng, M. P., Wang, Y. H., Chen, H. H., and Pan, C. C. (1992) *Solid State Electron.* **35**, 67.

Joyce, B. A. (1985) in *Molecular Beam Epitaxy and Heterostructures*, eds. Chang, L. L. and Ploog, K. (Martinus Nijhoff, Boston) p. 37.

Jung, H., Döhler, G. H., Göbel, E. O., and Ploog, K. (1983) *Appl. Phys. Lett.* **43**, 40.

Kazarinov, R. F. and Luryi, S. (1981) *Appl. Phys. Lett.* **38**, 810.

Kennedy, D. P. and O'Brien (1969) *IBM J. Res. Dev.* **13**, 212.

Kennedy, D. P., Murley, P. C., and Kleinfelder, W. (1968) *IBM J. Res. Dev.* **12**, 399.

Kobayashi, N., Makimoto, T., and Horikoshi, Y. (1986) *Jpn. J. Appl. Phys.* **25**, L746.

Koenraad, P. M., de Lange, W., Blom, F. A. P., Leys, M. R., Perenboom, J. A., Singleton, J., van der Vleuten, W. C., and Wolter, J. (1990a) *Materials Science Forum* **65 & 66**, 461.

Koenraad, P. M., Voncken, A. P. J., Singleton, J., Blom, F. A. P., Langerak, C. J. G., Leys, M. R., Perenboom, J. A., Spermon, S. J. R. M., van der Vlenten, W. C., and Wolter, J. H. (1990b) *Surface Science* **228**, 538.

Lanzillotto, A.-M., Santos, M., and Shayegan, M. (1989) *Appl. Phys. Lett.* **55**, 1445.

Lee, H., Schaff, W. J., Wicks, G. W., Eastman, L. F., and Calawa, A. R. (1985) *Inst. Phys. Conf. Ser.* **74**, 321.

Levi, A. F. J., McCall, S. L., and Platzman, P. M. (1989) *Appl. Phys. Lett.,* **54**, 940.

Luy, J.-F. (1990) *Thin Solid Films* **184**, 185.

Maguire, J., Murray, R., Newman, R. C., Beall, R. B., and Harris, J. J. (1987) *Appl. Phys. Lett.* **50**, 516.

Makimoto, T., Kobayashi, N., and Horikoshi, Y. (1986) *Jpn. J. Appl. Phys.* **25**, L513.

Malik, R. J. and Dixon, S. (1982) *IEEE Elec. Dev. Lett.* **3**, 205.

Malik, R. J., AuCoin, T. R., Ross, R. L., Board, K., Wood, C. E. C., and Eastman, L. F. (1980b) *Electron. Lett.* **16**, 836.

Malik, R. J., Board, K., Eastman, L. F., Wood, C. E. C., AuCoin, T. R., and Ross, R. L. (1980a) *Gallium Arsenide and Related Compounds, Inst. Phys. Conf. Ser.* **56**, 697.

Malik, R. J., Lunardi, L. M., Walker, J. F., and Ryan, R. W. (1988a) *IEEE Electron Dev. Lett.* **9**, 7.

Malik, R. J., Nottenburg, R. N., Schubert, E. F., Walker, J. R., and Ryan, R. W. (1988b) *Appl. Phys. Lett.* **53**, 2661.

Masselink, W. T. (1991a) *Phys. Rev. Lett.* **66**, 1513.

Masselink, W. T. (1991b) *Appl. Phys. Lett.* **59**, 694.

Nagle, J., Malik, R. J., and Gershoni, D. (1991) *J. Cryst. Growth* **111**, 264.

Ovsyannikov, M. I., Romanov, Y. A., Shabanov, V. N., and Loginova, R. G. (1971) *Sov. Phys. Semicond.* **4**, 1919.

Ploog, K. and Döhler, G. H. (1983) *Adv. Phys.* **32**, 285.

Proetto, C. R. (1990) *Phys. Rev. B***41**, 6036.

Reboredo, F. A. and Proetto, C. R. (1992) *Solid State Commun.* **81**, 163.

Romanov, Y. A. (1972) *Sov. Phys. Semicond.* **5**, 1256.

Romanov, Y. A. and Orlov, L. K. (1973) *Sov. Phys. Semicond.* **7**, 182.

Sasa, S., Muto, S., Kondo, K., Ishikawa, H., and Hiyamizu, S. (1985) *Jpn. J. Appl. Phys. Lett.* **24**, L602.

Saxon, D. S. (1968) *Elementary Quantum Mechanics* (Holden-Day, San Francisco).

Schrieffer, J. R. (1957) in *Semiconductor Surface Physics*, ed. by R. H. Kingston (Univ. of Pennsylvania Press, Philadelphia) pp. 55–69.

Schubert, E. F. (1990a) *Optical and Quantum Electronics* **22**, S141.

Schubert, E. F. (1990b) *J. Vac. Sci. Technol.* A**8**, 2980.

Schubert, E. F. (1993) in *Semiconductor and Semimetals*, ed. Gossard, A. C. (Academic, New York).

Schubert, E. F. and Cunningham, J. E. (1988) *Electron. Lett.* **24**, 980.

Schubert, E. F. and Ploog, K. (1985) *Jpn. J. Appl. Phys. Lett.* **24**, L608.

Schubert, E. F., Cunningham, J. E., and Tsang, W. T. (1986a) *Appl. Phys. Lett.* **49**, 1729.

Schubert, E. F., Cunningham, J. E., and Tsang, W. T. (1987a) *Solid State Commun.* **63**, 591.

Schubert, E. F., Cunningham, J. E., and Tsang, W. T. (1987b) *Appl. Phys. Lett.* **51**, 817.

Schubert, E. F., Cunningham, J. E., Tsang, W. T., and Chiu, T. H. (1986b) *Appl. Phys. Lett.* **49**, 292.

Schubert, E. F., Cunningham, J. E. and Tsang, W. T. and Timp, G. L. (1987c) *Appl. Phys. Lett.* **51**, 1170.

Schubert, E. F., Fischer, A., Horikoshi, Y. and Ploog, K. (1985a) *Appl. Phys. Lett.* **47**, 219.

Schubert, E. F., Fischer, A. and Ploog, K. (1985b) *Electron Lett.* **21**, 411.

Schubert, E. F., Fischer, A., and Ploog, K. (1986c) *IEEE Transactions on Electron Devices,* **33**, 625.

Schubert, E. F., Harris, T. D., Cunningham, J. E., and Jan, W. (1989a) *Phys. Rev. B***39**, 11011.

Schubert, E. F., Horikoshi, Y. and Ploog, K. (1985c) *Phys. Rev. B***32**, 1085.

Schubert, E. F., Kopf, R. F., Kuo, J. M., Luftman, H. S., and Garbinski, P. A. (1990a) *Appl. Phys. Lett.* **57**, 497.

Schubert, E. F., Kuo, J. M., Kopf, R. F., Jordan, A. S., Luftman, H. S., and Hopkins, L. C. (1990b) *Phys. Rev. B***42**, 1364.

Schubert, E. F., Kuo, J. M., Kopf, R. F., Luftman, H. S., Hopkins, L. C., and Sauer, N. J. (1990c) *J. Appl. Phys.* **67**, 1969.

Schubert, E. F., Luftman, H. S., Kopf, R. F., Headrick, R. L., and Kuo, J. M. (1990d) *Appl. Phys. Lett.* **57**, 1799.

Schubert, E. F., Ploog, K., Fischer, A., and Horikoshi, Y. (1984) Europ. Patent No. 0183146 A2.

Schubert, E. F., Stark, J. B., Ullrich, B., and Cunningham, J. E. (1988a) *Appl. Phys. Lett.* **52**, 1508.

Schubert, E. F., Tu, C. W., Kopf, R. F., Kuo, J. M., and Lunardi, L. M. (1989b) *Appl. Phys. Lett.* **54**, 2592.

Schubert, E. F., Tu, L. W., Zydzik, G. J., Kopf, R. F., Benvenuti, A., and Pinto, M. R. (1992) *Appl. Phys. Lett.* **60**, 466.

Schubert, E. F., Ullrich, B., Harris, T. D., and Cunningham, J. E. (1988b) *Phys. Rev. B***38**, 8305.

Schubert, E. F., van der Ziel, J. P., Cunningham, J. E., and Harris, T. D. (1989c) *Appl. Phys. Lett.* **55**, 757.

Shockley, W. (1961) *Solid-State Electron.* **2**, 35.

Stern, F. (1983) *Appl. Phys. Lett.* **43**, 974.

Stern, F. and Das Sarma, S. (1984) *Phys. Rev. B***30**, 840.

Sun, C.-Y. and Liu, W.-C. (1991) *Appl. Phys. Lett.* **59**, 2823.

Tokumitsu, E., Dentai, A. G., Joyner, C. H., and Chandrasekhar, S. (1990) *Appl. Phys. Lett.* **57**, 2841.

Vinter, B. (1984) *Appl. Phys. Lett.* **44**, 307.

Vojak, B. A., Zajac, G. W., Chambers, F. A., Meese, J. M., Chumbley, P. E., Kaliski, R. W., Holonyak, N., Jr., and Nam, D. W. (1986) *Appl. Phys. Lett.* **48**, 251.

Wang, Y. H., Houng, M. P., Chen, H. H., and Wei, H. C. (1991) *Electron. Lett.* **27**, 1667.

Webb, C. (1989) *Appl. Phys. Lett.* **54**, 2091.

Wood, C. E. C., Metze, G. M., Berry, J. D., and Eastman, L. F. (1980) *J. Appl. Phys.* **51**, 383.

Yang, G. M., Park, S. G., Seo, K. S., and Choe, B. D. (1992) *Appl. Phys. Lett* **60**, 2380.

Chapter 12

Abernathy, C. R., Pearton, S. J., Manasreh, M. O., Fischer, D. W., and Talwar, D. N. (1990) *Appl. Phys. Lett.* **57**, 294.

Akatsuka, T., Miyake, R., Nozaki, S., Yamada, T., Konagai, M., and Takahashi, K. (1990) *Jpn. J. Appl. Phys.* **29**, L537.

Almassy, R. J., Reynolds, D. C., Litton, C. W., Bajaj, K. K., and McCoy, G. L. (1981) *Solid State Commun.* **38**, 1053.

Amron, I. (1964) *Electrochem. Tech.* **2**, 337.

Amron, I. (1967) *Electrochem. Tech.* **5**, 94.

Ashen, D. J., Dean, P. J., Hurle, D. T., Mullin, J. B., White, A. M., and Greene, P. D. (1975) *J. Phys. Chem. Solids* **36**, 1041.

Benninghoven, A. (1970) *Z. Phys.* **230**, 403.

Benninghoven, A., Rüdenauer, F. G., and Werner, H. W. (1987) *Secondary Ion Mass Spectrometry: Basic Concepts, Instrumental Aspects, Applications and Trends* (Wiley, New York).

Blood, P. (1986) *Semicond. Sci. Technol.* **1**, 7.

Blood, P. and Orton, J. W. (1978) *Rep. Prog. Phys.* **41**, 157.

Bohm, D. (1951) *Quantum Theory* (Prentice-Hall, Englewood Cliffs).

Bube, R. H. (1960) *J. Appl. Phys.* **31**, 315.

Buehler, M. G. and Phillips, W. E. (1976) *Solid State Electron,* **19**, 777.

Burstein, E. (1954) *Phys. Rev.* **93**, 632.

Calleja, E., Munoz, E., and Garcia, F. (1983) *Appl. Phys. Lett.* **42**, 528.

Carballes, J. C. and Lebailly, J. (1968) *Solid State Commun.* **6**, 167.

Cardona, M. and Güntherodt, G. (1984) *Light Scattering in Solids Vol. IV* (Springer, Berlin).

Cardona, M. and Güntherodt, G. (1989) *Light Scattering in Solids Vol. V* (Springer, Berlin).

Castaing, R. and Slodzian, G. (1962) *Compt. Rend.* **255**, 1893.

Castaing, R., Jouffrey, B., and Slodzian, G. (1960) *Compt. Rend.* **251**, 1010.

Chakravarty, S., Subramanian, S., Sharma, D. K., and Arora, B. M. (1989) *J. Appl. Phys.* **66**, 3955.

Chase, L. L., Haynes, W., and Ryan, J. F. (1977) *J. Phys. C***10**, 2957.

Clegg, J. B. and Beall, R. B. (1989) *Surf. Interface Anal.* **14**, 307.

Davydov, A. S. (1965) *Quantum Mechanics*, 2nd ed. (Pergamon, Oxford).

Dean, P. J. (1973) in *Progress in Solid State Chemistry* **8**, eds. McCaldin, J. O. and Somorjai, G. (Pergamon, New York).

Dean, P. J. (1983) *Proc. Int. Conf. Phys. Semicond., Physica B***117**, 140.

Dean, P. J. and Herbert, D. C. (1979) *Excitons*, ed. Cho, K. (Springer, Berlin) p. 55.

Dean, P. J., Skolnick, M. S., Cockayne, B., MacEwan, W. R., and Iseler, G. W. (1984a) *J. Cryst. Growth* **67**, 486.

Dean, P. J., Skolnick, M. S., and Taylor, L. L. (1984b) *J. Appl. Phys.* **55**, 957.

de Lyon, T. J., Woodall, J. M., Goorsky, M. S., and Kirchner, P. D. (1990) *Appl. Phys. Lett.* **56**, 1040.

Downey, S. W. and Hozack, R. S. (1989) in *Secondary Ion Mass Spectrometry (SIMS VII)*, eds. Benninghoven, A., Evans, C. A., McKeegan, K. D., Storm, H. A., and Werner, H. W. (Wiley, New York), p. 283.

Downey, S. W., Emerson, A. B., Kopf, R. F., and Kuo, J. M. (1990) *Surf. and Interf. Anal.* **15**, 781.

Driver, M. C. and Wright, G. T. (1963) *Proc. Phys. Soc. (London)* **81**, 141.

Duenas, S., Izpura, I., Arias, J., Enriquez, L., and Barbolla, J. (1991) *J. Appl. Phys.* **69**, 4300.

Eagles, D. M. (1960) *J. Phys. Chem. Solids* **16**, 76.

Faktor, M. M., Ambridge, T., Elliott, C. R., and Regnault, J. C. (1980) in *Current Topics in Materials Science* **6**, ed. Kaldis E. (North Holland, New York).

Gammon, D., Merlin, R., Masselink, W. T., and Morkoc, H. (1986) *Phys. Rev. B***33**, 2919.

George, T., Weber, E. R., Nozaki, S., Yamada, T., Konagai, M., and Takahashi, K. (1991) *Appl. Phys. Lett.* **59**, 60.

Göbel, E. O. (1982) in GaInAsP *Alloy Semiconductors*, ed. Pearsall, T. P. (Wiley, New York) p. 313.

Greiner, M. E. and Gibbons, J. F. (1984) *Appl. Phys. Lett.* **44**, 750.

Hall, E. H. (1879) *Am. J. Math.* **2**, 287.

Harris, T. D. (1992) personal communication.

Harris, T. D., Skolnick, M. S., Parsey, Jr., J. M. and Bhat, R. (1988) *Appl. Phys. Lett.* **52**, 389.

Haynes, J. R. (1960) *Phys. Rev. Lett.* **4**, 361.

Haynes, J. R. (1966) *Phys. Rev. Lett.* **17**, 860.

Henry, C. H. and Lang, D. V. (1977) *Phys. Rev. B***15**, 989.

Henry, C. H., Hopfield, J. J., and Luther, L. C. (1966) *Phys. Rev. Lett.* **17**, 1178.

Herzog, R. K. and Liebl, H. (1963) *J. Appl. Phys.* **34**, 2893.

Hickmott, T. W. (1991) *Phys. Rev. B***44**, 13487.

Hoke, W. E., Lemonias, P. J., Weir, D. G., Hendricks, H. T., and Jackson, G. S. (1991) *J. Appl. Phys.* **69**, 511.

Holtz, M., Zallen, R., Geissberger, A. E., and Sadler, R. A. (1986) *J. Appl. Phys.* **59**, 1946.

Hopfield, J. J. (1964) in *Proc. 7th Int. Conf. on the Phys. of Semicond.* (Academic, New York) p. 725.

Hornstra, J. and Bartels, W. J. (1978) *J. Cryst. Growth* **44**, 513.

Kennedy, D. P. and O'Brien (1969) *IBM J. Res. Develop.* **13**, 212.

Kennedy, D. P., Murley, P. C., and Kleinfelder, W. (1968) *IBM J. Res. Develop.* **12**, 399.

Kimerling, L. C. (1974) *J. Appl. Phys.* **45**, 1839.

Kitagawara, Y., Itoh, I., Noto, N., and Takenaka, T. (1986) *Appl. Phys. Lett.,* **48**, 788.

Knox, R. S. (1963) *Theory of Excitons* (Academic, New York).

Kroemer, H., Chien, W.-Y., Harris, Jrs. J. S., and Edwall, D. D. (1980) *Appl. Phys. Lett.* **36**, 295.

Lang, D. V. (1974) *J. Appl. Phys.* **45**, 3023.

Lang, D. V. (1979) in *Thermally Stimulated Relaxation in Solids*, ed. Bräunlich, P. (Springer, New York).

Lang, D. V., Cho, A. Y., Gossard, A. C., Ilegems, M., and Wiegmann, W. (1976) *J. Appl. Phys.* **47**, 2558.

Leigh, R. S. and Newman, R. C. (1982) *J. Phys. C: Solid State Phys.* **15**, L1045.

Levinson, M. and Temkin, H. (1983) *Appl. Phys. Lett.* **42**, 605.

Losee, D. L. (1972) *Appl. Phys. Lett.* **21**, 54.

Losee, D. L. (1975) *J. Appl. Phys.* **46**, 2204.

Macrander, A. T., Schwartz, B., and Focht, M. W. (1984) *J. Appl. Phys.* **55**, 3595.

Maguire, J., Murray R., Newman, R. C., Beall, R. B., and Harris, J. J. (1987) *Appl. Phys. Lett.* **50**, 516.

Martin, G. M., Mitonneau, A., and Mircea, A. (1977) *Electron. Lett.* **13**, 191.

Meyer, N. I. and Guldbrandson, T. (1963) *Proc. IEEE* **51**, 1631.

Miller, G. L. (1972) *IEEE Trans. Electron Devices* **19**, 1103.

Miller, G. L., Lang, D. V., and Kimerling, L. C. (1977) *Ann. Rev. Mater. Sci.* **7**, 377.

Miller, G. L., Ramirez, J. V., and Robinson, D. A. H. (1975) *J. Appl. Phys.* **46**, 2638.

Mircea, A., Mitonneau, A., and Hallais, J. (1977) *Phys. Rev. B***16**, 3665.

Mitonneau, A., Martin, G. M., and Mircea, A. (1977) *Electron. Lett.* **13**, 666.

Moll, J. L. (1964) *Physics of Semiconductors*, (McGraw-Hill, New York).

Moss, T. S. (1954) *Proc. Phys. Soc. (London) B***76**, 775.

Murray, R., Newman, R. C., Sangster, M. J. L., Beall, R. B., Harris, J. J., Wright, P. J., Wagner, J., and Ramsteiner, M. (1989) *J. Appl. Phys.* **66**, 2589.

Nakamura, T. and Katoda, T. (1985) *J. Appl. Phys.* **57**, 1084.

Newman, R. C. (1974) *Infrared Studies of Crystal Defects* (Taylor and Francis, London).

Newman, R. C., Thompson, F., Hyliands, M., and Peast, R. F. (1972) *Solid State Commun.* **10**, 505.

Oechsner, H., ed. (1984) *Thin Film and Depth Profile Analysis* (Springer, New York).

Omling, P., Samuelson, L., and Grimmeiss, H. G. (1983) *J. Appl. Phys.* **54**, 5117.

Pankove, J. I. (1965) *Phys. Rev.* **140**, A2059.

Perry, T. A., Merlin, R., Shanabrook, B. V., and Comas, J. (1985a) *J. Vac. Sci. Technol.* **B3**, 636.

Perry, T. A., Merlin, R., Shanabrook, B. V., and Comas, J. (1985b) *Phys. Rev. Lett.* **54**, 2623.

Pickering, C., Tapsten, P. R., Dean, P. J., and Ashen, D. J. (1983) *Inst. Phys. Conf. Ser.* **65**, 469.

Ramsteiner, M., Wagner, J., Emnen, H., and Maier, M. (1988) *Phys. Rev. B***38**, 10 669.

Rashba, E. I. and Sturge, M. D. (1982) *Excitons* (North-Holland, Amsterdam).

Redfield, D. (1963) *Phys. Rev.* **130**, 916.

Reynolds, D. C., Litton, C. W., Smith, E. B., Bajaj, K. K., Collins, T. C., and Pilkuhn, M. H. (1983) *Phys. Rev. B***28**, 1117.

Sah, C. T. (1967) *Proc. IEEE* **55**, 654, 672.

Sah, C. T. (1976) *Solid-State Electron.* **19**, 975.

Sah, C. T., Chan, W. W., Fu, H. S., and Walker, J. W. (1972) *Appl. Phys. Lett.* **20**, 193.

Sah, C. T., Forbes, L., Rosier, L. L., and Tasch Jr., A. F. (1970) *Solid State Electron.* **13**, 759.

Schubert, E. F., Göbel, E. O., Horikoshi, Y., Ploog, K., and Queisser, H. J. (1984) *Phys. Rev. B***30**, 813.

Schubert, E. F., Kopf, R. F., Kuo, J. M., Luftman, H. S., and Garbinski, P. A. (1990a) *Appl. Phys. Lett.* **57**, 497.

Schubert, E. F., Luftman, H. S., Kopf, R. F., Headrick, R. L., and Kuo, J. M. (1990b) *Appl. Phys. Lett.* **57**, 1799.

Shanabrook, B. V., Moore, W. J., and Bishop, S. G. (1986) *Phys. Rev. B***33**, 5943.

Sinai, J. J. and Wu, S. Y. (1989) *Phys. Rev. B***39**, 1856.

Skolnick, M. S., Dean, P. J., Groves, S. H., and Kuphal, E. (1984) *Appl. Phys. Lett.* **45**, 962.

Smith, R. A. (1978) *Semiconductors* (Cambridge University Press, Cambridge) p. 123.

Spitzer, W. (1971) in *Advances in Solid State Physics Vol. XI*, ed. Madelung, O. (Pergamon-Vieweg, Braunschweig).

Stillman, G. E., Wolfe, C. M., and Dimmock, J. O. (1977) in *Semiconductor and Semimetals Vol. 12*, eds. Willardson, R. K. and Beer, A. C. (Academic, New York).

Sturge, M. D. (1962) *Phys. Rev.* **127**, 768.

Takanohashi, T., Komiya, S., Yamazaki, S., Kishi, Y., and Umebu, I. (1984) *Jpn. J. Appl. Phys. (Lett.)* **23**, L849.

Takanohashi, T., Tanahashi, T., Sugawara, M., Kamite, K., and Nakajima, K. (1988) *J. Appl. Phys.* **63**, 1961.

Tasch, A. F. and Sah, C. T. (1970) *Phys. Rev. B*1, 800.

Theis, W. M., Bajaj, K. K., Litton, C. W., and Spitzer, W. G. (1982) *Appl. Phys. Lett.* **41**, 70.

Thomas, C. O., Kahng, D., and Manz, R. C. (1962) *J. Electrochem. Soc.* **109**, 1055.

Thomas, D. G., Gershenzon, M., and Trumbore, F. A. (1964) *Phys. Rev.* **133**, A269.

Urbach, F. (1953) *Phys. Rev.* **92**, 1324.

Van der Pauw, L. J. (1958) *Philips Res. Rep.* **13**, 1.

Vincent, G., Chantre, A., and Bois, D. (1979) *J. Appl. Phys.* **50**, 5484.

Wagner, J. and Ramsteiner, M. (1986) *Appl. Phys. Lett.* **49**, 1369.

Wagner, J. and Ramsteiner, M. (1989) *IEEE J. Quant. Electron.* **25**, 993.

Wagner, J. and Seelewind, H. (1988) *J. Appl. Phys.* **64**, 2761.

Wagner, J., Ramsteiner, M., Seelewind, H., and Clark (1988) *J. Appl. Phys.* **64**, 802.

Wagner, J., Seelewind, H., and Kaufmann, U. (1986a) *Appl. Phys. Lett.* **48**, 1054.

Wagner, J., Seelewind, H., and Koidl, P. (1986b) *Appl. Phys. Lett.* **49**, 1080.

Wan, K. and Bray, R. (1985) *Phys. Rev. B32*, 5265.

Wiley, J. D. and Miller, G. L. (1975) *IEEE Trans. Electron Devices* **22**, 265.

Williams, E. W. and Bebb, H. B. (1972) in *Semiconductors and Semimetals* **8**, eds. Willardson, R. K. and Beer, A. C. (Academic, New York) p. 181

Wilson, R. G., Stevie, F. A., and Magee, C. W. (1989) *Secondary Ion Mass Spectrometry* (Wiley, New York).

Wolf, P. (1962) *Phys. Rev.* **126**, 405.

Wolfe, C. M. and Stillman, G. E. (1971) in *Gallium Arsenide and Related Compounds 1970*, (Inst. Phys. and Phys. Soc., London) p. 3.

Wright, G. B. and Mooradian, C. (1968) *Proc. 9th Inst. Conf. Phys. Semicon., Moscow, 1968* (Nauka, Leningrad) p. 1067.

Yacobi, B. G. and Holt, D. B. (1986) *J. Appl. Phys.* **59**, R1.

Appendixes

Adachi, S. (1985) *J. Appl. Phys.* **58**, R1.

Adachi, S. (1992) *Physical Properties of III-V Semiconductor Compounds* (Wiley, New York).

Blakemore, J. S. (1982) J. Appl. Phys. **53**, R123.

Casey, H. C. Jr., and Panish, M. B. (1978) *Heterostructure Lasers, Part A and B* (Academic, San Diego).

Landolt-Bornstein (1982) *Numerical Data and Functional Relationships in Science and Technology*, New Series, Hellwege, K.-H. (ed), *Group III: Crystal and Solid State Physics, Vol. 17, Semiconductors* (Springer, Berlin).

Landolt-Bornstein (1987) *Numerical Data and Functional Relationships in Science and Technology*, New Series, Madelung, O, (ed), *Group III: Crystal and Solid State Physics, Vol. 22, Semiconductors* (Springer, Berlin).

Neuberger, M. (1971) III-V semiconductor-data tables in *Handbook of Electronic Materials* **2**, (Plenum, New York).

Neuberger, M. (1972) III-V semiconductor-data tables in *Handbook of Electronic Materials* **7**, (Plenum, New York).

Pearsall, T. P. (ed) (1982) *GaInAsP Alloy Semiconductors* (Wiley, New York).

Properties of Gallium Arsenide, 2nd edition (INSPEC, IEE, London, 1990).

Properties of Indium Phosphide (INSPEC, IEE, London, 1991).

Shur, M. (1990) *Physics of Semiconductor Devices* (Prentice Hall, Englewood Cliffs).

Index

79.95